Statistics for Biology and Health

Series Editors
K. Dietz, M. Gail, K. Krickeberg, A. Tsiatis, J. Samet

Statistics for Biology and Health

Borchers/Buckland/Zucchini: Estimating Animal Abundance: Closed Populations.

Burzykowski/Molenberghs/Buyse: The Evaluation of Surrogate Endpoints.

Everitt/Rabe-Hesketh: Analyzing Medical Data Using S-PLUS.

Ewens/Grant: Statistical Methods in Bioinformatics: An Introduction. 2nd ed.

Hougaard: Analysis of Multivariate Survival Data.

Keyfitz/Caswell: Applied Mathematical Demography, 3rd ed.

Klein/Moeschberger: Survival Analysis: Techniques for Censored and Truncated Data, 2nd ed.

Kleinbaum/Klein: Survival Analysis: A Self-Learning Text.

Kleinbaum/Klein: Survival Analysis: A Self-Learning Text. 2nd ed.

Kleinbaum/Klein: Logistic Regression: A Self-Learning Text, 2nd ed.

Lange: Mathematical and Statistical Methods for Genetic Analysis, 2nd ed.

Manton/Singer/Suzman: Forecasting the Health of Elderly Populations.

Nielsen: Statistical Methods in Molecular Evolution.

Moyé: Multiple Analyses in Clinical Trials: Fundamentals for Investigators.

Parmigiani/Garrett/Irizarry/Zeger: The Analysis of Gene Expression Data: Methods and Software.

Salsburg: The Use of Restricted Significance Tests in Clinical Trials.

Simon/Korn/McShane/Radmacher/Wright/Zhao: Design and Analysis of DNA Microarray Investigations.

Sorensen/Gianola: Likelihood, Bayesian, and MCMC Methods in Quantitative Genetics.

Stallard/Manton/Cohen: Forecasting Product Liability Claims: Epidemiology and Modeling in the Manville Asbestos Case.

Therneau/Grambsch: Modeling Survival Data: Extending the Cox Model.

Vittinghoff/Glidden/Shiboski/McCulloch: Regression Methods in Biostatistics: Linear, Logistic, Survival, and Repeated Measures Models

Zhang/Singer: Recursive Partitioning in the Health Sciences.

Rasmus Nielsen
Editor

Statistical Methods
in Molecular Evolution

 Springer

Rasmus Nielsen
Dept. of Biological Sciences and
    Computational Biology
Cornell University
Ithaca, NY 14853
USA

Series Editors

K. Dietz
Institut für Medizinische Biometrie
Universität Tübingen
Westbahnhofstraße 55
D-72070 Tübingen
Germany

M. Gail
National Cancer Institute
Rockville, MD 20892
USA

K. Krickeberg
Le Châtelet
F-63270 Manglieu
France

A. Tsiatis
Department of Statistics
North Carolina State University
Raleigh, NC 27695
USA

J. Samet
Department of Epidemiology
School of Public Health
Johns Hopkins University
615 Wolfe Street
Baltimore, MD 21205
USA

Library of Congress Cataloging-in-Publication Data
On file.

ISBN 0-387-22333-9   Printed on acid-free paper.

Printed in the United States of America.   (MVY)

9 8 7 6 5 4 3 2 1   SPIN 10945506

springeronline.com

To Hannah, Milla, and Dawn

Preface

The field of molecular evolution is devoted to elucidating the processes generating variation within and between species at the molecular level. It addresses the fundamental question of why all life on Earth looks as it does from the perspective of molecular and evolutionary biology. Molecular evolution arose as a scientific field in the 1960s after protein sequences from multiple species first became available. In the 1970s, the first journal exclusively devoted to this field arose, and today it molecular evolution dominates the literature on evolutionary biology. Since the appearance of large-scale genomic data, the field of molecular evolution has emerged as one of the major scientific pillars in the analysis of genomic data, especially when it comes to data from multiple species.

The field of molecular evolution relies heavily on statistical theory. Usually, researchers only have access to DNA data, or other molecular data, from extant species. From such data they try to make inferences regarding past evolutionary processes. This inference problem is fundamentally statistical in nature, and it is not surprising that a large body of literature on statistical methods in molecular evolution has emerged. The statistical problems encountered in molecular evolution are often rather non-standard because the underlying statistical models usually involve superimposing stochastic processes along the edges of a tree. Several interesting and peculiar algorithmic and statistical problems arise from these models. While many books contain excellent coverage of the specialized, but important, area of phylogenetic inference (estimation of trees) from molecular data, there are no books that provide an introduction to the more general area of statistical methods in molecular evolution. With the publication of this book, we hope to rectify this problem.

The first four chapters of the book provide a general introduction to the area. The first chapter, by Galtier and his colleagues, introduces the models of DNA sequence change usually applied in molecular evolution. Evolution does not remember past states except by current form, and Markov models have therefore been the natural choice for statistical models to describe the evolution of DNA sequences. The first chapter provides an introduction to these models and sets the stage for the subsequent chapters describing how the models are used for statistical inferences. The second chapter, by Buschbom and von Haeseler, provides an introduction to the use of the likelihood function in molecular evolution. In addition to introducing some basic statistical concepts for the uninitiated biologist, it also provides an introduction to the computational aspects involved in calculating sampling distributions based on the Markov models discussed in the first chapter. The third chapter, by Larget, provides an introduction to the use of Markov chain Monte Carlo (MCMC) methods in molecular evolution. Methods based on MCMC are receiving an increasing amount of attention and the third chapter provides an introduction to this area. Chapter 4, by Bustamante, provides an introduction to population genetic theory with special emphasis on areas of relevance for molecular evolution. The classical mathematical theory underlying studies of molecular evolution is population genetic, and much research in molecular evolution cannot be fully understood without an appreciation of population genetic theory.

The second section in the book contains four chapters written by the authors of some of the most important statistical computer packages used in the study of molecular evolution. These chapters discuss practical statistical approaches for analyzing DNA sequences and molecular data. Chapter 5, by Bielawski and Yang, discusses methods for detecting natural Darwinian selection using the program *Paml*. *Paml* is possibly the most commonly used computer program for analyzing models of molecular evolution. Chapter 6, by Pond and Muse, discusses a recently developed versatile computer package, *HyPhy*, for analyzing models of molecular evolution. Readers interested in developing and analyzing new models of molecular evolution may want to take advantage of this computer package. Chapter 7, by Huelsenbeck and Ronquist discusses Bayesian inference in molecular evolution based on the popular computer program *MrBayes*. The use of Bayesian methods in molecular evolution is quite new but has already had a tremendous impact on the field, largely due to the availability of *MrBayes*. Chapter 8 is written by the authors of the computer program *Multidivtime*, Kishino and Thorne, and it discusses statistical issues relating to the molecular clock. The molecular clock assumption is that the rate of molecular change has been constant through evolutionary time. This assumption has been used extensively in the literature to date evolutionary events, but on numerous occasions is has also been shown to be not invalid. The main focus of Chapter 8 is to discuss methods for dating evolutionary events when the molecular clock assumption is not met.

The third section introduces other models of molecular evolution beyond the basic nucleotide-based Markov chain models that were the main focus of Chapter 1. Chapter 9, by Dimmic, discusses models of protein evolution. Such models are important not only because they can be used for making inferences regarding protein evolution and function but also because the substitution matrices estimated using these models are important in alignment algorithms. The issue of estimating substitution matrices is revisited in Chapter 15, and the issue of alignment is dealt with in depth in Chapter 14. In Chapter 10, Calabrese and Sainudiin discuss models of microsatellite evolution and statistical inferences based on these models. Microsatellites are small repeated patterns of DNA and are used extensively in many genetic studies because they are highly variable. The models used to analyze this type of data are fundamentally different from most other models used to analyze DNA sequence data. Chapter 11, by Durrett, discusses methods and models for analyzing whole-genome evolution incorporating rearrangements such as inversions and translocations. Although these models are still in their infancy, they have become highly relevant with the recent availability of large-scale genomic data. Chapter 12, by Siepel and Haussler, provides an introduction to the use of Hidden Markov models (HMMs) in the study of molecular evolution. Such models are very important when certain properties of the evolutionary process are thought to vary among positions along the DNA sequence. The use of HMMs for statistical alignment is also discussed in Chapter 14.

The last section of the book contains five chapters that further detail methods of inference in molecular evolution. In Chapter 13, McVean relates the Markov models of molecular evolution discussed in most chapters of this book to the population genetic models discussed in Chapter 4 in the context of variation of nucleotide composition among species. The frequency of different nucleotides is known to vary among species. This observation is interesting from the perspective of evolutionary biology, and it is also highly relevant to our choice of models for analyzing molecular evolution. In Chapter 14 on statistical alignment, Lunter, Drummond, Mikls, and Hein explore the relationship among evolutionary models, trees and the problem of aligning DNA or protein sequences. In Chapter 15, Yap and Speed discuss the estimation of substitution matrices for use in alignment problems. The last two chapters, Chapter 16 by Bollback and Chapter 17 by Shimodaira and Hasegawa, discuss issues related to hypothesis testing and model choice in molecular evolution. Bayesian methods have recently gained much popularity in the area of molecular evolution, leading to a debate regarding choice of statistical methodology not dissimilar to the discussions that have occurred in many other areas of applied statistics. Chapter 17 argues for the use of certain frequentist procedures for the tree estimation problem and discusses problems with Bayesian procedures, while other chapters of the book (e.g., Chapter 7) provide more optimistic views of the use of Bayesian methods in molecular evolution. Chapter 16 discusses the use of posterior predictive distributions for statistical inferences, in addition to providing a review of recent methods for estimating the history of mutations from DNA sequence data.

This book provides a comprehensive review of the many interesting statistical problems arising in molecular evolution provided by leading researchers in the field. It is intended for researchers and students from the statistical and biological sciences alike. For the statistician, the book will provide an introduction to an exciting area of application that often has been overlooked by statisticians. For the biologist, the book provides an introduction to the theory underlying many of the methods they use in their daily research. Several of the chapters, including the four introductory chapters, are also highly suitable as texts for advanced undergraduate or graduate-level courses in molecular evolution.

Copenhagen, April 2004   *Rasmus Nielsen*

Contents

Part I

Introduction

Markov Models in Molecular Evolution

Nicolas Galtier[1], Olivier Gascuel[2], and Alain Jean-Marie[3]

[1] Génome, Populations, Interactions, Adaptation, Université Montpellier, Montpellier, France, `galtier@univ-montp2.fr`
[2] Laboratoire d'Informatique, Robotique et Microélectronique de Montpellier, Université Montpellier, Montpellier, France, `gascuel@lirmm.fr`
[3] Laboratoire d'Informatique, Robotique et Microélectronique de Montpellier, Université Montpellier, Montpellier, France, `ajm@lirmm.fr`

1.1 Introduction to Markov Models in Molecular Evolution

Markov chains (or Markov processes) are memoryless stochastic processes. Formally, a stochastic process is a collection $X(t)$ of random variables, where t is typically time, and the Markovian property is defined by (for a discrete-time process):

$$\Pr(X(t+1) = x_{t+1} \mid X(t) = x_t, X(t-1) = x_{t-1}, \ldots, X(1) = x_1, X(0) = x_0)$$
$$= \Pr(X(t+1) = x_{t+1} \mid X(t) = x_t)$$

for all states $x_0, x_1, \ldots, x_{t-1}, x_t, x_{t+1}$ of the process and any time t. More intuitively, this means that the future of the process (that is, the various states possibly reached and their probabilities of occurrence) depends only on the present state, not on past states (i.e. the pathway followed to reach the current state). Markov processes can be in discrete time, when states are assigned to successive "steps," or "generations," or in continuous time, when the time to next event is an exponential random variable. The space of states can be discrete (finite or infinite) or continuous. Branching processes (discrete state, discrete time), random walks (continuous state, discrete time), Poisson processes (discrete state, continuous time), and Brownian motion (continuous state, continuous time) are well-known instances illustrating this variety of stochastic processes [32]. Markov chains have been widely used in a variety of scientific fields, including physics, chemistry, networks, and, of course, evolutionary biology.

The reasons why Markov chains are useful to model biological evolution are obvious: evolution is very generally memoryless. Some examples of non-Markovian evolutionary processes can be thought of, however. The future size of a population of current size N, for instance, depends somewhat on past population sizes. This is because population history determines the genetic diversity of the current population, which might itself influence the growth rate. It should be noted, however, that the joint evolutionary process of population size and population genetic diversity is Markovian: the future of the population depends only on current size and genetic resources. The potentially non-Markovian nature of the size process therefore appears to be due to an incomplete representation of the system rather than true evolutionary memory.

Markov models are routinely used in several domains of evolutionary biology. We have already introduced population dynamics (that is, the evolution of census population size), a field in which stochastic processes are central. Branching processes, for example, are used for estimating demographic parameters (birth rate, mortality) and extinction risks, with applications in species management and conservation [14]. In the case of structured populations, in which individuals are assigned to classes (e.g. age classes) and can switch between classes, Markov processes are used to predict the proportion of each class at equilibrium [2]. Markov chains are also widely used for representing the evolution of quantitative traits (e.g. morphology, behavior, growth rate), modeled as Brownian motion when neutral (e.g. [17]) or using more complex continuous-time Markov chains when selected (e.g. [24]). The evolution of genomic data is also typically modeled as Markov chains, as we now discuss in more detail.

Genes and genomes are made with DNA, a polymer of four distinct monomers called adenine (A), cytosine (C), guanine (G), and thymine (T). DNA sequences are therefore naturally represented as words in the {A,C,G,T} alphabet, where letters of the alphabet are called "nucleotides" or "bases." DNA sequences evolve according to a two-level process: sequence transmission and sequence change. Genes are transmitted from parents to offspring (at a short timescale) or from ancestral to descendant species (at a larger timescale), so that the history of a gene will typically be represented by a tree, called a genealogy or phylogeny. A gene is lost if not transmitted, resulting in the extinction of one of the evolving lineages. This is called a birth-death process. A gene, when transmitted, can undergo a change of its DNA sequence. Many kinds of changes have been reported, including base replacement, insertion, deletion, inversion, tandem duplication, translocation, recombination, and gene conversion. (The last two events involve two sequences.) The process of sequence change is superimposed on the process of gene transmission: changes occur along the branches of an underlying genealogy.

A model that would aim at representing the way sequences actually evolve should therefore incorporate both sequence reproduction and sequence change. This is achieved by the coalescent model of population genetics, in which the genealogy of a sample of genes is considered as a random tree whose distribution and characteristics are determined by certain parameters of interest (e.g. population history). The pattern of sequence variability in the sample is used to infer the plausible shapes (topology, branch lengths) of the underlying unknown genealogy, and assess the likelihood of various hypotheses about the birth-death process of gene transmission. This is usually done by assuming that the genealogy is independent of sequence changes, i.e., that the probability of gene transmission is the same for any state of the sequence space, (the neutrality assumption). See [26] for a recent review of the coalescent theory.

Objectives and methods are somewhat different when sequences from distinct species, rather than from individuals of a single species, are sampled. In this case, the underlying phylogenetic tree can be known, or can be what we want to reconstruct. It is generally taken as a parameter or a known quantity, not a realization of a certain random process. As far as sequence changes are concerned, most models for between-species data focus on base replacement, the prevalent process in the long-term evolution of coding sequences. Under these assumptions, the evolution of a DNA sequence of length n is represented by n Markov processes running along the branches of a common tree. Each of these processes takes values on $\mathcal{E} = \{A,C,G,T\}$, the so-called state space. They will be considered as continuous-time Markov chains. We now recall some of the major mathematical properties of these models in relationship with the underlying biological assumptions. Then we examine popular Markov models of sequence evolution aimed at representing the specificities of sequence evolutionary processes. Finally, we review and discuss the various uses of Markov chains in phylogenetic analyses.

1.2 Modeling DNA Sequence Evolution: Mathematical Background

1.2.1 Continuous-Time Transition Rates

Consider a DNA sequence of fixed length n evolving in time by the process of base replacement. Assume that the processes followed by the n sites (positions) are Markovian, independent, identically distributed, and constant in time (we shall discuss these assumptions later). Let

$$\mathbf{F}(t) = {}^t(f_A(t), f_C(t), f_G(t), f_T(t))$$

be the column vector of the probabilities of states A, C, G, and T, respectively, for a certain site at time t. Let $\mu_{xy}(y \neq x)$ be the transition rate from state $x \in \mathcal{E}$ to state $y \in \mathcal{E}$, and let $\mu_x = \sum_{y \neq x} \mu_{xy}$. The evolutionary dynamics is described by the differential equations

$$f_A(t + dt) = f_A(t) - f_A(t)\, \mu_A dt + \sum_{x \neq A} f_x(t)\, \mu_{xA}\, dt,$$

$$f_C(t + dt) = f_C(t) - f_C(t)\, \mu_C dt + \sum_{x \neq C} f_x(t)\, \mu_{xC}\, dt,$$

$$f_G(t + dt) = f_G(t) - f_G(t)\,\mu_G dt + \sum_{x \neq G} f_x(t)\,\mu_{xG}\,dt,$$

$$f_T(t + dt) = f_T(t) - f_T(t)\,\mu_T dt + \sum_{x \neq T} f_x(t)\,\mu_{xT}\,dt,$$

where the summations are over $\mathcal{E}$.

The first of the four equations above states that the frequency of A at time $t + dt$ equals the frequency of A at time t minus the frequency of lost A's, plus the frequency of newly arisen A's. This set of equations has a compact matrical form

$$\mathbf{F}(t + dt) = \mathbf{F}(t) + \mathbf{MF}(t)dt$$

or

$$\frac{d\mathbf{F}(t)}{dt} = \mathbf{MF}(t), \qquad (1.1)$$

where $\mathbf{M}$ is the 4×4 matrix defined as

$$M = \begin{pmatrix} -\mu_A & \mu_{CA} & \mu_{GA} & \mu_{TA} \\ \mu_{AC} & -\mu_C & \mu_{GC} & \mu_{TC} \\ \mu_{AG} & \mu_{CG} & -\mu_G & \mu_{TG} \\ \mu_{AT} & \mu_{CT} & \mu_{GT} & -\mu_T \end{pmatrix}.$$

$\mathbf{M}$ is called the rate matrix, or generator, of the process. It is such that column entries sum to zero. Entries of $\mathbf{M}$ are expressed in (time unit)$^{-1}$. They are homogeneous to the rate of a Poisson process; $\mu_{xy}dt$ tends to the probability of being in state y at time $t + dt$ given state x at time t as dt tends to zero. The time to next change, given current state x, is exponentially distributed with rate μ_x. Given that the process leaves state x, it will enter state $y \neq x$ with probability μ_{xy}/μ_x.

1.2.2 Stationary Distribution

If all μ_{xy} rates are positive, so that all states "communicate", then the Markov chain has a stationary distribution $\{\pi_x, x \in \mathcal{E}\}$: an equilibrium (or steady state) is reached when t tends to infinity, at which any state x has a nonzero probability of occurrence, π_x, that does not depend on the initial state of the process. Such a Markov chain is called ergodic . Stationary frequency π_x is the expected proportion of time spent in state x after the Markov process has run infinitely long. In the case of DNA, under the assumption of a common process for every site, the π_x's correspond to the equilibrium base composition (that is, the proportions of A, C, G, and T) of the evolving sequence. A Markov process is said to be stationary when its current distribution is the stationary distribution, (i.e. when $\mathbf{F}(t) = \mathbf{\Pi}$). By definition, the stationary distribution is such that

$$\frac{d\mathbf{\Pi}}{dt} = \mathbf{M\Pi} = \mathbf{0},$$

where the first equality follows from equation (1.1). This implies that $\mathbf{\Pi}$ is an eigenvector of $\mathbf{M}$ for eigenvalue zero.

1.2.3 Time Reversibility

A stationary Markov process is said to be time-reversible if, for every pair (x, y) of states, we have

$$\pi_x \mu_{xy} = \pi_y \mu_{yx} \tag{1.2}$$

Reversibility means that in steady state the amount of change from state x to state y is equal to amount of change from state y to state x (although the two states need not to be equally frequent). Not every stationary process is reversible. Reversibility, however, is a convenient, reasonable assumption made by virtually every model of DNA sequence evolution. Under the reversibility assumption, transition rates μ_{xy} can be expressed as:

$$\mu_{xy} = s_{xy} \pi_y \tag{1.3}$$

where $s_{xy} = s_{yx}$ is a symmetric term sometimes called "exchangeability" between x and y. The twelve nondiagonal entries of rate matrix $\mathbf{M}$ can therefore be described by just nine independent parameters under the assumption of reversibility, namely six exchangeability terms s_{xy} and three stationary frequencies π_x (remember that the π_x's have to sum to one).

1.2.4 Calculating Transition Probabilities

The piece of theory above has to do with the instantaneous dynamics (equation (1.1)) and the long-run behavior (stationarity, reversibility) of a continuous-time Markov process. In sequence data analysis, however, we will typically compare sequences that have diverged during a finite amount of time. To get some insight about sequence evolutionary processes from Markov models, we need to address the transient behavior of Markov chains. This is achieved by solving differential equation (1.1):

$$\mathbf{F}(t) = e^{\mathbf{M}t} \mathbf{F}(0). \tag{1.4}$$

Equation (1.4) relates the distribution of the process at time t to its initial distribution $\mathbf{F}(0)$. Let $\mathbf{P}(t) = e^{\mathbf{M}t}$. Entry $p_{xy}(t)$ of $\mathbf{P}(t)$ is the probability of state y after evolution according to process $\mathbf{M}$ during time t given initial state x. The $\mathbf{P}$ matrix is defined as the exponential of matrix $\mathbf{M}t$:

$$\mathbf{P}(t) = e^{\mathbf{M}t} = \sum_{i=0}^{\infty} (\mathbf{M}t)^i / i! \quad .$$

There are numerous ways to calculate the exponential of a matrix [25]. In the case of DNA sequence Markov models, and since we want to calculate the $p_{xy}(t)'s$ for many t values but constant $\mathbf{M}$ (see below), the appropriate calculation involves diagonalising $\mathbf{M}$. If $\mathbf{M} = \mathbf{Q}\mathbf{D}\mathbf{Q}^{-1}$ with $\mathbf{D}$ diagonal then:

$$\mathbf{P}(t) = e^{\mathbf{M}t} = e^{(\mathbf{Q}\mathbf{D}t\mathbf{Q}^{-1})}$$

$$= \sum_{i=0}^{\infty} (\mathbf{Q}\mathbf{D}t\mathbf{Q}^{-1})^i / i!$$

$$= \sum_{i=0}^{\infty} \mathbf{Q}(\mathbf{D}t)^i \mathbf{Q}^{-1} / i!$$

$$= \mathbf{Q}\left(\sum_{i=0}^{\infty} (\mathbf{D}t)^i / i!\right)\mathbf{Q}^{-1}$$

$$= \mathbf{Q}e^{\mathbf{D}t}\mathbf{Q}^{-1},$$

which is easily calculated since the exponential of a diagonal matrix is obtained by replacing its diagonal terms by their exponentials.

1.2.5 Trees and Likelihood

Up to now, we have considered the evolution of a single sequence in a one-dimensional time space. But biological sequences reproduce and die, as indicated above, so that the process of sequence change should be regarded as running along the branches of a binary rooted tree, called phylogeny. The generalization is obtained simply by stating that when the process reaches a node of the tree, the current state is duplicated, and two independent processes restart along the two child branches. The generalized process models the evolution of a set of DNA sequences sharing a common ancestor at the root of the tree. Such a model has three kinds of parameters, namely the tree topology, branch lengths (that is, the amount of time during which the process runs in each branch), and entries of the rate matrix.

Molecular phylogeny essentially aims at estimating these parameters from a data set of extant homologous sequences. This is typically achieved using the likelihood function. The likelihood L of a certain set of parameter values θ is defined as the probability of the data Y conditional on these parameter values:

$$L(\theta) = \Pr(Y \mid \theta). \tag{1.5}$$

In the case of DNA sequence data, Y corresponds to a set of (aligned) DNA sequences, each of length n, associated to the tips of the tree. Let Y_i $(1 \leq i \leq n)$ be the ith site of Y, defined as the set of bases at position i in the various sequences of the data set (ith column of the alignment). Each Y_i is the outcome of a distinct Markov process. Under the assumption of independent sites, we have

$$L(\theta) = \Pr(Y \mid \theta) = \prod_{i=1}^{n} \Pr(Y_i \mid \theta). \tag{1.6}$$

Now the probability of a site Y_i given the rate matrix, rooted tree topology, and branch lengths can be calculated recursively, as shown by Felsenstein [5].

Let k be an internal node of the tree, and let $L_i^k(x)(x \in \mathcal{E})$ be the partial conditional likelihood defined as

$$L_i^k(x) = \Pr(Y_i^k \mid \theta, k_i \equiv x),$$

where Y_i^k is the restriction of site Y_i to the sequences associated to tips descending from node k and where $k_i \equiv x$ means that the ancestral state for site i at node k was x. $L_i^k(x)$ is the likelihood at site i for the subtree underlying node k conditional on state x at k. The likelihood at site i can be expressed as

$$\Pr(Y_i \mid \theta) = \sum_{x \in \mathcal{E}} \Pr(r_i \equiv x) L_i^r(x), \tag{1.7}$$

where r is the root node. The recurrence on $L_i^k(x)$ is

$$L_i^k(x) = \sum_{y_1 \in \mathcal{E}} p_{xy_1}(t_1) L_i^{k_1}(y_1) \sum_{y_2 \in \mathcal{E}} p_{xy_2}(t_2) L_i^{k_2}(y_2), \tag{1.8}$$

where k_1 and k_2 are the two child nodes to internal node k, and where t_1 and t_2 are the lengths of the (k, k_1) and (k, k_2) branches, respectively. Equation (1.8) results from the independence of the processes in the two subtrees underlying node k. This equation holds if k is an internal node. The recurrence closes at leaves (terminal nodes) of the tree. Let l be a leaf

$$L_i^l(x) = \begin{cases} 1 & \text{if } l_i \equiv x \\ 0 & \text{otherwise,} \end{cases}$$

where the state at node l is determined by the base observed at position i in the corresponding sequence. This calculation is achieved in a time linear in the number of sequences and in the number of sites. Usually, the logarithm of the likelihood is computed rather than the likelihood itself. The product in equation (1.6) becomes a summation if the log-likelihood is computed.

Note that equation (1.7) requires knowledge of the base composition of the ancestral sequence, (i.e., the probabilities of states A, C, G, and T at the root node). Under the stationarity assumption, these probabilities correspond to the stationary distribution of the process. The calculation above was defined on a rooted tree. For a reversible process, however, the location of the root does not matter: the likelihood value is unchanged whatever the position of the root [5].

The likelihood function is used in the first place for estimation purposes. The parameters of the model (tree, branch lengths, and rate matrix) can be estimated jointly by the maximum-likelihood method: the maximum-likelihood estimator is defined as the parameter value $\hat{\theta}$ that maximizes $L(\theta)$. Alternatively, the likelihood function can be used in the Bayesian framework to calculate the posterior probabilities of parameters or other unknown quantities (see Chapters 3 and 7). Likelihood is also useful for comparing alternative models and testing hypotheses (see section 1.4). See [4] for an introduction to the likelihood theory and Chapter 2 of the present volume for applications in molecular evolution.

1.3 Popular Markov Models of Sequence Evolution

Many different Markov models of sequence evolution embedded in the mathematical background above have been proposed in the literature and applied to sequence data. The reason for this diversity is that genomic evolutionary processes vary between genomes and between regions of a genome. Different evolutionary forces apply to coding and noncoding regions or to the mitochondrial and nuclear genomes, for example. Models essentially differ in the parametrization of the rate matrix and in the modeling of rate variations.

1.3.1 Specifying the Rate Matrix

In its more general form, the rate matrix $\mathbf{M}$ is described by 12 parameters μ_{xy} corresponding to the 12 rates of base change. From a statistical point of view, 12 parameters can be too many. More economical parametrizations have been proposed. Inversely, some of the assumptions made by standard models of sequence evolution (e.g. stationarity, independent sites) were found inappropriate for specific data sets, leading to the development of more general models. So far, we have considered evolution at the DNA level. The evolution of protein-coding sequences, however, can also be modeled at the protein or codon level, requiring specific state spaces and specific rate matrices. We now review these various topics.

DNA models

The first proposed Markov model of DNA sequence evolution, called the Jukes-Cantor model, assumed a constant rate for every possible change [21]:

$$\text{JC model (one parameter):} \quad \mu_{xy} = \mu \quad \forall (x, y) \in \mathcal{E}^2. \tag{1.9}$$

This is a strong assumption that turned out to fit virtually no sequence data set. A very general feature of DNA sequence evolutionary processes is that *transitions* are more frequent than transversions. *Transitions* are changes within the purine {A,G} or pyrimidine {C,T} state subsets, while transversions are changes from purine to pyrimidine or from pyrimidine to purine. There are four *transitions* and eight transversions. The latter are more frequent for biochemical reasons: pyrimidines (respectively, purines) share a similar molecular structure. A change from C to T, for example, only requires one methylation and one deamination, while switching from, say, C to A is a much more complex chemical pathway. Kimura [22] amended Jukes and Cantor's model to distinguish these two kinds of changes:

$$\text{K2 model (two parameters):} \quad \mu_{xy} = \begin{cases} \alpha & \text{for } transitions \\ \beta & \text{for transversions.} \end{cases}$$

Note that the use of the word *transition* to define a subset of the possible changes of state is awkward since transition is usually defined as a synonym state change (as it is in Section 1.2 of this chapter). This confusion results from the collision between two bodies of literature (molecular evolution and stochastic processes) within which the term has an unambiguous meaning. In this chapter, italicized *transition* will refer to the specific category of base changes defined above, while regular transition will be used in its generic sense.

The JC and K2 models both have a balanced stationary distribution: proportion 0.25 is expected for the four bases at equilibrium under these models. But many DNA sequence data sets show unbalanced base composition, requiring the introduction of additional parameters. Equation (1.3) suggests the use of parameter π_x's, explicitly controlling the stationary distribution, as in the HKY model [18]:

$$\text{HKY model (five parameters):} \quad \mu_{xy} = \begin{cases} \alpha\, \pi_y & \text{for } transitions \\ \beta\, \pi_y & \text{for transversions.} \end{cases}$$

It should be noted that there are good biological reasons for having the proportions of C and G (respectively, A and T) in a genome equal. This comes from the fact that the DNA molecule is double-stranded, and made only with C:G and A:T pairs. If one assumes that the evolutionary processes followed by the two strands are identical, then the rate of change from (to) C and from (to) G (respectively, A and T) in one strand should be equal (because a G in the plus strand must change as soon as a C in the minus strand changes [23]). Very similar G and C contents (and A and T contents) are actually observed in most genomic sequences [23]. This suggests adding the $\pi_G = \pi_C$ and $\pi_A = \pi_T$ assumptions to the HKY model, resulting in the so-called T92 model [34]. The T92 model has three parameters, namely α, β, and $\theta = \pi_C + \pi_G$.

The REV or GTR (general time-reversible) model was defined above when introducing the concept of reversibility. It has nine parameters, namely six exchangeability terms s_{xy} and three stationary frequencies π_x (equation (1.3)). The HKY model is a special case of the REV model in which the s_{xy}'s are constrained: *transitions* (respectively, transversions) have to share a common s_{xy}. The T92 model is a special case of HKY in which $\pi_G = \pi_C$ and $\pi_A = \pi_T = 0.5 - \pi_G$. The K2 model is a special case of T92 in which $\pi_A = \pi_C = \pi_G = \pi_T$. The JC model is a special case of K2 in which $\alpha = \beta$. All these models are special cases of the most general, twelve-parameter model.

Amino acid models

The evolution of protein-coding genomic sequences can be considered at the protein, rather than DNA, level. Proteins are made of 20 distinct amino acids, so that the state space in protein models will have size 20, not 4. The mathematical background introduced in the previous section is still valid apart from this difference.

The protein rate matrix has 380 nondiagonal entries (to be compared with 12 in the DNA case), making the modeling effort more complex. Like DNA models, most protein models assume reversibility. The π_x's and s_{xy}'s are typically defined from the analysis of protein databases, resulting in generic models of protein evolution subsequently applied to any data set. There is, indeed, little hope to estimate the many entries of the protein rate matrix from a single data set. The so-called PAM, JTT, and WAG models are such database-defined models of protein evolution. See Chapter 9 of this volume for a detailed description of protein models.

Codon models

Protein-coding genes can also be analyzed at the codon level. A codon is a triplet of bases encoding for a certain amino acid. The information contained in codon sequences exceeds that of protein sequences since a given amino acid can be encoded by more than one codon. There are 61 sense codons, classi-fied in 20 groups of synonymous codons. Every codon of a group encodes the same amino acid. The group size ranges from one to six. Codon changes within groups are called synonymous. Such changes do not modify the sequence of the encoded protein. They have no (or weak) functional consequences. Between-group codon changes are called nonsynonymous. Nonsynonymous changes can affect the function of proteins, and the fitness of organisms: their fate is influ-enced by natural selection.

The relative rate of synonymous and nonsynonymous change occurring in a gene therefore provides some information about the selective forces apply-ing to that gene. Synonymous changes are essentially neutral (or under weak selection) and accumulate at a rate equal (close) to the mutation rate. Non-synonymous changes, in contrast, can be strongly selected, either negatively (if deleterious) or positively (if advantageous). Most genes are undergoing negative (i.e. purifying) selection, resulting in a synonymous/nonsynonymous rate ratio lower than one. Neutrally evolving genes (e.g. pseudogenes) have roughly equal synonymous and nonsynonymous rates. Genes showing a higher nonsynonymous than synonymous evolutionary rate are of biological interest: this pattern suggests that they have been recurrently adapting to some envi-ronmental change.

Goldman and Yang [13] introduced the first codon model in the early 1990s, and all the subsequent developments in the field were based on this contribution. This model has 63 parameters, namely 60 stationary frequencies π_{xyz}, where (xyz) is a codon, one *transition* rate α, one transversion rate β, and the nonsynonymous/synonymous ratio, ω. Entries of the 61×61 codon rate matrix are defined by the GY codon model (63 parameters):

$$\mu_{(x_1x_2x_3)(y_1y_2y_3)} = \begin{cases} \alpha\,\pi_{y_1y_2y_3} & \text{for synonymous } transitions \\ \beta\,\pi_{y_1y_2y_3} & \text{for synonymous transversions} \\ \omega\,\alpha\,\pi_{y_1y_2y_3} & \text{for nonsynonymous } transitions \\ \omega\,\beta\,\pi_{y_1y_2y_3} & \text{for nonsynonymous transversions.} \end{cases}$$

The formula above applies if the two codons $(x_1x_2x_3)$ and $(y_1y_2y_3)$ differ by exactly one base. The instantaneous rate of codon changes involving more than one base change is assumed to be zero. The probability of occurrence of such events after a finite amount of time is nonzero, however.

Of the 63 parameters of the GY model, the 60 codon equilibrium frequencies are usually not estimated by the maximum-likelihood method. Rather, they are estimated a priori as the observed frequency of every codon in the data set, or from the observed frequencies of bases at each codon position. In the latter case, the frequency π_{xyz} of codon (xyz) is estimated by the $\pi_x^1\pi_y^2\pi_z^3$ product, where π_w^i is the observed frequency of base w at codon position i in the data set. This greatly simplifies the likelihood maximization. This trick is also used for protein and DNA models.

Nonhomogeneous models

One fundamental assumption of the Markov models above is stationarity: the base (amino acid, codon) composition is assumed to be at equilibrium throughout the tree. This implies that the base composition is the same in every sequence of the data set. Some data sets, however, depart from this assumption. Observing significantly different base compositions between sequences implies that distinct evolutionary processes, with distinct stationary distributions, have been followed in distinct lineages. Simulation studies have shown that neglecting base composition variation between sequences when effective leads to biased phylogeny estimates in which sequences of similar base composition tend to be grouped irrespective of their true phylogenetic relationship [10].

To accommodate this peculiarity of some data sets, Galtier and Gouy developed a nonhomogeneous, nonstationary model of DNA sequence evolution [11]. Under this model, every branch of the tree follows a T92 process (see above). The α and β parameters are shared by all branches (common transition/transversion ratio), but the θ parameter is branch-specific: each branch has its own stationary G+C content. This accounts for variable G+C contents between sequences at the cost of a large increase in the number of parameters (one θ per branch versus one θ for the whole tree in the T92 model). The model is called nonhomogeneous because the rate matrix is not constant in time, nor between lineages. Yang and Roberts had previously proposed a nonhomogeneous version of the HKY model [43], but their implementation appeared to be limited to a small number of sequences.

A remarkable property of nonhomogeneous, nonstationary models is that ancestral base composition is a free parameter, whereas it is deducible from the rate matrix under the stationarity assumption. Past base compositions can therefore be estimated under these models. Computer simulations showed that reliable estimates of ancestral base composition can be obtained from data sets of reasonable size [11]. Also note that the likelihood becomes dependent on the location of the root under such nonstationary, nonreversible processes.

Nonindependence between sites

The vast majority of Markov models for sequence evolution make the assumption of independence between sites. This has the desirable property of validating equation (1.6), greatly simplifying the likelihood calculation. This assumption, however, is quite probably violated by most coding sequence data sets. This is because sites in a protein (or an RNA) interact to determine the selected tridimensional structure (and function) of the molecule, and the evolutionary processes of interacting sites are not independent. Two major attempts were made to relax the independence assumption.

The first relies on the (plausible) idea that most molecular interactions involve neighbor (or nearly neighbor) amino acids (bases) in a protein (DNA, RNA) sequence. Yang, followed by Felsenstein and Churchill, introduced an autocorrelation parameter measuring how much the evolutionary rate of a site is correlated with that of neighboring sites ([41], [6]). Goldman and colleagues extended this view for the specific case of proteins. They propose that sites belong to a small number of structural categories (helices, sheets, loops, turns), neighboring sites having a higher probability than random sites to be in the same category. Each site category has a distinct rate matrix. The assignment of sites to categories is not known but modeled by a hidden Markov chain running along the sequence ([35], [31]). The likelihood is calculated by conditioning on possible assignments of sites to categories, the probability of certain assignment being controlled by the hidden Markov chain. Evolutionary modes, not rates, are correlated between neighbor sites in this model.

Pollock and co-workers tackled the problem differently, without a relationship to the site neighborhood. Consider the joint evolutionary process of any two sites of a protein. The state space for the joint process is $\mathcal{E} \times \mathcal{E}$. Under the assumption of independent sites, the rate matrix for the joint process is deductible from that of the single-site process (assume reversibility),

$$
\begin{aligned}
\bar{\mu}_{xx',yx'} &= \mu_{xy} = s_{xy}\pi_y, \\
\bar{\mu}_{xx',xy'} &= \mu_{x'y'} = s_{x'y'}\pi'_y, \\
\bar{\mu}_{xx',yy'} &= 0,
\end{aligned}
\tag{1.10}
$$

for $x \neq y$ and $x' \neq y'$, where $\bar{\mu}_{xx',yy'}$ is the rate of change from x to y at site 1 and from x' to y' at site 2 (in $\mathcal{E} \times \mathcal{E}$), and where μ_{xy}, s_{xy}, and π_y are the above-defined transition terms for the single-site process (in $\mathcal{E}$). Modeling nonindependence between the two sites involves departing from equations (1.10). This is naturally achieved by amending stationary frequencies. It is easy to show that the stationary frequency $\bar{\pi}_{xx'}$ of state $(x, x') \in \mathcal{E}$ is equal to the $\pi_x \pi'_x$ product under the independence assumption. Nonindependence can be introduced by rewriting the equation above as:

$$\bar{\mu}_{xx',yx'} = s_{xy}\bar{\pi}_{yx'},$$
$$\bar{\mu}_{xx',xy'} = s_{x'y'}\bar{\pi}_{xy'}, \tag{1.11}$$
$$\bar{\mu}_{xx',yy'} = 0,$$

where $\bar{\pi}_{xx'}$'s are free parameters (possibly some function of π_x's). This formalization accounts for the existence of frequent and infrequent combinations of states between the two sites, perhaps distinct from the product of marginal site-specific frequencies. Pollock and co-workers applied this idea (first introduced by Pagel for quantitative characters [27]) in a simplified, two-state model of protein evolution, with the aim of detecting pairs of co-evolving amino acid sites in vertebrate myoglobin [29]. Duret and Galtier somewhat combined the two approaches to model the evolution of (overlapping) pairs of successive bases in the human genome [3].

1.3.2 Modeling Variations of Evolutionary Rate

Molecular evolutionary rates are of primary interest because they reflect to some extent the way natural selection applies to molecules. If one assumes that the mutation rate (that is, the rate of random occurrence of changes in individuals of a population) is more or less constant–a reasonable assumption for many data sets–then only natural selection (that is, the force determining the chances of eventual fixation of mutations in the population) can explain differences in evolutionary rates between lineages, molecules, or sites. Functionally important sequences will evolve at a slower rate (because most changes are deleterious and therefore eliminated by natural selection) than nonfunctional DNA. This idea was introduced above (codon models) when defining the synonymous and nonsynonymous evolutionary rates of protein-coding sequences.

Rate variation between lineages

Consider a homogeneous Markov process with constant transition rate μ (assume a JC model for simplicity) running along the branches of a phylogenetic tree whose lengths t_i are measured in some time unit. Present-day sequences (leaves) are equidistant from the ancestral sequence (root)–such trees are called ultrametric. A sequence that would evolve this way is said to be consistent with the so-called molecular clock hypothesis [46]. It is a fact that many data sets depart from the molecular clock assumption, sometimes spectacularly (e.g. [28]): some lineages evolve faster than others, possibly because the selective pressure applied to the sequence varies in time and between lineages.

A simple way to account for a departure from the clock is to fully relax the homogeneity assumption by letting each branch of the tree have its own freely varying transition rate μ (again in the JC case). Note that an equivalent model would be reached by assuming a constant μ but unconstrained branch lengths t_i. The relevant recoverable parameters are the $\mu\, t_i$ products (i.e. the amount of evolution between connected nodes). Allowing unconstrained branch lengths is the most flexible way to represent departure from the clock. It is the default option of most phylogenetic methods and programs.

Alternatively, one might want to model the way the transition rate varies over time, maybe with the aim of reconstructing past events of rate change. This can be achieved by assuming that transition rate μ is itself evolving according to some stochastic process upon which the process of sequence evolution is dependent ([36], [20]). Rate changes are assumed to occur at nodes of the tree in [36], continuously in [20]. These models are less parameter-rich than the standard, one-parameter-per-branch model. Full-likelihood calculation, however, is computationally difficult under these models because sites are correlated: a putative event of rate change would affect all sites simultaneously, making equation (1.6) incorrect. These models have been used in the Bayesian framework, where the integration over all possible scenarios of rate change is achieved through Monte Carlo Markov chains.

Rate variation between sites

The distinct sites of a molecule do not evolve at the same rate: functional sites are mostly conserved, showing little or no variation between sequences, while unimportant sites are free to evolve. This is a strong determinant of coding sequence variation patterns, with important implications with respect to the molecular structure/function link.

Yang first introduced likelihood calculation under the hypothesis of variable rates ([39], [40]). He proposed to model the variation of evolutionary rates across sites by a Gamma (rather than constant) distribution. The likelihood for site Y_i is therefore integrated over all possible rates,

$$\Pr(Y_i \mid \theta) = \int_0^\infty f_\Gamma(u)\, \Pr(Y_i \mid r(Y_i) = u, \theta)\, du, \qquad (1.12)$$

where f_Γ is the probability density of the Gamma distribution and where $\Pr(Y_i \mid r(Y_i) = u)$ is the likelihood for site Y_i conditional on rate u for this site. The latter term is easily calculated by applying recurrence (1.8) after having multiplied branch lengths by u. The variance (and shape) of the Gamma distribution is determined by a parameter that can be estimated from the data. The continuous Gamma distribution is usually discretized to avoid the integration of equation (1.12). Equation (1.12) becomes

$$\Pr(Y_i \mid \theta) = \sum_{j=1}^{g} \Pr(j)\, \Pr(Y_i \mid r(Y_i) = r_j, \theta), \qquad (1.13)$$

where g is the assumed number of rate classes and $\Pr(j)$ the probability of class j ($1/g$ for an equiprobable discretization). The complexity of the likelihood calculation under the discrete-Gamma model of rate variation is therefore g times the complexity of the equal-rate calculation. Waddell et al. explored other distributions for variable rates among sites [38]. In the case of codon models, Yang and co-workers used many different distributions to model the variation between sites in ω, the ratio of nonsynonymous to synonymous substitution rates [42].

Note that sites are not assigned to rate classes in this calculation. Rather, all possible assignments are considered and the conditional likelihoods averaged. Sites can be assigned to rate classes posterior to the calculation. The posterior probability of class j for site Y_i can be defined as

$$\Pr(Y_i \text{ in class } j) = \frac{\Pr(j)\Pr(Y_i \mid r(Y_i) = r_j)}{\Pr(Y_i)}, \qquad (1.14)$$

where the calculation is achieved using the maximum-likelihood estimates of parameters (tree, branch lengths, rate matrix). This equation does not account for the uncertainty on unknown parameters, an approximate procedure called "empirical Bayesian" [45].

"Covarion" models

In models of rate variation between sites, the (relative) rate of a site is constant in time: a slow site is slow in every lineage of the tree. There are biological reasons, however, why the specific rate of a site could vary. The rate of a site essentially reflects its level of structural constraint: sites important for the tridimensional structure (and therefore the function) of a protein cannot change much. But tridimensional structures evolve in the long run. The level of constraint applying to a certain site might therefore vary in time.

The notion that the evolutionary rate of a site can evolve was first introduced by Fitch [8] and subsequently modeled by Tuffley and Steel [37] and Galtier [9]. This process has been called covarion (for COncomitantly VARIable codON [8]), heterotachy, or site-specific rate variation. The covarion model is close, in spirit, to models of the relaxed molecular clock [20]. The rate of a site evolves in time according to some continuous-time process. The process of sequence change is defined conditionally on the outcome of the rate process. In the covarion model, each site runs its own specific rate process, so that not all sites are simultaneously rapid or slow, in contrast with the relaxed-clock model. The covarion model is an instance of the so-called Markov-modulated Markov chains [7].

Likelihood calculation is tractable under the covarion model. Sites are independent, allowing the use of equation (1.6). Recursion (1.8) must be modified to account for the underlying rate process. This is achieved by considering the compound process as a single process taking values on $\mathcal{E} \times \mathcal{G}$, where $\mathcal{G}$ is the set of possible rates. The rate matrix of this process is deductible from those of the rate and base (amino acid, codon) processes (see [9] for details).

It is worth noting that Fitch's initial definition of covarion implied both the notions of both site-specific rate change and nonindependent sites. In Fitch's model, only a subset of sites can evolve (covary) during a given period of time, but this set varies in time, the reasons for this variation being interactions between sites. Modern literature has separated "covariation" (that is, nonindependent sites) and site-specific rate variation. Unfortunately, the word "covarion" has been misleadingly kept as a synonym for "site-specific rate variation" so that the modern meaning of covarion has nothing to do with covariation.

1.4 Use of Markov Models for Phylogenetic Analyses

The many models presented above can be used and combined to represent the peculiarities of various genomic sequence data sets. This is done in the first place with the goal of reconstructing the past: testing evolutionary hypotheses, and estimating evolutionary parameters. This requires fitting the model to data, typically in the maximum-likelihood framework. Another use of Markov models in molecular evolution is simulation, which we now examine.

1.4.1 Simulations

Simulating data means running a statistical model on a computer using arbitrary (controlled) parameter values and storing the outcome. In the case of Markov models for sequence evolution, simulations are typically achieved by (i) randomly drawing an ancestral sequence at the root of a given tree (typically from the stationary distribution of the simulated process) and (ii) making this sequence evolve by recursively drawing states at child nodes from the $p_{xy}(t_i)$ probability distribution, where the recursion stops at leaves. The number and length of sequences, the tree, branch lengths, and process have to be chosen prior to simulation either arbitrarily or randomly drawing in some distribution on the tree space.

Simulations have been used during the 1980s and 1990s to compare the efficiency of competing tree-building methods: simulate data sets using a certain model tree $\mathcal{T}$, and ask how often methods will reconstruct $\mathcal{T}$ from the simulated data. Likelihood-based methods are now consensually considered as optimal. Some of their statistical properties (consistency, accuracy) are known theoretically or can be derived analytically. Simulations can still be useful to compare algorithms of likelihood maximization [16] or to assess the robustness of phylogenetic estimates: simulate data under model $\mathcal{M}_1$ and estimate parameters using model $\mathcal{M}_2$ to check how problematic it is to use a wrong model.

Simulations are also used to calculate confidence intervals around parameter estimates. The procedure is as follows: (i) estimate parameters of a model from the data by the maximum-likelihood method, (ii) simulate many data sets using parameter values equal to the ML estimates, (iii) for each simulated data set, reestimate parameters by ML, and (iv) seek an interval including 95% (or 99%) of these estimates. This is the very definition of a confidence interval; simulation is often the only way to calculate confidence intervals in phylogeny. The technique above is often called "parametric bootstrap" [19].

1.4.2 Hypothesis Testing

The likelihood framework provides tools for hypothesis testing, the so-called likelihood-ratio test (LRT). Let $\mathcal{M}_0$ (p_0 parameters) and $\mathcal{M}_1$ ($p_1 > p_0$ parameters) be two models, and assume that $\mathcal{M}_0$ is nested into (i.e., it is a special case of) $\mathcal{M}_1$. For example, $\mathcal{M}_0$ could be the JC model and $\mathcal{M}_1$ the K2 model. Now fit the two models to some data, call L_0 and L_1 the corresponding maximum likelihoods, and define the LR statistics as

$$\text{LR} = 2 \ln\left(\frac{L_1}{L_0}\right). \tag{1.15}$$

It can be shown that this statistic is asymptotically (that is, for an infinite amount of data) χ^2 distributed with $p_1 - p_0$ degrees of freedom under the null $\mathcal{M}_0$ hypothesis. L_1 must be higher than (or equal to) L_0 since $\mathcal{M}_0$ is a special case of $\mathcal{M}_1$. The LRT quantifies the expected increase in log-likelihood obtained by switching from $\mathcal{M}_0$ to $\mathcal{M}_1$ if data had been generated under $\mathcal{M}_0$. A data set showing an excessive increase in log-likelihood would lead to rejection of $\mathcal{M}_0$.

LRT between Markov models has been used for testing various evolutionary hypotheses. The molecular clock hypothesis, for instance, can be tested by comparing clock and relaxed-clock models. This is important for the purpose of dating events of speciation (i.e., internal nodes of the tree): divergence time is proportional to sequence divergence if and only if the molecular clock hypothesis holds (see below). LRT has also been used to test selective hypotheses from codon sequence data. Selective scenarios are modeled by making assumptions about the distribution of ω (nonsynonymous/synonymous rate ratio) across sites [42]. The neutral model, for example, assumes a constant $\omega = 1$ for all sites. A model involving purifying selection would let one class of sites have $\omega < 1$. Recurrent adaptation at some sites is modeled by incorporating an additional class with $\omega > 1$. LRT is used to compare these competing models, usually with the goal of detecting adaptation (that is, having a significant increase in log-likelihood when adding a class of sites with $\omega > 1$).

LRT can also be used at the level of the site. Pupko and Galtier, for example, proposed an LRT for the detection of covarion-like sites [30] (i.e., sites having a high substitution rate in certain subtrees but a low rate in the remaining part of the tree). This approach is useful to detect functional shifts in the history of a molecule, as illustrated by the analysis of primate mitochondr-ial proteins [30]. Such tests are typically applied to every site of an alignment so that the statistical problem of multiple testing must be addressed.

1.4.3 Parameter Estimation

If you believe in some model of sequence evolution, you might want to esti-mate parameters of this model from the data (i.e., recovering the past from the present). This is achieved by the maximum-likelihood (ML) method: the ML estimate (MLE) of the parameters is the set of parameter values that maximizes the likelihood; that is, the probability of the data. Parameters of interest include the tree topology, branch lengths, parameters of the rate matrix, and ancestral sequences and base compositions. This is the most vo-luminous body of literature in the field. Fast and accurate methods are now available for parameter estimation (e.g., [16]).

Reconstructing trees is a goal pursued by most users of Markov model-based phylogenetic methods. Trees are useful because they are the basis of systematics, the field of biology aiming at understanding biological diversity and its origins. Branch lengths are also of interest, either for dating purposes and for subsequent links between molecular phylogenies and paleontology or for understanding the dynamics of molecular evolutionary rates (how they change and why). With regard to the rate matrix, some parameters are of little interest (e.g., the *transition*/transversion ratio), but others have a clear bio-logical meaning. The ω parameter of codon models, for example, measures the amount and nature of the selective pressure applying to sequences or codons and is worth estimating (e.g., [44]). Nonstationary models allow estimation of ancient base composition, as indicated above. For specific genes, base compo-sition can be related to ecological features (e.g., growth temperature in bac-teria), and Markov models can be used to infer life-history traits of ancestral species [12]. Ancestral sequences can also be estimated from Markov models, such as using the empirical Bayesian approach [43]. This lead two spectacular studies in which the reconstructed ancestral proteins were synthesized in vitro and their biochemical properties compared with that of extant proteins [1].

1.4.4 Model Choice

The question of which model to choose for a given data set is an important one that we did not yet address. We have, however, introduced the LRT, a technique for comparing nested models. LRT will favor complex model $\mathcal{M}_1$ over simpler model $\mathcal{M}_0$ if the increase in log-likelihood yielded by switching from $\mathcal{M}_0$ to $\mathcal{M}_1$ is higher than expected under $\mathcal{M}_0$. The so-called Akaike information criterion (AIC) is a related likelihood-based measure of model appropriateness applicable for nonnested competing models. AIC is defined as

$$\text{AIC} = -2\ln(L) + 2p,$$

where L and p are the maximum likelihood and number of parameters of the model considered. The model minimizing AIC will be considered the most suitable. LRT and AIC are commonly used in current molecular evolution literature when the problem of choosing a model occurs.

We would argue, however, that this usage of LRT or AIC is a bit too systematic. Markov models of sequence evolution are built to address a variety of biological questions, as we just discussed. It is unclear that, for a given data set, these many purposes will require a common model. Of central interest is the problem of model choice in molecular phylogeny. That a certain model is favored by the AIC or LRT techniques does not guarantee that it is optimal for the purpose of estimating the tree. AIC and LRT favor models optimizing the balance between accuracy (fit) and number of parameters, something not directly related to the desirable properties of an estimator, namely small bias and small sampling variance. Empirical results from distance-based tree-building methods suggest that using a model simpler (i.e., less parameter-rich) than the true model can improve phylogenetic estimates ([33], [15]). We hope that forthcoming work in this area will clarify the relationship between the bias/variance and fit/parameter-richness balances and maybe amend current recommendations about AIC/LRT-based model choice in molecular phylogeny.

1.5 The Future of Markov Models for Sequence Evolution

A considerable amount of work has been done on Markov models for sequence evolution, from the theoretical basis to the use of highly specific models for inference purposes. One may wonder whether this is a nearly closed body of literature or whether important advances may be expected in the near future. We would speculate that the answer is double-faced, again because of the multiplicity of uses for Markov chains.

As far as the problem of tree reconstruction is concerned, we do not expect major advances from the Markov chain literature in the future. Building new models that would fit the data more accurately would not clearly lead to an improvement of phylogenetic estimators, as discussed above. Perspectives in this field have more to do with data management (e.g., dealing with conflicting data sets, that is, detecting the existence of distinct trees for distinct genes) than with improvements of models of sequence evolution, in our opinion.

Evolutionary genomics (that is, understanding the way genes and genomes evolve) should, in contrast, benefit from further developments of Markov models. Models explicitly aiming at representing the various evolutionary forces applying to genomic sequences, and especially natural selection, have just started to be built. These include the various codon models, models for non-independent sites, and the covarion model, for instance. Further refinements of these models are to be expected, especially in the context of Monte Carlo

Markov chain (MCMC) Bayesian analysis, a technique that overcomes many computational limitations induced by complex models (see Chapter 3 of this volume). Improving Markov models of sequence evolution should help in understanding the way genomes evolve and how their diversity originated–one of the big current issues in biology.

References

[1] B. S. W. Chang and M. J. Donoghue. Reconstructing ancestral proteins. *Trends in Ecology and Evolution*, 15:109–113, 2000.

[2] B. Charlesworth. *Evolution in Age-Structured Populations*. Cambridge University Press, Camebridge, 1994.

[3] L. Duret and N. Galtier. The covariation between TpA deficiency, CpG deficiency, and G+C content of human isochores is due to a mathematical artifact. *Molecular Biology and Evolution*, 17:1620–1625, 2000.

[4] A. W. F. Edwards. *Likelihood.* Cambridge University Press, Camebridge, 1972.

[5] J. Felsenstein. Evolutionary trees from DNA sequences: a maximum likelihood approach. *Journal of Molecular Evolution*, 17:368–376, 1981.

[6] J. Felsenstein and G. A. Churchill. A Hidden Markov Model approach to variation among sites in rate of evolution. *Molecular Biology and Evolution*, 13:93–104, 1996.

[7] W. Fischer and K. Meier-Hellstern. The Markov-modulated Poisson process (MMPP) cookbook. *Performance Evaluation*, 18:149–171, 1992.

[8] W. M. Fitch. Rate of change of concomitantly variable codons. *Journal of Molecular Evolution*, 1:84–96, 1971.

[9] N. Galtier. Maximum-likelihood phylogenetic analysis under a covarion-like model. *Molecular Biology and Evolution*, 18:866–873, 2001.

[10] N. Galtier and M. Gouy. Inferring phylogenies from DNA sequences of unequal base compositions. *Proceedings of The National Academy of Sciences of the USA*, 92:11317–11321, 1995.

[11] N. Galtier and M. Gouy. Inferring pattern and process: maximum-likelihood implementation of a nonhomogeneous model of DNA sequence evolution for phylogenetic analysis. *Molecular Biology and Evolution*, 15:871–879, 1998.

[12] N. Galtier, N. Tourasse, and M. Gouy. A nonhyperthermophilic common ancestor to extant life forms. *Science*, 283:220–221, 1999.

[13] N. Goldman and Z. Yang. A codon-based model of nucleotide substitution for protein-coding DNA sequences. *Molecular Biology and Evolution*, 11:725–736, 1994.

[14] F. Gosselin and J. D. Lebreton. *The Potential of Branching Processes as a Modeling Tool for Conservation Biology*, pages 199–225. Springer, New York, 2000.

[15] S. Guindon and O. Gascuel. Efficient biased estimation of evolutionary distances when substitution rates vary across sites. *Molecular Biology and Evolution*, 19:534–543, 2002.

[16] S. Guindon and O. Gascuel. A simple, fast and accurate algorithm to estimate large phylogenies by maximum likelihood. *Systematic Biology*, 52:696–704, 2003.

[17] P. H. Harvey and A. Purvis. Comparative methods for explaining adaptations. *Nature*, 351:619–624, 1991.

[18] M. Hasegawa, H. Kishino, and T. Yano. Dating of the human-ape splitting by a molecular clock of mitochondrial DNA. *Journal of Molecular Evolution*, 22:160–174, 1985.

[19] J. P. Huelsenbeck, D. M. Hillis, and R. Jones. *Parametric Bootstrapping in Molecular Phylogenetics: Applications and Performances*, pages 19–45. Wiley-Liss, New York, 1996.

[20] J. P. Huelsenbeck, B. Larget, and D. Swofford. A compound Poisson process for relaxing the molecular clock. *Genetics*, 154:1879–1892, 2000.

[21] T. H. Jukes and C. R. Cantor. *Evolution of Protein Molecules*, pages 21–132. Academic Press, New York, 1969.

[22] M. Kimura. A simple method for estimating evolutionary rates of base substitutions through comparative studies of nucleotide sequences. *Journal of Molecular Evolution*, 16:111–120, 1980.

[23] J. R. Lobry. Properties of a general model of DNA evolution under no-strand-bias conditions. *Journal of Molecular Evolution*, 40:326–330, 1995.

[24] G. Meszéna, E. Kisdi, U. Dieckmann, S. A. H. Geritz, and J. A. J. Metz. Evolutionary optimization models and matrix games in the unified perspective of adaptive dynamics. *Selection*, 2:193–210, 2001.

[25] C. Moler and C. Van Loan. Nineteen dubious ways to compute the exponential of a matrix. *SIAM Review*, 20:801–836, 1978.

[26] M. Nordborg. *Coalescent Theory*, pages 179–212. Wiley, Chichester, 2001.

[27] M. D. Pagel. Detecting correlated evolution on phylogenies: a general method for the comparative analysis of discrete characters. *Proceedings of the Royal Society of London, Series B: Biological Sciences*, 255:37–45, 1994.

[28] J. Pawlowski, I. Bolivar, J. F. Fahrni, C. De Vargas, M. Gouy, and L. Zaninetti. Extreme differences in rates of molecular evolution of foraminifera revealed by comparison of ribosomal DNA sequences and the fossil record. *Molecular Biology and Evolution*, 14:498–505, 1997.

[29] D. D. Pollock, W. R. Taylor, and N. Goldman. Coevolving protein residues: maximum likelihood identification and relationship to structure. *Journal of Molecular Biology*, 287:187–198, 1999.

[30] T. Pupko and N. Galtier. A covarion-based method for detecting molecular adaptation: application to the evolution of primate mitochondrial

genomes. *Proceedings of the Royal Society of London, Series B: Biological Sciences*, 269:1313–1316, 2002.

[31] D. M. Robinson, D. T. Jones, H. Kishino, N. Goldman, and J. L. Thorne. Protein evolution with dependence among codons due to tertiary structure. *Molecular Biology and Evolution*, 20:1692–1704, 2003.

[32] S. Ross. *Stochastic Processes*. John Wiley and Sons, New York, 1996.

[33] A. Rzhetsky and T. Sitnikova. When is it safe to use an oversimplified substitution model in tree-making? *Molecular Biology and Evolution*, 13:1255–1265, 1996.

[34] K. Tamura. Estimation of the number of nucleotide substitutions when there are strong transition-transversion and G+C-content biases. *Molecular Biology and Evolution*, 9:678–687, 1992.

[35] J. L. Thorne, N. Goldman, and D. T. Jones. Combining protein evolution and secondary structure. *Molecular Biology and Evolution*, 13:666–673, 1996.

[36] J. L. Thorne, H. Kishino, and I. S. Painter. Estimating the rate of evolution of the rate of molecular evolution. *Molecular Biology and Evolution*, 15:1647–1657, 1998.

[37] C. Tuffley and M. A. Steel. Modeling the covarion hypothesis of nucleotide substitution. *Mathematical Biosciences*, 147:63–91, 1998.

[38] P. J. Waddell, D. Penny, and T. Moore. Hadamard conjugations and modeling sequence evolution with unequal rates across sites. *Molecular Phylogenetics and Evolution*, 8:33–50, 1997.

[39] Z. Yang. Maximum-likelihood estimation of phylogeny from DNA sequences when substitution rates differ over sites. *Molecular Biology and Evolution*, 10:1396–1401, 1993.

[40] Z. Yang. Maximum likelihood phylogenetic estimation from DNA sequences with variable rates over sites: Approximate methods. *Journal of Molecular Evolution*, 39:306–314, 1994.

[41] Z. Yang. A space-time process model for the evolution of DNA sequences. *Genetics*, 139:993–1005, 1995.

[42] Z. Yang, R. Nielsen, N. Goldman, and A. M. Pedersen. Codon-substitution models for heterogeneous selection pressure at amino acid sites. *Genetics*, 155:431–449, 2000.

[43] Z. Yang and D. Roberts. On the use of nucleic acid sequences to infer early branchings in the tree of life. *Molecular Biology and Evolution*, 12:451–458, 1995.

[44] Z. Yang, W. J. Swanson, and V. D. Vacquier. Maximum-likelihood analysis of molecular adaptation in abalone sperm lysin reveals variable selective pressures among lineages and sites. *Molecular Biology and Evolution*, 17:1446–1455, 2000.

[45] Z. Yang and T. Wang. Mixed model analysis of DNA sequence evolution. *Biometrics*, 51:552–561, 1995.

[46] E. Zuckerkandl and L. Pauling. Molecules as documents of evolutionary history. *Journal of Theoretical Biology*, 8:357–366, 1965.

2

Introduction to Applications of the Likelihood Function in Molecular Evolution

Jutta Buschbom[1] and Arndt von Haeseler[2,3]

[1] Institut für Bioinformatik, Heinrich-Heine Universität, Düsseldorf, Germany,
jbuschbom@cs.uni-duesseldorf.de
[2] Heinrich-Heine Universität, Düsseldorf, Germany,
haeseler@cs.uni-duesseldorf.de
[3] Neumann Institute for Computing, FZ Jülich, Germany,
haeseler@cs.uni-duesseldorf.de

2.1 Introduction

This chapter is about the likelihood function in the context of molecular evolution. We will introduce the concept of likelihood and try to illuminate the flexibility of the likelihood approach in terms of modeling, inference, and test statistics.

From its beginning, molecular evolution has dealt with the analysis of data that are amenable to mathematical description and statistical testing. Composed of building blocks representing a limited alphabet, molecular data are–ideally–the product of comparatively simple recurring processes with predictable outcomes.

Zuckerkandl and Pauling [23] took advantage of this situation by proposing one of the now classical null hypotheses in evolutionary biology. Their molecular clock hypothesis states that the rate of change for a given protein is more or less constant over time and in different evolutionary lineages. Upon publication, this hypothesis created tumult among classical biologists since it seemed to contradict the traditional view of evolution that some organisms are evolved "further," while others might represent relics. Zuckerkandl and Pauling's hypothesis, being based on an explicit model of protein substitution, however, presents the advantage that it can be tested and potentially rejected. Moreover, the explicit model representing the hypothesis provides the opportunity to estimate actual values for evolutionary rates and thus gain further, more detailed information on the underlying processes.

Nowadays molecular evolution is a flourishing field of scientific research that profits from the advances in molecular biology and statistics and last but not least from the increasing power of modern computers, so that it is impossible to cover all the areas where the likelihood function comes into play. Thus, we will only be able to define the likelihood function and sketch upon some central ideas through examples. The fine details of the varying and wide-ranging applications are left to subsequent chapters.

2.1.1 Terminology

To set the stage, we introduce some terminology that will be used throughout this chapter. We observe *individuals* out of a population. Based on these observations, we ask what the characteristics of the whole *population* might be; that is, we want to know the values of *variables* that describe its characteristics (e. g., genetic composition, demography, historical events, and population processes). Generally, mathematical *models* need to be employed to arrive at values for the characteristics of interest. These models summarize background information and describe defined interactions among several variables.

If we collected information on all individuals of a population, we would obtain a *census* of the population. However, without such a drastic and often impossible measure, the only chance to obtain information about the population is to draw a random *sample* and analyze the data contained in the sample with regard to the variables that interest us. If the sample is representative of the whole population, we can use it to infer the characteristics of the original population. This we will do by specifying evolutionary models for various questions of interest to a molecular biologist with some knowledge of statistics.

2.2 The Likelihood

2.2.1 The Likelihood Function

Likelihood has become one of the central concepts in statistical inference [7]. It provides the means through which the information supplied by a sample can be incorporated into the process of statistical inference; that is, in arriving at a conclusion about characteristics of the underlying population.

A typical textbook introduces the likelihood as a function L of a hypothesis H, given a set of observations O and assuming a specific interaction model or set of model parameters. The likelihood $L(H|O)$ is proportional to the conditional probability $P(O|H)$ of observing the data given that hypothesis H applies. More formally,

$$L(H|O) = C \cdot P(O|H), \tag{2.1}$$

where C denotes an arbitrary constant.

To be a bit more formal, we regard the observation O as the realization of a random variable $\mathbf{X} = (X_1, \ldots, X_m)^T$ with an unknown probability distribution (with respect to an appropriate measure) that is defined by a probability density function $p(x) = p_{\mathbf{X}}(x)$. Moreover, the unknown density is restricted to an appropriate family of distributions. We will consider densities that involve a finite number of unknown parameters $\boldsymbol{\theta} = (\theta_1, \ldots, \theta_k)^T$. Finally, the *parameter space* defines the region of possible values of $\boldsymbol{\theta}$. The notation $p_{\mathbf{X}}(\mathbf{x}, \boldsymbol{\theta})$ indicates the dependency of the density on $\boldsymbol{\theta}$. In other words, the model function, $p_{\mathbf{X}}(\mathbf{x}, \boldsymbol{\theta})$, describes the model we have in mind. When studying the model function, one may study the effect on the function on $\mathbf{x}$ for each fixed $\boldsymbol{\theta}$ (i.e, on the probability density determined by $\boldsymbol{\theta}$). On the other hand, we may switch the viewpoint and study the model function as a function of $\boldsymbol{\theta}$ for fixed $\mathbf{x}$. For an $\mathbf{x}$ actually observed, we obtain the *likelihood function*

$$L(\boldsymbol{\theta}) = L(\mathbf{x}|\boldsymbol{\theta}) = p_{\mathbf{X}}(\mathbf{x}|\boldsymbol{\theta}). \tag{2.2}$$

Sometimes, we will also discuss the random variable

$$L_{\mathbf{X}}(\mathbf{X}|\boldsymbol{\theta}) = p_{\mathbf{X}}(\mathbf{X}|\boldsymbol{\theta}). \tag{2.3}$$

The definitions above were given for arbitrary multidimensional random variables. In most applications, we will treat the components $X_1, X_2, \ldots, X_n$ of $\mathbf{X}$ as mutually independent and identically distributed with $p_X(X|\boldsymbol{\theta})$. Then

$$L(X_1, \ldots, X_n|\boldsymbol{\theta}) = \prod_{i=1}^{n} p(X_i \mid \boldsymbol{\theta}) \tag{2.4}$$

defines the likelihood function of the sample of size n. In some applications, it is convenient to study the natural logarithms of the likelihood function, which will be denoted by

$$\ell(X_1, \ldots, X_n|\boldsymbol{\theta}) = \sum_{i=1}^{n} \log[p(X_i \mid \boldsymbol{\theta})]. \tag{2.5}$$

Example 2.1 Binomial Distributions

Somebody catches a fish from a pond and considers an experiment as successful if a red individual is caught. The color of the fish is noted and the fish is returned to the pond. Then X is a random variable assuming values "red," "not red" with a certain probability $\theta \in [0, 1]$ (i.e., $p(X = \text{red}|\theta) = \theta$ and $p(X = \text{not red}|\theta) = 1 - \theta$). If $n = 10$ fish were caught and k were red, then the likelihood function according to equation (2.4) is defined as

$$L(X_1, \ldots, X_{10}|\theta) = \prod_{i=1}^{10} p(X_i \mid \theta) \tag{2.6}$$

$$= \binom{10}{k} \theta^k (1 - \theta)^{10-k}, \tag{2.7}$$

the binomial distribution with parameter θ and ten realizations, where $\binom{10}{k}$ denotes a factor that depends on the realization of the data only. Figure 2.1

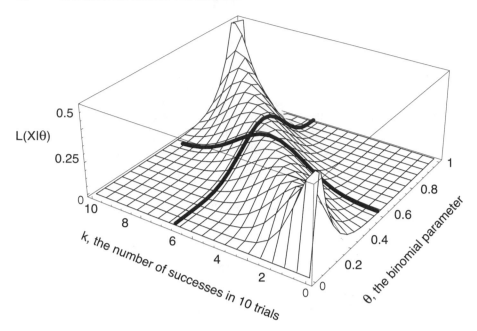

Fig. 2.1. Plot of the binomial distribution showing the relationship between probability and likelihood. The thick line running from "east to west" represents the continuous version of the binomial distribution with parameter 1/2. The line running from "south to north" displays the likelihood function when $k = 6$ successful experiments were observed.

displays this example, for all possible realizations of k. Figure 2.1 displays also the dual way of looking at the model function $p(\mathbf{X}, \boldsymbol{\theta})$. The thick line running from left to right represents the probability distribution for a fixed θ, whereas the highlighted line running from front to back illustrates the probability of observing $k = 6$ successful outcomes as a function of θ (i.e., represents one realization of $L(X|\theta)$).

2.2.2 Maximum Likelihood Estimation

One of the questions we want to address when dealing with likelihoods is the construction of parameter estimates. That is, we think of estimates as some plausible values $\hat{\boldsymbol{\theta}}$. Once we observe some data as in Example 2.1, the likelihood function depends only on $\boldsymbol{\theta}$. For any admissible value of $\boldsymbol{\theta}$, the likelihood function gives the (a priori) probability of observing what has actually been observed. This explains the name *likelihood*. Please notice that the likelihood function $L(\boldsymbol{\theta})$ is not a probability distribution, whereas all values of the likelihood function are probabilities. This interpretation then leads to the concept of *maximum likelihood estimate* (MLE), where we select the value of $\boldsymbol{\theta}$ that maximizes the likelihood function for a given realization $\mathbf{x}$ of the random variable $\mathbf{X}$. More formally,

$$\boldsymbol{\theta}^{\mathrm{ml}} = \operatorname{argmax}\{L(\mathbf{x}|\boldsymbol{\theta})|\boldsymbol{\theta}\} \qquad (2.8)$$

is the MLE of the likelihood function.

The MLE $\boldsymbol{\theta}^{\mathrm{ml}}$ provides the best explanation for observing the data $\mathbf{X}$ given the model. This does not mean that $\boldsymbol{\theta}^{\mathrm{ml}}$ is the value of $\boldsymbol{\theta}$ that maximizes the probability as n increases. E. L. Lehmann and G. Casella showed that under certain conditions for n independent observations drawn from $p(X|\boldsymbol{\theta}_0)$, the probability that

$$\prod_{i=1}^{n} p(X_i \mid \boldsymbol{\theta}_0) > \prod_{i=1}^{n} p(X_i \mid \boldsymbol{\theta})$$

approaches 1 as the sample size n tends to infinity, where θ_0 is the true value of the parameter. Since, by definition,

$$L(\boldsymbol{\theta}^{\mathrm{ml}}) \geq L(\mathbf{x}|\boldsymbol{\theta}),$$

for all θ, the combination of both inequalities suggests that in a large sample the MLE is close to the true value. This statement is made more precise by the definition of *consistency*, which states that as $n \to \infty$ the series $(\boldsymbol{\theta}_n^{\mathrm{ml}})$ tends to the true value θ_0 (for details see [14]).

Computing the MLE

Equation (2.4) defines the joint distribution of $X_1, \ldots, X_n$. Again, for a given realization $(x_1, \ldots, x_n)$ we consider $L(x_1, \ldots, x_n|\boldsymbol{\theta})$ as a function of $\boldsymbol{\theta}$. The MLE $\boldsymbol{\theta}^{\mathrm{ml}}$ is found by differentiation with respect to $\boldsymbol{\theta}$ and using standard calculus to prove maximality. If differentiation is used, then the equation

$$\frac{\partial L(X_1, \ldots, X_n|\boldsymbol{\theta})}{\partial \boldsymbol{\theta}} = 0 \qquad (2.9)$$

must be solved. Equation (2.9) is referred to as the *likelihood equation(s)*.

Sometimes it is easier to work with the natural logarithm of the likelihood function $\log[L]$, the so-called *support*. Thus, equation (2.9) becomes

$$\frac{\partial \log[L(X_1, \ldots, X_n|\boldsymbol{\theta})]}{\partial \boldsymbol{\theta}} = 0. \qquad (2.10)$$

Obviously, a solution of equation (2.9) or equation (2.10) is not necessarily the global maximum of the likelihood function. The derivatives in equation (2.10) are called scores. Global maximality can be difficult to establish. In certain instances, one can show that the likelihood function is concave, which implies that the solution to equation (2.9) is indeed the MLE. For the one-dimensional parameter case, it is sufficient to show that the second derivative is negative at $\boldsymbol{\theta}^{\mathrm{ml}}$; namely

$$\frac{\partial^2 \log[L(\boldsymbol{\theta})]}{\partial^2 \boldsymbol{\theta}}\bigg|_{\boldsymbol{\theta}=\boldsymbol{\theta}^{\mathrm{ml}}} < 0. \qquad (2.11)$$

If $\boldsymbol{\theta}$ is multidimensional, then it is sometimes possible to show that the matrix in equation (2.11) is negative definite and thus maximal locally but not necessarily globally.

Unfortunately, for many complex problems, such as the optimization of many parameters, analytical solutions cannot be obtained. In this case, numerical iterations are employed to arrive at approximations of the maximum likelihood estimator. Because these methods cannot guarantee actually finding the global maximum, it is advisable to search the parameter space for all solutions.

Example 2.1 (Continued)

The likelihood function (equation (2.6)) is an instance where the maximum likelihood estimate can be analytically obtained. To this end, consider the random variable X that counts the number of successful events in n trials, where each successful event has probability θ. Then the score equals

$$S(X, \theta) = \frac{X}{\theta} - \frac{n - X}{1 - \theta}. \qquad (2.12)$$

Setting equation (2.12) equal to zero and solving it for θ, we obtain the maximum likelihood estimator

$$\theta^{\mathrm{ml}} = \frac{X}{n}, \qquad (2.13)$$

which is the global maximum because the second derivative

$$-n\frac{\left(\theta^2 + \frac{X}{n}\right)}{\theta^2(1 - \theta)^2} \qquad (2.14)$$

is certainly less than zero at $\theta = \theta^{\mathrm{ml}}$.

Example 2.2: MLEs for Multinomial Distributions

In the context of molecular genetics, the method of maximum likelihood is frequently applied to questions involving the multinomial distribution. We assume that the random variables $X_1, \ldots, X_m$ count the number of elements in m distinct cells or categories, where the total number of elements equals n and the cell probabilities are $p_1, \ldots, p_m$ such that $p_i > 0$ and $p_1 + \ldots + p_m = 1$. The likelihood function for $\mathbf{x} = (x_1, \ldots, x_m)$ is then

$$L(x_1, \ldots x_m | p_1, \ldots, p_m) = \binom{n}{x_1 \ldots x_m} \prod_{i=1}^{m} p_i^{x_i}. \qquad (2.15)$$

A little calculus then shows that the MLEs are given by

$$p_i^{\text{mle}} = \frac{x_j}{n} \quad \text{for} \quad j = 1, \dots, m. \tag{2.16}$$

We will later encounter situations where the cell probabilities are functions of unknown parameters $\boldsymbol{\theta}$; that is, $p_i = p_i(\boldsymbol{\theta})$ for $i = 1, \ldots, m$. Then the log-likelihood function of $\boldsymbol{\theta}$ is

$$\ell(\mathbf{x}|\boldsymbol{\theta}) = \log n! - \sum_{i=1}^{m} \log x_i! + \sum_{i=1}^{m} x_i \log p_i(\boldsymbol{\theta}). \tag{2.17}$$

To find the MLE, we need to solve equation (2.10), which reduces to

$$\frac{\partial S(\theta)}{\partial \theta} = \sum_{i=1}^{m} x_i \frac{p_i'(\theta)}{p_i(\theta)} = 0 \tag{2.18}$$

if we assume that the dimension of $\boldsymbol{\theta}$ equals 1 and $p_i'(\theta)$ is the derivative with respect to θ. To find an analytical solution in this case can get quite tedious.

Before we conclude this subsection, it is worthwhile to point out several features of the maximum likelihood estimate that are relevant for practical applications.

1. It is not necessary to confine the parameter space to $\mathbb{R}^d$. In fact, we will later consider the branching pattern of a tree as a parameter.
2. Sometimes the MLE does not exist.
3. The MLE need not be unique. However, in most cases the maximum likelihood exists and is unique.
4. The likelihood function is maximized over the parameter space defined by the model and not over the set of mathematically admissible values.
5. In most realistic applications in molecular biology, the MLE has no closed-form expression. Thus, it must be computed numerically for the observed data $\mathbf{X}$. This leads to interesting numerical approaches. We will give some examples later.

2.2.3 Large Samples

In the previous section, we outlined approaches to estimate the parameter $\boldsymbol{\theta}^{\text{ml}}$ that provides the best estimate of the observed data. This so-called *point estimate* is thus a good guess about the unobservable parameter $\boldsymbol{\theta}_0$. In the following, we simply state some relevant results that give a recipe to estimate variances of the MLE or confidence intervals.

For the sake of illustration, we assume a one-dimensional parameter θ. Using the so-called *expected information*

$$\mathbb{I}(\theta) = -\mathbb{E}\left[\frac{\partial^2}{\partial \theta^2} \log L(\mathbf{X}|\theta)\right], \tag{2.19}$$

one can show that the large sample distribution is approximately normal with mean θ_0 and variance $1/(n\mathbb{I}(\theta_0))$. To indicate that this is only true as n approaches infinity, we typically say that the MLE is asymptotically unbiased and that $1/(n\mathbb{I}(\theta_0))$ is the asymptotic variance of the MLE.

Moreover, one can prove that the distribution of $\sqrt{\mathbb{I}(\theta_0))}(\theta^{\mathrm{mle}} - \theta_0)$ is approximately equal to the standard normal distribution $N(0,1)$. Replacing the unknown $\mathbb{I}(\theta_0)$ by $\mathbb{I}(\theta^{\mathrm{mle}})$, the approximation is still valid and we obtain

$$\theta^{\mathrm{mle}} \pm \frac{z(\alpha/2)}{\sqrt{n\mathbb{I}(\theta^{\mathrm{mle}})}} \tag{2.20}$$

as the approximate $100(1 - \alpha)\%$ confidence interval.

Example 2.1 (Continued)

We illustrate the procedure on Example 2.1. First, we need to compute the expected information (equation (2.19)), which becomes according to equation (2.14)

$$
\begin{aligned}
\mathbb{I}(\theta) &= -\mathbb{E}\left[-\frac{X}{\theta^2} + \frac{n - X}{(1 - \theta)^2}\right] \\
&= \frac{\mathbb{E}(X)}{\theta^2} + \frac{\mathbb{E}(X) - n}{(1 - \theta)^2} \\
&= \frac{n\theta}{\theta^2} + \frac{n\theta - n}{(1 - \theta)^2} \\
&= \frac{n}{\theta(1 - \theta)}.
\end{aligned}
$$

By substituting the MLE X/n and applying equation (2.20), we get

$$\frac{X}{n} \pm z(\alpha/2)\sqrt{\frac{X(n - X)}{n^3}}$$

as the approximate confidence interval for the accuracy of the estimation.

In the example above it was relatively straightforward to compute the approximate confidence interval. In molecular evolution, matters are more complicated and one must resort to *bootstrap estimates* for finding approximate confidence intervals [8].

We explain the bootstrap principle for Example 2.1. Because the true parameter and the distribution $\theta^{\mathrm{ml}} - \theta_0$ are not known, we use the MLE θ^{mle} to generate many samples, B, from a binomial distribution with parameters n (sample size) and θ^{ml}. For each randomly generated sample, we compute the MLE $\theta_b^{\mathrm{ml}}, b = 1,\ldots,B$. The unknown distribution $\theta^{\mathrm{ml}} - \theta_0$ is then approximated by the simulated distribution $\delta_b = \theta_b^{\mathrm{ml}} - \theta^{\mathrm{ml}}, b = 1,\ldots,B$. This distribution can then be used to compute the corresponding quantiles.

Such approaches are easily implemented and are widely distributed in the literature on molecular evolution.

2.2.4 Efficiency

Besides the MLE, the statistical literature provides a collection of parameter estimates, such as the sample mean or the method of moments. The question arises as to which one should be chosen. To aid a decision, statistics introduces the *efficiency* of two estimates $\hat{\theta}$ and $\tilde{\theta}$ as

$$\text{eff}(\hat{\theta}, \tilde{\theta}) = \frac{\text{Var}(\hat{\theta})}{\text{Var}(\tilde{\theta})}. \tag{2.21}$$

If the efficiency is smaller than 1, then $\tilde{\theta}$ has a larger variance than $\hat{\theta}$. The Cramer-Rao inequality then states [3] that the variance of an unbiased estimate of the unknown parameter θ is greater than or equal to $1/(n\mathbb{I}(\theta))$(assuming some condition on the distribution). From this inequality and the fact that the asymptotic variance of the maximum likelihood estimate is equal to the lower bound, we say that MLEs are *asymptotically efficient*. Note that the asymptotic efficiency does not allow a conclusion about the efficiency for finite sample sizes. Sometimes other estimators may be more efficient in the sense of equation (2.21).

2.2.5 Hypothesis Testing and Adequacy of Models

We have seen that the likelihood depends on an underlying model. To base our biological conclusions on solid grounds, it is necessary to have confidence in the models. We will outline some theoretical aspects to check whether models or hypotheses are appropriate. However, it is beyond the scope of this chapter to expound the full theory of testing hypotheses, and we will focus on the likelihood ratio tests.

General remarks

In the classical hypothesis setting one tests a *null hypothesis* H_0 against an *alternative hypothesis* H_A. In the first step, one specifies both hypotheses. This should be done before the data are actually observed. The statistical literature distinguishes between simple and composite hypotheses. In the former case, the numerical values of all unknown parameters in the probability distribution of interest are specified, whereas in the latter case not all unknown parameters are declared.

Following Neyman and Pearson, a decision whether or not to reject H_0 is made on the so-called **test statistic**, which is computed for the observed data $\mathbf{x}$. The choice of the test statistic is a feat in itself. Based on the test statistic, one defines the *acceptance region* (i.e., the set of values of the test statistic which accept H_0) and the *rejection region* (i.e., the set of values that reject H_0).

Because we are dealing with a random outcome of an experiment, H_0 may be rejected when it is true. This is the so-called *type I error*. The probability for this event is denoted by α. If H_0 is simple, then α is called the *significance level*. Not surprisingly, one also deals with a *type II error*, which is the probability β of accepting H_0 when it is false.

In an ideal world, one would like to have $\alpha = \beta = 0$, but this is almost always impossible. Thus, a pragmatic procedure is to determine a small significance level α in advance and then to construct a test with a small type II error.

The generalized likelihood ratio test

One important tool for gaining insight into different hypotheses in molecular evolution is the so-called *generalized likelihood ratio test*, which applies if the hypotheses are not simple. In the following, we outline the test statistic and then show that the null distribution of the appropriately scaled statistics is approximated by the chi-square distribution.

Assume that the model function $p_{\mathbf{X}}(\mathbf{x}|\boldsymbol{\theta})$ is given. To specify the null hypothesis, we constrain the parameter space to some subset ω_0 where we assume that the entire parameter space Ω is admissible for the alternative hypothesis. We then compute the test statistic

$$\Lambda = \frac{\max\{L(\mathbf{X}|\boldsymbol{\theta})|\boldsymbol{\theta} \in \omega_0\}}{\max\{L(\mathbf{X}|\boldsymbol{\theta})|\boldsymbol{\theta} \in \Omega\}}. \tag{2.22}$$

The null hypothesis is rejected if Λ is small. Now the following theorem holds. Under certain regularity conditions, the distribution

$$-2\log\Lambda \tag{2.23}$$

is for large sample sizes n approximately chi-square distributed with $m = \dim(\Omega) - \dim(\omega_0)$ degrees of freedom. For the sake of completeness, the density of the chi-square distribution with m degrees of freedom is

$$f(y) = \frac{1}{2^{m/2}\Gamma(m/2)}y^{\frac{m}{2}-1}\exp(-y/2) \tag{2.24}$$

for $y \geq 0$. Since $\omega_0 \subset \Omega$, the hypotheses are nested. If the test statistic in equation (2.23) is large for the observed data $\mathbf{x}$, then H_0 is rejected. Unfortunately, the chi-square approximation cannot always be used in applications of molecular evolutionary problems because typically large samples are required, and more importantly it is sometimes difficult to determine the degrees of freedom (see [11]). Sometimes the models are not nested. In such situations, it is useful to apply statistical tests based on Monte Carlo procedures.

Example 2.2 (Continued)

We consider the multinomial model outlined before. Our null hypothesis states that the cell probabilities depend on the k-dimensional parameter θ, where the alternative hypothesis is the full multinomial model. Then the unrestricted maximum likelihood estimator is given by equation (2.16), while $p_i(\theta^{\mathrm{ml}})$ are the cell probabilities under H_0. Then

$$-2\log(\Lambda) - 2\sum_{i=1}^{m} x_i \log\left(\frac{p_i(\theta^{\mathrm{ml}})}{p_i^{\mathrm{ml}}}\right) \approx \chi^2_{m-1-k} \qquad (2.25)$$

is approximately chi-square distributed with $m-1-k$ degrees of freedom.

2.2.6 Bayesian Inference

The Bayesian approach to statistics is quite different from the approaches we have explained. However, in recent years it has become quite popular in molecular evolution (for a recent review see [2]). So far, we have assumed that the parameter θ is an unknown and fixed quantity. From the observed data $\mathbf{x}$, we obtained some knowledge about θ; for example, by computing the MLE. In the "Bayesian world", θ is considered a random variable with a known probability distribution, the *prior distribution*, which needs to be specified in advance, hence the name prior. The prior distribution reflects our subjective knowledge about the plausibility of certain parameter values or hypotheses. Once the data have been observed, the prior distribution is "updated" using the information (i.e., the probability of the sample given the data). Thus the computation of the "update" probabilities is a simple exercise using the Bayes formula. Going back to equation (2.1), we compute the posterior probability of a collection of hypothesis models $\theta_i, i = 1, \ldots, k$ given some data $\mathbf{x}$ as

$$P(\theta_i|\mathbf{x}) = \frac{L(\mathbf{x}|\theta_i)P(\theta_i)}{\sum_{j=1}^{k} L(\mathbf{x}|\theta_j)P(\theta_j)}, \qquad (2.26)$$

where we require that $\sum P(\theta_j) = 1$. Without entering the discussion about the theoretical controversies of the Bayesian approach, we simply point out that the success of the method is due to an enormous increase in computing power. Before the development of fast computers, it was a major obstacle in the field to actually compute the posterior probability since explicit formulas were rarely available. Computers allow via stochastic simulation the computation of the denominator, which is typically the hard part. Markov chain Monte Carlo methods especially allow an efficient sampling of huge parameter spaces (see Chapters 3 and 7).

2.3 Applications of the Likelihood Function in Molecular Evolution

The theory outlined in the preceding paragraphs has many applications in molecular evolution. In the following, we will start with an almost classical example that illustrates the "easy" application of the likelihood example. The following examples will get increasingly complicated and can only be solved by computational approaches.

2.3.1 The Hardy-Weinberg Equilibrium

The Hardy-Weinberg principle states that the gene frequencies in a stationary population determine the frequencies of the genotypes of a population. The most simple example deals with a diploid population and a two-allele gene locus. Let A and a denote the corresponding alleles, and θ denote the frequency of a. Then, in Hardy-Weinberg equilibrium, the frequencies of genotypes AA, Aa, and aa are specified by

$$
\begin{aligned}
p_{AA}(\theta) &= (1-\theta)^2, \\
p_{Aa}(\theta) &= 2\theta(1-\theta), \\
p_{aa}(\theta) &= \theta^2.
\end{aligned}
\tag{2.27}
$$

With X_{AA}, X_{Aa}, and X_{aa}, we denote the genotype counts of a sample of size n drawn from a population. One may ask whether the population is in Hardy-Weinberg equilibrium. To test this, we first compute the MLE θ^{mle} and then apply the likelihood ratio test. Equation (2.27) is an example of the MLE for the multinomial distribution where the cell probabilities are functions of the allele frequency θ. The solution of equation (2.18) provides the maximum likelihood estimator

$$
\theta^{\mathrm{mle}}_{\mathrm{HW}} = \frac{2X_3 + X_2}{2n},
\tag{2.28}
$$

which agrees with intuition. If we want to test the null hypothesis that the population is in Hardy-Weinberg equilibrium, we apply the statistic in equation (2.25), which is approximately chi-square distributed with 1 degree of freedom. Thus, if $-2\log(\Lambda)$ exceeds the critical value c_α for a significance level α, one rejects the null hypothesis.

This was a simple example. The procedure gets more complicated when we deal with more than two alleles and when we cannot observe the genotypes directly.

2.3.2 Models of Sequence Evolution

With the advent of molecular data, models of DNA and amino acid sequence evolution represent the work horses in analyses of molecular evolution. These

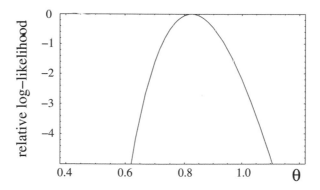

Fig. 2.2. Relative log-likelihood function (i.e., $\ell(\theta) - \ell(\theta^{\mathrm{ml}})$) for the Jukes and Cantor model of sequence evolution. Sequences were 400 base pairs long with 200 identical nucleotide pairs. $\theta^{\mathrm{ml}} = 0.824$.

substitution models have been widely explored and expanded. In Chapter 1, models of sequence evolution were described in detail. The simplest such model was introduced by T. H. Jukes and C. R. Cantor [13], which distinguishes between two categories for a pair of DNA sequences, the fraction of identical nucleotide pairs $(p_0(\theta))$ and the fraction of nonidentical pairs $(p_1(\theta))$, where $\theta = \mu t$ is the product of mutation rate μ and the appropriately scaled total amount of time t that elapsed between the two sequences under study.

From the theory outlined in Chapter 1, one readily computes

$$p_1(\theta) = \frac{3}{4}\left(1 - \exp\left(-\frac{4\theta}{3}\right)\right). \tag{2.29}$$

Let X_0, X_1 be the number of identical base pairs and of nonidentical base pairs, respectively, where $X_0 + X_1 = n$ denotes the total number of nucleotide pairs. Then, the log-likelihood function is

$$\ell(X_0, X_1 \mid \theta) = X_0 \log\left(1 - p_1(\theta)\right) + X_1 \log\left(p_1(\theta)\right) \tag{2.30}$$

and one computes the maximum likelihood estimator as

$$\theta_{\mathrm{JC}}^{\mathrm{mle}} = -\frac{3}{4}\log\left(1 - \frac{4}{3}\frac{X_1}{X_0 + X_1}\right). \tag{2.31}$$

Figure 2.2 displays an example for a pair of sequences with 200 identical nucleotide pairs and 200 nonidentical pairs of nucleotides. We notice that the maximum likelihood estimate for a pair of sequences is nothing but the famous Jukes-Cantor [13] correction formula for multiple hits. Thus, $\theta_{\mathrm{JC}}^{\mathrm{mle}}$ is the number of substitutions that occurred between the two sequences. From the preceding, it follows that if the model (Jukes-Cantor evolution) were correct and if n were large, then we could infer with high accuracy the number of

substitutions that actually occurred with an asymptotic large sample variance
[21]

$$\text{var}(\theta) = \frac{p_1(1 - p_1)}{n} \frac{1}{(1 - \frac{4p_1}{3})^2}, \tag{2.32}$$

where $p_1 = X_1/n$ is estimated from the data.

As in the Hardy-Weinberg example, this model of sequence evolution is extremely simple. Biological sequences almost never evolve according to the Jukes-Cantor model. However, in the likelihood framework outlined here, we are actually in a position to apply the likelihood ratio test to check the hypothesis whether the Jukes-Cantor model is plausible or not.

Comparing two DNA sequences of length n, we observe the following counts $n_{AA}, n_{AC}, \ldots n_{TT}$, in the cells AA, AC, ... TT, that sum up to n. The Jukes-Cantor model predicts equal counts for the 12 possible cells of different nucleotide pairs, and equal counts for the cells of identical pairs. Therefore the alternative hypothesis is defined by the multinomial distribution with $m = 16$ cells, whereas the Jukes-Cantor model has only one parameter. Thus according to equation (2.25) the statistic is chi-square distributed with $m - 1 - 1 = 14$ degrees of freedom. If the Jukes Cantor model is rejected as too simple, one can move on to more complicated models (see Chapter 6).

As models of sequence evolution get more complex, the complexity of the likelihood function grows. For the general reversible [21] model, the likelihood function depends on six parameters for the substitution process and three parameters for the stationary base composition. Although the computations are more involved, it is still possible to do the computations on a computer.

2.3.3 The Likelihood Function in Phylogenetics

When two sequences are compared, the likelihood function can get very complex, as we have outlined in the preceding sections. Now we extend the complexity by considering n sequences arranged in a multiple alignment $\mathbf{X} = (X_1, \ldots, X_l)$ consisting of l aligned sites. Note the X_i's are n-dimensional words from the alphabet $\mathcal{A} = \{A, C, G, T\}$. We assume that sequence positions are evolving independently of one another according to the same model M. Thus an alignment constitutes a random sample of size l, where each sample is drawn from the same distribution, in our world, following the same evolutionary scenario. This scenario will now be made more specific.

Besides the evolutionary model, we introduce the tree that relates the n sequences as an additional parameter.

A *tree* $T = \{V, E\}$ is a cycle-free graph, where V represents the nodes (vertices) of the tree and $E \subset \{\{u, v\} | u, v, \in V, u \neq v\}$ the branches (edges). We furthermore specify a length function $s : E \to \mathbb{R}^+$ that assigns the number of substitutions to each branch of the tree. $s(e), e \in E$ is given by the evolutionary model M. We will call $s_c = s(e)$ the *branch length*. To keep the computation tractable, we assume that the model M is identical for all branches

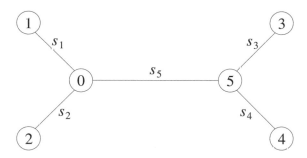

Fig. 2.3. An example topology with four leaves (e.g., sampled sequences and two internal nodes).

and that M belongs to the class of time-homogeneous, time-continuous, and reversible models (see Chapter 1).

Thus, the parameter $\boldsymbol{\theta}$ comprises a tree T, substitutions on the branches $s(e)$, and an evolutionary model M. Because we think of sequences as being obtained from contemporary organisms–that is, the leaves (end nodes) of the tree are labelled with the sequences–for convenience the labels are $\{1, \ldots, n\}$. Therefore, $p_X(X|\boldsymbol{\theta})$ specifies for a fixed $\boldsymbol{\theta}$ the probability of observing $X \in \mathcal{A}^n$. Thus, an essential point in our model is an alignment that is nothing more than a sample from a multinomial distribution with 4^n categories (the words of length n), where the probability of observing a category is specified by $\boldsymbol{\theta}$. Thus the likelihood of observing the alignment $\mathbf{X}$ is

$$L(\mathbf{X}|\boldsymbol{\theta}) = L(X_1, \ldots, X_l|\boldsymbol{\theta}) = \prod_{i=1}^{l} p(X_i|\boldsymbol{\theta}). \qquad (2.33)$$

Computing the likelihood function

According to equation (2.33), it suffices to compute the probability for a single alignment site. The joint probability of all l sites is then the probability of the alignment. Contrary to the examples given so far, an analytical formula is not available, so we will give a nontrivial example. From this example, one can readily conclude how to evaluate more complex trees. To this end, consider the tree in Figure 2.3 for $n = 4$ sequences $1, 2, 3, 4$. Because M is assumed to be generally reversible (see Chapter 1), we suppose that evolution "starts" in node "0" and proceeds along the branches to generate the pattern $x = (x_1, x_2, x_3, x_4)$. If the nucleotides for internal nodes 0 and 5 are known, (i.e., x_0, x_5), then

$$P(x|\boldsymbol{\theta}, y_0, y_5) = P_{y_0 x_1}(s_1) \cdot P_{y_0 x_2}(s_2) \cdot P_{y_0 y_5}(s_5) \cdot P_{y_5 x_3}(s_3) \cdot P_{y_5 x_4}(s_4), \quad (2.34)$$

where $P_{z_u z_v}(s)$ denotes the probability of substituting nucleotide z_u with nucleotide z_v if s substitutions occurred along the branch leading from u to v.

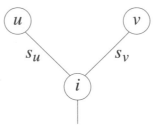

Fig. 2.4. Graph showing the relationship between an internal node (i) and its two offspring nodes (u, v).

Typically, the ancestral states are not known, thus, one sums over all possible assignments,

$$P(x|\boldsymbol{\theta}) = \sum_{y_0 \in \mathcal{A}} \sum_{y_5 \in \mathcal{A}} \pi_{y_0} P(x|\boldsymbol{\theta}, y_0, y_5), \tag{2.35}$$

where $(\pi_A, \pi_C, \pi_G, \pi_T)$ is the stationary distribution of the evolutionary model (see Chapter 1). In the example (Figure 2.3), $16 = 4^{(\text{number of internal nodes})}$ assignments are possible, thus equation (2.35) looks like one has to evaluate a sum with exponentially many summands.

However, Joseph Felsenstein [9] suggested a "dynamic programming" solution that computes the solution in $O(nl^2)$ times.

To this end, consider the pattern $X = (x_1, \dots, x_n) \in \{A, C, G, T\}^n$ as a realization of evolution, where x_i is the nucleotide at sequence i. With $y_i, i = n+1, \dots, 2n-2$, we denote the assignments of nucleotides to the internal nodes. We furthermore assume that the tree T is arbitrarily rooted at node $2n - 2$ and then each internal node has two offsprings (u, v) (see Figure 2.4). We define $\mathbf{L}^w = (L_A^w, L_C^w, L_G^w, L_T^w)$ as the vector of partial likelihoods for the subtree descending from node w. The computation of the partial likelihoods $\mathbf{L}^i$ for the internal nodes $i = n + 1, \dots n - 2$ proceeds recursively according to the already computed partial likelihoods of its offsprings u and v,

$$L_y^i = \left(\sum_{z \in \{A,C,G,T\}} P_{y,z}(s_u) L_z^u \right) \cdot \left(\sum_{z \in \{A,C,G,T\}} P_{y,z}(s_v) L_z^v \right), \tag{2.36}$$

with $y \in \{A, C, G, T\}$.

The partial likelihood vectors are initialized at the end nodes of the tree. One simply sets $L_{x_i}^i = 1$, and the remaining components are equal to zero, more formally

$$\mathbf{L}_z^i = \begin{cases} 1 & \text{if } z = x_i \\ 0 & \text{otherwise} \end{cases} \text{ for } i = 1, \dots n. \tag{2.37}$$

For the root node $2n - 2$ with three offsprings, we modify equation (2.36) accordingly. The probability of the full model is then

$$P(X|\boldsymbol{\theta}) = \sum_{z \in \{A,C,G,T\}} \pi_z L_z^{2n-2}. \tag{2.38}$$

The likelihood of the alignment $\mathbf{X}$ follows from equation (2.33).

Maximizing the likelihood for a given tree

We have described a computational approach to compute the likelihood of an alignment for fixed $\boldsymbol{\theta}$. Equation (2.33) shows that the resulting probability distribution is very complicated. It involves the branching pattern of the tree, the number of substitutions on each branch, and a model of sequence evolution.

We assume for the rest that the model of sequence evolution M is fixed. Now, $\mathbf{x}$ is observed and we want to find $\boldsymbol{\theta}^{\mathrm{ml}}$.

This task is divided into two parts. In the first part, one fixes the branching pattern of the tree and wants to estimate branch lengths $(s_e)_{e \in E}$ to maximize the likelihood. Even for a fixed tree it is generally not possible to obtain an analytical estimator. Thus one resorts to numerical optimization methods. Newton's method is one instance to find an optimal assignment of the number of branch lengths. However, often other numerical routines, or simply hill-climbing techniques, which stop when a local maximum is found, are also applied. Numerical methods are typically time-consuming, and sometimes the result depends on the numerical method applied.

In recent years, yet another obstacle has been observed where sequence alignments were found that generate multiple optima on the same tree. Chor et al. [4] even found sequence alignments with a continuum of optimal points, such as the following alignment:

$$\begin{array}{ll} 1 & AAAACCCAC \\ 2 & AAAACCCCA \\ 3 & AAAACCACC \\ 4 & AAAACCAAA. \end{array} \tag{2.39}$$

Thus, the likelihood surface is more complicated than originally expected. However, one should note that the alignment in equation (2.39) is not very tree-like. J. S. Rogers and David L. Swofford [16] asked "*Is it generally true that the trees of highest maximum likelihood for a given data set have only a single optimum?*" Based on intuition and the hill-climbing method implemented in PAUP* [20], they found in each case that the maximum likelihood point was the unique global optimum. In other words, if data are "close" to a "true" tree, then it is hard to find multiple maxima. At this point, a final conclusion seems impossible, and further work is necessary to detect alignments (also in real data) that give rise to multiple optimal solutions.

As complicated as the likelihood function can get, in some simple cases it is possible to get the maximum likelihood estimator as an analytical function. Yang [22] gave a solution for the simple two-state model and three sequences, and Chor et al. [5] extended this result to the case when a molecular clock holds true. Recently, Chor et al. [6] gave results for four sequences.

Finding the tree topology that maximizes the likelihood

In the previous chapter, we saw that the computation of the maximum likelihood assignments of branch lengths to a given tree τ poses already some complications. In molecular evolution, however, the branching pattern of the tree is also unknown, and a typical goal is to find the tree $\tau^{\max}$ that maximizes the likelihood function over *all* trees. This problem, like most of the phylogenetic approaches that optimize an objective function (maximum parsimony, distance-based methods), is even harder. No efficient algorithms are known that guarantee to pick the best tree(s) from the vast number of possible trees. The naive method to compute the maximum likelihood for each of the

$$t_n = \frac{(2n-2)!}{2^{n-3}(n-3)!} = 1 \cdot 2 \cdot 3 \ldots (2n-5) \qquad (2.40)$$

possible unrooted, binary, leaf-labelled trees is impossible already for $n = 10$ or 11 sequences. To overcome this problem, various heuristics are employed and implemented in, for example, PHYLIP [10], MOLPHY [1], NJ [17], PAUP* [20], and TREE-PUZZLE [19].

2.4 Outlook

We have tried to give an introduction to the application of the likelihood function in molecular evolution. Since this chapter has only introductory character, we could not give a full introduction to the flexibility of a likelihood approach in molecular evolution.

We have focussed on reconstructing the phylogenetic history of evolution, which applies to the million-year timescale. The coalescence framework (see reviews in [12, 15, 18]) provides a powerful approach to investigate genealogical processes within species. Here, however, the focus shifts from the reconstruction of a single most likely tree to the estimation of the population parameters that govern the genealogical process through the integration over all possible genealogies. In such analyses, the tree is no longer a parameter but a random variable with a well-defined prior probability specified by the model. In this field of molecular evolution, Bayesian inference has become popular in recent years.

With a further increase in computing power, we will be able to refine our models of evolution and will certainly integrate more realistic aspects of evolution. Thus, applications of the likelihood function in molecular evolution are a continuously interesting and flowering field of research.

References

[1] J. Adachi and M. Hasegawa. PROTML (maximum likelihood inference of protein phylogeny), 1995.

[2] M. A. Beaumont and B. Rannala. The Bayesian revolution in genetics. *Nature Review Genetics*, 5:251–261, 2004.

[3] G. Casella and R. L. Berger. *Statistical Inference.* Wadsworth & Brooks/Cole, Pacific Grove, CA, 1990.

[4] B. Chor, M. D. Hendy, B. R. Holland, and D. Penny. Multiple maxima of likelihood in phylogenetic trees: An analytical approach. *Molecular Biology and Evolution*, 17:1529–1541, 2000.

[5] B. Chor, M. D. Hendy, and D. Penny. Analytic solutions for three-taxon MLMC trees with variable rates across sites. In O. Gascuel and B. M. E. Moret, editors, *Algorithms in Bioinformatics: First International Workshop, WABI 2001, Aarhus, Denmark, August 28–31, 2001, Proceedings*, pages 204–214. Springer Verlag, Heidelberg, 2001.

[6] B. Chor, A. Khetan, and S. Snir. Maximum likelihood on four taxa phylogenetic trees: Analytic solutions. In RECOMB, to appear, 2003.

[7] A. W. F. Edwards. *Likelihood.* Cambridge University Press, Cambridge, expanded edition.

[8] B. Efron and R. J. Tibshirani. *An Introduction to the Bootstrap.* Chapman & Hall, New York and London, first edition, 1993.

[9] J. Felsenstein. Evolutionary trees from DNA sequences: A maximum likelihood approach. *Journal of Molecular Evolution*, 17:368–376, 1981.

[10] J. Felsenstein. PHYLIP (phylogeny inference package), 1993.

[11] N. Goldman. Statistical tests of models of DNA substitution. *Journal of Molecular Evolution*, 36:182–198, 1993.

[12] R. R. Hudson. Gene genealogies and the coalescent process. In D. J. Futuyma and J. Antonovics, editors, *Oxford Surveys in Evolutionary Biology*, volume 7, pages 1–43. Oxford University Press, Oxford, 1990.

[13] T. H. Jukes and C. R. Cantor. Evolution of protein molecules. In H. N. Munro, editor, *Mammalian Protein Metabolism*, pages 21–132. Academic Press, New York, 1969.

[14] E. L. Lehmann and G. Casella. *Theory of Point Estimation.* Springer-Verlag, New York, 1998.

[15] M. Nordborg. Coalecent theory. In D. J. Balding, M. Bishop, and C. Cannings, editors, *Handbook of Statistical Genetics*. Oxford University Press, Chichester, 2001.

[16] J. S. Rogers and D. L. Swofford. Multiple local maxima for likelihoods of a phylogenetic tree: A simulation study. *Molecular Biology and Evolution*, 16:1079–1085, 1999.

[17] N. Saitou and M. Nei. The neighbor-joining method: A new method for reconstructing phylogenetic trees. *Molecular Biology and Evolution*, 4:406–425, 1987.

[18] M. Stephens. Inference under the coalescent. In D. J. Balding, M. Bishop, and C. Cannings, editors, *Handbook of Statistical Genetics*, pages 213–238. John Wiley & Sons, Chichester, 2001.

[19] K. Strimmer and A. von Haeseler. Quartet puzzling: A quartet maximum-likelihood method for reconstructing three topologies. *Molecular Biology and Evolution*, 7:964–969, 1996.

[20] D. L. Swofford. PAUP*. Phylogenetic analysis using parsimony (* and other methods), 1998.

[21] S. Tavaré. Some probabilistic and statistical problems in the analysis of DNA sequences. In M. S. Waterman, editor, *Some Mathematical Questions in Biology: DNA Sequence Analysis*, pages 57–86. The American Mathematical Society, Providence, RI, 1986.

[22] Z. Yang. Complexity of the simplest phylogenetic estimation problem. *Proceedings of the Royal Society of London, Series B: Biological Sciences*, 267:109–119, 2000.

[23] E. Zuckerkandl and L. Pauling. Evolutionary divergence and convergence in proteins. In V. Bryson and H. J. Vogel, editors, *Evolving Genes and Proteins*, pages 97–166. Academic Press, New York, 1965.

3

Introduction to Markov Chain Monte Carlo Methods in Molecular Evolution

Bret Larget

Departments of Statistics and Botany,
University of Wisconsin, Madison, WI, USA, brlarget@wisc.edu

3.1 Introduction

Markov chain Monte Carlo (MCMC) is a general computational technique for evaluating sums and integrals, especially those that arise as probabilities or expectations under complex probability distributions. *Monte Carlo* implies that the method is based on using chance (in the form of a pseudo-random number generator). *Markov chain* indicates a *dependent* sampling scheme with the probability distribution of each sampled point depending on the value of the previous one. Due to this dependence, MCMC samplers typically require sample sizes that are substantially larger than the sizes of independent samples produced by Monte Carlo integration methods to be able to achieve similar accuracy. However, independent sampling methods often require detailed knowledge of characteristics of the function being integrated, as their computational efficiency relies upon having a close approximation of this function. MCMC has proved to be highly useful because of its great flexibility and its success at solving many high-dimensional integration problems where other methods are computationally prohibitive.

3.1.1 A Brief History of MCMC Methods

The primary ideas behind MCMC were created by physicist Nicholas Metropolis and colleagues over fifty years ago at Los Alamos National Laboratory in the years after the Manhattan Project as part of a solution to a problem in statistical physics [22]. Hastings provided an important generalization to this pioneering work [12]. Hastings' foundational paper was ahead of its time in the statistics literature, and it took more than a decade (and the start of a personal computing revolution) before MCMC methods attracted additional attention in the statistics community. Their first use was in the form of the Gibbs sampler applied to image analysis [8]. Interest in MCMC exploded in the 1990s as it proved to be a powerful and flexible technique for solving a variety of previously unsolvable computational problems, especially those arising in Bayesian analyses. Refinement and extension of MCMC methods and their application to new problems continues to be an area of active research. MCMC methodology has completely transformed the Bayesian approach to statistics and its application to large-scale complicated modeling problems. For a more extensive description of the historical development of MCMC methods, please see the article by Hitchcock [13].

MCMC approaches in molecular evolution

MCMC approaches to problems in molecular evolution first appeared in the mid-1990s as several authors developed various methods to calculate posterior probabilities of phylogenies on the basis of aligned DNA sequence data [27, 20, 21, 17, 18]. Bayesian approaches using MCMC have since been applied to a growing number of problems in molecular evolution [16, 6]. This volume includes several applications of MCMC, including relaxation of the molecular clock assumption, the detection of positive selection, Bayesian analysis of aligned molecular sequences, models of protein evolution, evolution by genome rearrangement, and the calculation of predictive distributions and posterior mappings [25, 1, 14, 4, 5, 2]. The remainder of this chapter describes the theory behind MCMC methodology and illustrates the methods using examples in molecular evolution.

3.2 Bayesian Inference

The Bayesian approach to statistical inference in molecular evolution most often fits into the following general framework. (In what follows, I use p to stand for a generic probability density and let the arguments distinguish them.) A likelihood function $p(D \mid \theta)$ describes the probability (or probability density) of data D given the values of parameters θ. The prior distribution $p(\theta)$ expresses the uncertainty in the parameters prior to observation of the data. Bayes' theorem provides the form of the posterior distribution $p(\theta \mid D)$, the probability distribution that describes uncertainty in the parameters after observing the data and the object of all Bayesian inference

$$p(\theta \mid D) = \frac{p(D \mid \theta)p(\theta)}{p(D)}. \tag{3.1}$$

The denominator $p(D)$ is the marginal probability of the data, averaged over all possible parameter values weighted by their prior distribution. Formally, we can write

$$p(D) = \int_\Theta p(D \mid \theta)p(\theta)\, d\theta, \tag{3.2}$$

where Θ is the parameter space for θ. In almost all problems of practical interest, it is not tractable to compute $p(D)$ directly, the normalizing constant of the posterior distribution. MCMC offers a means to make Bayesian inferences without the need to compute this normalizing constant.

In a typical application in Bayesian inference in molecular evolution, the parameter θ contains both discrete and continuous components. For example, θ might include the discrete tree topology and the continuous branch lengths and nucleotide substitution model parameters. The single integral in (3.2) represents a multiple sum over discrete parameters and a multiple integral over continuous parameters.

Calculating expected values

Usually, the posterior distribution $p(\theta \mid D)$ is a complicated function over a large parameter space that cannot be described adequately in full. We typically are interested in various summaries of the posterior distribution, all of which are posterior expectations of some function of the parameters. For example, the posterior probability of a tree topology is the expected value of the indicator variable for that tree topology, and the posterior density of a branch length can be summarized in part by its mean.

To simplify notation by eliminating the explicit dependence on the observed data, define the unnormalized posterior distribution to be $h(\theta) \equiv p(D \mid \theta)p(\theta)$. With this notation, the posterior expected value of a function of the parameter space is defined to be

$$\mathbb{E}[g(\theta)] = \frac{\int_{\Theta} g(\theta)h(\theta)\,\mathrm{d}\theta}{\int_{\Theta} h(\theta)\,\mathrm{d}\theta}. \tag{3.3}$$

The idea behind MCMC is to take a (dependent) random sample of points $\{\theta_i\}$ from the unnormalized target function $h(\theta)$ by simulating a Markov chain whose stationary distribution is proportional to $h(\theta)$. We can then approximate expectations with simple arithmetic averages,

$$\mathbb{E}[g(\theta)] \approx \frac{1}{n}\sum_{i=1}^{n} g(\theta_i). \tag{3.4}$$

We note that while most applications of MCMC to problems in molecular evolution have been part of Bayesian analyses, computations in the form of (3.4) can arise in non-Bayesian approaches as well. MCMC is a general-purpose tool.

3.3 The Metropolis-Hastings Algorithm

The most common form of MCMC is the Metropolis-Hastings algorithm. The idea is to create a *proposal distribution* q on the parameter space Θ. Instead of using q to generate a sequence of points sampled from Θ, we use q to generate a *candidate* for the next sampled point that is either accepted or rejected with some probability. If the candidate is rejected, the current point is sampled again. The random acceptance of proposals effectively changes the transition probabilities. A clever choice of acceptance probabilities results in a "metropolized" Markov transition matrix q' whose stationary distribution is proportional to the unnormalized target distribution h. Remarkably, the choice of q is nearly arbitrary. It suffices for q to be *irreducible*–from any points $x, y \in \Theta$, it should be possible to get from x to y through a finite number of transitions under q.

The initial sample point θ_0 may be arbitrary. If the current state is $\theta_i = x$, the Metropolis-Hastings algorithm generates candidate y from the distribution $q(\cdot \,|\, x)$. The acceptance probability is

$$r(y \,|\, x) = \min \left\{ 1, \frac{h(y)q(x \,|\, y)}{h(x)q(y \,|\, x)} \right\} . \tag{3.5}$$

With probability $r(y \,|\, x)$, we set $\theta_{i+1} = y$; otherwise $\theta_{i+1} = x$. In the special case where $q(x \,|\, y) = q(y \,|\, x)$ for each x and y, the proposal density drops out of (3.5). The original method in Metropolis et al. [22] assumed this symmetry, and Hastings [12] made the generalization that allowed nonsymmetric proposal distributions. Notice as well that the target distribution only needs to be known up to a normalizing constant, as it is only necessary to be able to compute the ratio of the target distribution evaluated at any two points. The ratio $q(x \,|\, y)/q(y \,|\, x)$ is known as the *Hastings ratio* or the *proposal ratio*. The *target ratio* $h(y)/h(x)$ is the *posterior ratio* in a Bayesian setting where it is the product of a *likelihood ratio* and a *prior ratio*.

3.3.1 Why Does the Metropolis-Hastings Algorithm Work?

The stationary distribution π of a Markov chain with transition function $q'(y \,|\, x)$ satisfies

$$\int_{x \in \Theta} \pi(x) q'(y \,|\, x) \, dx = \pi(y) \qquad \text{for each } y \in \Theta. \tag{3.6}$$

A stronger condition is for the chain to satisfy *detailed balance*,

$$\pi(x) q'(y \,|\, x) = \pi(y) q'(x \,|\, y) \qquad \text{for all } x, y \in \Theta. \tag{3.7}$$

Markov chains that satisfy detailed balance are *time-reversible*. The rate of transition from x to y is the same as that from y to x for each x and y, so the probability of any sequence of transitions would be the same in forward and backward time. Detailed balance of the target distribution h is easy to check for the Metropolis-Hastings algorithm. First, notice that the actual transition probability density is $q'(y \,|\, x) = q(y \,|\, x) r(y \,|\, x)$ for $x \neq y$. Therefore, we have

$$h(x)q'(y \mid x) = h(x)q(y \mid x) \min \left\{ 1, \frac{h(y)q(x \mid y)}{h(x)q(y \mid x)} \right\}$$
$$= \min \left\{ h(x)q(y \mid x), h(y)q(x \mid y) \right\}.$$

The last expression is symmetric in x and y, which implies that the first expression must have the same value if x and y are switched, so detailed balance is satisfied.

The Gibbs sampler

The Gibbs sampler is a special case of the Metropolis-Hastings algorithm in which the proposal distributions are the full conditional distributions of some part of the parameter space conditional on the rest. Suppose that the parameter vector $\theta = (\theta_{[1]}, \theta_{[2]}, \ldots, \theta_{[d]})$ is partitioned into d blocks. The idea behind the Gibbs sampler is to propose new values of a block of parameters $\theta_{[k]}$ from their full conditional distribution given the current values of all other parameters, denoted $p(\cdot \mid \theta_{[-k]})$, where $\theta_{[-k]}$ includes all of θ except for the kth block. The proposed values are always accepted. The *systematic-scan* Gibbs sampler updates blocks in a fixed order, cycling through them all. The *random-scan* Gibbs sampler randomly picks a block of parameters to estimate repeatedly.

We can understand why the Gibbs sampler works by checking the Metropolis-Hastings acceptance probability for one step of the Gibbs sampler. In updating the kth block given the current state θ, the candidate is

$$\theta^* = (\theta_{[1]}, \ldots, \theta_{[k-1]}, \theta^*_{[k]}, \theta_{[k+1]}, \ldots, \theta_{[d]}).$$

The posterior ratio is $h(\theta^*)/h(\theta) = p(\theta^*)/p(\theta)$ and the proposal ratio is $p(\theta_{[k]} \mid \theta_{[-k]})/p(\theta^*_{[k]} \mid \theta_{[-k]})$. Conditioning on parameters outside the kth block leads to $p(\theta^*) = p(\theta_{[-k]} \cap \theta^*_{[k]}) = p(\theta_{[-k]})p(\theta^*_{[k]} \mid \theta_{[-k]})$ with a similar expression for $p(\theta)$. The acceptance probability is then

$$r = \min \left\{ 1, \frac{p(\theta^*)}{p(\theta)} \times \frac{p(\theta_{[k]} \mid \theta_{[-k]})}{p(\theta^*_{[k]} \mid \theta_{[-k]})} \right\}$$
$$= \min \left\{ 1, \frac{p(\theta_{[-k]})p(\theta^*_{[k]} \mid \theta_{[-k]})}{p(\theta_{[-k]})p(\theta_{[k]} \mid \theta_{[-k]}))} \times \frac{p(\theta_{[k]} \mid \theta_{[-k]})}{p(\theta^*_{[k]} \mid \theta_{[-k]})} \right\}$$
$$= 1.$$

The advantage of the Gibbs sampler is that proposals are always accepted. While one might think that this feature would invariably lead to a sampler that moves through the parameter space rapidly, this is not always the case. It is well-known that the Gibbs sampler can mix slowly if highly correlated parameters are in different blocks. The other practical difficulty is that the flexibility of the Metropolis-Hastings approach in choosing a proposal distribution is lost. Candidates from the full conditional distributions are often not easy to simulate, which can make the problem difficult. In the case where full conditional distributions are available and easy to simulate, Gibbs sampling will be a good choice. Experience indicates that the more general Metropolis-Hastings approaches are often a more practical solution for many statistical problems in molecular evolution.

3.3.2 An Example in Bayesian Phylogenetic Inference

The Bayesian approach to estimating phylogenies from aligned DNA sequence data as implemented in the programs BAMBE [24] and MrBayes [15] uses MCMC to sample from the joint posterior probability distribution of phylogenies and nucleotide substitution model parameters. The state space for the Markov chain takes the form $\theta = (\tau, t, Q)$, where τ is the tree topology, t is a vector branch length, and Q is the generator of the continuous-time nucleotide substitution process. The MCMC samplers used in both BAMBE and MrBayes are actually *hybrid samplers* that combine several Metropolis-Hastings samplers, each of which samples from only part of the parameter space. BAMBE has a proposal distribution q_1 that updates the tree (both τ and t) while leaving Q fixed and a second proposal q_2 that updates Q leaving the tree fixed. BAMBE cycles back and forth between q_1 and q_2 proposals. Effectively, the hybrid sampler in BAMBE is a systematic-scan Gibbs sampler with a Metropolis-Hastings proposal at each step. In contrast, MrBayes has a collection of proposals to update parts of Q and another collection of proposals to update the tree. At each stage, one of these proposals is selected at random. Running only one chain, MrBayes uses a hybrid sampler that is a random-scan Gibbs sampler with a Metropolis-Hastings update at each step. Tierney [26] provides further examples and theoretical justifications of the use of hybrid MCMC samplers.

Description of the Local algorithm

BAMBE and MrBayes both use a local update method first described by Larget and Simon [17] to update unrooted trees. A description of this algorithm and the associated acceptance probability serves to illustrate the ideas of this section on an application of MCMC in molecular evolution. The acceptance probability originally reported in [17] was, in fact, incorrect. I am extremely grateful to Mark Holder, Paul Lewis, and David Swofford, who reported this to me quite recently.

The LOCAL algorithm begins by selecting an internal branch of the tree at random. (Please see Figure 3.1 for a graphical description of this algorithm.) The nodes at the ends of this branch are each connected to two other branches. One of each pair is chosen at random. Imagine taking these three selected adjacent edges and stringing them like a clothesline from left to right, where the direction is also selected at random. The two endpoints of the first branch selected will each have a subtree hanging like a piece of clothing strung to the line. The algorithm proceeds by multiplying the three selected branches by a common random amount, akin to stretching or shrinking the clothesline. Finally, the leftmost of the two hanging subtrees is disconnected and reattached to the clothesline at a location selected uniformly at random. This is the candidate tree.

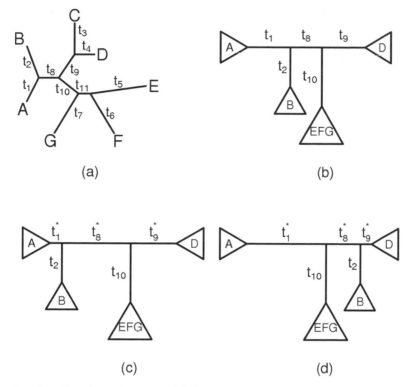

(a)   (b)

(c)   (d)

Fig. 3.1. Local update algorithm. (a) A seven-taxon unrooted tree. (b) A randomly chosen local neighborhood of the tree. Triangles represent subtrees. (c) A candidate tree with the same tree topology. (d) A candidate tree with a different tree topology.

Next, we then ask with what probability the candidate should be accepted. See Figure 3.1 (a), which displays a sample seven-taxon unrooted tree for the definition of the parameters in the following description. Suppose that we began by selecting the internal branch with length t_8 that separates taxa A and B from the rest, that we selected branches with lengths t_1 and t_9 from each side, and that we oriented these branches as shown in Figure 3.1 (b). The probability of this part of the proposal is $(1/b) \times (1/2)^3$ because there are $b = 4$ internal branches and we made three binary choices.

Let $m = t_1 + t_8 + t_9$ be the current length of the clothesline. We select the new length to be $m^* = m \exp(\lambda(U_1 - 0.5))$, where U_1 is a Uniform$(0, 1)$

random variable independent of everything else. It is straightforward to show that given m, m^* has density

$$q(m^* \mid m) = \frac{1}{\lambda m^*} \qquad \text{for } me^{-\lambda/2} < m^* < me^{\lambda/2}. \tag{3.8}$$

Suppose as well that $u = t_1$ is the current distance from the left endpoint of the clothesline to the B subtree. Given m^*, we pick a new distance $u^* = U_2 m^*$, where U_2 is another independent Uniform$(0,1)$ random variable. The distance from the left end point to the EFG subtree changes proportionally from $v = t_1 + t_8$ to $v^* = m^* v/m$. There are now two cases. If $u^* < v^*$ (see Figure 3.1 (c)), the tree topology does not change and the updated branch lengths are $t_1^* = u^*$, $t_8^* = v^* - u^*$, and $t_9^* = m^* - v^*$. Otherwise (see Figure 3.1 (d)), $v^* < u^*$ and the tree topology does change. The new branch lengths are $t_1^* = v^*$, $t_8^* = u^* - v^*$, and $t_9^* = m^* - u^*$. The probability density of this part of the proposal is the density of u^* given m^*, which is uniform, $q(u^* \mid m^*) = 1/m^*$ on $(0, m^*)$.

    The joint proposal density given the local choice of the subtree to update is

$$q(u^*, v^*, m^* \mid m, u, v) = q(m^* \mid m) q(u^* \mid m^*) \delta_{\{v^* = vm^*/m\}}$$
$$= \frac{\delta_{\{v^* = vm^*/m\}}}{\lambda (m^*)^2}, \tag{3.9}$$

where δ is Dirac's delta function. If x is the original tree and y is the candidate tree, the acceptance probability for the LOCAL proposal is

$$\min \left\{ 1, \frac{h(y) \left(\frac{1}{6}\right) \left(\frac{1}{2}\right)^3 \times \frac{\delta_{\{v = v^* m/m^*\}}}{(\lambda m^2)}}{h(x) \left(\frac{1}{6}\right) \left(\frac{1}{2}\right)^3 \times \frac{\delta_{\{v^* = vm^*/m\}}}{(\lambda (m^*)^2)}} \right\} = \min \left\{ 1, \frac{h(y)}{h(x)} \times \left(\frac{m^*}{m}\right)^3 \right\}$$

since

$$\frac{\delta_{\{v^* m/m^*\}}}{\delta_{\{v^*\}}} = \frac{\delta_{\{v^*\}}/(m/m^*)}{\delta_{\{v^*\}}} = \frac{m^*}{m}$$

The incorrect acceptance probability published previously [17] had a power of 2 rather than the correct power of 3.

3.4 Reversible Jump MCMC

In all of the examples we have considered to this point, the state space has had a fixed number of parameters. One can imagine a number of problems arising in molecular evolution where this need not be the case. For example, consider a Bayesian approach to phylogeny estimation from aligned DNA sequence data in which there is a prior distribution on the class of likelihood model. Specifically, suppose we think, for example, that the HKY85 and TN93 models are equally likely. The HKY85 model has one fewer free parameter than the TN93 model. We could define different Metropolis-Hastings samplers to update the Q matrix separately within each model, but we would also need to be able to switch between models. In this example, the number of parameters is itself a parameter of the model. *Reversible Jump MCMC* describes how to extend the Metropolis-Hastings approach to allow jumps between subspaces of different dimensions [11].

A typical situation is that we want a set of proposal distributions between subspaces Θ_1 and Θ_2 where the kth subspace has m_k parameters and $m_1 \neq m_2$. The key idea to make this work is *dimension matching*. The basic idea is to supplement each set of parameters with different numbers of random variables so that the dimensions match and then to transform one set into the other with a bijection. Let θ_1 and θ_2 be two states in Θ_1 and Θ_2, respectively. Then the vectors $\phi^{(1)} = (\theta_1, u_1)$ and $\phi^{(2)} = (\theta_2, u_2)$ each have length $d = m_1 + n_1 = m_2 + n_2$, where u_k is an n_k-vector and n_k are chosen so that the dimensions match. (It is often the case that $n_k = 0$ for the larger subspace.) Suppose that T_1 is a bijection so that $\phi^{(2)} = T_1(\phi^{(1)})$ and $T_1^{-1} = T_2$.

The proposal from $\theta_x \in \Theta_x$ to $\theta_y \in \Theta_y$ follows this procedure.

1. Generate random vector u_x, which has length n_x.
2. Let $\phi^{(x)} = (\theta_x, u_x)$.
3. Evaluate $\phi^{(y)} = T_x(\phi^{(x)})$.
4. Project $\phi^{(y)} = (\theta_y, u_y)$ into first m_y coordinates to determine θ_y.

3.4.1 Acceptance probability

The acceptance probability for this proposal includes a Jacobian in addition to the usual ratios. The Jacobian for the transformation is $|\det J_x|$, where J_x is a $d \times d$ square matrix whose i, j entry is

$$\{J_x\}_{ij} = \frac{\partial \phi_i^{(y)}}{\partial \phi_j^{(x)}} .$$

The acceptance probability is

$$r(\theta_y \,|\, \theta_x) = \min \left\{ 1, \frac{h(\theta_y)q(\theta_x \,|\, \theta_y)}{h(\theta_x)q(\theta_y \,|\, \theta_x)} \times |\det J_x| \right\} .$$

Example

We illustrate these ideas with the example of modeling the nucleotide substitution process in which we have equal prior probabilities that Q is from either an HKY85 or a TN93 class of models. Each of these models has three free parameters for the base composition that do not need to change in moving between models. HKY85 has a single transition/transversion parameter κ, while TN93 allows two different transition rates, κ_1 for purines and κ_2 for pyrimidines. In a proposal from TN93 to HKY85, assume we set κ to be the mean of κ_1 and κ_2. To attain detailed balance, we need for the proposal density given κ to have support on all positive (κ_1, κ_2) such that $\kappa = (\kappa_1 + \kappa_2)/2$. We could do this by letting $u \mid \kappa$ be a Uniform$(-\kappa, \kappa)$ random variable. We have this bijection:

$$T_1(\kappa, u) = (\kappa - u, \kappa + u) \qquad \text{and} \qquad T_2(\kappa_1, \kappa_2) = \left(\frac{\kappa_1 + \kappa_2}{2}, \frac{\kappa_2 - \kappa_1}{2} \right).$$

We have

$$J_1 = \begin{bmatrix} \dfrac{\partial(\kappa - u)}{\partial \kappa} & \dfrac{\partial(\kappa - u)}{\partial u} \\[2ex] \dfrac{\partial(\kappa + u)}{\partial \kappa} & \dfrac{\partial(\kappa + u)}{\partial u} \end{bmatrix} = \begin{bmatrix} 1 & -1 \\ 1 & 1 \end{bmatrix}$$

so that $|\det J_1| = 2$. By a similar calculation, $|\det J_2| = 1/2$.

Suppose that the unnormalized posterior distribution is h and we are interested in calculating the acceptance probabilities for a proposal from $\theta_1 = \kappa$ to $\theta_2 = (\kappa_1, \kappa_2)$. If a_1 is the probability that we propose that a TN93 Q matrix given the current model is HKY85 and a_2 is the probability of the reverse situation, the acceptance probability is determined as

$$r(\theta_2 \mid \theta_1) = \min \left\{ 1, \frac{h(\theta_2)}{h(\theta_1)} \times \frac{a_2}{a_1/(2\kappa)} \times 2 \right\}$$

provided that $(\kappa_1 + \kappa_2)/2 = \kappa$. The acceptance probability of a proposal in the other direction is

$$r(\theta_1 \mid \theta_2) = \min \left\{ 1, \frac{h(\theta_1)}{h(\theta_2)} \times \frac{a_1/(2\kappa)}{a_2} \times \frac{1}{2} \right\}.$$

3.5 Assessing Convergence

The theoretical justification of MCMC as a computational tool is that sample averages converge to their expected values. However, this result is asymptotic and, in practice, no chain can be run forever. We must therefore address the following practical questions: How long should a chain be run? Should we discard the initial portion of a sample? Should we subsample the Markov chain? How do we assess the accuracy of the MCMC estimates? How can we compare MCMC samplers? None of these questions has a definitive answer, and rarely can we have absolute trust in the MCMC calculations. Despite this, there are steps that will increase confidence in the results.

We will illustrate these ideas in the context of a very simple example. Suppose that we are interested in estimating a branch length from a two-taxon tree under the Jukes-Cantor model for a data set in which n_1 sites are unvaried and n_2 sites are variable. We will assume an exponential prior distribution with rate λ. The density is $p(t) = \lambda e^{-\lambda t}$. The probabilities of the possible site patterns are $\frac{1}{4}\left(\frac{1}{4} + \frac{3}{4}e^{-\frac{4}{3}t}\right)$ for unvaried sites and $\frac{1}{4}\left(\frac{1}{4} - \frac{1}{4}e^{-\frac{4}{3}t}\right)$ for varied sites. Putting these two probabilities together, the unnormalized posterior distribution is as follows.

$$h(t) = \left(\frac{1}{4}\right)^{n_1+n_2} \left(\frac{1}{4} + \frac{3}{4}e^{-\frac{4}{3}t}\right)^{n_1} \left(\frac{1}{4} - \frac{1}{4}e^{-\frac{4}{3}t}\right)^{n_2} \left(\lambda e^{-\lambda t}\right).$$

Consider updating the branch length by choosing a new value uniformly at random from a window of half-width w centered at the current value, reflecting off the origin when a negative branch length is proposed. Specifically, $t^* = |t + U|$, where U is uniformly distributed between $-w$ and w. It is straightforward to show that the proposal ratio is one. Acceptance probabilities are then $\min\{1, h(t^*)/h(t)\}$.

In a specific numerical example, suppose that $n_1 = 70$, $n_2 = 30$, and $\lambda = 5$. We will compare results for two choices of w, namely $w = 0.1$ and $w = 0.5$. In each case, we will begin with an initial edge length of 5.0 (a poor choice) and update the edge length 2000 times (much shorter than we might typically do). Figure 3.2 displays summaries of the MCMC samples.

3.5.1 Burn-in

The initial portion of an MCMC sample is often discarded as *burn-in*. The logic behind this practice is that the initial portion of a run will typically be highly dependent on the starting value of the Markov chain, and if this value is not likely under the stationary distribution, the sample would be biased toward the initial point. The estimate

$$\sum_{i=m+1}^{n} g(\theta_i)/(n - m),$$

which discards the first m sample points, is typically a more accurate estimate of the expectation of g under the target distribution when m is substantially larger than one in the usual case of an atypical initial state.

3.5.2 Trace Plots

This then begs the question of how one should determine the portion of a sample to discard. *Trace plots* of one-dimensional summaries of the state space are a crude but often effective way of determining burn-in. For Bayesian MCMC sampling from a posterior distribution of trees, both BAMBE and MrBayes produce trace plots of the log likelihood. When beginning runs at random initial trees, it is typical for the log-likelihood to increase dramatically as the chain rapidly approaches an area of the state space of relatively high posterior probability before changing behavior dramatically and reaching a plateau around which the likelihood fluctuates for the remainder of the run.

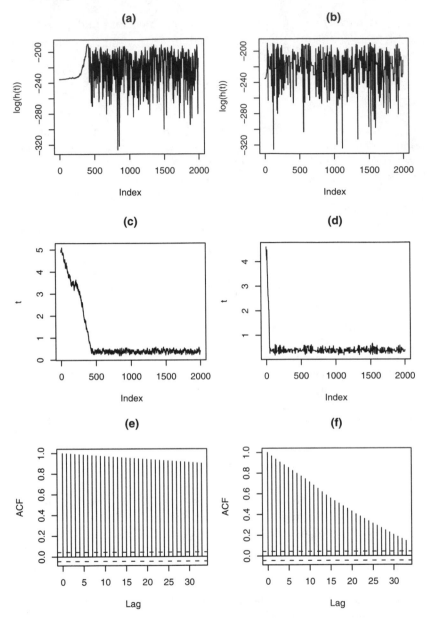

Fig. 3.2. Trace plots in the Jukes-Cantor example. (a) Unnormalized posterior versus index with window size $w = 0.1$. (b) Unnormalized posterior versus index with window size $w = 0.5$. (c) Branch length versus index with window size $w = 0.1$. (d) Branch length versus index with window size $w = 0.5$. (e) Autocorrelation plot of sampled edge lengths with $w = 0.1$. (f) Autocorrelation plot of sampled edge lengths with $w = 0.5$.

Trace plots of the log-likelihood are good indicators of minimum values for burn-in but are insufficient on their own to assess convergence. If the chain were stuck in a local minimum, the behavior exhibited in the trace plot would be indistinguishable from the trace plot behavior of a well mixing chain. Trace plots of other one-dimensional summaries of the state space, such as parameter values in the substitution models or the sum of all branch lengths of the tree, should also be examined for visual evidence that after burn-in the initial portion of the sample looks similar to the end portion.

The trace plots of h and t displayed in Figure 3.2 for the Jukes-Cantor example provide a means to informally assess convergence. The trace plots of the edge length in each run more clearly indicate the necessary time to convergence. In the run with $w = 0.1$, we need to discard at least the first 500 sample points, and I would discard a few more to be safe, say at least the first 10% of the sample after apparent convergence. Discarding the first $m = 700$ points of each run suffices for this example.

For the run with $w = 0.1$, a 95% credibility region for the edge length is $(0.24, 0.52)$. The post-burn-in credible region for the run with $w = 0.5$ is quite similar, $(0.25, 0.52)$. Had we not discarded the initial part of the run, the 95% credible region would have been either $(0.25, 4.48)$ or $(0.26, 0.62)$, with right endpoints substantially too large in both cases. Of course, we could have lessened the bias due to burn-in by either running the chains for many more iterations or by using an initial edge length closer to the center of the posterior distribution.

Figure 3.2 also displays the autocorrelation function of the sampled branch lengths for both window sizes. Notice that in this example mixing is significantly faster using the larger window size. With $w = 0.5$, the Markov chain has reached approximate independence after about 40 steps. Dependence decreases much more slowly in the case with a smaller window. Acceptance probabilities can offer a clue about convergence speed. In this example, updates with $w = 0.1$ were accepted 73% of the time as opposed to only 23% of the time for $w = 0.5$. Acceptance probabilities between 0.15 and 0.40 often indicate chains that mix relatively well. This simple example suffices to show that adjustment of tuning parameters can have a large effect on mixing properties; running slowly mixing chains for a long time can compensate. Notice also in Figure 3.2 as well that the trace plots of the edge lengths are more informative about burn-in than are the trace plots of the posterior distribution. With larger trees, larger models, and longer sequences, it is highly advisable to examine trace plots of many posterior summaries and to complete several very long MCMC simulations.

3.5.3 How Many Chains?

While there is no consensus on how many chains should be run, I advocate running several long chains from widely disparate starting values. The advantage is that if the post-burn-in summaries of important characteristics of the target distribution are similar, there is evidence that the Markov chains are successfully mixing. In contrast, summaries from independent chains that are wildly different are a certain indicator that one or more chains has not reached stationarity or that the chains are mixing so slowly that substantially longer runs are needed to obtain more accurate calculations. If one has access to several processors, the real time to take several long samples is the same as the time to complete a single run on one machine. The other advantage to having several independent estimates of posterior characteristics is that simple and accurate estimates of Monte Carlo error are easily computed. Estimates of Monte Carlo error from single runs depend on estimates of the dependence in a single chain. Such estimates can vary considerably with the method used to estimate the dependence.

3.5.4 How Often Should the Markov Chain Be Subsampled?

From a purely statistical perspective, there is nothing to gain from subsampling–a loss of data represents a potential loss of information. However, from a practical sense, because chains tend to be highly dependent, regular subsamples of the Markov chain output will typically be just as accurate as if the entire post-burn-in sample were saved and summarized. Practical issues involving the ease of the storage and analysis of the output of a long MCMC run often outweigh the negligible potential loss of information from subsampling.

3.5.5 How Long Should a Chain Be Run?

There are formal methods to decide upon chain convergence that are based on running a number of chains in parallel and stopping when variability in the chains' estimates of a number of scalar posterior summaries between chains is small relative to the variability within each chain [7]. A cruder yet effective approach is to learn from preliminary runs how much time is required to run a chain a specified number of steps, extrapolate this to the time available (such as overnight), run several independent chains in parallel in that time, and calculate the Monte Carlo standard error of each important scalar posterior characteristic from the estimates in each independent chain. If this Monte Carlo error estimate is too big for the problem at hand, then it may be that longer runs are necessary (or that a better proposal distribution is required).

3.6 Metropolis-Coupled MCMC

There are many strategies for improving the sampling properties of MCMC approaches. One of the most useful is Metropolis-coupled MCMC, or MCMCMC [9]. The idea is to run several simultaneous chains on the state space Θ.

Only one of these chains, the *cold chain*, needs to have the correct stationary distribution. The other *heated* chains are typically selected to have stationary distributions that are flatter than the stationary distribution of the cold target chain. The heated chains are able to move more easily between regions where the target is relatively high.

After some number of steps of each chain, a move that swaps the states of two of the chains is proposed and accepted or rejected according to a Metropolis-Hastings rule. This type of proposal can effectively jump the cold chain to a different portion of the parameter space. Only the sampled points from the cold chain are saved as a sample from the target distribution. Suppose that the chains have unnormalized target distributions $\{h_i\}$ for $i = 1, \ldots, m$. If the current states in chains i and j are x_i and x_j, respectively, the probability of accepting a proposed swap of the two states is

$$\min\left\{1, \frac{h_i(x_j)h_j(x_i)}{h_i(x_i)h_j(x_j)}\right\}.$$

Generally speaking, running m chains requires m times the computational effort that running a single chain would require. This trade-off can be worth-while if the cold chain is very slow-mixing.

Figure 3.3 illustrates these ideas in a small artificial example. The target function (solid line in Figure 3.3(a)) contains two separate modes of relatively high probability separated by a region of very low probability. We are using a proposal chain that proposes new values in a small uniform window extending one unit below and above the current position. Crossing the valley between the two peaks in the cold chain requires an unlikely proposal and acceptance of several consecutive steps through the low region between the modes. The dashed line is a single heated distribution. The same proposal distribution will more easily cross between the two modes. In a simulation, both chains began at the value $x = 20$, are updated by Metropolis-Hastings individually, and are then followed by a proposed swap after each set of updates. The chains ran for 100,000 cycles of updates. Figure 3.3(b) shows a histogram of the sampled values that matches the target quite well. Figure 3.3(c) shows the same sampled values in a trace plot versus the iteration number. It is clear that the sampled chain jumped between modes many times during the simulation. Figure 3.3(d) shows the sampled values from a regular Metropolis-Hastings MCMC simulation in a trace plot versus iteration number. This particular realization jumped between modes only once. Simulation-based sample esti-mates of target characteristics will likely be inaccurate and will be highly sensitive to the decision on when to stop the chain. A substantially longer simulation in which the sampled chain crossed the low region several times would be required for accurate estimation.

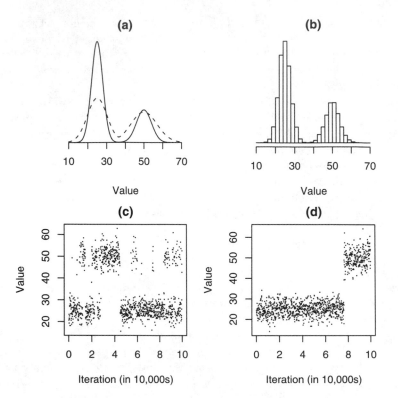

Fig. 3.3. Illustration of MCMCMC. (a) The solid line shows the target function, h. The dashed line is proportional to $h^{1/3}$ (rescaled to have a similar normalizing constant). A *heated* chain run under the dashed line will have the incorrect stationary distribution but will move more freely about the region. (b) Histogram of sampled points from the MCMCMC run. (c) Plot of the sampled points in the MCMCMC run versus iteration number. (d) Plot of the sampled points in a regular Metropolis-Hastings run versus iteration number.

3.7 Discussion

MCMC has become an indispensable tool for statistical computing, with special importance to the Bayesian approach. MCMC is especially useful for the high-dimensional calculation problems that arise in statistical models of molecular evolution. As evolutionary biologists address problems in molecular evolution of increasing complexity (larger trees, genome-scale data of varied type, more realistic and parameter-rich models of molecular evolution, accounting for additional forms of biological interaction), most of the tools that will be successful in providing answers to these questions are likely to be based on MCMC computation.

3.7.1 Other References about MCMC

MCMC is an important topic that is described in much greater detail in many other sources and is an area of much continuing active research. Gilks et al. have written an entire book on the topic of MCMC [10]. The books by Robert and Cassela and by Liu on Monte Carlo methods each include several chapters on MCMC [23, 19]. The books on Bayesian statistics by Gelman et al. and by Carlin and Louis include extensive descriptions of MCMC [7, 3]. Joe Felsenstein's recent book includes a chapter on Bayesian approaches to phylogenetic inference using MCMC as well as a chapter on using MCMC to make likelihood calculations on coalescent trees [6].

References

[1] J. P. Bielawski and Z. Yang. Maximum likelihood methods for detecting adaptive protein evolution. Chapter 5, this volume.

[2] J. P. Bollback. Posterior mappings and posterior predictive distributions. Chapter 16, this volume.

[3] B. P. Carlin and T. A. Louis. *Bayes and Empirical Bayes Methods for Data Analysis*. Chapman and Hall/CRC, Boca Raton, second edition, 2000.

[4] M. Dimmic. Markov models of protein sequence evolution. Chapter 9, this volume.

[5] R. Durrett. Genome rearrangement. Chapter 11, this volume.

[6] J. Felsenstein. *Inferring Phylogenies*. Sinauer Associates, Inc., Sunderland, MA, 2004.

[7] A. Gelman, J. B. Carlin, H. S. Stern, and D. B. Rubin. *Bayesian Data Analysis*. Chapman and Hall/CRC, Boca Raton, 1995.

[8] S. Geman and D. Geman. Stochastic relaxation, Gibbs distributions, and the Bayesian restoration of images. *IEEE Transactions on Pattern Analysis and Machine Intelligence*, 6:721–741, 1984.

[9] C. J. Geyer. Markov chain Monte Carlo maximum likelihood. In E. M. Kerimidas, editor, *Computing Science and Statistics: Proceedings of the 23rd Symposium on the Interface*, pages 156–163. Interface Foundation, Fairfax Station, VA, 1991.

[10] W. R. Gilks, S. Richardson, and D. J. Spiegelhalter, editors. *Markov Chain Monte Carlo in practice*. Chapman and Hall/CRC, Boca Raton, 1996.

[11] P. J. Green. Reversible jump MCMC computation and Bayesian model determination. *Biometrika*, 82:711–732, 1995.

[12] W. K. Hastings. Monte Carlo sampling methods using Markov chains and their applications. *Biometrika*, 57:97–109, 1970.

[13] D. Hitchcock. A history of the Metropolis-Hastings algorithm. *The American Statistician*, 57:254–257, 2003.

[14] J. P. Huelsenbeck and F. Ronquist. Bayesian analysis of molecular evolution using MrBayes. Chapter 8, this volume.

[15] J. P. Huelsenbeck and F. Ronquist. MRBAYES: Bayesian inference of phylogenetic trees. *Bioinformatics*, 17:754–755, 2001.

[16] J. P. Huelsenbeck, F. Ronquist, R. Nielsen, and J. P. Bollback. Bayesian inference of phylogeny and its impact on evolutionary biology. *Science*, 294:2310–2314, 2001.

[17] B. Larget and D. L. Simon. Markov chain Monte Carlo algorithms for the Bayesian analysis of phylogenetic trees. *Molecular Biology and Evolution*, 16:750–759, 1999.

[18] S. Li, H. Doss, and D. Pearl. Phylogenetic tree reconstruction using Markov chain Monte Carlo. *Journal of the American Statistical Society*, 95:493–508, 2000.

[19] J. S. Liu. *Monte Carlo Strategies in Scientific Computing*. Springer, New York, 2001.

[20] B. Mau and M. A. Newton. Phylogenetic inference for binary data on dendograms using Markov chain Monte Carlo. *Journal of Computational and Graphical Statistics*, 6:122–131, 1997.

[21] B. Mau, M. A. Newton, and B. Larget. Bayesian phylogenetic inference via Markov chain Monte Carlo methods. *Biometrics*, 55:1–12, 1999.

[22] N. Metropolis, A. W. Rosenbluth, M. N. Rosenbluth, A. H. Teller, and E. Teller. Equation of state calculations by fast computing machines. *Journal of Chemical Physics*, 21:1087–1092, 1953.

[23] C. P. Robert and G. Casella. *Monte Carlo Statistical Methods*. Springer, New York, 2002.

[24] D. Simon and B. Larget. Bayesian analysis in molecular biology and evolution (BAMBE). http://www.mathcs.duq.edu/larget/bambe.html, 2001.

[25] J. L. Thorne and H. Kishino. Estimation of divergence times from molecular sequence data. Chapter 9, this volume.

[26] L. Tierney. Markov chains for exploring posterior distributions (with discussion). *Annals of Statistics*, 22:1701–1762, 1994.

[27] Z. Yang and B. Rannala. Bayesian phylogenetic inference using DNA sequences: A Markov chain Monte Carlo method. *Molecular Biology and Evolution*, 14:717–724, 1997.

4

Population Genetics of Molecular Evolution

Carlos D. Bustamante

Department of Biological Statistics and Computational Biology, Cornell University, 422 Warren Hall, Ithaca, NY 14850, USA, cdb28@cornell.edu

Summary

The aim of this chapter is to provide an introduction to aspects of population genetics theory that are relevant to current research in molecular evolution. We review the roles of mutation rates, natural selection, ancestral polymorphism, and linkage among sites in molecular evolution. We also discuss why it is possible to detect the workings of natural selection from comparing rates of substitution for different classes of mutations along a branch in the phylogeny. The problem of estimating the distribution of selective effects among newly arising mutations is given considerable treatment, as are neutral, nearly neutral, and selective population genetics theories of molecular evolution. The chapter does not aim to be an exhaustive description of the field but rather a selective guide to the literature and theory of the population genetics of molecular evolution.

4.1 Introduction

Evolution is the outcome of population-level processes that transform genetic variation within species into genetic differences among species in time and space. Two central goals of evolutionary biology are to describe both the branching order of the history of life (phylogeny) and the evolutionary forces (selective and nonselective) that explain why species differ from one another. Since the 1980s there has been an explosion in the number and complexity of probabilistic models for tackling the first problem, with the motivation that to understand evolution at any level one needs to get the history right (or at least integrate over one's uncertainty in the matter) (for a review, see [113]). Current Markov chain models of evolution deal with the complexities of DNA [48, 60, 40, 119]), RNA [69, 92], codon [39, 70], and protein evolution (see [104] for a review), as well as rate variation among sites [120, 26] and diverse complex dependencies such as tertiary structure [85] and CpG mutational effects [94]. Likewise, there has been tremendous growth in using probabilistic models for hypothesis testing and model selection. For example, it is currently possible to exploit rate variation among codons [72, 124] and among lineages and codons [123] to detect amino acid sites that are likely to be involved in adaptive evolution assuming silent sites evolve neutrally and codons evolve independently of one another.

The purpose of this chapter is to introduce population genetics concepts relevant to the study of molecular evolution, with particular emphasis on understanding how natural selection affects rates and patterns of molecular evolution. Some effort is also made to discuss how population genetics models relate to continuous-time discrete-space Markov chain models of molecular evolution. For example, if the transformation of genetic variation is mostly governed by genetic drift acting on evolutionarily neutral mutations that evolve independently of one another, the outcome will be a Poisson process with constant rate that is independent of the species size [81, 88, 51]. A Markov chain model of evolution (perhaps with rate variation among sites) is a quite appropriate model to capture the dynamics of such a system since the exponential distribution of times among transitions corresponds to an underlying Poisson process. If mutations are not neutral but sites evolve independently of one another, the substitution process can remain a Poisson process that differs among lineages depending on population size and the strength of selection. Under such a model, it is possible to use variation in the rates of substitution among sites to infer the distribution of selective effects among new mutations [25, 73, 90]. Alternatively, if mutations are linked and either slightly deleterious or advantageous (e.g., [81, 77, 78, 79, 59]), or if the fitness effects of mutations vary randomly with the environment (e.g., [100, 30, 31]), the observed patterns of molecular evolution can depart greatly from the expectations of a Poisson process with constant rate [31, 32, 33, 34, 17, 18].

We will begin with a brief historical overview of the population genetics of molecular evolution (Subsection 4.1.1). In Section 4.2, we discuss some of the major predictions of neutral and nearly neutral models of molecular evolution. In Section 4.3, we demonstrate how the classical Wright-Fisher models of population genetics give rise to the neutral theory of molecular evolution. Next will follow a discussion on how ancestral polymorphism can cause departures from the expectations of the neutral independence-among-sites model (Section 4.4). We will then discuss natural selection and demonstrate how comparing the rate of substitution of a putatively selected class of mutations to a neutrally evolving class can be used to infer the signature of natural selection from sequence data (Section 4.5). A discussion will follow on the effects of a distribution of selection coefficients among new mutations on rates and patterns of molecular evolution. Lastly, we investigate the effects of linkage and selection on rates of molecular evolution. A definitive and more mathematical treatment of the subject of theoretical population genetics can be found in Warren Ewens' excellent work *Mathematical Population Genetics*, which has just been published in a second edition by Springer [23].

4.1.1 Setting the Stage

To understand the relationship between population genetics and the study of molecular evolution, one must begin at the point in history where the two became intertwined. In their seminal paper, Zukerkandl and Pauling [126] proposed that the preferred characteristic for inferring the evolutionary relationships among organisms ought to be similarity at the level of DNA or protein sequences. Their paper, while deeply philosophical and contentious, was rooted in the observation that the rate of amino acid evolution in hemoglobin-α and cytochrome-c per year was roughly constant for various vertebrate species. If DNA and protein sequences ("informational macromolecules") accrued substitutions at a near constant rate, then the changes along the phylogeny represented a "molecular clock" that could be used for dating species divergence. Since these changes are more plentiful and presumably subject to less scrutiny by natural selection than morphological characters, the authors reasoned that DNA and protein changes provide better markers for inferring evolutionary relationships. Their paper provided a simple stochastic model of molecular evolution whereby each site had equal probability of being substituted and the number of substitutions that occur along a branch was proportional to the length.

The theoretical foundation for this model (and thus for the molecular clock hypothesis and ultimately for modern-day methods) was provided by the "neutral-mutation drift" theory of molecular evolution, which posited that the vast majority of molecular evolution was due to the stochastic fixation of selectively neutral mutations [55, 63, 57, 62]. The theory concerns both variation within and between species and is summed up most elegantly by the title of Kimura and Ohta's seminal paper: "Protein polymorphism as a phase of molecular evolution" [62]. In other words, the neutral theory arises from considering the evolutionary implications of genetic drift operating on neutral variation [55, 62, 58]. As we will see, the theory predicts (among other things) that the rate of molecular evolution ought to be independent of the population size. In many ways, the true concern of the theory is the distribution of selective effects among newly arising mutations since everything else follows from this premise. The neutral theory is predicated upon the notion that almost all mutations are either highly deleterious or evolutionarily neutral. Highly deleterious mutations contribute little to variation within species and nothing to the genetic differences among species. Adaptive mutations are assumed to be very rare and to fix quickly, thus leaving neutral mutations as the only real source of genetic variation within species that can lead to fixed differences among species. It is important to note that the mature theory says little about the proportion of all mutations that are neutral; rather, it states that most mutations that go on to contribute to differences among species and variation within species are neutral. In this sense, even very constrained molecules such as histones can evolve neutrally. Their molecular clock just ticks at a much lower rate than that of unconstrained molecules such as, perhaps, noncoding DNA. Present-day rate-variation models [120, 26] allow this constraint parameter to vary among sites.

While the neutral theory arises as an extension of population genetics theory, it is not the only population genetics theory of molecular evolution (e.g., [81, 100, 79, 30, 59, 82, 35, 89, 90]). In fact, the field of population genetics has had a long-standing debate over the relative contribution of competing evolutionary forces (mutation, migration, genetic drift, and natural selection) to patterning genetic differences among species. Much of this debate has focused on the question of how much genetic variation within species is maintained by natural selection as well as how much of the molecular differences that we observe among species are due to adaptive molecular evolution [64, 61, 31].

One of the most important critiques of the neutral theory has been put forth by John Gillespie in *The Causes of Molecular Evolution* [31]. He used two lines of evidence to argue that most amino acid substitutions are adaptive. The first is specific examples of adaptive molecular evolution in response to environmental stress. The second is a thorough analysis of variation in the index of dispersion (ratio of the variance to the mean) for amino acid substitutions among mammalian and *Drosophila* species. As mentioned above, a major prediction of the neutral model is that the pattern of substitutions along different branches in a phylogeny ought to be Poisson-distributed with constant rate [81]. Gillespie conclusively demonstrated that the index of dispersion is, on average, much greater than 1 for both sets of species (i.e., it is overdispersed) and that the observations cannot easily be accounted for by neutral or nearly neutral models. He concludes that amino acid evolution occurs due to natural selection in "response to environmental factors, either external or internal, that are changing through time/or space." While the specific model Gillespie espoused [30] may not explain the overdispersed molecular clock (see [34, 35, 17, 18]), the data are certainly not consistent with the strict neutral model.

In fact, recent genome-wide analyses suggest quite an important role for both adaptive and weak negative natural selection in patterning molecular evolution in *Drosophila* (e.g., [91, 24, 90, 75, 98, 5, 8, 90, 38, 93, 6, 84]), *Arabidopsis* (e.g., [8, 67, 4, 110, 84]), maize (e.g., [103, 14, 47]), mouse [96], HIV (e.g., [118, 115, 121, 125, 68, 12, 19]), mammalian mitochondrial genomes [73, 112], and humans (e.g., [46, 87, 83, 1, 41, 13, 97, 29, 50, 114]). While many agree selection is important, there is still considerable debate as to the relative contribution of negative versus positive selection in patterning molecular evolution. As we will see in Section 4.6, the key to the debate rests on rates of recombination and the distribution of selective effects among newly arising mutations. In the next section, we will delve into the specifics of neutral and nearly neutral models before turning to the underlying population genetics machinery.

4.2 The Neutral Theory of Molecular Evolution

It is Darwin [20], of course, who posited that evolution occurs as the result of natural selection by which heritable differences that alter the probability of survival and reproduction of organisms are passed on from generation to generation. Sir Ronald Fisher [27, 28] and Sewall Wright [116] provided the first mathematical models of "the Darwinian evolution of Mendelian populations" by treating genetic drift (i.e., fluctuations in allele frequencies at a given locus due to finite population size) as analogous to the diffusion of heat along a metal bar. In these works, Wright and Fisher also provided the first genetic theories of evolution by deriving a formula for the probability that a mutation subject to natural selection would become fixed in the population (a result we will derive in Section 4.3). What they showed is that if a mutation alters the expected number of offspring a haploid individual (chromosome) contributes to the next generation by a small amount s so that those carrying the mutation leave on average $1 + s$ offsprings and those that do not carry the mutation leave 1 offspring on average, then the probability that a new mutation eventually becomes fixed in the population is roughly

$$\Pr(\text{fixation}) \approx \frac{2s}{1 - e^{-4Ns}}, \tag{4.1}$$

where N is the effective population size of the species, $2N$ is the number of chromosomes in the population, and s is on the order of N^{-1}. If $s > 0$, we say the mutation is *selectively favored* and that there is *positive selection* operating on the mutation since as the magnitude of s increases above 0 so does the probability of fixation (4.1). Likewise, if $s < 0$, we say the mutation is *selectively disfavored* and there is *negative selection* operating on the mutation since as s becomes more negative, the probability of eventual fixation becomes smaller and smaller. In the neutral case ($s \approx 0$), we can see by applying L'Hopital's rule that the probability of eventual fixation is simply the initial frequency of the mutation $p = \frac{1}{2N}$ (the mutation must have occurred in a heterozygous form).

While Fisher and Wright laid out a great deal of the foundation, it is Motoo Kimura who built up much of the population genetics theory of molecular evolution. His neutral theory of molecular evolution [55, 57, 58, 61] arises from a beautifully simple cancellation of terms: if mutations enter the population at some rate μ per locus per generation, some fraction f_0 are neutral, and $1 - f_0$ are completely lethal, then the rate of evolution k_0 would equal the neutral mutation rate:

$$k_0 = E(\# \text{ of neutral mutations entering per generation.}) \tag{4.2}$$
$$\times \Pr(\text{neutral mutation becomes fixed})$$
$$= 2N f_0 \mu \frac{1}{2N}$$
$$= f_0 \mu. \tag{4.3}$$

Three major predictions or consequences arise from (4.3):

1. Neutral molecular evolution is independent of the population size and depends only on the *per generation* rate of input of neutral mutations.
2. Neutral molecular evolution is linear in time, thus providing a "molecular clock" by which the relative divergence time of different populations can be dated.
3. Since neutral evolution occurs more rapidly in regions of low selective constraint (high f_0) and more slowly in regions of high selective constraint (low f_0), differences in rates of substitution can be used to infer functional constraint [63].

Furthermore, it is often assumed that the number of neutral mutations that fix in some interval of t generations (substitutions) is Poisson-distributed with rate $k_0 t$.

Our goal in Section 4.3 is to understand the population genetics theory behind equation (4.3) and, more importantly, to understand when this simple neutral model holds and when it does not hold. For example, the assertion that the substitution process is a Poisson process only holds if sites evolve independently of one another [51, 108]. This will be true only if there is free recombination among sites or if there is a sufficiently low mutation rate that only 1 or 0 nucleotides vary at a given point in time for a non-recombining region. High mutation rates and linkage among neutral sites can have a pronounced effect, leading to the fixation of "bursts" of mutations that are approximately geometrically distributed [108, 109, 32].

It is important to mention at this point that population genetics models of molecular evolution differ in some regards from discrete-space continuous-time models [48, 40, 60, 119]. For example, the Poisson assertion above ignores the possibility of multiple substitutions at the same site. The reason many population genetics models make such an assumption is that the timescale on which they operate is relatively short compared with the timescale on which phylogenetic reconstruction of distantly related species is usually carried out. Likewise, much of the theory is based on the behavior of single-locus two-allele models, where the goal is to understand the probability of fixation of a new mutation under various scenarios. Such a model is not rooted in the actual A, C, T, and G of DNA but rather on the fact that at a given nucleotide site the probability of having more than two nucleotides segregating is very low. Likewise, if the population size and mutation rates are small, there will be few linked polymorphic sites. Therefore, the independently evolving single-locus model with two alleles is a reasonable place to start in modeling molecular evolution.

4.2.1 Nearly Neutral Models of Molecular Evolution

From the beginning, it was evident that the great power of the neutral theory of molecular evolution lay in its quantitative predictions regarding rates and patterns of molecular evolution. In Kimura's original paper [55], the problems the neutral theory solved were the inordinately high rate of nucleotide evolution inferred from patterns of amino acid evolution [126] as well as the plentiful amounts of amino acid variation within species [43, 65]. According to Kimura's calculations, Darwinian evolution would produce too high a genetic load on the population to account for these patterns; therefore, most of the changes were likely neutral. Likewise, King and Jukes [63] set out to demonstrate that "most evolutionary change in proteins may be due to neutral mutations and genetic drift" by testing some of the predictions of a neutral molecular evolution theory using almost all of the available data in the world on protein, RNA, and DNA sequence variation.[1] One key prediction of the neutral theory was that if proteins were more constrained than genomic DNA, then proteins should evolve at a slower rate. If, on the other hand, proteins were constantly being refined by positive natural selection, then the rate of evolution of proteins would be faster than that of genomic DNA. Using early DNA hybridization experiments coupled with protein sequence information, King and Jukes concluded (rightly) that most proteins evolve at a much slower rate than most regions of genomic DNA. Another key argument they used was a near Poisson fit to the number of substitutions per site across the gene trees of various molecules (globins, cytochrome-c, and immunoglobulin-G light chains).

It was soon pointed out that if the neutral theory of molecular evolution was strictly true, then the rate of amino acid evolution should be proportional to generation time and not chronological time. Kimura and Tomoko Ohta [81] countered with the first "nearly neutral" model of molecular evolution. This model posits that newly arising nonlethal mutations are not strictly neutral ($s \approx 0$) but rather have selection coefficients drawn from a distribution such that the mean selective effect is slightly deleterious and most mutations are in the interval $(-\frac{1}{N} \leq s \leq \frac{1}{N})$.[2] Under such a scheme, the evolutionary fate

---

[1]King and Jukes had independently proposed a neutral theory of molecular evolution, but their paper was initially rejected by *Science*. In the interim, Kimura's paper appeared, and Kimura's results were added to the revised King and Jukes manuscript [99].

[2]The definition of "nearly neutral" is somewhat of a moving target and context-dependent. In their original paper, Ohta and Kimura [81, p.22] implicitly considered nearly neutral those mutations in the interval $(-\frac{2}{N} \leq s' \leq \frac{2}{N})$, where $s' = 2s$. In Ohta and Kimura's later work [77, 78, 79, 59], the emphasis was on explaining how slightly deleterious mutations could be considered an engine for nonadaptive molecular evolution. Likewise, Gillespie [31] has argued that nearly neutral should only refer to mutations in the interval $(-\frac{1}{N} \leq s' < 0)$ since slightly advantageous mutations are helped along by selection. Ohta [80] (not surprisingly) has explicitly reclaimed the "slightly advantageous" as nearly neutral ground by arguing that the fate of slightly advantageous mutations is very much governed by both selection and drift. Unless otherwise noted, we will adopt Ohta's view and consider nearly neutral mutations as those that are in the interval $-2 \leq \gamma \leq 2$, where $\gamma = 2Ns$.

of mutations is mostly governed by genetic drift. One implication of near-neutrality is an inverse relationship between population size N and the rate of molecular evolution at selected sites k_s. Letting f_s be the fraction of mutations that are selected, under the assumption that selected mutations evolve independently of one another, the rate of evolution for a selected mutation k_s is given by

$$k_s = E(\text{\# of selected mutations entering per generation.}) \qquad (4.4)$$
$$\times \Pr(\text{selected mutation becomes fixed})$$
$$= 2N f_s \mu \frac{2s}{1 - e^{-4Ns}}$$
$$= f_s \mu \frac{4Ns}{1 - e^{-4Ns}} . \qquad (4.5)$$

We see from (4.5) that for a fixed $s < 0$

$$\lim_{N \to \infty} k_s = 0.$$

The interpretation of this equation is that if mutations are slightly deleterious, a species with a large population size will evolve at a slower rate than a species with a small population size. Ohta and Kimura [81] posited that since population size is roughly inversely proportional to body size and body size is roughly inversely proportional to generation time (i.e., big animals have long times between generations but also live at low densities), these two factors cancel each other out to produce a rate of evolution that is close to linear in chronological time. Kimura [59] later argued that if $-s$ follows a Gamma distribution with mean 1 and shape parameter $\beta = 0.5$, then the rate of evolution will be proportional to $\sqrt{N}$.

A very useful way of studying the consequences of natural selection on rates of molecular evolution is by comparing the relative rate of substitution for selected mutations (4.5) to neutral mutations (4.3)

$$\omega = \frac{k_s}{k_0} = \frac{f_s}{f_0} \frac{2\gamma}{1 - e^{-2\gamma}}$$

letting $\gamma = 2Ns$. We will refer to γ as the scaled selection coefficient, and it will reappear when we derive (4.5) from an approximation to the Wright-Fisher process (Section 4.5). We note that ω can be interpreted as the expected dn/ds ratio assuming silent mutations are neutral, replacement mutations have the same selective effect, and mutations evolve independently of one another. Assuming $f_s = f_0$, if $s = -1 \times 10^{-4}$ and the population size is small ($N = 1000$), the rate of evolution at selected sites is $\omega = 0.81$, the rate of evolution at neutral sites, which we might refer to as a modest reduction. On the other hand, if s does not change and the population size is large ($N = 10,000$), then $\omega = 0.074$ and we would observe a large reduction in the substitution rate. In Figure 4.1, we plot the rate of substitution for selected mutations as compared with neutrality assuming $f_s = f_0$.

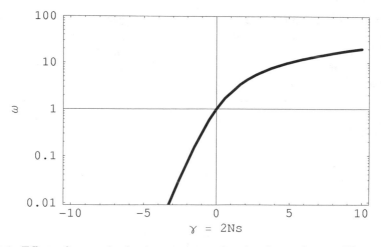

Fig. 4.1. Effect of natural selection on rates of molecular evolution. The x-axis is the scaled selection coefficient for new mutations, and the y-axis is the relative rate of substitution as compared with neutrality. Note that the y-axis is on a log-scale.

4.3 Wright-Fisher Model

4.3.1 No Mutation, Migration, or Selection

Consider a diploid population of constant size N (i.e., a population of $2N$ chromosomes) with discrete nonoverlapping generation [116, 28]. The population in the next generation is produced by randomly pairing gametes from an infinitely large pool of gametes produced by the current population. Focus on a neutrally evolving locus A with two alleles A_1 and A_2, and assume that there is no mutation between A_1 and A_2. Let $X(t)$ be the number of chromosomes in the population that carry the A_1 allele at generation t. The collection of random variables $\{X(t)\}$ for $t = 0, 1, \ldots$ is a discrete-time discrete-space Markov chain with state space $\{0, 1, \ldots, 2N\}$. The transition probability P_{ij} for going from state i to state j comes from binomial sampling:

$$P_{ij} \equiv \Pr(X(t+1) = j \mid X(t) = i) = \binom{2N}{j} \left(\frac{i}{2N}\right)^j \left(1 - \frac{i}{2N}\right)^{2N-j} . \quad (4.6)$$

This model is known as the Wright-Fisher model of population genetics, and the stochastic sampling of gametes from generation to generation is known as *genetic drift*. It is easy to verify that $X(t) = 0$ and $X(t) = 2N$ are *absorbing* states (i.e., $P_{00} = P_{2N\,2N} = 1$), corresponding to loss ($X(t) = 0$) or fixation ($X(t) = 2N$) of the A_1 allele. It is also relatively easy to show that all other states $(1, 2, \ldots, 2N - 1)$ are transient. This conforms with our biological intuition that if a population has 0 copies of allele A_1 in generation t_0, $\Pr(X(t) = 0) = 1$ for all $t > t_0$.

An implication of the Wright-Fisher model is that *each segregating neutral mutation in a population is eventually fixed or lost*. The stochastic fixation of neutral mutations (along with the fixation of selected mutations) thus underpins molecular evolution. It is then of immediate interest to find the probability that a mutation initially at frequency $p = \frac{X(0)}{2N}$ is eventually fixed in the population. The expected gene frequency in generation $t + 1$ given the gene frequency in generation t comes directly from the binomial model for gametic sampling:

$$E\left(\frac{X(t+1)}{2N} \mid X(t)\right) = \frac{\sum_{j=0}^{2N} j P_{ij}}{2N} = \frac{X(t)}{2N}.$$

Similarly, the variance in gene frequency is

$$V\left(\frac{X(t+1)}{2N} \mid X(t)\right) = \frac{X(t)(1 - X(t))}{2N}.$$

The first result implies that for the Wright-Fisher model without mutation, the expected change in allele frequency from generation to generation is zero (i.e., the $X(t)$ process is a Martingale). We can thus think of the change in gene frequency as a random walk without bias. As a result, we might intuit from symmetry alone that the probability of eventually fixing the A_1 allele should equal the initial frequency of the A_1 allele in the population (i.e., p).

A more rigorous approach is to set up a set of linear recurrence equations that the Wright-Fisher process must satisfy [74, p. 15]. Let p_j be the probability that a population that starts with j copies of the A_1 allele ($X(0) = j$) eventually fixes the A_1 allele (i.e., the probability that the process reaches $2N$ before it reaches 0). Clearly, $p_0 = 0$ and $p_{2N} = 1$. By exploiting the Markov property of the system, we can write down the following set of equations:

$$p_i = \sum_{j=0}^{2N} p_j P_{ij}, \quad \text{for } i = 1, \dots, 2N - 1. \tag{4.7}$$

The reason our model must satisfy these equations is that once the process enters state j, it "forgets" that it had previously been in state i and the process is restarted. The probability of reaching state $2N$ before state 0 is p_j, and by weighing the p_j's by the probability of transitioning from state i into state j, we obtain a set of $2N - 1$ equations (4.7) for $2N - 1$ unknowns $(p_1, p_2, \dots, p_{2N-1})$. By substituting (4.6) into (4.7), we verify that $p_j = \frac{j}{2N}$ is the non-negative solution to the system of equations. Therefore, the probability of eventual fixation of a neutral mutation is

$$p_1 = \frac{1}{2N}. \tag{4.8}$$

4.3.2 Rate of Fixation of Neutral Mutations

Now consider a process whereby in each generation a Poisson number of mutations occurs at a rate $\frac{\theta}{2} = 2N f_0 \mu$, where $f_0 \mu$ is the generation neutral mutation rate per locus. It is assumed that each mutation occurs at a previously invariant DNA site [58, 107]. We will now consider the rates and patterns of neutral molecular evolution under two assumptions: (a) complete independence among sites [58, 21, 22, 89] and (b) complete linkage among sites [107].

Independence among sites

Following [21, 89], model the mutation process as starting a Poisson number of new Wright-Fisher processes each generation. Let $X_j(t)$ be the state of the process (frequency) at site j at time t, where t is measured as the time since the mutation at site j originated in the population (i.e., $X_j(0) = \frac{1}{2N}$ for all j). It is assumed that mutations $\{i = 1, 2, \ldots\}$ evolve independently of one another so that X_j processes are i.i.d. Considering some absolute interval of time $(0, T]$, let M_i for $i = 1, 2, \ldots, T$ be the number of mutations that enter the population in generation i that are destined to be fixed. The time of entry of mutations that eventually fix in the population is known as the *origination* process [33, 88, 51]. Since each mutation has probability $p_1 = \frac{1}{2N}$ of eventually fixing in the population and the trajectories $X_1, X_2, \ldots$ are independent of each other, M_i for $i = 1, 2, \ldots, T$ are i.i.d. filtered Poisson random variables with rate

$$\mathbb{E}(M_i) = \frac{\theta}{2} p_1 = 2N \mu f_0 \frac{1}{2N} = \mu f_0 \ .$$

Furthermore, the total number of mutations $K = \sum_{i=1}^{T} M_i$ that enter the population during $(0, T]$ and eventually fix is also a Poisson random variable with rate $\mathbb{E}(K) = \mu f_0 T$ by the additivity property of independent Poisson random variables.

It is important to note that K is not the *actual* number of mutations that fix during the given interval of T generations (known as the *fixation process* [33]) but rather the number of mutations that enter during this interval and *eventually* become fixed. In the case of independently evolving sites, the origination process and the fixation process will have the same distribution as long as the time intervals are exchangeable. An example of when the time intervals would not be exchangeable is a difference in mutation rates for different time intervals.

Complete linkage among sites

Birky and Walsh [7] showed that the expected substitution rate for neutral mutations is not affected by linkage to neutral, deleterious, or advantageous mutations. Here we follow Cutler's discussion of the problem [16] closely to show that the distribution of the number of mutations that ultimately fix in the population remains a filtered Poisson process with rate μf_0 [81]. This was originally shown using reversibility arguments by Sawyer [88] and Kelly [51, p. 158].

Assume that mutations enter at a Poisson process rate $\frac{\theta}{2} = 2N f_0 \mu$, and write $X_j(t)$ for $j = 1, 2, \ldots$ to denote the frequency of the j process at time t since the origination of mutation j. Assume complete linkage among sites and write $f_j(x \mid t)dt$ to denote the $\Pr(X_j(t) = x)$. Let us introduce an indicator variable that tracks whether a given mutation becomes fixed in the population:

$$I_j = \begin{cases} 1 & \text{if mutation } j \text{ fixes in the population} \\ 0 & \text{otherwise.} \end{cases}$$

Since the number of neutral mutations on a chromosome does not alter the probability of fixation, $\mathbb{E}(I_j) = p_1$ for all j. Likewise, since the expected change in frequency from generation to generation is 0, the expected frequency of the j process is

$$\mathbb{E}(X_j(t)) = \int_0^1 x f_j(x \mid t)dx = \mathbb{E}(X_j(0)) = p_1.$$

Now consider two mutations, which we arbitrarily label $j = 1$ and $j = 2$, and assume mutation 1 is older than mutation 2. Consider the probability that both mutations become fixed ($\mathbb{E}(I_1 I_2)$). For this to happen, mutation 2 must occur on a background that contains mutation 1. The probability of this occurring is the frequency of the first mutation at the time the second mutation originates, $X_1(t)$. The marginal probability that mutation 2 fixes is simply its initial frequency $X_2(0) = p_1$. Therefore, the probability that both mutation 1 and mutation 2 fix in the population is given by

$$\mathbb{E}(I_1 I_2) = \Pr(\text{mutation 2 fixes}) \Pr(\text{mutation 1 fixes} \mid \text{mutation 2 fixes})$$
$$= \Pr(\text{mutation 2 fixes}) \cdot$$
$$\Pr(\text{mutation 2 arose on a chromosone containing mutation 1})$$
$$= p_1 \int_0^1 x f_1(x \mid t)dx$$
$$= p_1^2 \ .$$

Since the probability that both mutations fix is shown to be the product of the probability that each mutation fixes alone, the random variables $X_1(t)$ and $X_2(t)$ must be independent. This implies that linkage among neutral mutations does not affect the neutral rate of evolution. Likewise, since X_1 and X_2 are independent, the origination process remains a filtered Poisson process. The fixation process, on the other hand, does not remain a Poisson process in the presence of linkage. Informally, one can reason that the time intervals are no longer exchangeable. As has been discussed by Gillespie [31, 33] and

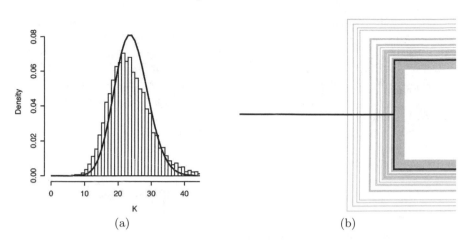

Fig. 4.2. Population dynamics can influence of molecular evolution. Two populations are split, evolved for $t = 10N$ generations, and a random chromosome from each is compared. (a) Distribution of the number of differences between a pair of random sequence from two populations that separated $10N$ generations ago and accrue mutations at rate $\mu = \frac{1}{N}$. The solid line is the expected distribution from a Poisson model. (b) Variation in branch length for the process due to random coalescence in ancestral population for $t = 10N$. The black line is the expected branch length (measured along the horizontal axis), and the grey lines are 100 replicates of the process.

Watterson [108, 109], the fixation process for the neutral infinite-sites model is a "burst" process whereby a geometric number of mutations fix when a chromosome reaches frequency 1 in the population. The effect of correlation in the substitution process is to reduce the efficiency of statistical methods for phylogenetic reconstruction [45].

4.4 Ancestral Polymorphism and Neutral Molecular Evolution

The analysis in Section 4.3 is predicated upon being able to follow the history of the entire population. The purpose of this Section is to derive the mean and variance of the sampling distribution for the number of nucleotide differences K between a sample of two DNA sequences drawn from a pair of populations that diverged t generations in the past. The full distribution for a sample of size $n = 2$ can be found in [102].

Measuring time into the past so that 0 is the present day, let

$$K = K_1 + K_2 + K_A,$$

where K_1 and K_2 are the number of mutations that accumulate on the first and second sequences since time t and K_A is the number of fixed differences due to ancestral polymorphism. Assuming a molecular clock, K_1 and K_2 are Poisson with rate $f_0 \mu t$. Without loss of generality, assume $f_0 = 1$. It will be shown that K_A is a geometrically distributed random variable so that the sampling distribution of K is not Poisson (see Figure 4.2). We will also see that the degree to which K will differ from a Poisson random variable with the same mean will depend on the parameters t and N_A, where N_A is the ancestral population size.

We will begin by considering the distribution of the number of differences for a sample of two chromosomes drawn from a panmictic population. This is equivalent to deriving the distribution of heterozygosity under an infinite-sites model and is a well-studied problem in population genetics (e.g., [56, 58, 107]). We will use the machinery of coalescent theory [44] to address the issue.

4.4.1 Average Pairwise Distance

Consider a sample of size $n = 2$ chromosomes drawn from a randomly mating population of size $2N$ chromosomes. Let S_2 be the number of nucleotide differences between two sequences at our locus of interest.

The probability that a random pair of chromosomes find a common ancestor in the previous generation is $\frac{1}{2N}$. Therefore, the distribution of the number of generations M until the two chromosomes find a common ancestor is a "first success" distribution with mean $2N$:

$$\Pr(M = m) = \left(1 - \frac{1}{2N}\right)^{m-1} \left(\frac{1}{2N}\right). \tag{4.9}$$

If N is large, (4.9) can be approximated using an exponential distribution. Measuring time in units of $2N$ generations, the random variable $T_2 = \frac{M}{2N}$ follows the exponential distribution with rate 1,

$$\Pr(M \leq 2Nx) = Pr(T_2 \leq x) \approx 1 - e^{-x}.$$

The random variable T_2 is known as the *coalescent* time for a sample of size $n = 2$ and describes the waiting time until two random chromosomes from a population coalesce (or merge) in a common ancestor. As one follows the two sequences back in time until the coalescent event, each accrues mutations independently at a rate $\frac{\theta}{2} = 2N\mu$ per unit of time assuming a Poisson model of mutation. This assumption implies that the waiting time until a mutation (T_M) occurs along either chromosome is exponential with rate θ. By the usual result for competing exponentials

$$\Pr(T_M < T_2) = \frac{\theta}{\theta + 1}.$$

Likewise, because of the memoryless property of the exponential distribution, once a mutation event occurs along either chromosome, the coalescent process is restarted. Therefore, the distribution of the number of mutations before a coalescent event for $n = 2$ is geometric:

$$\Pr(S_2 = k) = \left(\frac{\theta}{\theta + 1}\right)^k \frac{1}{\theta + 1}. \tag{4.10}$$

The expected value and variance of S_2 are easily shown to be

$$\mathbb{E}(S_2) = \theta, \quad \mathbb{V}(S_2) = \theta^2 + \theta. \tag{4.11}$$

Equations (4.10) and (4.11) were first derived by Watterson [107] when he found the distribution of the number of segregating sites S_i in a sample of size i. Li [66] also derived these results while finding the transient distribution of S_2. For our problem, $K_A = S_2$ with N replaced by N_A.

Recall that K is the sum of two independent Poisson random variables, each with mean μt, and a geometric random variable with mean $\theta_A = 4N_A\mu$, where N_A is the size of the ancestral population. This implies that

$$E(K) = 2\mu(t + 2N_A), \quad V(K) = 2\mu(t + 2N_A + 8N_A^2\mu) . \tag{4.12}$$

The index of dispersion (the ratio of the variance to the mean) is one way to assess the concordance between K and a Poisson random variable with the same mean [81, 31]. For K it is easy to show that

$$R(K) = 1 + \frac{8N_A^2\mu}{t + 2N_A} = 1 + \frac{\theta_A}{1 + \tau} ,$$

where $\tau = t/2N_A$. Figure 4.2 illustrates that ancestral polymorphism can lead to deviations from the Poisson expectations. In this figure, we have simulated 10,000 comparisons of $n = 2$ sequences drawn from a pair of populations that diverged $t = 10N_A$ generations ($\tau = 5$) in the past. Mutations occur in each daughter population as a Poisson process with rate $\mu = \frac{1}{N_A}$ per chromosome per generation ($\theta_A = 4$). Note that the distribution of K has a much larger variance than expected from the Poisson prediction ($E(K) = 24$) with $R(K) = 1.666$.

4.4.2 Lineage Sorting

Ancestral polymorphism can also lead to the phenomenon of "lineage sorting", where the genealogical tree for a sample of DNA sequences has a different branching order than the tree relating the history of population-splitting events. That is, if we have a sample of three sequences from three species $\{A, B, C\}$ and the tree relating our three populations is $((A, B), C)$, there is some probability of recovering *discordant* gene trees that are of the form $(A, (B, C))$ and $((A, C), B)$. (For an excellent discussion of the problem from a population genetics perspective, see [86]). The probability of recovering discordant trees in the three-taxon case is relatively easy to calculate using coalescent theory.

Assume that the population size N of three species is the same and has been constant for the history of $\{A, B, C\}$. Let t_1 be the time in the past in units of $2N$ generations when populations A and B split and let t_2 be the time in the past when the ancestral populations of A and B split from C. Write T_{AB} to denote the coalescent time of the sequence from species A and from species B and define T_{AC} and T_{BC} analogously. The probability that a gene tree will be concordant is the probability that A and B coalesce with each other before either coalesces with C. That is, the probability of concordance is given by $\Pr(\min(T_{AB}, T_{AC}, T_{BC})) = T_{AB}$.

The first coalescent event in the history of $\{A, B, C\}$ cannot occur before t_1. Between times t_1 and t_2, only coalescent events between A and B are allowed, and after t_2 all three lineages are equally likely to coalesce with one another. Letting $t = t_2 - t_1$, we can write

$$
\begin{aligned}
T_{AB} &= t_1 + X_1 \,, \\
T_{BC} &= t_1 + t + X_2 \,, \\
T_{AC} &= t_1 + t + X_3 \,,
\end{aligned}
\tag{4.13}
$$

where X_1, X_2, and X_3 are i.i.d. exponentially distributed random variables with rate 1. The justification for (4.13) comes from the results derived above that for large N the coalescent time for a sample of two sequences is exponential with rate 1. Recalling that the minimum of k independent exponential random variables is exponentially distributed with the sum of the k rates, we can also write

$$
\min(T_{BC}, T_{AC}) = t_1 + t + Y \,,
$$

where Y is an exponential random variable with rate 2 that is independent of X_1. Therefore,

$$
\begin{aligned}
\Pr(\text{concordance}) &= \Pr(\min(t + Y, X_1) = X_1) \\
&= \Pr(\min(t + Y, X_1) = X_1 \mid X_1 \leq t) \times \Pr(X_1 \leq t) + \\
&\qquad \Pr(\min(X_1, Y) = X_1 \mid X_1 > t) \times \Pr(X_1 > t) \\
&= 1 \times (1 - e^{-t}) + \frac{1}{3} \times e^{-t} \\
&= 1 - \frac{2}{3} e^{-t} \,.
\end{aligned}
$$

This simple example illustrates that to understand molecular evolutionary patterns on relatively short timescales, one must model the population genetics dynamics.

The question of estimating ancestral population genetics parameters has a rich history. Equations (4.12) were first derived by Takahata and Nei [101]. The full distribution of K in the case of one sequence from each of a pair as well as each of a triplet of species is given in Takahata, Satta, and Klein [102, eqs. (3), (6)]. As they discuss, these probabilities can be used for maximum likelihood estimates of the species divergence time and ancestral population size from multilocus data. Likewise, Yang [122] and Wall [106] have developed methods that incorporate rate variation among loci as well as recombination. The effects of population growth and differences in population size on levels of variation within and between a pair of species are taken up by Wakeley and Hey [105]. Likewise, a Bayesian method for distinguishing migration from isolation using within- and between-species sequence data is presented by Nielsen and Wakeley [71].

4.5 Natural Selection

The Wright-Fisher machinery can be adapted for modeling other evolutionary forces by specifying the joint effects of all forces on the change in gene frequency per generation. This is usually done in a two-step process. First an infinite gamete pool is assumed such that the frequency of the A_2 allele changes in the gamete pool deterministically due to mutation, selection, and other factors from some value $p = \frac{i}{2N}$ to p'. The effect of genetic drift is modeled using an equation analogous to (4.6), where p' depends on i and the evolutionary forces being considered:

$$P_{ij} \equiv \Pr(X(t+1) = j \mid X(t) = i) = \binom{2N}{j} (p')^j (1-p')^{2N-j} . \qquad (4.14)$$

In modeling natural selection, one needs to specify the fitness of the three relevant genotypes. Let the expected relative contribution of the $A_1 A_1$, $A_1 A_2$, and $A_2 A_2$ genotypes to the next generation be 1, $1 + 2sh$, and $1 + 2s$. (Note that h is known as the dominance parameter and summarizes the effect of selection on the heterozygote fitness.) The effect of natural selection is to bias the chance of picking an allele A_2 at random from the next generation. The expected proportion of offspring left by each of the three genotypes is

$$A_1 A_1 : \frac{(1-p)^2}{\overline{w}}, \quad A_1 A_2 : \frac{2p(1-p)(1+2sh)}{\overline{w}}, \quad A_2 A_2 : \frac{(1+2s)p^2}{\overline{w}},$$

where $\overline{w} = (1-p)^2 + 2(1+2sh)p(1-p) + p^2(1+2s)$.

Therefore, the frequency of the A_2 allele after one round of natural selection is

$$p_{t+1} = \frac{p_t^2(1+2s) + (1+2sh)p_t(1-p_t)}{\overline{w}} .$$

As we will see below, the number of selected mutations that fix in the history of a population under the assumption of recurrent mutation and selection is also Poisson and depends on the parameter $\gamma = 2Ns$ and h.

4.5.1 Diffusion Approximation

To study the Wright-Fisher model with selection (and other complicated population genetics models), it is often more convenient to work with a continuous-time continuous-space approximation to a discrete process. The natural state space is the frequency of a mutation ($0 \leq x = \frac{X(\cdot)}{2N} \leq 1$), and the natural time scaling is in units of $2N$ generations. Fisher [27] first noted that the action of genetic drift on a locus could be modeled using the same differential equations used to model the diffusion of heat. The classical problem of finding the stationary distribution of allele frequencies visited by a mutation under a variety of selective, mutation, and demographic models was taken up by Fisher in *The Genetical Theory of Natural Selection* [28] as well as by Sewall Wright [116, 117]. The time-dependent solution of what was later recognized as the Fokker-Planck or Kolmogorov forward equation was given in [52]. A definitive treatment of the subject is given in Kimura's classic paper [54]. We will now proceed to derive the stationary distribution, omitting many technical details that can be found by the interested reader in [54, 49, 23].

As discussed in Karlin and Taylor [49, p. 180], as $N \to \infty$, the Wright-Fisher process has a limiting diffusion that depends on the mean $M_{\delta x}$ and variance $V_{\delta x}$ of the change of gene frequency per generation. $M_{\delta x}$ will usually depend on the specifics of the model that produces the change in the gamete pool (mutation, migration, selection, etc.), while $V_{\delta x}$ is almost always given by the effects of binomial sampling. It is important to note that neither $M_{\delta x}$ nor $V_{\delta x}$ depend on time.

Write $\phi(x \mid p, t)dx$ to represent the conditional probability that a mutation at frequency p goes to frequency x in time t. In this equation, p is fixed and x is a random variable. When $dx = \frac{1}{2N}$ is substituted, $f(x \mid p, t) = \phi(x \mid p, t)\frac{1}{2N}$ gives the approximate frequency of mutations in the interval $x + dx$ for $0 < x < 1$ [54]. As discussed in [54], $\phi(x \mid p, t)$ is the solution to the Kolmogorov forward equation

$$\frac{\partial \phi(x \mid p, t)}{\partial t} = \frac{1}{2}\frac{\partial^2}{\partial x^2}\{V_{\delta x}\phi(x \mid p, t)\} - \frac{\partial}{\partial x}\{M_{\delta x}\phi(x \mid p, t)\} . \qquad (4.15)$$

A very useful consequence of (4.15) is that we can solve for the *stationary* or time-independent solution (if it exists) of $\phi(x \mid p)$ by setting $\frac{\partial \phi(x \mid p, t)}{\partial t} = 0$,

$$\phi(x) = \frac{C}{V_{\delta x}}\exp\left(-2\int\frac{M_{\delta x}}{V_{\delta x}}\right) , \qquad (4.16)$$

where C is a constant chosen so that $\int \phi(x)dx = 1$. The time-independent solution of (4.15) was first found by Sewall Wright [117].

Example 4.1: Reversible mutation neutral model

Consider a neutral model with reversible mutation so that $A_1 \to A_2$ at rate μ and $A_2 \to A_1$ at rate ν per generation. Let x_t represent the frequency of the A_1 allele at time t,

$$x_{t+1} = (1 - x_t)\nu - x_t\mu ,$$

implying that $M_{\delta x} = (1 - x)\nu - x(1 + \mu)$. The variance of the change in gene frequency is

$$V_{\delta x} = \frac{x(1-x)}{2N} .$$

Plugging M_{δ_x} and V_{δ_x} into (4.16), it is relatively straightforward to show that

$$\phi(x) = Cx^{4N\nu-1}(1-x)^{4N\mu-1} .$$

Recognizing that this is the density of a Beta distribution with parameters $4N\nu$ and $4N\mu$, the necessary constant is $C = \frac{\Gamma(4N\nu+4N\mu)}{\Gamma(4N\nu)\Gamma(4N\mu)}$.

4.5.2 Probability of Fixation

One of the most useful applications of the diffusion approximation is to calculate the probability of fixation of a mutation given its frequency in the population. To do so, we will follow [53] and use the Kolmogorov backwards equation to solve for $\phi(x \mid p, t)$. In this equation, we write the differential equation with respect to p varying, and the model is equivalent to running the process backwards in time (i.e., reversing the diffusion from x to p). The Kolmogorov backwards equation is

$$\frac{\partial \phi(x \mid p, t)}{\partial t} = \frac{V_{\delta p}}{2}\frac{\partial^2 \phi(p \mid x, t)}{\partial p^2} + M_{\delta p}\frac{\partial \phi(x \mid p, t)}{\partial p} . \tag{4.17}$$

If we substitute in $x = 1$, the solution to equation (4.17) gives us the probability of a mutation reaching fixation by time t given an initial frequency p. We will follow Kimura [54] and refer to this probability as $u(p, t)$. The boundary conditions for solving (4.17) are $u(0, t) = 0$ (i.e., probability of reaching 1 before 0 is 0 if $p = 0$) and $u(1, t) = 1$.

Again, following [54], by letting t tend towards infinity, we can find the probability of ultimate fixation:

$$u(p) = \lim_{t\to\infty} u(p, t) .$$

For the probability of ultimate fixation, $u(p)$, the left-hand side of (4.17) is 0, and thus the solution satisfies

$$0 = \frac{V_{\delta p}}{2}\frac{d^2 u(p)}{dp^2} + M_{\delta p}\frac{du(p)}{dp} .$$

Kimura [53] showed that the solution to this equation is

$$u(p) = \frac{\int_0^p G(x)dx}{\int_0^1 G(x)dx} ,$$

where

$$G(x) = \exp\left(-2\int \frac{M_{\delta x}}{V_{\delta x}}dx\right) .$$

4.5.3 No Selection

Recall that in the case of no mutation and no selection, $M_{\delta x} = 0$ and $V_{\delta x} = \frac{x(1-x)}{2N}$. This implies that $G(x) = 1$ and $u(p) = p$. This is the exact result we derived in a different way above, which states that the probability of ultimate fixation of a neutral mutation is given simply by its frequency.

4.5.4 Genic Selection

In the case of genic selection, $h = 0.5$ and the fitnesses of the individual genotypes are $\{1, 1 + s, 1 + 2s\}$. Letting x be the frequency of the selected allele,

$$M_{\delta x} = \frac{x^2(1 + 2s) + (1 + s)x(1 - x)}{\bar{w}} - x = \frac{sx(1 - x)}{1 + 2xs} .$$

If s is small, $M_{\delta x} \approx sx(1 - x)$, $G(x) = \exp(-4Nsx)$, and $u(p \mid s) = \frac{1 - \exp(-4Nsp)}{1 - \exp(-4Ns)}$. This implies that the probability of fixation of a new mutation is

$$u\left(\frac{1}{2N} \mid s\right) = \frac{1 - e^{-2s}}{1 - e^{-4Ns}} \approx \frac{2s}{1 - e^{-4Ns}}$$

using the fact that $e^x \approx 1 + x$ if x is small.

Since the mutation process for both selected and neutral mutations is Poisson, their relative substitution rates are given by the ratio of the probabilities of fixation *assuming independence among sites*. Let ω equal the ratio of the probability of fixation of a selected mutation per selected site relative to the probability of fixation of a neutral mutation per neutral site:

$$\omega = \frac{f_s u(p \mid s \neq 0)}{f_0 u(p \mid s = 0)} = \frac{f_s \frac{2s}{1 - e^{-4Ns}}}{f_0 \frac{1}{2N}} = \frac{f_s}{f_0} \frac{2\gamma}{1 - e^{-2\gamma}} .$$

As previously mentioned, ω can be interpreted as the expected dn/ds ratio assuming silent mutations are neutral. We will assume $f_0 = f_s$ for the remainder of the chapter (for coding DNA). As we see from Figures 4.1, 4.3, and 4.5, even modest amounts of natural selection can have a profoundly strong effect on rates of substitution. For example, it has been estimated that the historical effective population size of humans is close to $N = 10^5$ (for a review, see [106]). This implies that sites where a mutation would lower the expected number of offspring an individual contributes to the next generation by as little as 0.0025% ($\gamma = -5$) would not evolve at any appreciable rate ($\omega < 0.01$).

In the case of positive genic selection, as s becomes large, the probability of ultimate fixation for a new mutation is well-approximated by $u \approx 2s$ and the expected ratio of substitution rates for selected to neutral mutations by $\omega \approx 2\gamma$. This implies that if mutations at some class of sites increased the expected number of offspring by as little as 0.0025% ($\gamma = 5$), they would evolve at 10 times the rate of neutral mutations.

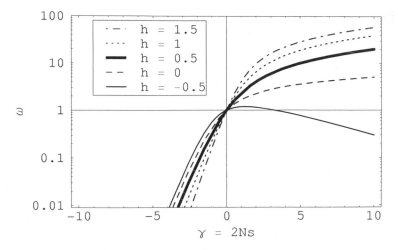

Fig. 4.3. Effect of dominance and selection on rates of molecular evolution.

4.5.5 Dominance

In the case of general selection, it follows directly from the Wright-Fisher model that $M_{\delta x} \approx s(h + (1 - 2h)x)x(1 - x)$ if s is small. This implies that $G(x) = \exp(-4Nshx + 2Ns(1 - 2h)x^2)$ and

$$
u(p) = \frac{\int_0^p e^{-2\gamma shx + \gamma(1-2h)x^2} dx}{\int_0^1 e^{-2\gamma shx + \gamma(1-2h)x^2} dx} .
$$

This integral can be evaluated numerically to investigate the effect of heterozygous fitness on rates of molecular evolution. As we see from Figure 4.3, the most profound effects occur when mutations are selectively favored ($\gamma > 0$) and produce heterozygote advantage ($h > 1$). This condition is known as overdominance and such a mutation is said to be subject to balancing selection. In an infinitely large population, overdominance leads to a stable equilibrium in gene frequency such that both alleles are maintained in the population indefinitely. In a finite population, though, higher heterozygote fitness translates into a higher substitution rate relative to neutrality as well as relative to genic selection ($h = 0.5$). The reason for these perplexing results is that having a high heterozygote fitness decreases the probability that a mutation will be lost from the population and thus increases the probability that it will ultimately become fixed in the population.

Another interesting case to consider is that of a mutation whose fitness relative to the wildtype depends on whether it is in heterozygous or homozygous form ($h = -0.50$). If the mutation is deleterious in homozygous form but advantageous in heterozygous form, the mutation will have a slightly higher rate of fixation relative to the case when the heterozygote has intermediate fitness ($h = 0.5$). Alternatively, a beneficial mutation in homozygous form that produces heterozygotes that are less fit than either homozygote will have a lower substitution rate. In interpreting these results, it is important to remember that in estimating ω we are assuming independence among sites. As we will see below, linkage among selected sites can cause interference effects that will counter the single-site dynamics illustrated in Figure 4.3. This is particularly true in the case of strong dominance.

4.6 Variation in Selection Among Sites

Understanding how the distribution of selection coefficients among newly arising mutations affects the rates and patterns of molecular evolution has been a focus of extensive research in theoretical population genetics. In a series of papers, Tomoko Ohta (along with Kimura) [81, 76, 77, 78] first investigated the molecular evolution of "nearly neutral" mutations and found that their behavior was quite different from that of strictly neutral mutations ($\gamma = 0$). In particular, she showed that if there is a high rate of input of slightly deleterious mutations ($-2 < \gamma < 0$) into a population, then this class of mutations can contribute significantly to the overall substitution rate even though these mutations are slightly less fit than the existing wildtype allele. As discussed in Section 4.2, Ohta and Kimura also demonstrated that a nearly neutral model would predict a negative correlation between population size and rate of molecular evolution since natural selection is more efficient in a larger population.

The original work of Ohta and Kimura went on to inspire a plethora of nearly neutral, nonneutral, and fluctuating-environment population genetics theories of molecular evolution. For example, Ohta proposed the exponential-shift model [79], where $-s$ follows an exponential distribution among new mutations (the term shift is used since s is relative to the wildtype allele and the distribution must shift after an allele fixes in the population). Likewise, Kimura [59] suggested a Gamma-shift model that conveniently had sufficient mass near $s = 0$ to account for several neutral and nearly neutral predictions [61, 31]. Ohta and Tachida [82] also proposed a fixed fitness model, where the distribution of s was Gaussian and independent of parental type (a so-called house-of-cards model). These models have been used to argue that if a substantial proportion of slightly deleterious mutations are input into the population, the rate of fixation contributes significantly to the proportion of mutations that fix in the population. It is important to note, though, that the conclusion comes directly from assumptions regarding the functional form of the distribution of selective effects among sites. Since there is no biological reason to favor one distribution over another a priori, in practical applications it is important to be catholic on the matter and consider several potential candidate distributions.

Recently, two methods have come on the market for estimating the distribution of selective effects among new mutations. Nielsen and Yang [73] have developed a likelihood-based method for use with divergence data that considers ten different models (e.g., constant, normal, Gamma, exponential, normal + invariant). (A similar method was suggested by Felsenstein [25] but to our knowledge not fully implemented.) Nielsen and Yang applied their model to a data set of eight mtDNA primate genomes and found that of the models considered, a normal or Gamma-shift model with some sites held invariant was the best fit to data (and significantly better than an exponential distribution [79]). Likewise, Stanley Sawyer and colleagues have developed a method for fitting a normal-shift model to polymorphism and divergence data [90] and applied it to 56 loci with polymorphism from *Drosophila simulans* and divergence data relative to a *D. melanogaster* reference strain. In these models, it is assumed that selection coefficients at a given site are constant in time and do not depend on the nucleotide present. Below we present a brief analysis of the normal-shift model and discuss the findings of Nielsen and Yang [73] and Sawyer et al. [90] in light of the analysis.

4.6.1 Normal Shift

Assume that we starts a Poisson number of Wright-Fisher processes at rate $2N\mu$ per generation and that these processes do not interfere with one another. The number of processes that fix for the selected mutation in some interval of time t will be Poisson with rate

$$
\begin{aligned}
\mathbb{E}(K \mid \gamma) &= 2N\mu t u(s) \\
&= \mu t \frac{2\gamma}{1 - e^{-2\gamma}} \\
&= \mu t k(\gamma).
\end{aligned}
$$

Likewise, if mutations have a distribution of selection coefficients such that the probability that a mutation has selection coefficient γ is governed by $f(\gamma)$, then the number of mutations that fix will be Poisson with rate

$$
\mathbb{E}(K) = \mu t \int_{-\infty}^{\infty} k(\gamma) f(\gamma) d\gamma . \tag{4.18}
$$

We can now calculate some statistics of interest. For example, the distribution of selection coefficients among fixed mutations (f is for "fixed") is

$$
p_f(\gamma) = \frac{k(\gamma) f(\gamma) d\gamma}{\int_{-\infty}^{\infty} k(\gamma) f(\gamma) d\gamma} . \tag{4.19}
$$

This implies that the average selection coefficient of substitutions can be easily computed as

$$
\mathbb{E}_f(\gamma) = \int_{-\infty}^{\infty} \gamma p_f(\gamma) d\gamma . \tag{4.20}
$$

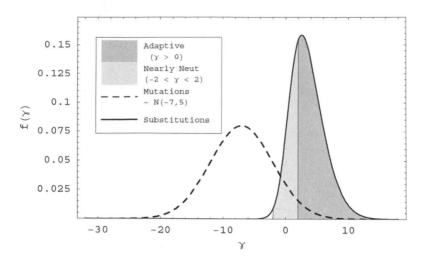

Fig. 4.4. A Gaussian model ("normal shift") for the distribution of selection co-efficients among mutations [90]. In this example, the selection coefficient of new mutations is normally distributed with mean $\mu = -7$ and standard deviation $\sigma = 5$. In this example, 68.9% of substitutions are adaptive (dark grey area), 30.7% are nearly neutral, and 0.4% are deleterious.

Likewise, the proportion of fixed differences that are nearly neutral (using the definition of nearly neutral as $-2 \leq \gamma \leq 2$) is

$$p_f(-2 \leq \gamma \leq 2) = \frac{\int_{-2}^{2} k(\gamma)f(\gamma)d\gamma}{\int_{-\infty}^{\infty} k(\gamma)f(\gamma)d\gamma} \qquad (4.21)$$

and the proportion of fixed differences that are positively selected (and not nearly neutral) is given by the tail probability

$$p_f(\gamma > 2 \mid \zeta) = \frac{\int_{2}^{\infty} k(\gamma)f(\gamma)d\gamma}{\int_{-\infty}^{\infty} k(\gamma)f(\gamma)d\gamma} . \qquad (4.22)$$

In Figures 4.4 and 4.5, we explore the effects of a Gaussian model for the distribution of selection coefficients among newly arising mutations. In Figure 4.4, mutations are assumed to follow a normal distribution with mean $\mu = -7$ and standard deviation $\sigma = 5$. Using (4.21) and (4.22), we can estimate the proportion of substitutions that are nearly neutral and adaptive via standard numerical integration (grey areas under the solid curve in Figure 4.4). We note that in this example the vast majority of mutations are deleterious ($> 91\%$ are below 0), while most of the substitutions (fixed differences) are positively selected: 92.8% are above $\gamma = 0$, and 68.3% have a selection coefficient above $\gamma = 2$. The average selection coefficient of fixed mutations is a (surprisingly) high $\gamma = 3.49$.

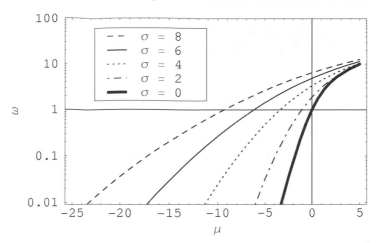

Fig. 4.5. Effect of variance in the distribution of selection coefficients among newly arising mutations on rates of molecular evolution. In this figure, μ is the mean of the distribution of selective effects.

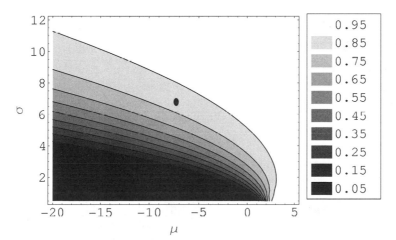

Fig. 4.6. Proportion of adaptive substitutions ($s > \frac{1}{N}$) as a function of the mean of the distribution of selection coefficients for new mutations μ and standard deviation σ. The black point represents the estimated mean and variance for a typical *Drosophila* gene [90].

The fact that mutations differ in their selective effects also has a strong implication for interpreting the ω ratio. In Figure 4.5, we plot the expected ω ratio for varying levels of selection (where the x-axis is the average selected effect of the new mutation) and variability among mutations assuming $f_s = f_0$, where σ corresponds to the standard deviation of selection coefficients among new mutations. In the case of moderate variance $\sigma = 6$, as long as the average selective effect of newly arising mutations is greater than -5, the ω ratio will

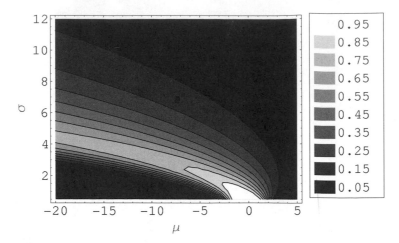

Fig. 4.7. Proportion of nearly neutral substitutions ($|s| \leq \frac{1}{N}$) as a function of the mean of the distribution of selection coefficients for new mutations μ and standard deviation σ. The black point represents the estimated mean and variance for a typical *Drosophila* gene [90].

be greater than 1 (even though most mutations are deleterious). This explains a perplexing phenomenon that is observed in day-to-day analysis of DNA sequence evolution: namely, how it is that one can detect positive selection in the first place if most of the amino acid sites in a protein are rather constrained. The answer is that natural selection is extremely efficient at fixing even slightly favored mutations, so that as long as there is some reasonable fraction of mutations that are adaptive, the average rate of fixation for selected sites (e.g., amino acid sites) may outstrip the neutral rate of evolution. In Figure 4.6, we plot the proportion of fixed differences that are adaptive as a function of both the average selective effect of new mutations (μ) and standard deviation (σ). We note that as long as the standard deviation among newly arising mutations is greater than 6, most of the substitutions will be adaptive even if, on average, mutations are extremely deleterious. The comparable contour plot for nearly neutral mutations is given in Figure 4.7. These simple results bolster the idea that comparing the rate of substitution for different types of sites in protein-coding genes is an effective way of detecting positively selected sites.

The results of Sawyer et al. [90] bear a strong resemblance to the pattern we have just described. They estimated the distribution of selective effects among new mutations in a typical *Drosophila* gene to have mean $\mu = -7.31$ and $\sigma = 6.79$. This implies that close to 97.1% of amino substitutions in a typical *Drosophila* nuclear gene are of positively selected mutations ($\gamma > 0$), with 84.7% being clearly adaptive, $\gamma \geq 2$; see (4.22). Furthermore, close to 15.2% of substitutions are of "nearly neutral" mutations ($-2 \leq \gamma \leq 2$), with only 2.7% being "slightly deleterious" ($-2 \leq \gamma \leq 0$) mutations while 12.4% are "slightly advantageous" $(0 \leq \gamma \leq 2)$. Lastly, the average selection coefficient of substituting mutations is 5.67; see (4.20). The black disks in Figures 4.6 and 4.7 correspond to the Sawyer et al. estimate of μ and σ for *Drosophila*. These results are consistent with previous findings of adaptive protein evolution in *Drosophila* (e.g., [95, 24, 8, 84]).

4.6.2 Linkage

One interpretation of the normal-shift model is that of "Darwin's wedge" at a molecular level [90]. As Darwin wrote in *The Origin of Species* [20, cp. 3]

> In looking at Nature, it is most necessary to keep the foregoing considerations always in mind never to forget that every single organic being around us may be said to be striving to the utmost to increase in numbers... . The face of Nature may be compared to a yielding surface, with ten thousand sharp wedges packed close together and driven inwards by incessant blows, sometimes one wedge being struck, and then another with greater force.

In this passage, Darwin views natural selection as competition for fixed resources leading to rapid turnover of species. That is, one wedge forces another out in order to fix its claim to a space in a cramped environment. At a molecular level, the metaphor works well: a slightly favored mutation sweeping through the population acts as a wedge to displace the existing alleles at a given locus. The efficacy of such a wedging scheme, of course, is predicated upon the frequency of favored wedges. If there are too many favored mutations competing for fixation at a given locus, they will knock each other out of competition and the efficacy of selection can be greatly reduced. In many ways, the fact that one can detect positive selection in the face of interference among selected sites is in fact *stronger* evidence for a selective model of molecular evolution. That is to say, if one estimates that the average selection coefficient of fixed mutations is $\gamma = 5.67$ in the presence of interference, the true selection coefficient on the mutation must be higher. There is relatively strong support for the view that linkage can affect rates and patterns of substitution for selected mutations [7, 42, 15, 36, 37].

For example, Birky and Walsh [7] have shown analytically and via simulation that linked selected mutation negatively interferes so as to increase the rate of substitution of deleterious mutations and to decrease the rate of substitution of advantageous mutations. They attribute this phenomenon to a reduction in the effective population size through an increase in the variance of offspring among individuals. As we saw in Section 4.5, if the effective population size of a species is reduced, genetic drift begins to play a more prominent role in determining the evolutionary fate of mutations.

The predictions of interference selection hypotheses have gained strong support in recent years. For example, Comeron and Kreitman used analytical, simulation, and genomic analyses to demonstrate that interference selection can explain patterns of codon usage and intron size in *Drosophila* [15]. Likewise, a prediction of the interference hypothesis is that rates of adaptive evolution should be reduced in regions of low recombination since the tighter the linkage among favored mutations, the stronger the interference effects. There is experimental evidence that regions of low recombination in *Drosophila* do, in fact, show a reduction in the rate of adaptive evolution [93, 5], as do nonrecombining mitochondria [111, 84]. Likewise, if we consider the analysis of Nielsen and Yang [73], they estimate a distribution of selective effects among mutations in primate mtDNA that has mean $\mu = -1.72$ and $\sigma = 0.72$. For such a model, the proportion of substitutions that have a selection coefficient greater than $\gamma = 0$ is a quite small 6%, consistent with the view that linkage limits the rate of adaptive evolution.

There is also important literature on the impact of linkage on rates of evolution in nearly neutral and fluctuating selection models [32, 33, 34, 35, 17, 18]. Much of it has focused on analytical and simulation work for describing which population genetics models lead to an overdispersed molecular clock. To summarize all of this work, Gillespie and Cutler have shown that the overdispersed molecular clock cannot readily be explained by overdominance, underdominance, a rapidly fluctuating environment, or the nearly neutral models presented above (although certain narrow parameter ranges can lead to an over-dispersed clock, the models do not, in general, lead to an overdispersed clock). Gillespie [32] has found that a slowly fluctuating environment can lead to an over-dispersed clock if the oscillations are on the same order as the mutation rate. Likewise, Cutler [17] has argued that a simple deleterious model that shifts between a favored and a deleterious allele is sufficient to explain the overdispersed clock.

Gillespie has also investigated the effects of linkage and selection on the relationship between population size and the rate of molecular evolution using extensive simulations. He has identified three domains, which he terms the Darwin domain ($k_s \propto N$), the Kimura domain ($k_s \approx \mu f_0$), and the Ohta domain ($k_s \propto \frac{1}{N}$). Not surprisingly, he finds that the nearly neutral models (exponential shift [79], Gamma-shift model [59], and house of cards [82]) all fall within the Ohta domain where the rate of evolution is inversely proportional to population size. He also finds that the normal-shift model with mean $\mu = 0$ (Darwin's wedge) appropriately falls in the Darwin domain, where the rate of substitution is proportional to the population size. He also notes that the rate of substitution for the normal-shift model is substantially reduced relative to the expectation under the independence-among-sites model (4.18) (as one might predict from [7]). Lastly, he finds, surprisingly, that the fluctuating selection, neutral, and overdominance models all lead to the Kimura domain, where the rate of molecular evolution is independent of the population size. A mechanism that Gillespie has proposed to explain this last observation is the theory of genetic drift, whereby positive selection on one locus leads to the reduction of effective population size at linked neutral loci even in an infinitely large population [36, 37].

Lastly, Brian Charlesworth and colleagues have also shown that linkage of neutral mutations to deleterious mutations ("background" selection) [9, 10, 11] leads to a chronic and pronounced reduction in the local effective population size of a chromosomal region. Recent experimental work on patterns of variation within the nonrecombining neo-sex chromosomes of *Drosophila miranda* [2, 3] has confirmed some theoretical predictions of the background selection model. Likewise, Cutler [18] has argued that the background selection hypothesis is consistent with the observed overdispersed molecular clock.

Acknowledgments

This work was supported by grants NSF-0319553 and USDA-HATCH(NYC-151411). I am indebted to Rasmus Nielsen and Ryan Hernandez for comments and suggestions on earlier drafts. Ryan also helped produce Figure 4.1. Likewise, Scott Williamson provided assistance in producing Figure 4.3. I would also like to thank the Institute for Pure and Applied Mathematics (IPAM) for administrative support and office space where some of this work was completed. Beatrix Johnson is also acknowledged for help in library research.

References

[1] J. M. Akey, G. Zhang, K. Zhang, L. Jin, and M. D. Shriver. Interrogating a high-density SNP map for signatures of natural selection. *Genome Res*, 12(12):1805–1814, Dec 2002.

[2] D. Bachtrog. Adaptation shapes patterns of genome evolution on sexual and asexual chromosomes in *Drosophila*. *Nat Genet*, 34(2):215–219, Jun 2003.

[3] D. Bachtrog. Protein evolution and codon usage bias on the neo-sex chromosomes of *Drosophila miranda*. *Genetics*, 165(3):1221–1232, Nov 2003.

[4] M. Barrier, C. D. Bustamante, J. Yu, and M. D. Purugganan. Selection on rapidly evolving proteins in the *Arabidopsis* genome. *Genetics*, 163(2):723–733, Feb 2003.

[5] A. J. Betancourt and D. C. Presgraves. Linkage limits the power of natural selection in *Drosophila*. *Proc Natl Acad Sci USA*, 99(21):13616–13620, Oct 2002.

[6] N. Bierne and A. Eyre-Walker. The genomic rate of adaptive amino-acid substitution in *Drosophila*. *Mol Biol Evol*, Mar 2004.

[7] C. W. Birky and J. B. Walsh. Effects of linkage on rates of molecular evolution. *Proc Natl Acad Sci USA*, 85(17):6414–6418, Sep 1988.

[8] C. D. Bustamante, R. Nielsen, S. A. Sawyer, K. M. Olsen, M. D. Purugganan, and D. L. Hartl. The cost of inbreeding in *Arabidopsis*. *Nature*, 416(6880):531–534, Apr 2002.

[9] B. Charlesworth. The effect of background selection against deleterious mutations on weakly selected, linked variants. *Genet Res*, 63(3):213–227, Jun 1994.

[10] B. Charlesworth, M. T. Morgan, and D. Charlesworth. The effect of deleterious mutations on neutral molecular variation. *Genetics*, 134(4):1289–1303, Aug 1993.

[11] D. Charlesworth, B. Charlesworth, and M. T. Morgan. The pattern of neutral molecular variation under the background selection model. *Genetics*, 141(4):1619–1632, Dec 1995.

[12] M. Choisy, C. H. Woelk, J. F. Guegan, and D. L. Robertson. Comparative study of adaptive molecular evolution in different human immunodeficiency virus groups and subtypes. *J Virol*, 78(4):1962–1970, Feb 2004.

[13] A. G. Clark, S. Glanowski, R. Nielsen, P. D. Thomas, A. Kejariwal, M. A. Todd, D. M. Tanenbaum, D. Civello, F. Lu, B. Murphy, S. Ferriera, G. Wang, X. Zheng, T. J. White, J. J. Sninsky, M. D. Adams, and M. Cargill. Inferring nonneutral evolution from human-chimp-mouse orthologous gene trios. *Science*, 302(5652):1960–1963, Dec 2003.

[14] R. M. Clark, E. Linton, J. Messing, and J. F. Doebley. Pattern of diversity in the genomic region near the maize domestication gene tb1. *Proc Natl Acad Sci USA*, 101(3):700–707, Jan 2004.

[15] J. M. Comeron and M. Kreitman. Population, evolutionary and genomic consequences of interference selection. *Genetics*, 161(1):389–410, May 2002.

[16] D. J. Cutler. Clustered mutations have no effect on the overdispersed molecular clock: A response to Huai and Woodruff. *Genetics*, 149(1):463–464, May 1998.

[17] D. J. Cutler. The index of dispersion of molecular evolution: Slow fluctuations. *Theor Popul Biol*, 57(2):177–186, Mar 2000.

[18] D. J. Cutler. Understanding the overdispersed molecular clock. *Genetics*, 154(3):1403–1417, Mar 2000.

[19] J. da Silva. The evolutionary adaptation of HIV-1 to specific immunity. *Curr HIV Res*, 1(3):363–371, Jul 2003.

[20] C. Darwin. *The Origin of Species*. Oxford University Press, Oxford, reissue edition, 1859.

[21] W. J. Ewens. A note on the sampling theory for infinite alleles and infinite sites models. *Theor Popul Biol*, 6(2):143–148, Oct 1974.

[22] W. J. Ewens. A note on the variance of the number of loci having a given gene frequency. *Genetics*, 80(1):221–222, May 1975.

[23] W. J. Ewens. *Mathematical Population Genetics*. Springer, New York, 2004.

[24] J. C. Fay, G. J. Wyckoff, and C. I. Wu. Testing the neutral theory of molecular evolution with genomic data from *Drosophila*. *Nature*, 415(6875):1024–1026, Feb 2002.

[25] J. Felsenstein. Taking variation of evolutionary rates between sites into account in inferring phylogenies. *J Mol Evol*, 53(4–5):447–455, Oct 2001.

[26] J. Felsenstein and G. A. Churchill. A hidden Markov model approach to variation among sites in rate of evolution. *Mol Biol Evol*, 13(1):93–104, Jan 1996.

[27] R. A. Fisher. On the dominance ratio. *Proc Roy Soc Edinburgh*, 42:321–341, 1922.

[28] R. A. Fisher. *The Genetical Theory of Natural Selection.* Clarendon Press, Oxford, 1st edition, 1930.

[29] Y. Gilad, C. D. Bustamante, D. Lancet, and S. Paabo. Natural selection on the olfactory receptor gene family in humans and chimpanzees. *Am J Hum Genet*, 73(3):489–501, Sep 2003.

[30] J. H. Gillespie. A general model to account for enzyme variation in natural populations. v. the sas-cff model. *Theor Popul Biol*, 14:1–45, 1978.

[31] J. H. Gillespie. *The Causes of Molecular Evolution.* Oxford University Press, Oxford, 1991.

[32] J. H. Gillespie. Substitution processes in molecular evolution. I. Uniform and clustered substitutions in a haploid model. *Genetics*, 134:971–981, 1993.

[33] J. H. Gillespie. Substitution processes in molecular evolution. II. Exchangeable models from population genetics. *Evolution*, 48:1101–1113, 1994.

[34] J. H. Gillespie. Substitution processes in molecular evolution. III. Deleterious alleles. *Genetics*, 138:943–952, 1994.

[35] J. H. Gillespie. The role of population size in molecular evolution. *Theor Popul Biol*, 55:145–156, 1999.

[36] J. H. Gillespie. Genetic drift in an infinite population. The pseudohitchhiking model. *Genetics*, 155(2):909–919, Jun 2000.

[37] J. H. Gillespie. The neutral theory in an infinite population. *Gene*, 261(1):11–18, Dec 2000.

[38] S. Glinka, L. Ometto, S. Mousset, W. Stephan, and D. De Lorenzo. Demography and natural selection have shaped genetic variation in *Drosophila melanogaster*: A multi-locus approach. *Genetics*, 165(3):1269–1278, Nov 2003.

[39] N. Goldman and Z. Yang. A codon-based model of nucleotide substitution for protein-coding DNA sequences. *Mol Biol Evol*, 11(5):725–736, Sep 1994.

[40] M. Hasegawa, H. Kishino, and T. Yano. Dating of the human-ape splitting by a molecular clock of mitochondrial DNA. *J Mol Evol*, 22(2):160–174, 1985.

[41] I. Hellmann, S. Zollner, W. Enard, I. Ebersberger, B. Nickel, and S. Paabo. Selection on human genes as revealed by comparisons to chimpanzee cDNA. *Genome Res*, 13(5):831–837, May 2003.

[42] W. G. Hill and A. Robertson. The effect of linkage on limits to artificial selection. *Genet Res*, 8(3):269–294, Dec 1966.

[43] J. L. Hubby and R. C. Lewontin. A molecular approach to the study of genic heterozygosity in natural populations. I. The number of alleles at different loci in *Drosophila pseudoobscura*. *Genetics*, 54(2):577–594, Aug 1966.

[44] R. R. Hudson. Gene genealogies and the coalescent process. In *Oxford Surveys in Evolutionary Biology 7*, pages 1–44. Oxford University Press, Oxford, 1990.

[45] J. P. Huelsenbeck and R. Nielsen. Effect of nonindependent substitution on phylogenetic accuracy. *Syst Biol*, 48(2):317–328, Jun 1999.

[46] G. A. Huttley, M. W. Smith, M. Carrington, and S. J. O'Brien. A scan for linkage disequilibrium across the human genome. *Genetics*, 152(4):1711–1722, Aug 1999.

[47] V. Jaenicke-Despres, E. S. Buckler, B. D. Smith, M. T. Gilbert, A. Cooper, J. Doebley, and S. Paabo. Early allelic selection in maize as revealed by ancient DNA. *Science*, 302(5648):1206–1208, Nov 2003.

[48] T. H. Jukes and C. R. Cantor. Evolution of protein molecules. In *Mammalian Protein Metabolism*, pages 21–132. Academic Press, New York, 1969.

[49] S. Karlin and H. M. Taylor. *A Second Course in Stochastic Processes*. Academic Press, New York, 1981.

[50] M. Kayser, S. Brauer, and M. Stoneking. A genome scan to detect candidate regions influenced by local natural selection in human populations. *Mol Biol Evol*, 20(6):893–900, Jun 2003.

[51] F. P. Kelly. *Reversibility and Stochastic Networks*. John Wiley & Sons, Chichester, 1979.

[52] M. Kimura. Solution of a process of random genetic drift with a continuous model. *Proc Natl Acad Sci USA*, 41:114–150, 1955.

[53] M. Kimura. On the probability of fixation of mutant genes in a population. *Genetics*, 47:713–719, 1962.

[54] M. Kimura. Diffusion models in population genetics. *J Appl Probab*, 1:177–232, 1964.

[55] M. Kimura. Evolutionary rate at the molecular level. *Nature*, 217:624–626, 1968.

[56] M. Kimura. The number of heterozygous nucleotide sites maintained in a finite population due to steady flux of mutations. *Genetics*, 61(4):893–903, Apr 1969.

[57] M. Kimura. The rate of molecular evolution considered from the standpoint of poulation genetics. *Proc Natl Acad Sci USA*, 63:1181–1188, 1969.

[58] M. Kimura. Theoretical foundation of population genetics at the molecular level. *Theor Popul Biol*, 2(2):174–208, Jun 1971.

[59] M. Kimura. Models of effectively neutral mutations in which selective constraint is incorporated. *Proc Nat Acad Sci USA*, 76:3440–3444, 1979.

[60] M. Kimura. A simple method for estimating evolutionary rates of base substitutions through comparative studies of nucleotide sequences. *J Mol Evol*, 16(2):111–120, Dec 1980.

[61] M. Kimura. *The Neutral Theory of Molecular Evolution*. Cambridge University Press, Camebridge, 1983.

[62] M. Kimura and T. Ohta. Protein polymorphism as a phase of molecular evolution. *Nature*, 229(5285):467–469, Feb 1971.

[63] J. L. King and T. H. Jukes. Non-Darwinian evolution. *Science*, 164:788–798, 1969.

[64] R. C. Lewontin. *The Genetic Basis of Evolutionary Change*. Columbia University Press, New York, 1974.

[65] R. C. Lewontin and J. L. Hubby. A molecular approach to the study of genic heterozygosity in natural populations. II. Amount of variation and degree of heterozygosity in natural populations of *Drosophila pseudoobscura*. *Genetics*, 54(2):595–609, Aug 1966.

[66] W. H. Li. Distribution of nucleotide differences between two randomly chosen cistrons in a finite population. *Genetics*, 85(2):331–337, Feb 1977.

[67] T. Mitchell-Olds and M. J. Clauss. Plant evolutionary genomics. *Curr Opin Plant Biol*, 5(1):74–79, Feb 2002.

[68] C. B. Moore, M. John, I. R. James, F. T. Christiansen, C. S. Witt, and S. A. Mallal. Evidence of HIV-1 adaptation to HLA-restricted immune responses at a population level. *Science*, 296(5572):1439–1443, May 2002.

[69] S. V. Muse. Evolutionary analyses of DNA sequences subject to constraints of secondary structure. *Genetics*, 139(3):1429–1439, Mar 1995.

[70] S. V. Muse and B. S. Gaut. A likelihood approach for comparing synonymous and nonsynonymous nucleotide substitution rates, with application to the chloroplast genome. *Mol Biol Evol*, 11(5):715–724, Sep 1994.

[71] R. Nielsen and J. Wakeley. Distinguishing migration from isolation: A Markov chain Monte Carlo approach. *Genetics*, 158(2):885–896, Jun 2001.

[72] R. Nielsen and Z. Yang. Likelihood models for detecting positively selected amino acid sites and applications to the HIV-1 envelope gene. *Genetics*, 148(3):929–936, Mar 1998.

[73] R. Nielsen and Z. Yang. Estimating the distribution of selection coefficients from phylogenetic data with applications to mitochondrial and viral DNA. *Mol Biol Evol*, 20(8):1231–1239, Aug 2003.

[74] J. R. Norris. *Markov Chains*. Cambridge University Press, Cambridge, 1997.

[75] D. Nurminsky, D. D. Aguiar, C. D. Bustamante, and D. L. Hartl. Chromosomal effects of rapid gene evolution in *Drosophila melanogaster*. *Science*, 291(5501):128–130, Jan 2001.

[76] T. Ohta. Evolutionary rate of cistrons and DNA divergence. *J Mol Evol*, 1:150–157, 1972.

[77] T. Ohta. Slightly deleterious mutant substitutions in evolution. *Nature*, 246:96–98, 1973.

[78] T. Ohta. Mutational pressure as the main cause of molecular evolution and polymorphism. *Nature*, 252:351–354, 1974.

[79] T. Ohta. Extension of the neutral mutation drift hypothesis. In M. Kimura, editor, *Molecular Evolution and Polymorphism*, pages 148–167. National Institute of Genetics, Mishima, Japan, 1977.

[80] T. Ohta. Near-neutrality in evolution of genes and gene regulation. *Proc Natl Acad Sci USA*, 99(25):16134–16137, Dec 2002.

[81] T. Ohta and M. Kimura. On the constancy of the evolutionary rate of cistrons. *J Mol Evol*, 1:18–25, 1971.

[82] T. Ohta and H. Tachida. Theoretical study of nearly neutrality. I. Heterozygosity and rate of mutant substitution. *Genetics*, 126:219–229, 1990.

[83] B. A. Payseur, A. D. Cutter, and M. W. Nachman. Searching for evidence of positive selection in the human genome using patterns of microsatellite variability. *Mol Biol Evol*, 19(7):1143–1153, Jul 2002.

[84] D. M. Rand, D. M. Weinreich, and B. O. Cezairliyan. Neutrality tests of conservative-radical amino acid changes in nuclear- and mitochondrially-encoded proteins. *Gene*, 261(1):115–125, Dec 2000.

[85] D. M. Robinson, D. T. Jones, H. Kishino, N. Goldman, and J. L. Thorne. Protein evolution with dependence among codons due to tertiary structure. *Mol Biol Evol*, 20(10):1692–1704, Oct 2003.

[86] N. A. Rosenberg. The probability of topological concordance of gene trees and species trees. *Theor Popul Biol*, 61(2):225–247, Mar 2002.

[87] P. C. Sabeti, D. E. Reich, J. M. Higgins, H. Z. Levine, D. J. Richter, S. F. Schaffner, S. B. Gabriel, J. V. Platko, N. J. Patterson, G. J. McDonald, H. C. Ackerman, S. J. Campbell, D. Altshuler, R. Cooper, D. Kwiatkowski, R. Ward, and E. S. Lander. Detecting recent positive selection in the human genome from haplotype structure. *Nature*, 419(6909):832–837, Oct 2002.

[88] S. A. Sawyer. On the past history of an allele now known to have frequency *p*. *J Appl Probab*, 14:439–450, 1977.

[89] S. A. Sawyer and D. L. Hartl. Population genetics of polymorphism and divergence. *Genetics*, 132(4):1161–1176, Dec 1992.

[90] S. A. Sawyer, R. J. Kulathinal, C. D. Bustamante, and D. L. Hartl. Bayesian analysis suggests that most amino acid replacements in *Drosophila* are driven by positive selection. *J Mol Evol*, 57 (Suppl 1):S154–S164, 2003.

[91] K. J. Schmid, L. Nigro, C. F. Aquadro, and D. Tautz. Large number of replacement polymorphisms in rapidly evolving genes of *Drosophila*: Implications for genome-wide surveys of DNA polymorphism. *Genetics*, 153(4):1717–1729, Dec 1999.

[92] M. Schoniger and A. von Haeseler. A stochastic model for the evolution of autocorrelated DNA sequences. *Mol Phylogenet Evol*, 3(3):240–247, Sep 1994.

[93] L. A. Sheldahl, D. M. Weinreich, and D. M. Rand. Recombination, dominance and selection on amino acid polymorphism in the *Drosophila* genome: Contrasting patterns on the x and fourth chromosomes. *Genetics*, 165(3):1195–1208, Nov 2003.

[94] A. Siepel and D. Haussler. Phylogenetic estimation of context-dependent substitution rates by maximum likelihood. *Mol Biol Evol*, 21(3):468–488, Mar 2004.

[95] N. G. Smith and A. Eyre-Walker. Adaptive protein evolution in *Drosophila*. *Nature*, 415(6875):1022–1024, Feb 2002.

[96] J. F. Storz and M. W. Nachman. Natural selection on protein polymorphism in the rodent genus *Peromyscus*: Evidence from interlocus contrasts. *Evol Int J Org Evol*, 57(11):2628–2635, Nov 2003.

[97] S. Sunyaev, F. A. Kondrashov, P. Bork, and V. Ramensky. Impact of selection, mutation rate and genetic drift on human genetic variation. *Hum Mol Genet*, 12(24):3325–3330, Dec 2003.

[98] W. J. Swanson, A. G. Clark, H. M. Waldrip-Dail, M. F. Wolfner, and C. F. Aquadro. Evolutionary EST analysis identifies rapidly evolving male reproductive proteins in *Drosophila*. *Proc Natl Acad Sci USA*, 98(13):7375–7379, Jun 2001.

[99] N. Takahata. *Population Genetics, Molecular Evolution, and the Neutral Theory*. University of Chicago Press, Chicago, 1994.

[100] N. Takahata, K. Ishii, and H. Matsuda. Effects of temporal fluctuation of selection coefficient on gene frequency in a population. *Proc Natl Acad Sci USA*, 72:4541–4545, 1975.

[101] N. Takahata and M. Nei. Gene genealogy and variance of interpopulational nucleotide differences. *Genetics*, 110(2):325–344, Jun 1985.

[102] N. Takahata, Y. Satta, and J. Klein. Divergence time and population size in the lineage leading to modern humans. *Theor Popul Biol*, 48(2):198 221, Oct 1995.

[103] M. I. Tenaillon, M. C. Sawkins, L. K. Anderson, S. M. Stack, J. Doebley, and B. S. Gaut. Patterns of diversity and recombination along chromosome 1 of maize (*Zea mays* L. ssp.). *Genetics*, 162(3):1401–1413, Nov 2002.

[104] J. L. Thorne. Models of protein sequence evolution and their applications. *Curr Opin Genet Dev*, 10(6):602–605, Dec 2000.

[105] J. Wakeley and J. Hey. Estimating ancestral population parameters. *Genetics*, 145(3):847–855, Mar 1997.

[106] J. D. Wall. Estimating ancestral population sizes and divergence times. *Genetics*, 163(1):395–404, Jan 2003.

[107] G. A. Watterson. On the number of segregating sites in genetical models without recombination. *Theor Popul Biol*, 7(2):256–276, Apr 1975.

[108] G. A. Watterson. Mutant substitutions at linked nucleotide sites. *Adv Appl Probab*, 14:206–224, 1982.

[109] G. A. Watterson. Substitution times for a mutant nucleotide. *J Appl Probab*, 19A:59–70, 1984.

[110] C. Weinig, L. A. Dorn, N. C. Kane, Z. M. German, S. S. Halldorsdottir, M. C. Ungerer, Y. Toyonaga, T. F. Mackay, M. D. Purugganan, and J. Schmitt. Heterogeneous selection at specific loci in natural environments in *Arabidopsis thaliana*. *Genetics*, 165(1):321–329, Sep 2003.

[111] D. M. Weinreich. The rates of molecular evolution in rodent and primate mitochondrial DNA. *J Mol Evol*, 52(1):40–50, Jan 2001.

[112] D. M. Weinreich and D. M. Rand. Contrasting patterns of nonneutral evolution in proteins encoded in nuclear and mitochondrial genomes. *Genetics*, 156(1):385–399, Sep 2000.

[113] S. Whelan, P. Lio, and N. Goldman. Molecular phylogenetics: state-of-the-art methods for looking into the past. *Trends Genet*, 17(5):262–272, May 2001.

[114] D. E. Wildman, M. Uddin, G. Liu, L. I. Grossman, and M. Goodman. Implications of natural selection in shaping 99.4% nonsynonymous DNA identity between humans and chimpanzees: Enlarging genus *Homo*. *Proc Natl Acad Sci USA*, 100(12):7181–7188, Jun 2003.

[115] S. Williamson. Adaptation in the env gene of HIV-1 and evolutionary theories of disease progression. *Mol Biol Evol*, 20(8):1318–1325, Aug 2003.

[116] S. Wright. Evolution in mendelian populations. *Genetics*, 160:97–159, 1931.

[117] S. Wright. The distribution of gene frequencies under irreversible mutation. *Proc Natl Acad Sci USA*, 24:253–259, 1938.

[118] W. Yang, J. P. Bielawski, and Z. Yang. Widespread adaptive evolution in the human immunodeficiency virus type 1 genome. *J Mol Evol*, 57(2):212–221, Aug 2003.

[119] Z. Yang. Estimating the pattern of nucleotide substitution. *J Mol Evol*, 39(1):105–111, Jul 1994.

[120] Z. Yang. A space-time process model for the evolution of DNA sequences. *Genetics*, 139(2):993–1005, Feb 1995.

[121] Z. Yang. Maximum likelihood analysis of adaptive evolution in HIV-1 gp120 env gene. In *Pacific Symposium on Biocomputing*, pages 226–237. World Scientific, Singapore, 2001.

[122] Z. Yang. Likelihood and Bayes estimation of ancestral population sizes in hominoids using data from multiple loci. *Genetics*, 162(4):1811–1823, Dec 2002.

[123] Z. Yang and R. Nielsen. Codon-substitution models for detecting molecular adaptation at individual sites along specific lineages. *Mol Biol Evol*, 19(6):908–917, Jun 2002.

[124] Z. Yang, R. Nielsen, N. Goldman, and A. M. Pedersen. Codon-substitution models for heterogeneous selection pressure at amino acid sites. *Genetics*, 155(1):431–449, May 2000.

[125] E. Yuste, A. Moya, and C. Lopez-Galindez. Frequency-dependent selection in human immunodeficiency virus type 1. *J Gen Virol*, 83(Pt 1):103–106, Jan 2002.

[126] E. Zuckerkandl and L. Pauling. Evolutionary divergence and convergence in proteins. In V. Bryson and H. Voge, editors, *Evolving Genes and Proteins*, pages 97–166. Academic Press, New York, 1965.

Part II

Practical Approaches for Data Analysis

Maximum Likelihood Methods for Detecting Adaptive Protein Evolution

Joseph P. Bielawski[1] and Ziheng Yang[2]

[1] Department of Biology, Dalhousie University, Halifax, Nova Scotia B3H 4J1, Canada, j.bielawski@dal.ca
[2] Department of Biology, University College London, Gower Street, London WC1E 6BT, United Kingdom, z.yang@ucl.ac.uk

5.1 Introduction

Proteins evolve; the genes encoding them undergo mutation, and the evolutionary fate of the new mutation is determined by random genetic drift as well as purifying or positive (Darwinian) selection. The ability to analyze this process was realized in the late 1970s when techniques to measure genetic variation at the sequence level were developed. The arrival of molecular sequence data also intensified the debate concerning the relative importance of neutral drift and positive selection to the process of molecular evolution [17]. Ever since, there has been considerable interest in documenting cases of molecular adaptation. Despite a spectacular increase in the amount of available nucleotide sequence data since the 1970s, the number of such well-established cases is still relatively small [9, 38]. This is largely due to the difficulty in developing powerful statistical tests for adaptive molecular evolution. Although several powerful tests for nonneutral evolution have been developed [33], significant results under such tests do not necessarily indicate evolution by positive selection.

A powerful approach to detecting molecular evolution by positive selection derives from comparison of the relative rates of synonymous and nonsynonymous substitutions [22]. Synonymous mutations do not change the amino acid sequence; hence their substitution rate (d_S) is neutral with respect to selective pressure on the protein product of a gene. Nonsynonymous mutations do change the amino acid sequence, so their substitution rate (d_N) is a function of selective pressure on the protein. The ratio of these rates ($\omega = d_N/d_S$) is a measure of selective pressure. For example, if nonsynonymous mutations are deleterious, purifying selection will reduce their fixation rate and d_N/d_S will be less than 1, whereas if nonsynonymous mutations are advantageous, they will be fixed at a higher rate than synonymous mutations, and d_N/d_S will be greater than 1. A d_N/d_S ratio equal to one is consistent with neutral evolution.

With the advent of genome-scale sequencing projects, we can begin to study the mechanisms of innovation and divergence in a new dimension. Undoubtedly, new examples of adaptive evolution will be uncovered; however, we will also be able to study the process of molecular adaptation in the context of the amount and nature of genomic change involved. Statistical tools such as maximum likelihood estimation of the d_N/d_S ratio [13, 24] and the likelihood ratio test for positively selected genes [26, 34] will be valuable assets in this effort. Hence, the objective of this chapter is to provide an overview of some recent developments in statistical methods for detecting adaptive evolution as implemented in the PAML package of computer programs.

5.1.1 The PAML Package of Programs

PAML (for Phylogenetic Analysis by Maximum Likelihood) is a package of programs for analysis of DNA or protein sequences by using maximum likelihood methods in a phylogenetic framework [36]. The package, along with documentation and source codes, is available at the PAML Web site (http://abacus.gene.ucl.ac.uk/software/paml.html). In this chapter, we illustrate selected topics by analysis of example datasets. The sequence alignments, phylogenetic trees, and the control files for running the program are all available at ftp://abacus.gene.ucl.ac.uk/pub/BY2004SMME/. Readers are encouraged to retrieve and analyze the example datasets themselves as they proceed through this chapter.

The majority of analytical tools discussed here are implemented in the codeml program in the PAML package. Data analysis using codeml and the other programs in the PAML package are controlled by variables listed in a "control file." The control file for codeml is called codeml.ctl and is read and modified by using a text editor. Options that do not apply to a particular analysis can be deleted from a control file. Detailed descriptions of all of codeml's variables are provided in the PAML documentation. Below we list a sample file showing the important options for codon-based analysis discussed in this chapter.

```
   seqfile = seqfile.txt    * sequence data filename
  treefile = tree.txt       * tree structure filename
   outfile = out.txt
   runmode = 0              * 0:user defined tree; -2:pairwise comparison
   seqtype = 1              * 1:codon models; 2: amino acid models
 CodonFreq = 2              * 0:equal, 1:F1X4, 2:F3X4, 3:F61
     model = 0              * 0:one-w for all branches; 2: w's for branches
   NSsites = 0              * 0:one-rtio; 1:neutral; 2:selection; 3:discrete;
                            * 7:beta; 8:beta&w
     icode = 0              * 0:universal code
 fix_kappa = 0              * 1:kappa fixed, 0:kappa to be estimated
     kappa = 2              * initial or fixed kappa
 fix_omega = 0              * 1:omega fixed, 0:omega to be estimated
     omega = 5              * initial omega
```

5.2 Maximum Likelihood Estimation of Selective Pressure for Pairs of Sequences

5.2.1 Markov Model of Codon Evolution

A Markov process is a simple stochastic process in which the probability of change from one state to another depends on the current state only and not on past states. Markov models have been used very successfully to describe changes between nucleotides, codons, or amino acids [10, 18, 13]. Advantages of a codon model include the ability to model biologically important properties of protein-coding sequences such as the transition to transversion rate ratio, the d_N/d_S ratio, and codon usage frequencies. Since we are interested in measuring selective pressure by using the d_N/d_S ratio, we will consider a Markov process that describes substitutions between the 61 sense codons within a protein- coding sequence [13]. The three stop codons are excluded because mutations to stop codons are not tolerated in a functional protein-coding gene. Independence among the codon sites of a gene is assumed, and hence the substitution process can be considered one codon site at a time. For any single codon site, the model describes the instantaneous substitution rate from codon i to codon j, q_{ij}. Because transitional substitutions are known to occur more often than transversions, the rate is multiplied by the κ parameter when the change involves a transition; the κ parameter is the transition/transversion rate ratio. Use of codons within a gene also can be highly biased, and consequently the rate of change from i to j is multiplied by the equilibrium frequency of codon j (π_j). Selective constraints acting on substitutions at the amino acid level affect the rate of change when that change represents a nonsynonymous substitution. To account for this level of selective pressure, the rate is multiplied by the ω parameter if the change is nonsynonymous; the ω parameter is the nonsynonymous/synonymous rate ratio (d_N/d_S). Note that only selection on the protein product of the gene influences ω.

The substitution model is specified by the instantaneous rate matrix, $Q = \{q_{ij}\}$, where

$$q_{ij} = \begin{cases} 0, & \text{if } i \text{ and } j \text{ differ at two or three codon positions,} \\ \mu\pi_j, & \text{if } i \text{ and } j \text{ differ by a synonymous transversion,} \\ \mu\kappa\pi_j, & \text{if } i \text{ and } j \text{ differ by a synonymous transition,} \\ \mu\omega\pi_j, & \text{if } i \text{ and } j \text{ differ by a nonsynonymous transversion,} \\ \mu\kappa\omega\pi_j, & \text{if } i \text{ and } j \text{ differ by a nonsynonymous transition.} \end{cases} \tag{5.1}$$

The diagonal elements of the matrix Q are defined by the mathematical requirement that the row sums be equal to zero. Because separate estimation of the rate (μ) and time (t) is not possible, the rate (μ) is fixed so that the expected number of nucleotide substitutions per codon is equal to one. This scaling allows us to measure time (t) by the expected number of substitutions per codon (i.e. genetic distance). The probability that codon i is substituted by codon j after time t is $p_{ij}(t)$, and $P(t) = p_{ij}(t) = e^{Qt}$. The above is a description of the basic codon model of Goldman and Yang [13]. A similar model of codon substitution was proposed by Muse and Gaut [24] and is implemented in codeml as well as in the program HyPhy (http://www.hyphy.org/).

5.2.2 Maximum Likelihood Estimation of the d_N/d_S Ratio

We can estimate ω by maximizing the likelihood function using data of two aligned sequences. Suppose there are n codon sites in a gene, and a certain site (h) has codons CCC and CTC. The data at site h, denoted $x_h = \{CCC, CTC\}$, are related to an ancestor with codon k by branch lengths t_0 and t_1 (Figure 5.1(a)). The probability of site h is

$$f(x_h) = \sum_k \pi_k p_{k,CCC}(t_0) p_{k,CTC}(t_1) = \pi_{CCC} p_{CCC,CTC}(t_0 + t_1). \qquad (5.2)$$

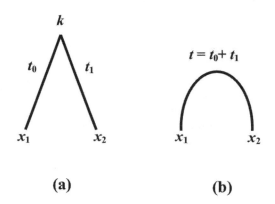

(a)   **(b)**

Fig. 5.1. Rooted (a) and unrooted (b) trees for a pair of sequences. Under reversible codon models, the root is unidentifiable; hence, only the sum of the branch lengths, $t = t_0 + t_1$, is estimable.

Since the ancestral codon is unknown, the summation is over all 61 possible codons for k. Furthermore, as the substitution model is time-reversible, the root of the tree can be moved around, say, to species 1, without changing the likelihood. Thus t_0 and t_1 cannot be estimated individually, and only $t_0 + t_1 = t$ is estimated (Figure 5.1(b)).

The log-likelihood function is a sum over all codon sites in the sequence

$$\ell(t, \kappa, \omega) = \sum_{h=1}^n \log f(x_h). \qquad (5.3)$$

Codon frequencies (π_i's) can usually be estimated by using observed base or codon frequencies. The ω parameter and parameters κ and t are estimated by maximizing the log- likelihood function. Since an analytical solution is not possible, numerical optimization algorithms are used.

5.2.3 Empirical Demonstration: Pairwise Estimation of the d_N/d_S Ratio for *GstD1*

In this section, we use a simple data set and the codeml program to illustrate maximum likelihood estimation of ω. The data set is *GstD1* genes of *Drosophila melanogaster* and *D. simulans*. The alignment has 600 codons. Our first objective is to evaluate the likelihood function for a variety of fixed values for the parameter ω. Codeml uses a hill-climbing algorithm to maximize the log-likelihood function. In this case, we will let codeml estimate κ (fix_kappa = 0 in the control file codeml.ctl) and the sequence distance t, but with parameter ω fixed (fix_omega = 1). All that remains is to run codeml several times, each with a different value for omega in the control file; the data in Figure 5.2 show the results for ten different values of ω. Note that the maximum likelihood value for ω appears to be roughly 0.06, which is consistent with purifying selection, and that values greater than 1 have much lower likelihood scores.

Our second objective is to allow codeml to use the hill-climbing algorithm to maximize the log-likelihood function with respect to κ, t, and ω. Thus we use fix_omega = 1 and can use any positive value for omega, which is used only as a starting value for the iteration. Such a run gives the estimate of ω of 0.067.

Alternatives to maximum likelihood estimates of ω are common [25, 15, 39]. Those methods count the number of sites and differences and then apply a multiple-hit correction, and they are termed the counting methods. Most of them make simplistic assumptions about the evolutionary process and apply ad hoc treatments to the data that can't be justified [23, 39]. Here we use the *GstD1* sequences to explore the effects of (i) ignoring the transition to transversion rate ratio (fix_kappa = 1; kappa = 1); (ii) ignoring codon usage bias (CodonFreq = 0); and (iii) alternative treatments of unequal codon frequencies (CodonFreq = 2 and CodonFreq = 3). Note that for these data transitions are occurring at higher rates than transversions, and codon frequencies are very biased, with average base frequencies of 6% (T), 50% (C), 5% (A), and 39% (G) at the third position of the codon. Thus, we expect estimates that account for both biases will be the most reliable.

Results of our exploratory analyses (Table 5.2.3) indicate that model assumptions are very important for these data. For example, ignoring the transition to transversion ratio almost always led to underestimation of the number of synonymous sites (S), overestimation of d_S, and underestimation of ω. This is because transitions at the third codon positions are more likely to be synonymous than are transversions [19]. Similarly, biased codon usage implies

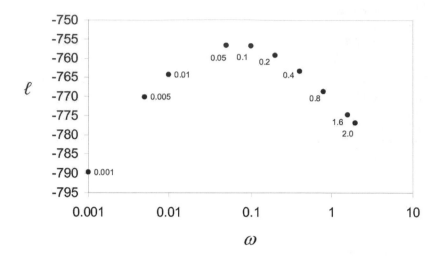

Fig. 5.2. Log-likelihood as a function of the ω parameter for a pair of *GstD1* genes from *Drosophila melanogaster* and *D. simulans*. The maximum likelihood estimate of ω is the value that maximizes the likelihood function. Since an analytical solution is not possible, the `codeml` program uses a numerical hill-climbing algorithm to maximize l. For these data, the maximum likelihood estimate of ω is 0.067, with a maximum likelihood of -756.57.

unequal substitution rates between the codons, and ignoring it also leads to biased estimates of synonymous and nonsynonymous substitution rates. In real data analysis, codon usage bias was noted to have an even greater impact than the transition/transversion rate ratio and is opposite to that of ignoring transition bias. This is clearly indicated by the sensitivity of S to codon bias, where S in this gene (45.2) is less than one-third the expected value under the assumption of no codon bias ($S = 165.8$). The estimates of ω differ by as much as 4.7-fold (Table 5.2.3). Note that these two sequences differed at just 3% of sites.

For comparison, we included estimates obtained from two counting methods. The method of Nei and Gojobori [25] is similar to ML ignoring transition bias and codon bias, whereas the method of Yang and Nielsen [39] is similar to ML accommodating transition bias and codon bias (F3×4). Note that estimation according to Nei and Gojobori [25] was accomplished by using the `codeml` program and according to Yang and Nielsen [39] by using the YN00 program of PAML. What is clear from these data is that when sequence divergence is not too great, assumptions appear to matter more than methods, with ML and the counting methods giving similar results under similar assumptions. This result is consistent with simulation studies examining the performance of

Table 5.1. Estimation of d_S and d_N between *Drosophila melanogaster* and *D. simulans GstD1* genes.

Method	κ	S	N	d_S	d_N	ω	ℓ
ML methods							
Fequal, $\kappa = 1$	1	152.9	447.1	0.0776	0.0213	0.274	-927.18
Fequal, κ estimated	1.88	165.8	434.2	0.0221	0.0691	0.320	-926.28
F3×4, $\kappa = 1$	1	70.6	529.4	0.1605	0.0189	0.118	-844.51
F3×4, κ estimated	2.71	73.4	526.6	0.1526	0.0193	0.127	-842.21
F61, $\kappa = 1$	1	40.5	559.5	0.3198	0.0201	0.063	-758.55
F61, κ estimated	2.53	45.2	554.8	0.3041	0.0204	0.067	-756.57
Counting methods							
Nei and Gojobori	1	141.6	458.4	0.0750	0.0220	0.288	
Yang and Nielsen (F3×4)	3.28	76.6	523.5	0.1499	0.0190	0.127	

different estimation methods [39]. However, as sequence divergence increases, ad hoc treatment of the data can lead to serious estimation errors [23, 8].

5.3 Phylogenetic Estimation of Selective Pressure

Adaptive evolution is very difficult to detect using the pairwise approach to estimating the d_N/d_S ratio. For example, a large-scale database survey identified less than 1% of genes (17 out of 3595) as evolving under positive selective pressure [9]. The problem with the pairwise approach is that it averages selective pressure over the entire evolutionary history separating the two lineages and over all codon sites in the sequences. In most functional genes, the majority of amino acid sites will be subject to strong purifying selection [31, 6], with only a small fraction of the sites potentially targeted by adaptive evolution [11]. In such cases, averaging the d_N/d_S ratio over all sites will yield values much less than one, even under strong positive selective pressure at some sites. Moreover, if a gene evolved under purifying selection for most of that time, with only brief episodes of adaptive evolution, averaging over the history of two distantly related sequences would be unlikely to produce a d_N/d_S ratio greater than one [4]. Clearly, the pairwise approach has low power to detect positive selection. Power is improved if selective pressure is allowed to vary over sites or branches [37, 40]. However, increasing the complexity of the codon model in this way requires that likelihood be calculated for multiple sequences on a phylogeny.

5.3.1 Likelihood Calculation for Multiple Sequences on a Phylogeny

Likelihood calculation on a phylogeny (Figure 5.3) is an extension of the calculation for two lineages. As in the case of two sequences, the root cannot be identified and is fixed at one of the ancestral nodes arbitrarily. For example, given an unrooted tree with four species and two ancestral codons, k and g, the probability of observing the data at codon site h, $x_h = \{x_1, x_2, x_3, x_4\}$ (Figure 5.3), is

$$f(x_h) = \sum_k \sum_g \left\{ \pi_k p_{kx_1}(t_1) p_{kx_2}(t_2) p_{kg}(t_0) p_{gx_3}(t_3) p_{gx_4}(t_4) \right\}. \tag{5.4}$$

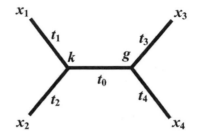

Fig. 5.3. An unrooted phylogeny for four sequences. As in the case of two sequences, the root cannot be identified. For the purpose of likelihood calculation, the root is fixed at one of the ancestral nodes arbitrarily, and t_0, t_1, t_2, t_3, and t_4 are estimable parameters in the model.

The quantity in the brackets is the contribution to the probability of observing the data by ancestral codons k and g at the two ancestral nodes. For an unrooted tree of N species, with $N - 2$ ancestral nodes, the data at each site will be a sum over 61^{N-2} possible combinations of ancestral codons. The log-likelihood function is a sum over all codon sites in the alignment

$$\ell = \sum_{h=1}^{n} \log\{f(x_h)\}. \tag{5.5}$$

As in the two-species case, numerical optimization is used to maximize the likelihood function with respect to κ, ω, and the $(2N - 3)$ branch-length parameters (t's).

5.3.2 Modelling Variable Selective Pressure among Lineages

Adaptive evolution is most likely to occur in an episodic fashion. For example, functional divergence of duplicated genes [43, 29, 5], colonization of a host by a parasitic organism [16], or colonization of a new ecological niche [21] all seem to occur at particular time points in evolutionary history. To improve detection of episodic adaptive evolution, Yang [37] (see also [24]) implemented models that allow for different ω parameters in different parts of a phylogeny. The simplest model, described above, assumes the same ω ratio for all branches in the phylogeny. The most general model, called the "free-ratios model," specifies an independent ω ratio for each branch in a phylogeny. In the codeml program, users can specify an intermediate model, with independent ω parameters for different sets of branches. Modelling variable selective pressure involves a straightforward modification of the likelihood computation [37]. Consider the example tree of fig. 5.4. Suppose we suspect selective pressure has changed in one part of this tree, perhaps due to positive selective pressure. To model this, we specify independent ω ratios (ω_0 and ω_1) for the two different sets of branches (Figure 5.4). The transition probabilities for the two sets of branches are calculated from different rate matrices (Q) generated by using different ω ratios. Under this model (Figure 5.4), the probability of observing the data at codon site x_h is

$$f(x_h) = \sum_k \sum_g \pi_k p_{kx_1}(t_1; \omega_0) p_{kx_2}(t_2; \omega_0) p_{kg}(t_0; \omega_0) p_{gx_3}(t_3; \omega_1) p_{gx_4}(t_4; \omega_1).$$

(5.6)

The log-likelihood function remains a sum over all sites but is now maximized with respect to ω_0 and ω_1, as well as branch lengths (t's) and κ. ω parameters for user-defined sets of branches are specified by model = 2 in the control file and by labelling branches in the tree, as described in the PAML documentation.

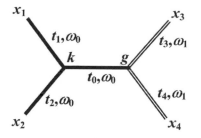

Fig. 5.4. Four-taxon phylogeny with variable ω ratios among its branches. The likelihood of this tree is calculated according to Yang [37], where the two independent ω ratios (ω_0 and ω_1) are used to calculate rate matrices (Q) and transition probabilities for the different branches.

5.3.3 Modelling Variable Selective Pressure among Sites

In practice, modelling variable selective pressure among sites appears to provide much greater gains in power than does modelling variable selective pressure among branches [38]. This is because adaptive evolution is generally restricted to a small subset of sites [6, 40], and the previous model for variation over branches effectively averages over all sites. Although differences in the relative rate of nonsynonymous substitution often can be detected among branches, averaging over sites means it is unlikely that estimated ω's will be greater than one. In fact, implementation of models with variable ω's among codon sites [26, 40, 41] has led to the detection of positive selection in many genes for which it had not previously been observed. For example, Zanotto et al. [42] used the models of Nielsen and Yang [26] to detect positive selection in the *nef* gene of HIV-1, a gene for which earlier studies had found no evidence for adaptive evolution [28, 7].

There are two approaches to modelling variation in ω among sites: (i) use a statistical distribution to model the random variation in ω over sites; and (ii) use a priori knowledge of a protein's structural and functional domains to partition sites in the protein and use different ω's for different partitions. Since structural and functional information are unknown for most proteins, a statistical distribution will be the most common approach. Collectively, Nielsen and Yang [26] and Yang et al. [40] implemented 13 such models, available in the `codeml` program. The continuous distributions are approximated by using discrete categories. In this approach, codon sites are assumed to fall into K classes, with the ω ratios for the site classes, and their proportions (p), estimated from the data. The number of classes (K) is fixed beforehand, and the ω's and p's are either treated as parameters or functions of parameters of the ω distribution [40]. We illustrate likelihood calculation by taking the discrete model (M3) as an example. M3 classifies codon sites into K discrete classes ($i = 0, 1, 2, \ldots, K - 1$), with d_N/d_S ratios and proportions given as:

$$\omega_0, \omega_1, \ldots, \omega_{K-1},$$
$$p_0, p_1, \ldots, p_{K-1}. \tag{5.7}$$

Equation (5.4) is used to compute the conditional probability $f(x_h|\omega_i)$ of the data at a site, h, for each site class. Since we do not know to which class site h belongs, we sum over both classes, giving the unconditional probability

$$f(x_h) = \sum_{i=0}^{K-1} p_i f(x_h|\omega_i). \tag{5.8}$$

In this way, the unconditional probability is an average over the site classes of the ω distribution. Still, assuming that the substitution process at individual codon sites is independent, the log-likelihood function is a sum over all sites in the sequence:

$$\ell = \sum_{h=1}^{n} \log\{f(x_h)\}. \tag{5.9}$$

The log-likelihood is now maximized as a function of the parameters of the ω distribution, branch-lengths (t), and κ.

With the second approach, we used knowledge of a protein's structural or functional domains to classify codon sites into different partitions with different ω's. Since we assume site independence, the likelihood calculation is straightforward; the transition probabilities in equation (5.4) are computed by using the appropriate ω parameter for each codon site. By taking this approach, we are effectively assuming our knowledge of the protein is without error; hence, we do not average over site classes for each site [41].

5.4 Detecting Adaptive Evolution in Real Data Sets

Maximum likelihood estimation of selective pressure is only one part of the problem of detecting adaptive evolution in real data sets. We also need the tools to rigorously test hypotheses about the nature of selective pressure. For example, we might want to test whether d_N is higher than d_S (*i.e.*, $\omega > 1$). Fortunately, we can combine estimation of selective pressure with a formal statistical approach to hypothesis testing, the likelihood ratio test (LRT). Combined with Markov models of codon evolution, the LRT provides a very general method for testing hypotheses about protein evolution, including: (i) a test for variation in selective pressure among branches; (ii) a test for variation in selective pressure among sites; and (iii) a test for a fraction of sites evolving under positive selective pressure. In the case of a significant LRT for sites evolving under positive selection, we use Bayes or empirical Bayes methods to identify positively selected sites in an alignment. In the following section, we provide an introduction to the LRT and Bayes' theorem and provide some empirical demonstrations of their use on real data.

5.4.1 Likelihood Ratio Test (LRT)

The LRT is a general method for testing assumptions (model parameters) through comparison of two competing hypotheses. For our purposes, we will only consider comparisons of nested models; that is, where the null hypothesis (H_0) is a restricted version (special case) of the alternative hypothesis (H_1). Note that the LRT only evaluates the differences between a pair of models, and any inadequacies shared by both models remain untested. Let ℓ_0 be the maximum log-likelihood under H_0 with parameters θ_0, and let ℓ_1 be the maximum log-likelihood under H_1 with parameters θ_1. The log-likelihood statistic is defined as twice the log likelihood difference between the two models,

$$2\Delta\ell = 2(\ell_1(\hat{\theta}_1) - \ell_0(\hat{\theta}_0)). \tag{5.10}$$

If the null hypothesis is true, $2\Delta\ell$ will be asymptotically χ^2 distributed with the degree of freedom equal to the difference in the number of parameters between the two models.

Use of the χ^2 approximation to the likelihood ratio statistic requires that certain conditions be met. First, the hypotheses must be nested. Second, the sample must be sufficiently large; the χ^2 approximation fails when too few data are used. Third, H_1 may not be related to H_0 by fixing one or more of its parameters at the boundary of parameter space. This is called the "boundary" problem, and the LRT statistic is not expected to follow a χ^2 distribution in this case [30]. When the conditions above are not met, the exact distribution can be obtained by Monte Carlo simulation [12, 1], although this can be a computationally costly solution.

5.4.2 Empirical Demonstration: LRT for Variation in Selective Pressure among Branches in *Ldh*

The *Ldh* gene family is an important model system for molecular evolution of isozyme multigene families [20]. The paralogous copies of lactate dehydrogenase (*Ldh*) genes found in mammals originated from a duplication near the origin of vertebrates (*Ldh-A* and *Ldh-B*) and a later duplication near the origin of mammals (Figure 5.5; *Ldh-A* and *Ldh-C*). Li and Tsoi [20] found that the rate of evolution had increased in mammalian *Ldh-C* sometime following the second duplication event. An unresolved question about this gene family is whether the increased rate of *Ldh-C* reflects (i) a burst of positive selection for functional divergence following the duplication event, (ii) a long-term change in selective pressure, or (iii) simply an increase in the underlying mutation rate of *Ldh-C*. In the following, we use the LRT for variable ω ratios among branches to test these evolutionary scenarios.

The null hypothesis (H_0) is that the rate increase in *Ldh-C* is simply due to an underlying increase in the mutation rate. If the selective pressure was constant and the mutation rate increased, the relative fixation rates of synonymous and nonsynonymous mutations (ω) would remain constant over the phylogeny, but the overall rate of evolution would increase in *Ldh-C*. One alternative to this scenario is that the rate increase in *Ldh-C* was due to a burst of positive selection following gene duplication (H_1). A formal test for variation in selective pressure among sites may be formulated as follows:

H_0: ω is identical across all branches of the *Ldh* phylogeny.

H_1: ω is variable, being greater than 1 in branch $C0$ of Figure 5.5.

Because H_1 can be transformed into H_0 by restricting ω_{C0} to be equal to the ω ratios for the other branches, we can use the LRT. The estimate of ω under the null hypothesis, as an average over the phylogeny in Figure 5.5, was 0.14, indicating that evolution of *Ldh-A* and *Ldh-C* was dominated by purifying selection. The LRT suggests that selective pressure in *Ldh-C* immediately following gene duplication (0.19) was not significantly different from the average over the other branches (Table 5.2). Hence, we found no evidence

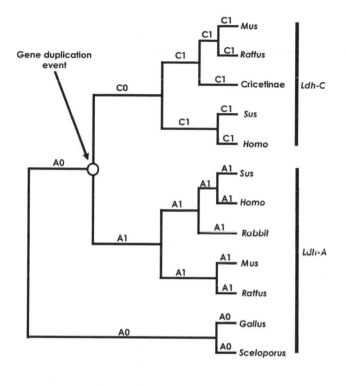

$$H_0: \omega_{A0} = \omega_{A1} = \omega_{C1} = \omega_{C0}$$
$$H_1: \omega_{A0} = \omega_{A1} = \omega_{C1} \neq \omega_{C0}$$
$$H_2: \omega_{A0} = \omega_{A1} \neq \omega_{C1} = \omega_{C0}$$
$$H_3: \omega_{A0} \neq \omega_{A1} \neq \omega_{C1} = \omega_{C0}$$

Fig. 5.5. A phylogenetic tree for the *Ldh-A* and *Ldh-C* gene families. The tree was obtained by a neighbor-joining analysis of a codon sequence alignment under the HKY85 substitution model [14] combined with a Gamma model of rate variation among sites [35]. Branch lengths are not to scale. The *Gallus* (chicken) and *Sccloporus* (lizard) *Ldh-A* sequences are pro-orthologs, as they predate the gene duplication event. The tree is rooted with the pro-orthologous sequences for convenience; all analyses were conducted by using the unrooted topology. The one ratio model (H_0) assumes uniform selective pressure over all branches. H_1 is based on the notion of a burst of positive selection in *Ldh-C* following the gene duplication event; hence the assumption of one ω for branch $C0$ and another for all other branches. H_2 is based on the notion of increased nonsynonymous substitution in all *Ldh-C* lineages following gene duplication; hence the assumption of one ω for the *Ldh-C* branches $(\omega_{C0} = \omega_{C1})$ and another for the *Ldh-A* branches $(\omega_{A0} = \omega_{A1})$. H_3 is based on the notion that selective pressure changed in both *Ldh-C* and *Ldh-A* following gene duplication, as compared with the pro-orthologous sequences; hence, one ω for the *Ldh-C* branches $(\omega_{C0} = \omega_{C1})$, one ω for the post-duplication *Ldh-A* branches (ω_{A1}), and one ω for the pro-orthologous branches (ω_{A0}).

for functional divergence of *Ldh-A* and *Ldh-C* by positive selection. It should be noted that if functional divergence of *Ldh-A* and *Ldh-C* evolved by positive selection for just one or a few amino acid changes, we would not observe a large difference in ω ratios among branches.

Table 5.2. Parameter estimates under models of variable ω ratios among lineages for the *Ldh-A* and *Ldh-C* gene families. (Note: The topology and branch-specific ω ratios are presented in Figure 5.5. The df is 1 for the comparisons of H_0 vs. H_1, H_0 vs. H_2, and H_2 vs. H_3.)

Models	w_{A0}	w_{A1}	w_{C1}	w_{C0}	ℓ
$H_0 : w_{A0} = w_{A1} = w_{C1} = w_{C0}$	$0.14 = w_{A0}$	$= w_{A0}$	$= w_{A0}$		-6018.63
$H_1 : w_{A0} = w_{A1} = w_{C1} \neq w_{C0}$	$0.13 = w_{A0}$	$= w_{A0}$	0.19		-6017.57
$H_2 : w_{A0} = w_{A1} \neq w_{C1} = w_{C0}$	$0.07 = w_{A0}$	0.24	$= w_{A1}$		-5985.63
$H_3 : w_{A0} \neq w_{A1} \neq w_{C1} = w_{C0}$	0.09	0.06	0.24	$= w_{A1}$	-5984.11

Using the same approach, we tested a second alternative hypothesis, where the rate increase in *Ldh-C* was due to an increase in the nonsynonymous substitution rate over all lineages of the *Ldh-C* clade (see H_2 in Figure 5.5). In this case, the LRT was highly significant, and the parameter estimates for the *Ldh-C* clade indicated an increase in the relative rate of nonsynonymous substitution by a factor of 3 (Table 5.2). Lastly, we tested the hypothesis that selective pressure differed in both *Ldh-A* and *Ldh-C* following gene duplication (see H_3 in Figure 5.5), and results of this test were not significant (Table 5.2). Collectively, these findings suggest selective pressure and mutation rates in *Ldh-A* were relatively unchanged by the duplication event, whereas the nonsynonymous rate increased in *Ldh-C* following the duplication event as compared with *Ldh-A*.

5.4.3 Empirical Demonstration: Positive Selection in the *nef* Gene in the Human HIV-2 Genome

The role of the *nef* gene in differing phenotypes of HIV-1 infection has been well-studied, including identification of sites evolving under positive selective pressure [42]. The *nef* gene in HIV-2 has received less attention, presumably because HIV-2 is associated with reduced virulence and pathogenicity relative to HIV-1. Padua et al. [27] sequenced 44 *nef* alleles from a study population of 37 HIV-2-infected people living in Lisbon, Portugal. They found that nucleotide variation in the *nef* gene, rather than gross structural change, was potentially correlated with HIV-2 pathogenesis. In order to determine whether the *nef* gene might also be evolving under positive selective pressure in HIV-2, we analyzed those same data here with models of variable ω ratios among sites [40].

Following the recommendation of Yang et al. [40] and Anisimova et al. [1], we consider the following models: M0 (one ratio), M1 (neutral), M2 (selection), M3 (discrete), M7 (beta), and M8 (beta & ω). Models M0 and M3 were described above. M1 (neutral) specifies two classes of sites: conserved sites with $\omega = 0$ and neutral sites with $\omega = 1$. M2 (selection) is an extension of M1 (neutral), adding a third ω class that is free to take a value > 1. Version 3.14 of `paml/codeml` introduces a slight variation to models M1 (neutral) and M2 (selection) in that $\omega_0 < 1$ is estimated from the data rather than being fixed at 0. Those are referred to as models M1a and M2a, also used here. Under model M7 (beta), ω varies among sites according to a beta distribution with parameters p and q. The beta distribution is restricted to the interval $(0, 1)$; thus, M1 (neutral), M1a (nearly neutral), and M7 (beta) assume no positive selection. M8 (beta & ω) adds a discrete ω class to the beta distribution that is free to take a value > 1. Under M8 (beta & ω), a proportion of sites p_0 is drawn from a beta distribution, with the remainder ($p_1 = 1 - p_0$) having the ω ratio of the added site class. We specified $K = 3$ discrete classes of sites under M3 (discrete), and $K = 10$ under M7 (beta) and M8 (beta & ω). We use an LRT comparing M0 (one ratio) with M3 (discrete) to test for variable selective pressure among sites and three LRTs to test for sites evolving by positive selection, comparing (i) M1 (neutral) against M2 (selection), (ii) M1a (nearly neutral) and M2a (positive selection), and (iii) M7 (beta) against M8 (beta & ω).

Maximum likelihood estimates of parameters and likelihood scores for the *nef* gene are presented in Table 5.3. Averaging selective pressure over sites and branches as in M0 (one ratio) yielded an estimated ω of 0.50, a result consistent with purifying selection. The LRT comparing M0 (one ratio) against M3 (discrete) is highly significant ($2\Delta\ell = 1087.2$, df $= 4$, $P < 0.01$), indicating that the selective pressure is highly variable among sites. Estimates of ω under models that can allow for sites under positive selection (M2, M2a, M3, M8) indicated a fraction of sites evolving under positive selective pressure (Table 5.3). To formally test for the presence of sites evolving by positive selection, we conducted LRTs comparing M1 and M2, M1a and M2a, and M7 and M8. All those LRTs were highly significant; for example, the test statistic for comparing M1 (neutral) and M2 (selection) is $2\Delta\ell = 223.58$, with $P < 0.01$, df $= 2$. These findings suggest that about 12% of sites in the *nef* gene of HIV-2 are evolving under positive selective pressure, with ω between 2 and 3. It is clear from Table 5.3 that this mode of evolution would not have been detected if ω were measured simply as an average over all sites of *nef*.

Models M2 (selection) and M8 (beta & ω) are known being multiple local optima in some data sets, often with ω_2 under M2 or ω under M8 to be < 1 on one peak and > 1 on another peak. Thus it is important to run these models multiple times with different starting values (especially different ω's) and then select the set of estimates corresponding to the highest peak. Indeed, the *nef* dataset illustrates this issue. By using different initial ω's, both the global and local optima can be found.

Table 5.3. Parameter estimates and likelihood scores under models of variable ω ratios among sites for HIV-2 *nef* genes. (Note: The number after the model code, in parentheses, is the number of free parameters in the ω distribution. The d_N/d_S ratio is an average over all sites in the HIV-2 *nef* gene alignment. Parameters in parentheses are not free parameters and are presented for clarity. PSS is the number of positive selected sites, inferred at the 50% (95%) posterior probability cutoff.)

Model	d_N/d_S	Parameter estimates	PSS	ℓ
M0: one ratio (1)	0.51	$\omega = 0.505$	none	-9775.77
M3: discrete (5)	0.63	$p_0 = 0.48, p_1 = 0.39, (p_2 = 0.13)$	31 (24)	-9232.18
		$\omega_0 = 0.03, \omega_1 = 0.74, \omega_2 = 2.50$		
M1: neutral (1)	0.63	$p_0 = 0.37, (p_1 = 0.63)$	not allowed	-9428.75
		$(\omega_0 = 0), (\omega_1 = 1)$		
M2: selection (3)	0.93	$p_0 = 0.37, p_1 = 0.51, (p_2 = 0.12)$	30 (22)	-9392.96
		$(\omega_0 = 0), (\omega_1 = 1), \omega_2 = 3.48$		
M1a: nearly neutral (2)	0.48	$p_0 = 0.55, (p_1 = 0.45)$	not allowed	-9315.53
		$(\omega_0 = 0.06), (\omega_1 = 1)$		
M2a: positive selection (4)	0.73	$p_0 = 0.51, p_1 = 0.38, (p_2 = 0.11)$	26 (15)	-9241.33
		$(\omega_0 = 0.05), (\omega_1 = 1), \omega_2 = 3.00$		
M7: beta (2)	0.42	$p = 0.18, q = 0.25$	not allowed	-9292.53
M8: beta & ω (4)	0.62	$p_0 = 0.89, (p_1 = 0.11)$	27 (15)	-9224.31
		$p = 0.20, q = 0.33, \omega = 2.62$		

5.4.4 Bayesian Identification of Sites Evolving under Positive Darwinian Selection

Under the approach described in this chapter, a gene is considered to have evolved under positive selective pressure if (i) the LRT is significant and (ii) at least one of the ML estimates of $\omega > 1$. Given that these conditions are satisfied, we have evidence for sites under positive selection but no information about which sites they are. Hence, the empirical Bayes approach is used to predict them [26, 40]. To do this, we compute, in turn, the posterior probability of a site under each ω site class of a model. Sites with high posterior probabilities under the class with $\omega > 1$ are considered likely to have evolved under positive selective pressure.

Say we have a model of heterogeneous ω ratios, with K site classes $(i = 0, 1, 2, \ldots, K - 1)$. The ω ratios and proportions are $\omega_0, \omega_1, \ldots, \omega_{K-1}$ and $p_0, p_1, \ldots, p_{K-1}$, with the proportions p_i used as the prior probabilities. The posterior probability that a site with data x_h is from site class i is

$$P(\omega|x_h) = \frac{P(x_h|\omega_i)p_i}{P(x_h)} = \frac{P(x_h|\omega_i)p_i}{\sum_{j=0}^{K-1} P(x_h|\omega_j)p_j}. \tag{5.11}$$

Because the parameters used in the equation above to calculate the posterior probability are estimated by ML (ω_i and p_i), the approach is called empirical Bayes. By using the ML parameters in this way, we ignore their

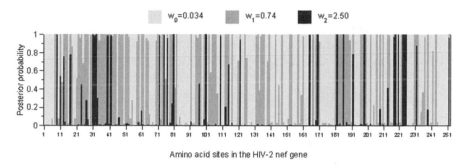

Fig. 5.6. Posterior probabilities for sites classes under M3 ($K = 3$) along the HIV-2 *nef* gene alignment.

sampling errors. The posterior probabilities will be sensitive to these parameter estimates, meaning that the reliability of this approach will be poor when the parameter estimates are poor, such as in small datasets or when obtained from a local optimum.

Because the *nef* dataset above is quite large, the parameter estimates are expected to be reliable [2]. Consistent with this, ML estimates of the strength and proportion of positively selected sites in *nef* are consistent among M2, M3, and M8 (Table 5.3). Figure 5.6 shows the posterior probabilities for the $K = 3$ site classes at each site of *nef* under model M3. Twenty-four sites were identified as having very high posterior probability ($P > 0.95$) of evolving under positive selection (site class with $\omega > 1$). Interestingly, none of these sites matched the two variable sites in a proline-rich motif that is strongly associated with an asymptomatic disease profile [27]. In fact, only four of the 24 sites were found in regions of *nef* considered important for function. Disruption of the important *nef* regions is associated with reduced pathogenicity in HIV-2-infected individuals [32, 27]. Our results suggest that selective pressure at such sites is fundamentally different from selection acting at the 24 positive selection sites predicted using the Bayes theorem. To be identified with such high posterior probabilities, the predicted sites must have been evolving under long-term positive selective pressure, suggesting that they are more likely subjected to immune-driven diversifying selection at epitopes [42, 34].

5.5 Power, Accuracy and Robustness

The boundary problem mentioned above applies to the LRT for variable selective pressure among sites and the LRT for positive selection at a fraction of sites [1]. The problem arises in the former because the null (M0) is equivalent to M3 ($K = 3$) with two of the five parameters (p_0 and p_1) fixed to 0, which is at the boundary of parameter space. In comparisons of M1 with M2, M1a with M2a, and M7 with M8, the null is equivalent to the alternative with a proportion parameter (p) fixed to 0. Therefore, the χ^2 approximation is not expected to hold. Anisimova et al. [1] used computer simulation to investigate the effect of the boundary problem on the power and accuracy of the LRT. Use of the χ^2 makes the LRT conservative, meaning that the false positive rate will be less than predicted by the specified significance level of the test [1]. Nevertheless, the test was found to be powerful, sometimes reaching 100% in data sets consisting of 17 sequences. Power was low for highly similar and highly divergent sequences but was modulated by the length of the sequence and the strength of positive selection. Note that simulation studies, both with and without the boundary problem, indicate that the sample size requirements for the χ^2 approximation are met with relatively short sequences in some cases as few as 50 codons [1].

Bayesian prediction of sites evolving under positive selection is a more difficult task than ML parameter estimation or likelihood ratio testing. The difficulty arises because the posterior probabilities depend on the (i) information contained at just a single site in the data set and (ii) the quality of the ML parameter estimates. Hence, a second study was conducted by Anisimova et al. [2] to examine the power and accuracy of the Bayesian site identification. The authors made the following generalizations: (i) prediction of positively selected sites is not practical from just a few highly similar sequences; (ii) the most effective method of improving accuracy is to increase the number of lineages; and (iii) site prediction is sensitive to sampling errors in parameter estimates and to the assumed ω distribution.

Robustness refers to the stability of results to changes in the model assumptions. The LRT for positive selection is generally robust to the assumed distribution of ω over sites [1]. However, as the LRT of M0 with M3 is a test of variable selective pressure among sites, caution must be exercised when only the M0–M3 comparison suggests positive selection. One possibility is to use M2, which tends to be more conservative than the other models [2]. Another approach is to select the subset of sites that are robust to the ω distribution [1, 34]. A third approach is to select sites that are robust to sampling lineages [34]. We believe that sensitivity analysis is a very important part of detecting positive selection, and we make the following recommendations: (i) multiple models should be used, (ii) care should be taken to identify and discard results obtained from local optima, and (iii) assumptions such as the ω distribution or the method of correcting for biased codon frequencies should be evaluated relative to their effects on ML parameter estimation and Bayesian site prediction.

All codon models discussed above ignore the effect of the physicochemical property of the amino acid being substituted. For example, all amino acid substitutions at a positively selected site are assumed to be advantageous, with $\omega > 1$. The assumption appears to be unrealistic; one can imagine that there might be a set of amino acid substitutions that are forbidden at a site because of physicochemical constraints, even though the site is subject to strong positive selection. Another limitation is that these methods are very conservative, only indicating positive selection when the estimate of ω is > 1. In cases where only one or a few amino acid substitutions result in a substantial change in phenotype, the methods will have little or no power because ω will be < 1. Another important limitation is the assumption of a single underlying phylogeny. When recombination has occurred, no single phylogeny will fit all sites of the data. A recent simulation study [3] found that the LRT is robust to low levels of recombination but can have a seriously high type I error rate when recombination is frequent. Interestingly, Bayesian prediction of positively selected sites was less affected by recombination than was the LRT. In summary, no matter how robust the results, they must be interpreted with these limitations in mind.

Acknowledgment

We thank an anonymous referee for many constructive comments. This work is supported by a grant from BBSRC to Z.Y.

References

[1] M. Anisimova, J. P. Bielawski, and Z. Yang. Accuracy and power of the likelihood ratio test in detecting adaptive molecular evolution. *Mol. Biol. Evol.*, 18:1585–1592, 2001.

[2] M. Anisimova, J. P. Bielawski, and Z. Yang. Accuracy and power of bayesian prediction of amino acid sites under positive selection. *Mol. Biol. Evol.*, 19:950–958, 2002.

[3] M. Anisimova, R. Nielsen, and Z. Yang. Effect of recombination on the accuracy of likelihood methods for detecting positive selection at amino acid sites. *Genetics*, 164:1229–1236, 2003.

[4] J. P. Bielawski and Z. Yang. The role of selection in the evolution of the daz gene family. *Mol. Biol. Evol.*, 18:523–529, 2001.

[5] J. P. Bielawski and Z. Yang. Maximum likelihood methods for detecting adaptive evolution after gene duplication. *J. Struct. Funct. Genomics*, 3:201–212, 2003.

[6] K. A. Crandall, C. R. Kelsey, H. Imanichi, H. C. Lane, and N. P. Salzman. Parallel evolution of drug resistance in HIV: Failure of nonsynonymous/synonymous substitution rate ratio to detect selection. *Mol. Biol. Evol.*, 16:372–382, 1999.

[7] J. Da Silva and A. L. Hughes. Conservation of cytotoxic t lymphocytes (CTL) epitopes as a host strategy to constrain parasitic adaptation: Evidence from the *nef* gene of human immunodeficiency virus 1 (HIV-1). *Mol. Biol. Evol.*, 15:1259–1268, 1998.

[8] K. D. Dunn, J. P. Bielawski, and Z. Yang. Rates and patterns of synonymous substitutions in *Drosophila*: Implications for translational selection. *Genetics*, 157:295–305, 2001.

[9] T. K. Endo, K. Ikeo, and T. Gojobori. Large-scale search for genes on which positive selection may operate. *Mol. Biol. Evol.*, 13:685–690, 1996.

[10] J. Felsenstein. Evolutionary trees from DNA sequences: A maximum likelihood approach. *J. Mol. Evol.*, 15:368–376, 1981.

[11] G. B. Golding and A. M. Dean. The structural basis of molecular adaptation. *Mol. Biol. Evol.*, 15:355–369, 1998.

[12] N. Goldman. Statistical tests of DNA substitution models. *J. Mol. Evol.*, 36:182–198, 1993.

[13] N. Goldman and Z. Yang. A codon based model of nucleotide substitution for protein-coding DNA sequences. *Mol. Biol. Evol.*, 11:725–736, 1994.

[14] M. Hasegawa, H. Kishino, and T. Yano. Dating the human-ape splitting by a molecular clock using mitochondrial DNA. *J. Mol. Evol.*, 22:160–174, 1985.

[15] Y. Ina. Pattern of synonymous and nonsynonymous substitutions: An indicator of mechanisms of molecular evolution. *J. Genet.*, 75:91–115, 1996.

[16] F. M. Jiggins, G. D. D. Hurst, and Z. Yang. Host-symbiont conflicts: Positive selection on the outer membrane protein of parasite but not mutualistic *Rickettsiaceae. Mol. Biol. Evol.*, 19:1341–1349, 2002.

[17] M. Kimura and T. Ohta. On some principles governing molecular evolution. *Proc. Natl. Acad. Sci. USA*, 71:2848–2852, 1974.

[18] H. Kishino, T. Miyata, and M. Hasegawa. Maximum likelihood inference of protein phylogeny and the origin of chloroplasts. *J. Mol. Evol.*, 31:151–160, 1990.

[19] W.-H. Li, C.-I. Wu, and C.-C. Luo. A new method for estimating synonymous and nonsynonymous rates of nucleotide substitutions considering the relative likelihood of nucleotide and codon changes. *Mol. Biol. Evol.*, 2:150–174, 1985.

[20] Y.-J. Li and C.-M. Tsoi. Phylogenetic analysis of vertebrate lactate dehydrogenase (ldh) multigene families. *J. Mol. Evol.*, 54:614–624, 2002.

[21] W. Messier and C.-B. Stewart. Episodic adaptive evolution of primate lysozymes. *Nature*, 385:151–154, 1997.

[22] T. Miyata and T. Yasunaga. Molecular evolution of mRNA: A method for estimating evolutionary rates of synonymous and amino acid substitutions from homologous nucleotide sequences and its applications. *J. Mol. Evol.*, 16:23–36, 1980.

[23] S. V. Muse. Estimating synonymous and non-synonymous substitution rates. *Mol. Biol. Evol.*, 13:105–114, 1996.

[24] S. V. Muse and B. S. Gaut. A likelihood approach for comparing synonymous and nonsynonymous nucleotide substitution rates, with applications to the chloroplast genome. *Mol. Biol. Evol.*, 11:715–725, 1994.

[25] M. Nei and T. Gojobori. Simple methods for estimating the numbers of synonymous and non-synonymous nucleotide substitutions. *Mol. Biol. Evol.*, 3:418–426, 1986.

[26] R. Nielsen and Z. Yang. Likelihood models for detecting positively selected amino acid sites and applications to the HIV-1 envelope gene. *Genetics*, 148:929–936, 1998.

[27] E. Padua, A. Jenkins, S. Brown, J. Bootman, M. T. Paixao, N. Almond, and N. Berry. Natural variation of the *nef* gene in human immunodeficiency virus type 2 infections in Portugal. *J. Gen. Virol.*, 84:1287–1299, 2003.

[28] U. Plikat, K. Nieselt-Struwe, and A. Meyerhans. Genetic drift can determine short-term human immunodeficiency virus type 1 *nef* quasispecies evolution in vivo. *J. Virol.*, 71:4233–4240, 1997.

[29] T. R. Schmidt, M. Goodman, and L. I. Grossman. Molecular evolution of the cox7a gene family in primates. *Mol. Biol. Evol.*, 16:619–626, 1999.

[30] S. Self and K. Y. Liang. Asymptotic properties of maximum likelihood estimators and likelihood ratio tests under non-standard conditions. *J. Am. Stat. Assoc.*, 82:605–610, 1987.

[31] P. M. Sharp. In search of molecular Darwinism. *Nature*, 385:111–112, 1997.

[32] W. M. Switzer and S. Wiktor et al. Evidence of *nef* truncation in human immunodeficiency virus type 2 infection. *J. Infect. Dis.*, 177:65–71, 1998.

[33] M. Wayne and K. Simonsen. Statistical tests of neutrality in an age of weak selection. *Trends Ecol. Evol.*, 13:236–240, 1998.

[34] W. Yang, J. P. Bielawski, and Z. Yang. Widespread adaptive evolution in the human immunodeficiency virus type 1 genome. *J. Mol. Evol.*, 57(2):212–221, 2003.

[35] Z. Yang. Maximum likelihood phylogenetic estimation from DNA sequences with variable rates over sites: Approximate methods. *J. Mol. Evol.*, 39:306–314, 1994.

[36] Z. Yang. PAML: A program package for phylogenetic analysis by maximum likelihood. *Appl. Biosci.*, 13:555–556, 1997.

[37] Z. Yang. Likelihood ratio tests for detecting positive selection and application to primate lysozyme evolution. *Mol. Biol. Evol.*, 15:568–573, 1998.

[38] Z. Yang and J. P. Bielawski. Statistical methods for detecting molecular adaptation. *Trends Ecol. Evol.*, 15:496–503, 2000.

[39] Z. Yang and R. Nielsen. Estimating synonymous and nonsynonymous substitution rates under realistic evolutionary models. *Mol. Biol. Evol.*, 17:32–43, 2000.

[40] Z. Yang, R. Nielsen, N. Goldman, and A.-M. K. Pedersen. Codon-substitution models for heterogeneous selective pressure at amino acid sites. *Genetics*, 155:431–449, 2000.

[41] Z. Yang and W. J. Swanson. Codon-substitution models to detect adaptive evolution that account for heterogeneous selective pressures among site classes. *Mol. Biol. Evol.*, 19:49–57, 2002.

[42] P. M. Zanotto, E. G. Kallis, R. F. Souza, and E. C. Holmes. Genealogical evidence for positive selection in the *nef* gene of HIV-1. *Genetics*, 153:1077–1089, 1999.

[43] J. Zhang, H. F. Rosenberg, and M. Nei. Positive Darwinian selection after gene duplication in primate ribonuclease genes. *Proc. Natl. Acad. Sci. USA*, 95:3708–3713, 1998.

6

HyPhy: Hypothesis Testing Using Phylogenies

Sergei L. Kosakovsky Pond[1] and Spencer V. Muse[2]

[1] Antiviral Research Center, University of California, San Diego, CA 92103, USA, spond@ucsd.edu.
[2] Bioinformatics Research Center, North Carolina State University, Raleigh, NC 27695-7566, USA, muse@stat.ncsu.edu

6.1 Introduction

The field of molecular evolution, though wide-reaching in its breadth, can be split into two types of investigations: studies of phylogeny and studies of the molecular evolutionary process. Of course, each of these two categories encompasses many different types of questions, and many investigations require studies of both phylogeny and evolutionary process, but the proposed binary classification is a useful construct. Software for molecular evolution is focused disproportionately on problems relating to phylogenetic reconstruction, with a number of outstanding comprehensive packages from which to choose. On the other hand, software for addressing questions of the molecular evolutionary process tends to be found in stand-alone programs that answer only one or two quite specific problems. The *HyPhy* system, available for download from www.hyphy.org, was designed to provide a unified platform for carrying out likelihood-based analyses on molecular evolutionary data sets, the emphasis of analyses being the molecular evolutionary process; that is, studies of rates and patterns of the evolution of molecular sequences.

HyPhy consists of three major components: a high-level programming language designed to facilitate the rapid implementation of new statistical methods for molecular evolutionary analysis; a collection of prewritten analyses for carrying out widely used molecular evolutionary methods; and a graphical user interface that allows users to quickly and interactively analyze data sets of aligned sequences using evolutionary models and statistical methods that they design using the software system. This chapter is intended to provide an overview of the key elements of each of the three system components, including both specific details of the basic functionality as well as a conceptual description of the potential uses of the software. The nature of the package prevents the creation of an exhaustive "cookbook" of available methods. Instead, we hope to provide a collection of fundamental tools and concepts that allow users to begin using *HyPhy* to carry out both existing and new methods of data analysis.

6.1.1 Standard Analyses

The second of the three enumerated *HyPhy* components was a collection of prewritten "standard" analyses. Since this section of the software is essentially just a collection of prepackaged analyses, we will not devote much time to a detailed discussion of it. However, we choose to describe it first in this chapter to illustrate the types of analyses that *HyPhy* has been designed to address. In Figure 6.1, we show the initial Standard Analyses menu invoked by ANALYSES:STANDARD ANALYSES... (note the use of SMALL CAPS to indicate menu items, with submenus or selections separated by a colon). Each of the nine major headings includes a collection of routines that can be selected by the user. For example, the POSITIVE SELECTION menu item expands to offer five different analyses relating to the task of identifying nucleotide sites undergoing positive selection. A total of 35 batch files are included in the collection, and most of these files include a variety of options enabling users to select items such as evolutionary models or topology search methods. Topics include molecular clock tests, positive selection analyses, phylogenetic reconstruction, and model comparison procedures. The authors frequently add new standard analyses to the package. *HyPhy* includes the ability to perform Web updates, which ensures that the distribution is kept up-to-date.

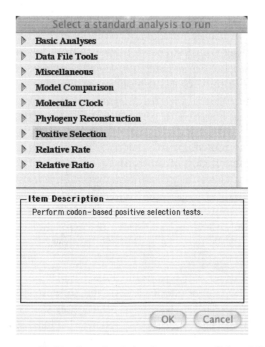

Fig. 6.1. *HyPhy* Standard Analyses menu (Mac OS X).

6.2 Using the *HyPhy* Graphical User Interface

6.2.1 Basic Analysis

The fundamental component of likelihood analyses of molecular evolutionary data is to fit a given phylogenetic tree with a specified model of evolution to an alignment and obtain maximum likelihood estimates (MLE) of all independent model parameters, which commonly include branch-length parameters and character substitution rates [3]. Before we demonstrate how to use *HyPhy* for simple model fitting, we will introduce the fundamental components required of virtually every *HyPhy* data analysis.

1. Data Set. A *data set* is a multiple-sequence alignment. *HyPhy* is able to read a variety of sequence formats, including NEXUS, PHYLIP, and FASTA.
2. Data Filter. A *data filter* specifies a part (or parts) of a data set. *HyPhy* provides powerful tools to select sites and sequences from a data set to analyze. The simplest data filter specifies the entire data set. Examples of nontrivial filters include every first and second position in a codon, exon-intron-exon arrangements, or alignment sites matching a particular motif, such as glycosylation sites. We will often refer to data filters as *partitions*.
3. Substitution Models. We also need to provide stochastic models describing how character substitutions occur along branches in a phylogenetic tree. *HyPhy* includes a multitude of standard "named" models and provides unparalleled flexibility for users to define their own models. A substitution model is specified by its *instantaneous rate matrix* and the vector of equilibrium character frequencies. For instance, one of the most commonly used nucleotide substitution models is the HKY85 model [5],whose instantaneous rate matrix is given by

$$
Q = \begin{array}{c} \\ A \\ C \\ G \\ T \end{array}
\begin{array}{cccc} A & C & G & T \end{array}
\left(\begin{array}{cccc}
\star & \kappa\pi_C & \pi_G & \kappa\pi_T \\
\kappa\pi_A & \star & \kappa\pi_G & \pi_T \\
\pi_A & \kappa\pi_G & \star & \kappa\pi_T \\
\kappa\pi_A & \pi_C & \kappa\pi_G & \star
\end{array} \right),
$$

where κ denotes the ratio of transversion and transition rates and π_i is the base frequency of nucleotide i, $i = A, C, G, T$. We use $\star$ as a notation to indicate that the diagonal elements of rate matrices are defined so that the sum of each row in the rate matrix is 0. This condition ensures that the transition probability matrix,

$$
P(t) = e^{Qt},
$$

defines a proper transition probability function (i.e., the sum of each row in P is 1).

4. Tree. A phylogenetic tree specifies the evolutionary history of extant sequences represented in the data set. It can either be given or can be inferred from the data/model combination. While most other software packages force the evolutionary process to follow the same model along every branch, in *HyPhy* the user can have multiple models, with different rate matrices at each branch. Therefore the notion of the tree in *HyPhy* is not just the evolutionary relationships but rather the *combination* of a tree topology and substitution models attached to tree branches. The distinction in *HyPhy* between a tree and a topology is an important one, as we will illustrate through later examples.

5. Likelihood Function. A combination of a data filter and a tree (which includes both topology and model information) is sufficient to define the probability of the observed data given model parameter values (i.e., the likelihood function). The likelihood function object in *HyPhy* is a convenient way to combine multiple data filter/tree objects (with shared or distinct model parameters) into a single likelihood function, which can then be maximized to obtain MLEs of all model parameters.

Example 6.1 Basic analysis

We are now conceptually prepared to set up the simplest nucleotide sequence analysis with the help of the *HyPhy* graphical user interface. Our example data set is the p51 subunit of the reverse transcriptase gene of HIV-1, obtained as one of the reference alignments from the Los Alamos HIV database, hiv-web.lanl.gov. This data set is included as an example file with *HyPhy* distribution.

Preparing the data

First we must load the sequence alignment. We accomplish this by starting *HyPhy* and selecting the FILE:OPEN:OPEN DATA FILE menu command from the *HyPhy* console window. The file we wish to open is named p51.nex and can be found in the data directory of the *HyPhy* standard installation. Alternatively, all example alignments used in this chapter can be downloaded from www.hyphy.org/pubs/HyphyBookChapter.tgz.

HyPhy will load the sequences and open a data Panel (fig. 6.2) We will explore some features of the data panel interface in later examples. For now, we wish to define a data filter (partition); in this case, the filter will simply be the entire alignment. Select all sites in the alignment by using the EDIT:SELECT ALL menu command, and then create a new partition by choosing DATA:SELECTION→PARTITION. The program creates a data filter with all the sites selected in the sequence viewer, assigns a default name and color to the partition, updates the navigation bar, and selects the newly created partition. One can edit the name and color of a partition by double clicking on the partition row in the "Analysis Setup" area or choosing DATA:PARTITION PROPERTIES, with the partition row selected. Rename the

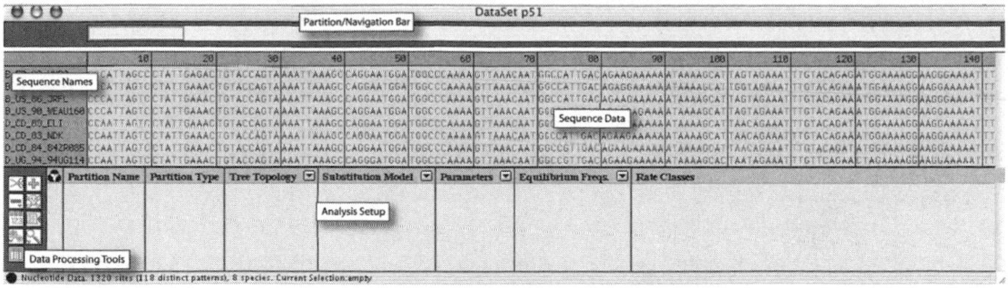

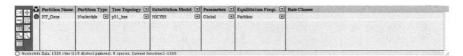

Fig. 6.2. *HyPhy* data panel (Mac OS X).

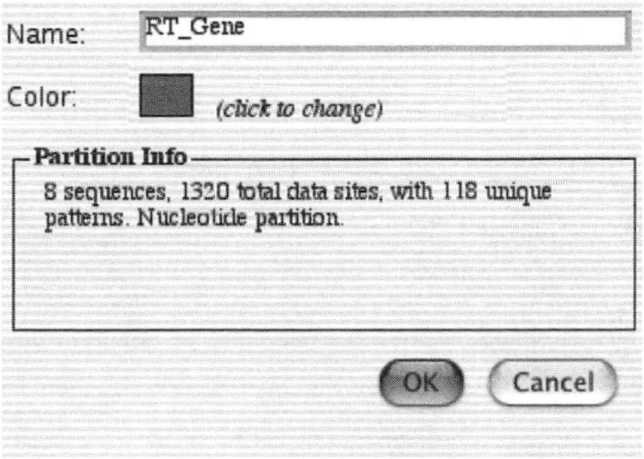

Name: RT_Gene

Color: ▨   *(click to change)*

┌─ **Partition Info** ─────────────────────────────
│   8 sequences, 1320 total data sites, with 118 unique
│   patterns. Nucleotide partition.
│
└──

OK   Cancel

Fig. 6.3. Partition properties dialog.

Fig. 6.4. Analysis Setup.

partition "RT_Gene" (for technical reasons, *HyPhy* doesn't allow spaces in the names of partitions) as shown in Figure 6.3.

Specifying the model

Once the data have been filtered, we may assign a tree topology and a model to the partition by clicking on the pulldown arrows in the appropriate columns of the "Analysis Setup" table (Figure 6.4). The data file `p51.nex` already included a tree topology, automatically loaded by *HyPhy* and made available in the "Tree Topology" pulldown list. For the model, let us choose substitution matrix HKY85, with global parameters (in this case meaning that there is a single transversion/transition ratio κ for every branch in the tree) and equilibrium frequencies gathered from the partition, so that entries of the frequency vector π are simply the frequencies of characters observed in the data. Once all the necessary analysis components have been successfully assigned to at least one data partition (RT_Gene in this case), the status light in the bottom left corner of the window will change from red to yellow, indicating that we are now ready to create a likelihood function.

Likelihood function

We will denote the likelihood function of the model parameters Θ, given a data set $\mathcal{D}$ and a tree $\mathcal{T}$, by

$$L(\Theta|\mathcal{D}, \mathcal{T}).$$

HyPhy is then able to obtain maximum likelihood parameter estimates $\hat{\Theta}$ by maximizing $L(\Theta|\mathcal{D}, \mathcal{T})$ over the possible values of Θ.

Let us now create and optimize the likelihood function. First, we select LIKELIHOOD:BUILD FUNCTION. *HyPhy* creates the likelihood function as requested and prints out some diagnostic messages to the console:

```
Created likelihood function 'p51_LF' with
 1  partition,
 1  shared parameters,
13 local parameters,
 0  constrained parameters.
Pruning efficiency 764 vs 1534 (50.1956 % savings)
```

The number of local parameters refers to the branch-length parameters, t. An unrooted binary tree on n sequences will have a total of $2N - 3$ branches. In our case, $N = 8$ and thus there are 13 branch-length parameters to estimate. Pruning efficiency numbers show the computational savings that *HyPhy* was able to realize using the column-sorting ideas of [6]. Now, choose LIKELI-HOOD:OPTIMIZE to instruct *HyPhy* to proceed with fitting selected models to the data and obtaining parameter MLEs.

Results

We are now ready to examine model-fitting results. For this example, *HyPhy* produces maximum likelihood estimates of 14 model parameters by numerical optimization of the likelihood function. The program reports a text summary to the console and also opens a graphical parameter table display, as shown in Figure 6.5. The status bar of the parameter table displays a one-line snapshot of the likelihood analysis: the maximum log-likelihood for our RT data set was -3327.25, and 14 parameters were estimated. Knowledge of these two quantities is sufficient to evaluate various information-theoretic criteria for relative goodness of fit, such as the Akaike information criterion [1], or to perform likelihood ratio tests for nested models.

Notice how *HyPhy* groups items in the parameter table by class: trees, global parameters (shared by all tree branches), and local parameters (those that affect a single branch); each item is labeled both by name and with an

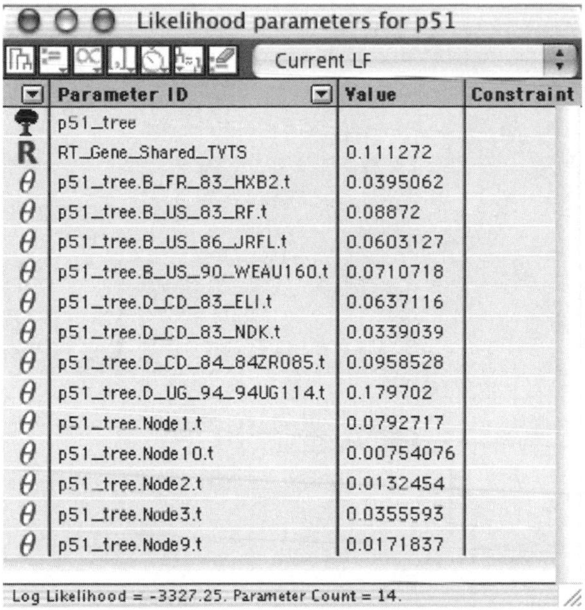

Fig. 6.5. Graphical parameter display.

appropriate icon. The single global parameter is the transversion:transition ratio, κ, of the HKY85 model and is labeled as RT Gene_Shared_TVTS. By default, each shared parameter is prefixed with the name of the data partition to which it is attached (RT_Gene in this case). While at first the names of local parameters may appear confusing, *HyPhy* uses a uniform naming scheme for all local model parameters: *tree name.branch name.parameter name*. For instance, p51_tree.B_FR_83_HXB2.t refers to a local parameter t along the branch ending in B_FR_83_HXB2 in the tree p51_tree. Leaf names in the tree correspond to sequence names in the data file, while *NodeN*, where N is an integer, are default names given to unlabeled internal nodes in the tree. (Users can give internal nodes custom names as well.) Parameter estimates can be exported in a variety of formats by invoking FILE:SAVE.

Let us now open a tree window to visualize the evolutionary distances between HIV-1 sequences in the example by double clicking on the tree row in the parameter table. *HyPhy* will open a tree viewer panel, as shown in Figure 6.6. A common measure used to assess evolutionary distances is the expected number of substitutions per site, E_{sub}, along a particular branch, equal to the weighted trace of the rate matrix:

$$E_{sub} = -t \sum_j \pi_j Q_{jj}. \tag{6.1}$$

The *HyPhy* tree viewer automatically scales branches on E_s, although the scaling may be changed by the user.

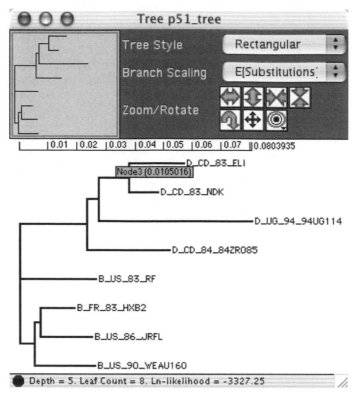

Fig. 6.6. *HyPhy* Tree Viewer for `p51.nex`, scaled on the expected number of substitutions per site inferred using the HKY85 model, with an example of a tooltip branch-length reporter.

Confidence intervals

All parameter estimates will be affected by sampling variations of various magnitudes. For instance, substitution-bias parameters often have large variances relative to those of branch-length estimates. *HyPhy* allows the user to obtain confidence intervals using the asymptotic normality of MLEs. Likelihood theory states that MLEs of model parameters are distributed asymptotically as multivariate normal around the true parameter values, and the covariance matrix of the normal distribution can be estimated by inverting the observed Fisher information matrix

$$\hat{I}\left(\hat{\Theta}\right) = \left(\frac{\partial^2 \log L(\Theta|\mathcal{D}, \mathcal{T})}{\partial \theta_i \partial \theta_j} \bigg|_{\Theta = \hat{\Theta}} \right).$$

The Fisher information matrix measures the curvature of the log-likelihood surface. Flat surfaces around the maximum do not inspire high confidence in estimated parameter values, while steep surfaces lead to sharp estimates.

HyPhy can be instructed to construct the covariance matrix as well as the confidence intervals for each parameter based on the estimated variance of the normal distribution, either for every parameter or for selected parameters (conditioned on the values of others). Select all the parameters in the table by choosing EDIT:SELECT ALL and then LIKELIHOOD: COVARIANCE AND CI, and set "Estimation Method" to "Asymptotic Normal[finer]" in the ensuing dialog box. "Crude" and "Finer" estimates differ in how *HyPhy* computes the Fisher Information Matrix (which must be done numerically because analytical derivatives of the likelihood function are not available in general). *HyPhy* will open two *chart windows*—the 95% confidence interval window for all selected parameters and the covariance matrix.

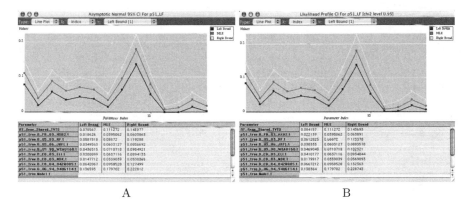

A   B

Fig. 6.7. *HyPhy* confidence interval estimates using (A) asymptotic normality of MLEs and (B) Profile plots using 95% levels of χ_1^2.

Likelihood profile

Confidence intervals based on asymptotic normality rely upon many assumptions that may be violated for short alignments or parameter-rich models. For example, such confidence intervals are always symmetric about the maximum likelihood estimate, and if the likelihood surface is skewed around the MLE, such intervals may be a poor representation of the real variance of parameter estimates. A second approach to determining statistical support for a parameter value estimate is to employ *likelihood profile* confidence intervals, obtained by inverting a likelihood ratio test.

Suppose we wish to compute a confidence interval CI_i^{α} of level α for a single model parameter θ_i. A common method is first to fix all other model parameters $\theta_{i'}, i' \neq i$ at their maximum likelihood estimates. We can now think of the likelihood function as a function of a single parameter θ_i. Thus, a restricted version of the full likelihood function is

$$\bar{L}(\theta_i) = L(\theta_i | \mathcal{D}, \mathcal{T}, \hat{\theta}_{i'}).$$

Clearly, the maximum likelihood estimate for θ_i using the restricted likelihood is the same as that given by the full likelihood function: $\hat{\theta}_i$.

Consider two hypotheses: $H_0 : \theta_i = x$ versus $H_A : \theta_i \neq x$. These hypotheses can be tested using the restricted likelihood function and a one degree of freedom likelihood ratio test (assuming that $\hat{\theta}_{i'}$ is not on the boundary of the parameter space)

$$2[\log \bar{L}(\hat{\theta}_i) - \log \bar{L}(x)] \sim \chi_1^2.$$

If $\hat{\theta}_j$ is on the boundary, then the asymptotic distribution changes to

$$2[\log \bar{L}(\hat{\theta}_i) - \log \bar{L}(x)] \sim \frac{\chi_1^2 + \chi_0^2}{2}.$$

Using this observation, a confidence region can be defined as all those values x for which we fail to reject H_0 (i.e., all those x for which the likelihood ratio statistic is less than the α percentile of the corresponding χ^2 or mixture distribution). If we also assume that the likelihood function is monotone (has no local maxima), then we find the boundaries of the confidence interval by tracing the log-likelihood function plot until the desired difference from the maximum is obtained in both directions (see Figure 6.8).

There are a couple of issues with this approach: (i) we assume sufficient data for the asymptotic likelihood distributions to be applicable, which may fail for short alignments or models that are too parameter-rich; and (ii) we are obtaining the confidence intervals for one parameter at a time rather than a confidence region for all parameters (which is mostly due to technical difficulties with finding such a region when there are many model parameters), thus ignoring covariation among parameter estimates.

The first issue may be resolved, to an extent, by accepting or rejecting H_0 using a non-LRT criterion, such as AIC [1]. The procedure is exactly the same, but the cutoff level is no longer determined by the asymptotic χ^2 distribution but rather by an information-theoretic parameter addition penalty. For AIC, $2[\log \bar{L}(\hat{\theta}_i) - \log \bar{L}(x)] \leq 2$ would place x in the confidence interval.

Also, to see how reasonable the asymptotic normality assumption is, one could check whether a quadratic approximation to the log-likelihood holds well. The quadratic approximation for the log restricted likelihood around the maximum likelihood estimate $\hat{\theta}_i$ can be derived from a Taylor series expansion:

$$\log \bar{L}(x) \approx \log \bar{L}(\hat{\theta}_i) + \frac{d}{d\theta_i} \log \bar{L}(\theta_i)\Big|_{\hat{\theta}_i} \left(x - \hat{\theta}_i\right) + \frac{d^2}{d\theta_i^2} \log \bar{L}(\theta_i)\Big|_{\hat{\theta}_i} \left(x - \hat{\theta}_i\right)^2.$$

Because $\hat{\theta}_i$ maximizes the likelihood function, the first derivative term vanishes, and we have the desired quadratic approximation:

$$\log \bar{L}(x) - \log \bar{L}(\hat{\theta}_i) \approx \frac{d^2}{d\theta_i^2} \log \bar{L}(\theta_i)\Big|_{\hat{\theta}_i} \left(x - \hat{\theta}_i\right)^2.$$

By plotting the likelihood profile and the quadratic approximation on the same graph, one can see how well the χ^2 approximation to the likelihood ratio test will work. *HyPhy* offers each of the confidence interval estimation techniques above via LIKELIHOOD:COVARIANCE AND CI and LIKELIHOOD:PROFILE PLOT from the parameter table window.

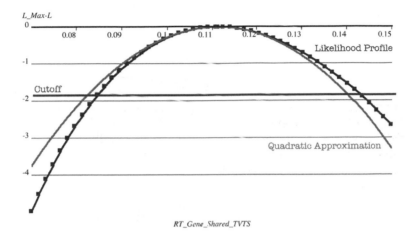
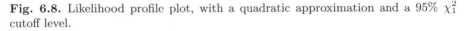

Fig. 6.8. Likelihood profile plot, with a quadratic approximation and a 95% χ_1^2 cutoff level.

Saving the analysis

HyPhy can store all the information needed to recreate the analysis we just performed in a single NEXUS file. This feature can be invoked by switching back to the data panel, selecting FILE:SAVE, and choosing the format option to include the data in the file. Let us save this simple analysis as p51_HKY85.bf in the "Saves" directory of the *HyPhy* installation.

6.2.2 Local Branch Parameters

Almost all treatments of likelihood analysis of molecular sequence data assume that there is only one parameter per branch in the phylogenetic tree—branch-length—and that other model parameters are shared by all branches. However, it may be often be desirable to relax this assumption. For example, to test whether a group of branches (such as a single lineage or a clade) have different substitution process parameters than the rest of the tree, it is necessary to compare likelihoods of constrained and unconstrained models. *HyPhy* provides a general mechanism for defining an arbitrary number of branch-specific and shared model parameters. Consider the HKY85 model discussed in the previous section. Rewrite the rate matrix as

$$Q = \begin{array}{c} \\ A \\ C \\ G \\ T \end{array} \begin{array}{cccc} A & C & G & T \\ \begin{pmatrix} \star & \beta\pi_C & \alpha\pi_G & \beta\pi_T \\ \beta\pi_A & \star & \beta\pi_G & \alpha\pi_T \\ \alpha\pi_A & \beta\pi_G & \star & \beta\pi_T \\ \beta\pi_A & \alpha\pi_C & \beta\pi_G & \star \end{pmatrix} \end{array}.$$

This may seem like a different matrix altogether, but if one sets $t = \alpha$ and $\kappa = \beta/\alpha$, we return to the previous parameterization if $\beta > 0$. In fact, this new parameterization allows the transition rate (α) to be 0 and transversion rate (β) to be nonzero, whereas the first (more common) parameterization does not. Even more importantly, we can now let each branch have a separate α and β, which is equivalent to allowing every branch to have its own transition/transversion ratio. We declare such a model to be fully local, as opposed to the fully global model of the previous section. Obviously, there is a range of intermediate models where some of the branches share transition/transversion ratios while others are free to vary.

To specify the fully local HKY85 model in *HyPhy* for our example data set, all that must be done differently is to select "Local" in place of "Global" in the "Parameters" column of the analysis setup table in Figure 6.4. You can either start a new analysis from scratch or continue from where we left off in the global analysis of the previous section. In the latter scenario, *HyPhy* will display a warning message because changing substitution models causes a fundamental change in the likelihood function (i.e., a different set of parameters and rate matrices). Next, invoke LIKELIHOOD:BUILD FUNCTION and observe that the resulting likelihood function has 26 local parameters (two per branch, as requested). Upon selecting LIKELIHOOD:OPTIMIZE, a parameter table is once again shown, and we observe that the log-likelihood has improved to -3320.84. A quick glance at the likelihood score improvement of seven units for 12 additional parameters suggests that there is insufficient evidence favoring the fully local model over the fully global model.

The rate parameter names in the parameter table for this analysis end with "trst" and "trsv," which hopefully mean "transition" and "transversion." *HyPhy* allows one to look at the rate matrix and map parameter names to what they actually stand for in case parameter names are less descriptive. To see how that is done, let us open the "Object Inspector" window (use WINDOW:OBJECT INSPECTOR on the Mac and FILE:OBJECT INSPECTOR in Windows). In the newly opened window (Figure 6.9(a)), select "Models" from the pulldown option list, and scroll through the rather long list of models until you find one in bold (meaning that this model is currently used in an active likelihood function) named "RT_Gene_HKY85_local." Again, the name of the data partition is incorporated in the model identifier for easy reference. Double click on that model and examine the rate matrix as shown in Figure 6.9b. The equilibrium frequencies for this model (π) are the actual proportions of A, C, G, and T in the RT gene alignment, and "trst" are indeed the rates for $A \leftrightarrow G$ and $C \leftrightarrow T$ substitutions, while "trsv" are the rates for all other substitutions. By default, *HyPhy* will automatically multiply rate matrix entries by the appropriate π, and hence there is no need to include them in the rate matrix explicitly.

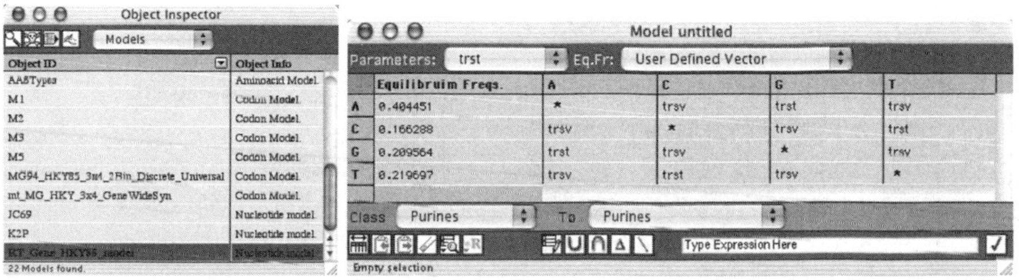

Fig. 6.9. (a) Models in the "Object Inspector"; (b) HKY85 local model for the RT gene.

Let us now open the tree window for the local model (Figure 6.10(a)). Recall that branch lengths are given by (6.1). The tree looks very similar to the global HKY85 tree from Figure 6.6. However, a more interesting comparison would be to see if the transition and transversion ratios vary from branch to branch. *HyPhy* allows scaling of the tree display on any local model parameter—"trst" and "trsv" in this instance.

Double click on the tree name in the parameter table once again to open another instance of the tree window— very useful for side-by-side comparison. Scale one of the trees on "trst" and another on "trsv" (Figure 6.10(b,c)). Notice that while the shapes are still similar, branch lengths are not quite proportional between trees, as they would be if all branch transition/transversion ratios were the same.

As a matter of fact, the *HyPhy* tree viewer allows scaling on any function of model parameters. Let us define the transversion/transition ratio parameter. For every branch, it is simply $ratio = trsv/trst$. To define this scaling parameter, switch to a tree window, select all branches (EDIT:SELECT ALL), and choose TREE:EDIT PROPERTIES. The dialog box that appears shows all available local branch parameters. Click on the "Add User Expression" button (the + icon), type in the formula for the expression, rename it "ratio," and select "OK" (Figure 6.11). *HyPhy* has added "ratio" to the list of local parameters (not estimable parameters but rather functions of other parameters). You can view the value of each branch ratio in the parameter table and scale

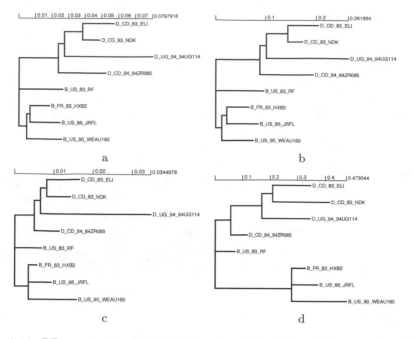

Fig. 6.10. RT gene tree under HKY85 local model scaled on (a) expected number of substitutions per site, (b) transition rates, (c) transversion rates, (d) transversion/transition ratios.

the tree on the transversion/transition ratio (Figure 6.10(d)). The differences in branch-to-branch ratios are quite striking.

The *HyPhy* tree viewer can automatically label each branch of the tree with any function of branch-model parameters. As an example, we will label each branch with the number of transitions E_t and transversions E_v per site, expected to occur along that branch. For the HKY85 local model,

$$E_t = 2\beta t(\pi_A\pi_G + \pi_C\pi_T), \quad E_v = 2\alpha t[(\pi_A + \pi_G)(\pi_C + \pi_T)].$$

Note that E_t and E_v add up to the total branch length and are linear functions of the rates. Substituting the actual values of π for our data set (Figure 6.9(b)), we get

$$E_t = 0.242583\beta t, \quad E_v = 0.474001\alpha t.$$

Employ the same process we did for adding the ratio parameter, and define $E_t = 0.242583 * trst$ and $E_v = 0.474001 * trsv$. Now use TREE:BRANCH LABELS:ABOVE BRANCHES and TREE:BRANCH LABELS:BELOW BRANCHES to label each branch with E_t and E_v, adjust fonts and alignments to your liking, and check "Scale tree by resizing window" in the dialog opened with TREE:TREE DISPLAY OPTIONS. The final display should look like Figure 6.12.

Parameter ID	Type	Value
trst	Parameter	*Multiple Values*
trsv	Parameter	*Multiple Values*
ratio	Expression	*Multiple Values*

Formula: `trsv/trst`

⊞ ▦ ✓   (OK)   (Cancel)

Fig. 6.11. New scaling parameter dialog.

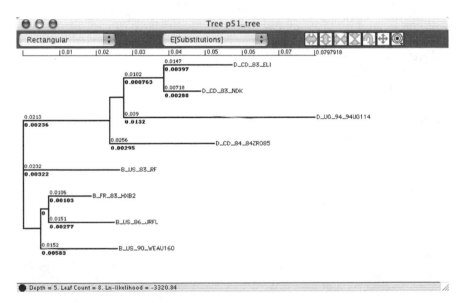

Fig. 6.12. RT tree scaled on expected number of substitutions per site and labeled with the expected number of transitions and transversions per site (above and below, respectively).

6.2.3 Multiple Partitions and Hypothesis Testing

Early attempts to model molecular evolution of protein-coding sequences used the observation that the evolution in the first and second positions of a codon differed markedly from that at the third position. Indeed, for the universal

genetic code, every substitution in the second codon position is nonsynonymous (i.e., it changes the protein encoded by the codon). For the first position,
all but eight possible (sense) substitutions are nonsynonymous. In contrast,
at the third position, 126 out of 176 substitutions are synonymous. Because
random nonsynonymous substitutions are likely to be deleterious, it is often
observed that the substitution rate for the third position is different (typically
much higher) than those in the first and second positions. Our next task is
to define a *HyPhy* analysis that treats the first and second codon positions
as one data partition and the third codon position as another, and then fit a
collection of models to the data. We will continue using the HIV-1 p51 subunit
of the RT gene data set from p51.nex.

First, open the data panel with p51.nex and select all the sites in the
alignment. Next, invoke one of the numerous data-filtering tools in *HyPhy*-
–the combing tool—by clicking on the comb tool button in the data panel
(Figure 6.13). To select the first two positions in every codon, we need a comb
of size 3 with first and second sites selected and the third omitted. In the
combing dialog, set the size of the comb to 3 and check the boxes next to
positions 1 and 2. Repeat the process to define the partition with every third
codon position (make sure that the first partition is *not* highlighted in the
analysis setup table while you are applying the second comb; otherwise *HyPhy*
will comb the partition again, effectively selecting every third column in the
data partition of the first and second positions we have just created). Rename
the partitions to "First_Second" and "Third", respectively. Assign the same
tree topology to both data partitions, the HKY85 model, global parameter
options, and equilibrium frequencies collected separately from each partition.
In the end, the data panel should resemble the one in Figure 6.13.

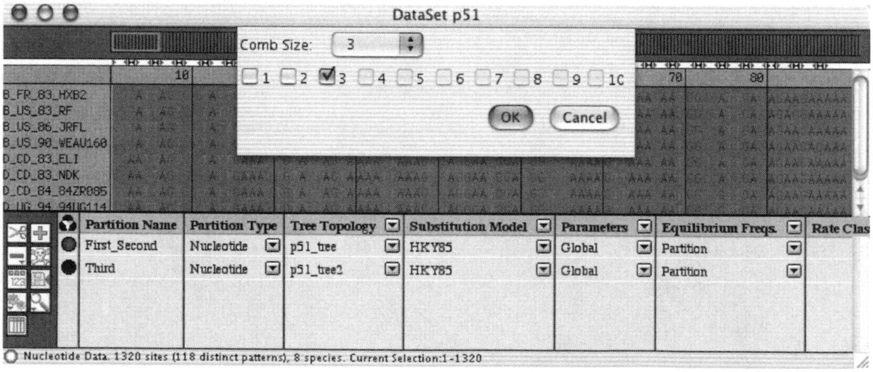

Fig. 6.13. Data panel with two data partitions and a comb filter dialog.

When we build the likelihood function, *HyPhy* prints out a message

`Tree topology p51_tree was cloned for partition Third.`

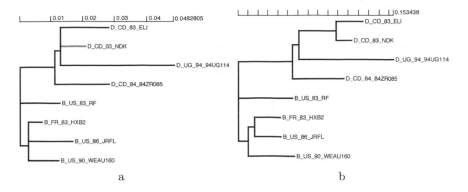

Fig. 6.14. HIV-1 RT scaled on the expected number of substitutions per site for (a) first and second codon positions and (b) third codon position.

It is important to understand that while both partitions share the tree *topology*, for *HyPhy* a tree means *both* topology and models/parameters. The two partitions need to have two trees with independent branch lengths and transversion/transition ratio parameters, κ_{12} and κ_3, assigned the names First_Second_Shared_TVTS and Third_Shared_TVTS by *HyPhy*.

After the models are fit to the data, we observe that both the shapes of the trees (Figure 6.14) and the transversion/transition ratios (0.198 versus 0.067) differ quite a lot between the partitions.

A careful reader might correctly point out that the analysis we have just performed could have been done by fitting HKY85 to each of the partitions separately. However, we will now illustrate what the joint likelihood function of both partitions can offer in terms of hypothesis testing.

Simple hypothesis testing

Consider the null hypothesis $H_0 : \kappa_{12} = \kappa_3$ versus the full-model alternative $H_A : \kappa_{12} \neq \kappa_3$. The analysis we just performed was for the full model, and before proceeding with the definition of the constraint in H_0, the MLEs for H_A must be saved. To do so, click on the pulldown menu in the parameter table (Figure 6.4) and choose SAVE LF STATE. A collection of parameter MLEs and constraints constitute a state (i.e., a hypothesis). Name the state "Full Model," and choose SELECT AS ALTERNATIVE from the same pulldown menu.

Now, the constraint for the null hypothesis must be defined, and a new set of MLEs for all independent model parameters must be calculated. To define the constraint, select both transversion/transition ratio parameters (shift-click to select multiple rows) and click on the constraint (second) button. Note that the parameter table updated to reflect that one of the ratios is no longer independent of the remaining parameters. Next, we calculate a new set of parameter MLEs by optimizing the likelihood function anew. Not surprisingly, $H_0 : \kappa_{12} = \kappa_3 = 0.11$, which is between the independently estimated values.

Save the set of MLEs for H_0 as "Constrained" and then choose SELECT AS NULL, which instructs *HyPhy* to treat "Constrained" as the null hypothesis. With all the components of a hypothesis test in place, choose LRT from the same pulldown menu. *HyPhy* computes the likelihood ratio statistic $2 \left(\log L_A - \log L_0 \right)$ and a *p*-value based on the asymptotic χ^2 distribution with (in this case) one degree of freedom:

```
Likelihood Ratio Test
     2*LR = 12.5286
     DF = 1
     P-Value = 0.000400774
```

The likelihood ratio test strongly rejects the null hypothesis of equal transversion/transition ratios between partitions.

Parametric bootstrap

The χ_1^2 distribution is the *asymptotic* distribution for the LRT statistic, and one would be well-advised to realize that it may not always apply directly. However, one can always verify or replace the results of a χ^2 test by the parametric bootstrap [2, 4]. *HyPhy* has a very general way of simulating sequence alignments parametrically – it can do so transparently for any likelihood function using current parameter values. For the purposes of this example, *HyPhy* simulates 1000 8-sequence alignments with 1320 sites each, using the model in the null hypothesis (i.e., constrained ratios). *HyPhy* then fits the models in H_0 and H_A to every simulated data set and tabulates the likelihood ratio test statistic. The resulting LRT distribution may then be used for obtaining significance values for the original LRT value or for verifying how well the LRT statistic follows the asymptotic χ^2 distribution.

The parametric bootstrap function can be accessed via the same pulldown menu in the parameter table window. Enter the number of data replicates to be simulated and choose whether or not *HyPhy* should save data and parameter estimates for every replicate. For the current data set, 1000 replicates should take $20 - 30$ minutes on a typical desktop computer. *HyPhy* opens a summary bootstrap table and adds simulated LRT statistic values as they become available, as well as keeping tabs on the current *p*-value. Replicates with larger values of the LRT than the original test are highlighted in bold. After bootstrapping has finished, you may open a histogram or cumulative distribution function plot for the LRT statistic, as shown in Figure 6.15. Your simulation results will differ from run to run, but you should still obtain a *p*-value very close to the asymptotic χ^2 *p*-value and an LRT histogram mirroring the shape of a χ^2 distribution with a single degree of freedom.

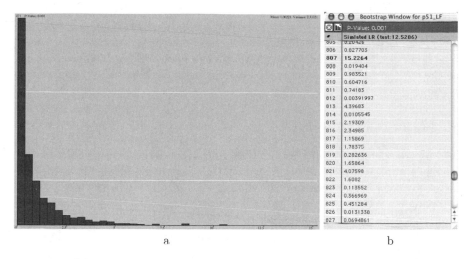

Fig. 6.15. (a) Simulated density for the likelihood ratio statistic and (b) bootstrapping window example.

Relative ratio test

It is clear from Figure 6.14 that the trees on the first and second positions $\mathcal{T}_{12}$ have much shorter branch lengths than the tree for the third position $\mathcal{T}_3$, which is to be expected. However, apart from a few internal branches, the overall shapes of the trees remain somewhat similar, suggesting that perhaps the only fundamental difference between nucleotide level substitution processes is the amount of change for the entire tree, while relative branch lengths $E_{sub}(b_i)$ are the same for both trees. Mathematically, this constraint can be expressed as

$$E_{sub}(b_i | \mathcal{T}_{12}) = R_R E_{sub}(b_i | \mathcal{T}_3), \text{ for all branches } b_i,$$

where the parameter R_R is the relative ratio. As we saw earlier, branch lengths for HKY85 are linear functions of the branch-length parameter t; thus it is sufficient to constrain t parameters to be proportional.

HyPhy has a built-in tool for easy specification of relative ratio constraints [13, 8] on trees or subtrees. To carry out the relative ratio test, select two trees (or two branches that root the subtrees; see below) and click on the relative ratio button (second from the right in the toolbar) in the parameter table. Name the ratio parameter, and then reoptimize the parameters. Use the technique from the previous example to save the full and constrained models and to carry out the likelihood ratio test using either the asymptotic distribution or the parametric bootstrap. The result from the chi-squared distribution is:

```
Likelihood Ratio Test
    2*LR = 24.0092
```

```
DF = 12
P-Value = 0.0202825
```

The relative ratio hypothesis can therefore be rejected at the 0.05 level but not at the 0.01 level. Application of the parametric bootstrap yields a comparable *p*-value.

Saving a complete analysis.

HyPhy is capable of saving an analysis and every hypothesis in a single file. Invoke FILE:SAVE from the *data panel*, and choose the format that includes sequence data in the resulting file dialog. If you later open the saved file by selecting FILE:OPEN:OPEN BATCH FILE, the analysis and all the hypotheses you have defined will be available.

6.2.4 Codon Models

The natural unit of evolution for stochastic models of protein-coding sequences is a codon. By modeling the substitutions on the level of codons rather than nucleotides, inherently different processes of synonymous and nonsynonymous substitutions can be handled adequately. By expanding the state space for the substitution process from four nucleotides to 61 nonstop codons in the universal genetic code, the computational cost increases dramatically, both when evaluating transition probability matrices and calculating the likelihood function itself. Modern computers can handle the added burden quite easily, though.

Consider a codon-based extension to the HKY85 model, which is similar to the model in [7]. We dub it MG94×HKY85_3×4. The 61×61 rate matrix for this model, which gives the probability of substituting codon x with codon y in infinitesimal time, is

$$
Q_{x,y}(\alpha, \beta, \kappa) = \begin{cases} \alpha \pi_{n_y} & x \to y \text{ 1-step synonymous transition,} \\ \alpha \kappa \pi_{n_y}, & x \to y \text{ 1-step synonymous transversion,} \\ \beta \pi_{n_y}, & x \to y \text{ 1-step nonsynonymous transition,} \\ \beta \kappa \pi_{n_y}, & x \to y \text{ 1-step nonsynonymous transversion,} \\ 0, & \text{otherwise.} \end{cases} \quad (6.2)
$$

As before, κ is the transversion/transition ratio. The parameter α denotes the synonymous substitution rate, while β provides the nonsynonymous substitution rate. The ratio of these two values, $\omega = \beta/\alpha$, can be used to measure the amount of selective pressure along a specific branch. The value π_{n_y} is the frequency of the "target nucleotide" for the substitution observed in the appropriate codon position in the data set. For instance, if $x = ATC$ and $y = AGC$, then π_{n_y} would be the frequency of nucleotide G observed at second codon positions in the alignment. The model only allows for one instantaneous nucleotide substitution between codons. For instance, $ATC \to AGG$

is not allowed to happen by two concurrent nucleotide substitutions because such events have negligibly small probabilities. However, such changes are allowed via multiple substitutions, as evidenced by the fact that all transition probabilities (entries in the matrix e^{Qt}) are nonzero for $t > 0$.

The specification of the model is completed by providing the equilibrium frequencies of the 61 codons. For a codon composed of three nucleotides i, j, k

$$\pi_{ijk} = \frac{\pi_i^1 \pi_j^2 \pi_k^3}{1 - \pi_T^1 \pi_A^2 \pi_A^3 - \pi_T^1 \pi_A^2 \pi_G^3 - \pi_T^1 \pi_G^2 \pi_A^3}, \tag{6.3}$$

where π_n^k denotes the observed frequency of nucleotide n at codon position k. The normalizing term accounts for the absence of stop codons TAA, TAG, and TGA from the state space and the model. Note that this model mixes local (α and β) and global (κ) parameters.

MG94×HKY85_3×4 applied to HIV-1 integrase gene

Following are the steps needed to apply a codon model to `integrase_BDA.nex`, found in the **Examples** directory of *HyPhy* standard distribution. This data file contains the integrase gene of six Ugandan subtype D, three Kenyan subtype A, and two subtype B (Bolivia and Argentina) HIV-1 sequences sampled in 1999. The integrase gene is relatively conserved and is appropriate for comparison between subtypes.

1. Open the data file via FILE:OPEN:OPEN DATA FILE.
2. Select all the data and define a partition—it will be created as a nucleotide partition at first.
3. Switch the partition type to "Codon." *HyPhy* will display a partition properties box. Rename the partition "Integrase," but keep all other default settings.
4. Assign "Integrase_BDA_tree" topology, "MG94×HKY85_3×4" model, and "Local" parameters option.
5. Build (LIKELIHOOD:BUILD FUNCTION) the likelihood function. Note that 38 local parameters (α and β for each of the 19 branches) and one global parameter (transversion/transition ratio) have been created.
6. Optimize (LIKELIHOOD:OPTIMIZE) the likelihood function. It should take a minute or so on a desktop computer. Open two tree displays, and scale one on synonymous rates and the other on nonsynonymous rates. Notice the radical differences between the trees, both in lengths and shapes, as shown in Figure 6.16.

Molecular clock tests

When reversible models of evolution are used, the rate parameters cannot be identified separately from the time parameters because only their products are

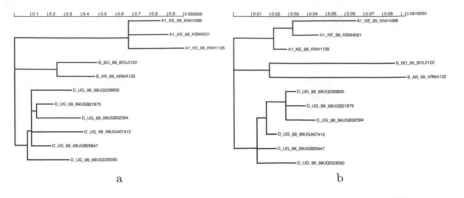

Fig. 6.16. HIV-1 integrase tree scaled on (a) synonymous rates α and (b) nonsynonymous rates β.

estimable. A set of sequences is said to have evolved under a molecular clock if the expected amount of evolution (measured in expected numbers of substitutions) from the most recent common ancestor to each of the descendent sequences is the same. Mathematically, we constrain the length of the paths between each sequence and the most recent common ancestor in the phylogenetic tree to be the same. For the tree in Figure 6.17, a molecular clock would be imposed by the following two constraints: $t_2 = t_1$ and $t_3 = t_1 + t_4$. Note that imposing a molecular clock typically requires a rooted tree. Thus, it is desirable to have a separate outgroup sequence (or groups of sequences) that can be used to establish the root of a tree. For instance, in the HIV-1 integrase example (Figure 6.16), subtype A sequences form an outgroup to both B and D subtype clades.

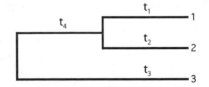

Fig. 6.17. Example of a molecular clock constraint.

For coding sequences, it is often useful to impose molecular clocks on synonymous substitutions only. Synonymous substitutions are assumed to be relatively free of selective constraints, whereas nonsynonymous substitutions will be heavily influenced by purifying and positive selection. *HyPhy* provides an easy way to impose molecular clock constraints on a subtree using some or all model parameters. For MG94×HKY85_3×4, it can be shown that the expected number of substitutions per site on a branch has the form

$$E_{sub} = t\alpha[f_1(\pi) + \kappa f_2(\pi)] + t\beta[g_1(\pi) + \kappa g_2(\pi)],$$

where f_1, f_2, g_1, and g_2 are functions determined by the nucleotide composition of the alignment. The first term in the sum corresponds to the contribution of synonymous substitutions and the second to the contribution of nonsynonymous substitutions. Since each is a multiple of the corresponding substitution parameter (α or β), imposing additive constraints on α and β will result in additivity of the corresponding expected substitution quantities. Note again that the time parameter t is not estimable alone, and the parameters actually being estimated (and constrained) are αt and βt.

Thus, three types of molecular clocks may be tested for local codon models: (i) synonymous only, (ii) nonsynonymous only, and (iii) full (both synonymous and nonsynonymous) rates.

Local clock tests on HIV-1 integrase

We now address the question of which, if any, of the three types of molecular clocks are supported for the D-subtype clade. We assume that the MG94×HKY85 model has been fit to the data as described above.

1. Save the likelihood function state as "Full Model." Select it to be the alternative hypothesis for our tests.
2. Select the branch that is the most recent common ancestor of the D clade in the tree viewer. Invoke TREE:SHOW PARAMETERS IN TABLE. This action will locate two rows in the parameter table, with the parameters attached to that branch—"Node9." This method is a general way for locating branch-specific model parameters in the table quickly—it also works for a multiple-branch selection. Highlight one of the two identified rows.
3. Click on the molecular clock button (fifth from the left) in the toolbar of the parameter table. A pulldown menu will appear with the parameters available for the molecular clock constraints. Choose to constrain "syn-Rate" for the synonymous rate clock.
4. Optimize the likelihood function, save the new likelihood function state as "Synonymous Clock," and set it to be the null hypothesis. Perform the likelihood ratio test. The test will report the likelihood ratio statistic of 9.52, which yields the p-value of 0.09 using the asymptotic χ^2 with 5 degrees of freedom. This value is reasonably close to rejecting the molecular clock hypothesis, so a bootstrap p-value verification may be desirable. For codon data, bootstrapping is a time-consuming process, so you may only choose to do 100 replicates. Our simulation yielded a p-value of 0.14, failing to reject the molecular clock.
5. Select "Full Model" from the pulldown menu in the toolbar of the parameter table, and then go back to step 3 and repeat steps 3 and 4, constraining

nonsynonymous rates first and then both rates. Likelihood ratio tests fail to reject either of the clocks.

6. Save the analysis from the data panel.

6.2.5 More General Hypothesis Testing

The hypotheses of the previous section are all examples of nested hypotheses, which can be obtained by constraining some of the model parameters in the more general hypothesis to reduce it to a particular case, the null hypothesis. Often, interesting biological questions cannot be framed as nested hypotheses. For example, the question of whether a particular phylogeny with certain taxa constrained to be monophyletic is significantly different from the unconstrained phylogeny is a nonnested question. Another example would be determining which of two competing models better explains the data when the models are nonnested. *HyPhy* includes a rather general mechanism for nonnested hypothesis testing based on the parametric bootstrap [2, 4]. All one needs to do is to define competing models (by models, we mean more than just the substitution matrices) on the same alignment and then test by parametric bootstrapping.

Consider the example data set of the p51 subunit of HIV-1 reverse transcriptase from the previous sections. As an illustration of testing nonnested hypotheses, we will consider whether there is enough evidence to suggest that the JTT model describes the data better than the Dayhoff model of amino acid evolution.

First, we must convert a codon alignment found in `p51.nex` into amino acids.

1. Open the data file `p51.nex`, select all alignment columns, and create a nucleotide partition.
2. Change the data type of the partition to "Codon," obeying the universal genetic code.
3. Select DATA:ADDITIONAL INFO:AMINO ACID TRANSLATION. Choose "All" in the ensuing dialog box. *HyPhy* will translate all the sequences in the codon partitions into amino acids, create a new data set, and open a new data panel displaying all the newly created amino acid sequences.
4. Let us now save the amino acid alignment to a separate data file. In the newly opened data panel with the amino acid alignment, create a partition with all the alignment sites and, with the partition row selected, click on the "Save Partition To Disk" button. Choose the "NEXUS Sequential[Labels]" format in the file save dialog, and save the file as `p51.aa` in the "data" directory of the *HyPhy* distribution.

Second, we evaluate the likelihood under the null hypothesis H_0: Dayhoff model:

1. Open the amino acid alignment `p51.aa`, select all alignment columns, and create a protein partition named "p51."

2. Assign the included p51 tree topology and the "Dayhoff" substitution model to the "p51" partition.
3. Build and optimize the likelihood function.

The null model has 13 estimable parameters and yields a log-likelihood of -2027.28.

Next, we set up the alternative hypothesis, H_A: JTT model:

1. Open the amino acid alignment p51.aa, *while the previous analysis is still open*. We need to keep both analyses in memory at the same time. Note how *HyPhy* renamed the new data panel "p512" to avoid a naming conflict with an already open window.
2. Assign the tree topology found in the data file and the "Jones" substitution model to the data partition.
3. Build and optimize the likelihood function.

The alternative model also has 13 adjustable parameters and yields a log-likelihood of -1981.61.

The JTT model provides a higher likelihood value, but since the models are not nested, we cannot simply compare the likelihoods to determine whether the difference is statistically significant. We can, however, use the parametric bootstrap to find a *p*-value for the test without relying on any asymptotic distributional properties.

1. Switch to either of the data panels, and invoke LIKELIHOOD:GENERAL BOOTSTRAP. *HyPhy* will display a bootstrap setup window, which is very similar to the window we have seen in nested bootstrap examples.
2. Set the appropriate null and alternative hypotheses by choosing the name of the data panel ("p51" should be the null, and "p512" should refer to the alternative, if you have followed the steps closely).
3. Click on the "Start Bootstrapping" button, select PARAMETRIC BOOTSTRAP from the pulldown, and enter 100 for the number of iterates.
4. *HyPhy* will perform the requested number of iterates (it should take five or ten minutes on a desktop computer), and report the *p*-value. In our simulation, we obtained a *p*-value of 0, suggesting that the data are better described by the JTT model.

6.2.6 Spatial Rate Heterogeneity: Selective Pressure and Functional Constraints

It is a well-documented fact that evolutionary rates in sequences vary from site to site. Good substitution models should be able to include such rate variation and offer ways to infer the rates at individual sites. Consider again the MG94×HKY85_3×4 codon model, but let us modify it to allow each codon s to have its own synonymous (α_s) and nonsynonymous (β_s) rates. The rate matrix for codon s must be modified as follows:

$$Q_{x,y} = \begin{cases} \alpha_s \pi_{n_y}, & x \to y \text{ 1-step synonymous transition,} \\ \alpha_s \kappa \pi_{n_y}, & x \to y \text{ 1-step synonymous transversion,} \\ \beta_s \pi_{n_y}, & x \to y \text{ 1-step nonsynonymous transition,} \\ \beta_s \kappa \pi_{n_y}, & x \to y \text{ 1-step nonsynonymous transversion,} \\ 0, & \text{otherwise.} \end{cases}$$

The most general estimation approach would be to estimate α_s and β_s separately for every codon, but that would require too many parameters and result in estimability issues. Another idea, first proposed in [11], is to treat the rate at a particular site as a random variable drawn from a specified distribution. Most work of this sort has considered only a single variable rate for each site, and the distribution of those rates has usually been assumed to follow a gamma distribution. We now extend the MG94×HKY85_3×4 model to have synonymous and nonsynonymous rates at codon s described by the bivariate distribution $F_\eta(\alpha_s, \beta_s)$ whose parameters η are either given or estimated. The likelihood for an alignment with S sites, tree $\mathcal{T}$, and the vector Θ of all model parameters can be written as

$$L(\Theta|\mathcal{T}, \mathcal{D}) = \prod_{s=1}^{S} E\left[L(\Theta|\mathcal{T}, \mathcal{D}_s, \alpha_s = a, \beta_s = b)\right].$$

The expectation is computed using the distribution specified by $F_\eta(\alpha_s, \beta_s)$. Site likelihoods, conditioned on the values of α_s and β_s, may be evaluated using Felsenstein's pruning algorithm [3]. Unless $F_\eta(\alpha_s, \beta_s)$ specifies a discrete distribution with a small number of classes, the expectation is computationally intractable. However, the approach of discretizing the continuous distribution of rates to obtain a computationally tractable formulation was introduced in [12].

If codon s in the alignment is following neutral evolution, then we expect to infer $\beta_s \approx \alpha_s$. For sites subject to functional constraints, nonsynonymous mutations are almost certain to be highly deleterious or lethal, leading to purifying selection and $\beta_s < \alpha_s$. If $\beta_s > \alpha_s$, the site s is likely to be evolving under positive selective pressure or undergoing adaptive evolution.

In contrast to existing methods that simply have sites varying according to their rates, *HyPhy* allows the user to identify multiple parameters that are free to vary over sites. In the following example, we allow both synonymous and nonsynonymous rates to be variable across sites, leading to the possibility, for instance, that a particular site might have a fast nonsynonymous rate but a slow synonymous rate. We will consider the case of MG94×HKY85_3×4 applied to a codon data set with α_s and β_s sampled independently from two separate distributions. Because only products of evolutionary rates and times can be estimated, we set the mean of the distribution of α_s to one. Widely used models of Nielsen and Yang [9] assume that $\alpha_s = 1$ for every site s; thus our approach is a natural extension. For our example, we choose to sample α_s from a gamma distribution $\gamma(\alpha_s; \mu_\alpha)$ with mean 1 and shape parameter μ_α

discretized into four rate classes by the method of [11]. The nonsynonymous rates β_s are assumed to come from a mixture of a general γ distribution and a point mass at 0 to allow for invariable sites (REF). The density of this distribution is

$$\beta_s \sim R\left[P_I\delta_0(\beta_s) + (1 - P_I)\gamma(\beta_s; \mu_\beta)\right], \tag{6.4}$$

where P_I is the proportion of (nonsynonymous) invariable sites, and R is the mean of the distribution and is the ratio of the means of the nonsynonymous and synonymous distribution (similar to dN/dS). The density of the unit mean gamma distribution with shape parameter μ_β is $\gamma(\beta_s, \mu_\beta)$. The gamma portion of the distribution is discretized into three rates, and, with the invariant rate class, the total number of nonsynonymous rate categories is four.

To perform a maximum likelihood fit of this model in *HyPhy* we follow these steps:

1. Open the data file `p51.nex`.
2. Select all data, create a single partition, and change its data type to codon and its name to RT_Gene.
3. Assign the tree and the model "MG94×HKY85×3_4×2_Rates" with "Rate Het" model parameters and four (per parameter) rate categories. The model we selected implements the extension to the MG94×HKY85_3×4 model we have just discussed.
4. Build the likelihood function and optimize it. Depending on the speed of your computer, this may take up to an hour.

Parameter estimates returned by the analysis are as follows:

```
RT_Gene_Shape_alpha    = 1.637
RT_Gene_Shape_beta_Inv = 0.708
RT_Gene_Shape_beta     = 1.174
RT_Gene_Shared_DNDS    = 0.527
```

HyPhy can also display the discretized distributions along with their continuous originals. This feature can be accessed via the pulldown in menu category variable rows in the parameter table (Figure 6.19). Density plots show the continuous density curve, the table of discrete rate classes, and their visual representations. Dotted lines depict the bounds for the intervals that each rate class (a solid vertical line) represents.

It is immediately clear that synonymous rates are not constant across sites. Indeed, the coefficient of variation for α_s, which is equal to $1/\sqrt{\mu}_\alpha$, is estimated to be 0.61, whereas we would expect a much smaller value were the synonymous rates equal among sites.

An especially interesting task is to determine which sites are conserved and which are evolving under selective pressure. An approach proposed in [14] is to employ the empirical Bayes technique. To do so, we fix all model parameter estimates (more on the validity of that later) and compute the posterior probability $p_{i,j}^s$ of observing rates a_i and b_j at site s. *HyPhy* can compute

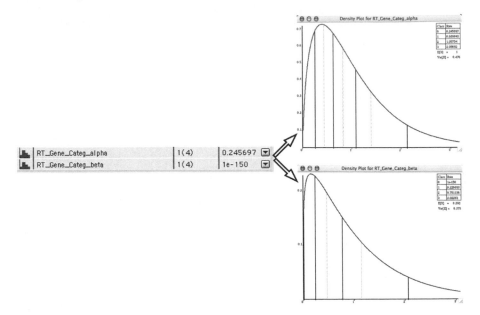

Fig. 6.18. Synonymous and nonsynonymous distributions for the analysis of the HIV-1 RT gene

the conditional likelihoods for every site (choose LIKELIHOOD:CATEGORIES PROCESSOR from the parameter table; see Figure 6.19) given that the rates come from the category i, j:

$$l_{i,j}^s = L(\Theta|\mathcal{T}, \mathcal{D}_s, \alpha_s = a_i, \beta_s = b_j).$$

Application of the Bayes rule yields

$$p_{i,j}^s = \Pr\{\alpha_s = a_i, \beta_s = b_j | \mathcal{D}_s\} = \frac{l_{i,j}^s \Pr\{\alpha_s = a_i, \beta_s = b_j\}}{\sum_{m,n} l_{m,n}^s}.$$

Consider two events at site s: positive selection, $PS_s = \{\alpha_s < \beta_s\}$, and negative or purifying selection, $NS_s = \{\alpha_s > \beta_s\}$. For any event, one can define the Bayes factor, which is simply the ratio of posterior and prior odds of an event. If the Bayes factor of an event is significantly greater than 1, then the data support the event.

Having opened the categories processor (Figure 6.20), we proceed to perform the posterior Bayes analysis as follows:

1. Create a new random variable $\beta_s - \alpha_s$. To do so, invoke CATEGORIES: DEFINE NEW VARIABLE and enter the expression (try to use the pulldown menu for quick access to category variables) $0.527 RT\_Gene\_Categ\_beta - RT\_Gene\_Categ\_alpha$. We multiply by the value of R (= 0.527) since in

Type: None ▼   X: Index ▼   Y: None ▼

Category Variables		Class 1	Class 2	Class 3	Class 4	Class 5	Class 6	Class 7
▽ RT_Gene_Categ_alpha	1	3.60832e-05	2.80549e-05	1.57008e-05	3.95586e-06	7.15948e-05	5.56654e-05	3.114
0.245697 (pr =0.250000)	2	0.0187765	0.0147323	0.00843288	0.00225548	0.0155176	0.0121754	0.006
0.609843 (pr =0.250000)	3	2.19122e-05	1.57329e-05	7.4111e-06	1.31907e-06	5.12999e-05	3.68033e-05	1.719
1.057537 (pr =0.250000)	4	0.00758236	0.00589534	0.00329882	0.000828205	0.00594012	0.00461849	0.002
2.086923 (pr =0.250000)	5	0.0187765	0.0147323	0.00843288	0.00225548	0.0155176	0.0121754	0.006
▽ RT_Gene_Categ_beta	6	0.000173276	0.000138801	8.3901e-05	2.65518e-05	0.000307939	0.000310632	0.000
0.000000 (pr =0.708158)	7	0.0139388	0.0112641	0.00689671	0.00215647	0.0109199	0.00882444	0.005
0.226033 (pr =0.097281)	8	0.0330851	0.0267265	0.0163531	0.0051144	0.0272804	0.0220374	0.013
	9	0.0169966	0.013215	0.00739485	0.00185787	0.0140146	0.0108964	0.006
	10							

Fig. 6.19. Conditional site likelihoods module of *HyPhy*.

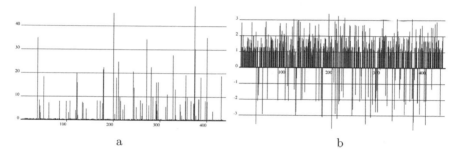

a   b

Fig. 6.20. (a) Bayes factor for the event of positive selection at a site. (b) Log of the Bayes factor for the event of negative selection at a site

the *HyPhy* parameterization RT_Gene_Categ_beta refers to the expression inside the brackets in (6.4)—you can check that by opening the model display in "Object Inspector."

2. Expand the view for the new difference variable by clicking on the arrow next to it, and choose (shift-click or drag select) the event for positive selection: all positive values of the difference variable.

3. Perform empirical Bayes analysis by selecting CATEGORIES:EVENT POSTERIORS. In the window that opens, select a type of "Bar Chart" and Y of "Bayes Factor." This display gives an easy overview of sites with large support for positive selection, say, with Bayes factor over 20.

4. Instruct *HyPhy* to find all the sites with the Bayes factor over 20. For this task, select the Bayes factor column (click on the column header), and choose CHART:DATA PROCESSING:SELECT CELLS BY VALUE. *HyPhy* will prompt for the selection criterion: type in "cell_value>20." The results are shown in Figure 6.21. According to this criterion, there are 12 positively selected codons: $35, 178, 179, 200, 211, 243, 272, 282, 329, 376, 377,$ and 403.

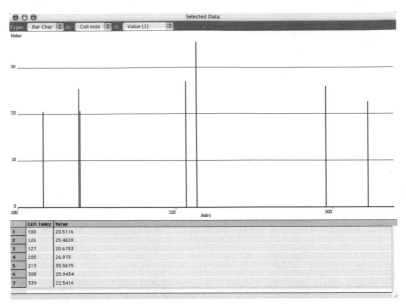

Fig. 6.21. Sites found to be under positive selection and supporting Bayes factors in the HIV-1 RT gene.

The weakness of empirical Bayes

It has been argued that maximum likelihood empirical Bayes methods for detecting rates at sites may yield many false positives. Alternatively, if very few sites in the alignment are under selective pressure, it is possible that the prior (and hence posterior) distributions will place zero probability on *any* site being positively selected, resulting in low power. The main shortcoming of empirical Bayes approaches is that parameter estimates are treated as correct values, and the uncertainties in estimation procedures are discounted altogether. If one were to compute 95% confidence intervals based on likelihood profiles with *HyPhy*, one would discover that

$$\mu_\alpha \in (0.759, 10.175), \mu_\beta \in (0.589, 3.467),$$

$$P_I \in (0.642, 0.779), R \in (0.405, 0.688).$$

That is quite a range of variation, and a change in any of those parameters would affect the conclusions of empirical Bayes methods. For instance, the most conservative (in terms of limiting false positives but also reducing power) estimates can be obtained by choosing the maximum possible values for μ_α, μ_β, and P_I and the minimum possible value for R. For this choice of parameters, the maximum Bayes factor at any site is a mere 17.914 and by our old criteria no sites are found to be under selective pressure. One should always realize that uncertainties in parameter estimates can greatly influence the conclusions of an empirical Bayes analysis, and it helps to compare various scenarios to assess inference reliability.

Further pointers

HyPhy can run analyses like the one just described in parallel on distributed systems using Message Passing Interface (MPI). For instance, if 16 processors are available, computations of $l_{i,j}^s$ for each of the 16 possible rate class combinations (i,j) are placed automatically on separate processors, achieving speeds similar to those of a single rate analysis on a single CPU system and making analyses with hundreds of sequences in an alignment feasible. Refer to www.hyphy.org for more details.

HyPhy also implements an ever-expanding collection of rapid positive/ negative selection analyses for data exploration loosely based on the counting method of [10], as well as site-by-site (and/or lineage-specific) likelihood ratio testing. It is accessible via standard analyses, and more details can be found in the *HyPhy* documentation.

6.2.7 Mixed Data Analyses

As more and more organisms are being fully sequenced, methods and tools for analyzing multigene sequence alignments and, ultimately, genome-wide data sets are becoming increasingly relevant. In the small example that follows, we will show how one can use *HyPhy* to begin to address such analytic needs.

We consider a sequence alignment of five sequences, each consisting of two introns and an exon, which can be found in intronexon.nex within the Examples directory. We must partition the data into introns and exons. As a first pass, it is appropriate to consider two partitions: coding and noncoding. For more complex data sets, one can easily define a separate partition for every gene, and so on. First, create a partition that includes all of the data (EDIT:SELECT ALL, followed by DATA:SELECTION->PARTITION).

The exon spans nucleotide positions 90 through 275. One of the ways to create the partition for the exon is to locate alignment column 90 in the data panel and select it, and then scroll to column 275 and shift-click on it (this selects the whole range). Note that the status line of the data panel was updated to reflect your current selection. Make sure it shows "Current Selection: 90–275." An alternative approach is to start at column 90 and then click-drag to column 275. Yet another possibility is to choose DATA:INPUT PARTITION and enter 89–274 (indices are 0-based).

Once the range has been selected, invoke DATA:SELECTION->PARTITION. We now have two partitions, overlapping over columns 90–275, as shown in the Navigation Bar. The final step is to "subtract" the partitions to create a new partition for the introns. To do this, we select both partition rows in the data panel table (shift-click selects multiple rows). Next, click on the "Subtract 2 Overlapping Partitions" button. Select the appropriate operation in the resulting pulldown menu. We have now specified two nonoverlapping partitions. Note that the intron partition is not contiguous. Rename the intron partition to "Introns" and the exon partition to "Exon." One could achieve this same partitioning scheme by defining three partitions, 1–89, 90–275, 276–552, and joining the first one and the third one.

There is one more filtering step left to do before we can begin analyzing the data. As often happens with smaller subsets extracted from larger alignments, there are several alignment columns consisting entirely of deletions. Such columns do not contribute informational content to likelihood analyses and should be removed. Select the "Exon" row in the partition table, click on the "Data Operations" button, and select SITES WITH ALL DELETIONS. *HyPhy* will locate all such sites *inside the selected partition only* and select them. Create a partition with those sites, subtract it from the exon partition as discussed above, and delete the partition with uninformative sites (select its row and click on the "Delete Partition" button).

Since introns are not subject to the functional constraints of coding sequences, it makes sense to model their evolution with a nucleotide model (HKY85 with global options). For the exon partition, a codon model is appropriate. Change the data type of "Exon" to "Codon" and apply the MG94×HKY85×3_4 model with local options. The end result should look like Figure 6.22 (a).

Next, build and optimize the likelihood function and open the parameter table. Our analysis includes two trees with the same topology (one for introns and the other for exons). The model for the intron tree has a single parameter per branch (branch length) and a shared transversion/transition ratio ($Exon\_Shared\_TVTS = 0.308$), whereas the model for the exon tree has two parameters per branch, synonymous and nonsynonymous rates, and a shared transversion/transition ratio ($Introns\_Shared\_TVTS = 0.612$). (Note that we could use previously discussed methods for testing hypotheses to decide whether the two transversion/transition ratios are different.)

One of the common assumptions made for analyses of molecular sequence data is that differences between coding and noncoding sequences can be explained by functional constraints and selective pressures on coding sequences, namely by changes in rates of nonsynonymous substitutions. In other words, synonymous substitutions in coding regions and nucleotide substitutions in neighboring noncoding stretches should have comparable relative rates. This assumption may be violated if mutation rates vary along the sequence or if there is selection operating in noncoding regions. We will now test this hypothesis of a relative ratio between the introns and the exon in our example data sets. In other words, we want to see if the exon tree scaled by synonymous rates has the same pattern of relative branch lengths as the intron tree. Mathematically, the set of relative ratio constraints is

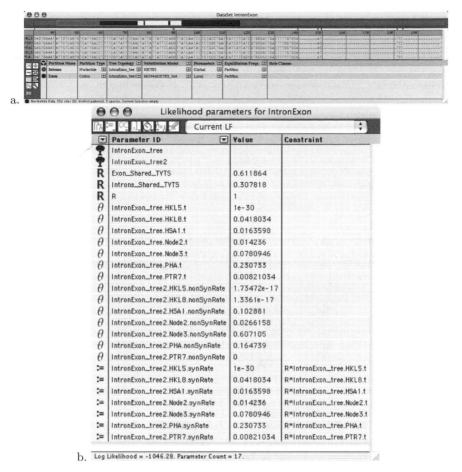

Fig. 6.22. Exon-intron mixed analysis. (a) Data panel setup and (b) parameter table with the relative ratio constraint.

$$exonTree.branch.synRate = R \times intronTree.branch.t,$$

where R is the (global) relative ratio, and the constraint is applied to every branch. For a small tree like ours, it is easy to use the proportional constraint tool in the parameter table interface module to define the constraints one at a time; however, this could become very tedious for larger trees. Luckily, *HyPhy* includes a command designed to traverse given trees and apply the same constraint to every branch. As you will learn from the next section, at the core *HyPhy* is a programming language (HBL), and all of the interface features we have discussed previously use HBL behind the scenes. If the interface does not include a built-in tool for a specific constraint, the user may tap directly into HBL to carry out the task at hand. We will do just that for our example.

Open the parameter table for the intron-exon analysis we have just set up (making sure none of the parameters are constrained). Invoke LIKELIHOOD: ENTER COMMAND. *HyPhy* will take any input from the dialog box that appears, parse the commands contained therein, and execute them. We need to invoke *ReplicateConstraint*, which is a powerful but somewhat complicated command. If we were to impose the constraints by hand at every branch, we would begin with

$$IntronExon\_tree2.HKL5.synRate = R \times IntronExon\_tree.HKL5.t$$

and repeat applying the same constraint, replacing "HKL5" with other branches in the tree. A single call using *ReplicateConstraint* will accomplish the same task:

```
global R = 1;
ReplicateConstraint("this1.?.synRate:=R*this2.?.t",
    IntronExon_tree2,IntronExon_tree);
```

The expression in quotation marks is the constraint *template*; "this1" is replaced with the first argument (IntronExon_tree2), "this2" with the second, and so on. The "?" is a wildcard meaning *match any branch name*. *ReplicateConstraint* is a very handy command to know, and we refer the reader to examples contained in the *HyPhy* distribution. The "global R=1" command is needed to declare R as a shared parameter and initialize it (further details are provided in the next section). Enter the commands above into the dialog box, and, if all went well, the parameter table will update and should look like Figure 6.22 (b). Optimize the likelihood function, define the null hypothesis, and perform the likelihood ratio test. The asymptotic p-value of the test is 0.023, rejecting the hypothesis of relative ratio. Since our data set is rather small, we would be wise to verify this result using the parametric bootstrap. We obtained a bootstrap p-value of 0.003 with 1000 replicates.

6.3 The *HyPhy* Batch Language

Underlying the *HyPhy* graphical user interface is a powerful interpreted programming language, HBL (*HyPhy* Batch Language). The authors originally developed HBL as a research tool to allow rapid development of molecular evolutionary analyses. The addition of the graphical interface is a more recent development and provides access to many common types of analyses. However, the underlying programming language is considerably more powerful and flexible (albeit with a steeper learning curve). The goal of this section is to provide readers with a basic understanding of the fundamentals of HBL programming and an appreciation of the power of the language. In doing so, we shall make use of a series of *HyPhy* batch files, which are available for download at www.hyphy.org/pubs/HyphyBookChapter.tgz. Complete documentation of the batch language is available in the Batch Language Command

Reference at www.hyphy.org and can also be accessed via the built-in command reference in the *HyPhy* console.

6.3.1 Fundamental Batch File Elements: *basics.bf*

The basic task shared by most *HyPhy* batch files is the optimization of a likelihood function for a given alignment/model/phylogeny combination. Therefore, almost every batch file will perform the following elementary tasks:

1. Input alignment data.
2. Describe an evolutionary model of sequence change.
3. Input or describe a phylogenetic tree.
4. Define a likelihood function based on the alignment, phylogeny, and model.
5. Maximize the likelihood function.
6. Print the results to the screen and/or an output file.

The simple batch file *basics.bf*, reproduced in its entirety below, illustrates the HBL code necessary to fit the F81 model of sequence evolution to an alignment of four sequences.

```
DataSet myData = ReadDataFile ("data/four.seq");
DataSetFilter myFilter = CreateFilter (myData,1);
HarvestFrequencies (obsFreqs, myFilter, 1, 1, 1);
F81RateMatrix =
                {{* ,mu,mu,mu}
                {mu,* ,mu,mu}
                {mu,mu,* ,mu}
                {mu,mu,mu,* }};
Model F81 = (F81RateMatrix, obsFreqs);
Tree myTree = ((a,b),c,d);
LikelihoodFunction theLikFun = (myFilter, myTree);
Optimize (MLEs, theLikFun);
fprintf (stdout, theLikFun);
```

Let us now explain how these nine statements accomplish the six key tasks enumerated above.

Input alignment data

The task of preparing data for analysis in *HyPhy* consists of two steps. First, the data must simply be read from a data file. After the data are read, they must be "filtered." The process of filtering involves selecting the precise taxa and alignment positions to be analyzed and identifying the "type" of the data (e.g., nucleotide, codon, dinucleotide).

```
DataSet myData = ReadDataFile ("data/four.seq");
DataSetFilter myFilter = CreateFilter (myData,1);
```

The first statement simply reads a sequence alignment into memory and names it *myData*. The HBL function automatically detects the sequence type (DNA) and the input format and then saves the data into a data structure of type *DataSet*, a predefined HBL data type. The second statement is the simplest version of the *CreateFilter* function. In this case, the function takes the alignment stored in *myData* and by default includes all of it in a structure named *myFilter*. The argument "1" indicates that the data should be treated as simple nucleotide data. Had we wanted the data to be interpreted as codons, the argument "3" would have been used instead. The *CreateFilter* command is quite powerful, and we will illustrate the use of some of its optional arguments in later examples. Multiple data filters may be created from the same data set.

Describe an evolutionary model of sequence change

The next task in our simple analysis is the definition of a model of sequence change. One of the unique strengths of *HyPhy* is its ability to implement any special case of a general time-reversible model (and, more generally, any continuous-time Markov chain model, not necessarily time-reversible), regardless of the dimensions of the character set. We rely on the fact that any special case of the general reversible model can be written in a form where entries in the substitution matrix are products of substitution parameters and character frequencies. Thus, we have adopted a convention of describing time-reversible models with two elements: a matrix consisting of substitution rate parameters, and a vector of equilibrium character frequencies.

```
F81RateMatrix =
                {{* ,mu,mu,mu}
                {mu,* ,mu,mu}
                {mu,mu,* ,mu}
                {mu,mu,mu,* }};
HarvestFrequencies (obsFreqs, myFilter, 1, 1, 1);
Model F81 = (F81RateMatrix, obsFreqs);
```

In our present example, the substitution parameter matrix of the F81 model is defined and named in an obvious fashion (the *HyPhy* matrix placeholder * is defined as "the negative sum of all nondiagonal entries on the row"). Next, the built-in function *HarvestFrequencies* tabulates the frequencies in *myFilter* and stores them in the newly created vector *obsFreqs*. The functions of the numerical arguments can be found in the Batch Language Command Reference. Finally, the matrix and frequencies are combined to form a valid substitution model using the *Model* statement.

For the F81 model, the instantaneous rate matrix is traditionally denoted

$$
\begin{array}{cccc}
& A & C & G & T \\
\begin{array}{c} A \\ C \\ G \\ T \end{array}
& \left(\begin{array}{c} -\mu(1-\pi_A) \\ \mu\pi_A \\ \mu\pi_A \\ \mu\pi_A \end{array} \right.
& \begin{array}{c} \mu\pi_C \\ -\mu(1-\pi_C) \\ \mu\pi_C \\ \mu\pi_C \end{array}
& \begin{array}{c} \mu\pi_G \\ \mu\pi_G \\ -\mu(1-\pi_G) \\ \mu\pi_G \end{array}
& \left. \begin{array}{c} \mu\pi_T \\ \mu\pi_T \\ \mu\pi_T \\ -\mu(1-\pi_T) \end{array} \right) .
\end{array}
$$

Observe the similarity between this matrix and the *HyPhy* syntax. By default, the *Model* statement multiplies each element of the rate matrix by the equilibrium frequency of an appropriate character, and hence the *HyPhy* declaration of F81 does not include the multiplication by elements of π. This behavior can be overridden by passing a third argument of 0 to the model statement (as is done, for example, for the original MG94 codon model).

Input or describe a phylogenetic tree

HyPhy uses standard (Newick) tree definitions. Thus, the statement

```
Tree myTree = ((a,b),c,d);
```

defines a tree named *myTree* with four OTUs, or taxa, named a, b, c, and d, corresponding to the names in the *HyPhy* data file. *HyPhy* will accept either rooted or unrooted trees; however, for most purposes, rooted trees are automatically unrooted by *HyPhy* because likelihood values for unrooted trees are the same as those for rooted trees.

The *Tree* data structure is much more complex than simply describing a tree topology. The *Tree* variable includes both topology information and evolutionary model information. The default behavior of a *Tree* statement is to attach the most recently defined *Model* to all branches in the tree. Thus, it is often *critical* that the *Model* statement appear before the *Tree* statement. We will discuss more advanced uses of the *Tree* statement later.

Define a likelihood function based on the alignment, phylogeny, and model

The likelihood function for phylogenetic trees depends on the data set, tree topology, and the substitution model (and its parameters). To define a likelihood function, we use a statement such as

```
LikelihoodFunction theLikFun = (myFilter,myTree);
```

We name the likelihood function *theLikFun*, and it uses the data in *myFilter* along with the tree topology and substitution model stored in *myTree*. Recall that the *Tree* structure *myTree* inherited the *Model F81* by default.

Maximize the likelihood function

Asking *HyPhy* to maximize the likelihood function is simple. The statement

```
Optimize (MLEs, theLikFun);
```

finds maximum likelihood estimates of all independent parameters and stores the results in the matrix named *MLEs*.

Print the results to the screen and/or an output file

The simplest way to display the results of a likelihood maximization step is simply to print the likelihood function:

```
fprintf(stdout,theLikFun);
```

This C-like command prints the structure *theLikFun* to the default output device *stdout* (stdout is typically the screen). The results of this statement are the following:

```
Log Likelihood = -616.592813234418;
Tree myTree=((a:0.0136035,b:0.0613344)Node1:0.0126329,
c:0.070388,d:0.0512889);
```

When asked to print a likelihood function, *HyPhy* first reports the value of the log-likelihood. It follows with a modified version of the Newick tree description as shown in the output above. Each of the branches in the unrooted phylogeny has an associated branch length, measured in units of expected number of nucleotide substitutions per site. Those values appear after the colon following the label for each branch. For example, the estimated branch length leading to the tip "b" is 0.0613344. Note that the internal node in the tree has been automatically named "Node1" by *HyPhy*, and its associated branch length is 0.0126329. Values of the estimated substitution parameters or base frequencies could be displayed by printing *MLEs* or *obsFreqs*. *HyPhy* also allows for more detailed user control of printed output using a C-like fprintf syntax. Later examples will illustrate this functionality.

6.3.2 A Tour of Batch Files

Defining substitution models

Simple nucleotide models: modeldefs.bf

One of the primary objectives of *HyPhy* is to free users from relying on the substitution models chosen by authors of software. While a relatively small set of model choices may be sufficient for performing phylogenetic analyses, having only a few potential models is often limiting for studies of substitution rates and patterns. To define a model in *HyPhy*, one needs only to describe the elements in a substitution rate matrix. If the characters being studied have n states, the rate matrix is $n \times n$. For example, nucleotide models are 4×4; models of amino acid change are 20×20; codon-based models might be 61×61. *HyPhy* can work properly with any member of the class of general time-reversible models, regardless of the number of character states. Instantaneous rate matrices in this class of models satisfy the condition $\pi_i Q_{ij} = \pi_j Q_{ji}$, where π_i is the equilibrium frequency of character i (for nucleotide data) and Q_{ij} is the ijth entry in the instantaneous rate matrix. *HyPhy* comes with many predefined rate matrices for commonly used substitution models. You can find examples in the *Examples* and *TemplateBatchFiles* directories of the *HyPhy* distribution.

To illustrate the basics of model definition, we discuss the batch file *modeldefs.bf*:

```
SetDialogPrompt("Select a nucleotide data file:");
DataSet myData   = ReadDataFile(PROMPT_FOR_FILE);
DataSetFilter myFilter = CreateFilter(myData,1);
HarvestFrequencies(obsFreqs,myFilter,1,1,1);
F81RateMatrix  = {{*,m,m,m}{m,*,m,m}{m,m,*,m}{m,m,m,*}};
Model F81 = (F81RateMatrix, obsFreqs); Tree myTree = ((a,b),c,d);
fprintf(stdout,"\n\n F81 Analysis \n\n");
LikelihoodFunction theLikFun = (myFilter, myTree);
Optimize(results,theLikFun);
fprintf(stdout,theLikFun);


fprintf(stdout,"\n\n HKY85 Analysis  \n\n");
HKY85RateMatrix = {{*,b,a,b}{b,*,b,a}{a,b,*,b}{b,a,b,*}};
Model HKY85 = (HKY85RateMatrix, obsFreqs);
Tree myTree = ((a,b),c,d);
LikelihoodFunction theLikFun = (myFilter, myTree);
Optimize(results,theLikFun);
fprintf(stdout,theLikFun);


fprintf(stdout,"\n\n Repeat F81 Analysis  \n\n");
UseModel(F81);
Tree myTree = ((a,b),c,d);
LikelihoodFunction theLikFun = (myFilter, myTree);
Optimize(results,theLikFun);
fprintf(stdout,theLikFun);
```

This batch file illustrates two new concepts. First, and most importantly, the lines

```
HKY85RateMatrix = {{*,b,a,b}{b,*,b,a}{a,b,*,b}{b,a,b,*}};
Model HKY85 = (HKY85RateMatrix, obsFreqs);
```

illustrate the definition of a new substitution matrix. In this case, we have defined the model of [5] and named the model HKY85. Those familiar with the HKY85 model will probably recognize the form of the matrix: transitions occur with rate a and transversions occur with rate b, with each of those substitution parameters multiplied by the appropriate nucleotide frequency to provide the final instantaneous rates. The second important point to note is that we must associate the model with a tree before we can do anything useful. In this case, we simply redefined the old tree to use the HKY85 model instead of the F81 model. (Recall that a tree consists of both the topology and the substitution matrices attached to its branches.) When the statement `Tree myTree = ((a,b),c,d);` is issued, the variable *myTree* is assigned the topology ((a,b),c,d) and the branches are assigned the HKY85 substitution model, which was the most recently defined *Model*. If we wanted to preserve the original variable *myTree*, we could simply have defined a new *Tree* structure using a command such as `Tree myNextTree = ((a,b),c,d);`.

Finally, for completeness, we created a new *Tree* and assigned it the F81 model and reproduced the original F81 analysis. Those final steps illustrate how predefined *Models* can be assigned to *Trees* using the *UseModel* command.

Note also the use of

```
SetDialogPrompt("Select a nucleotide data file:");
DataSet myData   = ReadDataFile(PROMPT_FOR_FILE);
```

to allow the user to locate the sequence file interactively instead of hard-coding it into the batch file.

More nucleotide models: models.bf

One of the most general models of nucleotide substitution is the general time reversible model (REV). The instantaneous rate matrix for the REV model is

$$Q_{\text{REV}} = \begin{array}{c} \\ A \\ C \\ G \\ T \end{array} \begin{pmatrix} \star & \theta_0 \pi_C & \theta_1 \pi_G & \theta_2 \pi_T \\ \theta_0 \pi_A & \star & \theta_3 \pi_G & \theta_4 \pi_T \\ \theta_1 \pi_A & \theta_3 \pi_C & \star & \theta_5 \pi_T \\ \theta_2 \pi_A & \theta_4 \pi_C & \theta_5 \pi_G & \star \end{pmatrix}.$$

It is simple to implement this model in *HyPhy*. The statements

```
REVRateMatrix  = {{*,a,b,c}{a,*,d,e}{b,d,*,f}{c,e,f,*}};
Model REV = (REVRateMatrix, obsFreq);
```

do the job.

To illustrate these notions in a more useful context, consider the batch file *models.bf*. In that batch file, models named F81, HKY85, REV, JC69, and K2P are defined, and each is fit to the same data set and tree topology. The batch file *models.bf* also demonstrates a few useful *HyPhy* features. First, notice the definition of a vector of frequencies for use by the equal-frequency models:

```
equalFreqs = {{0.25}{0.25}{0.25}{0.25}};
```

In a similar manner, we define the string constant *myTopology*:

```
myTopology = "((a,b),c,d)";
```

By changing the topology in the definition of *myTopology*, the entire analysis can be repeated using the new topology. This single step is faster than updating the topology for every *Tree* statement and is particularly useful for topologies with many taxa. Finally, note the reuse of the three substitution matrices and the two frequency vectors. The original matrix definitions are used as templates by the *Model* statements.

Global versus local parameters: *localglobal.bf*

Because the primary goal of *HyPhy* is to provide flexible modeling of the nucleotide substitution process, *HyPhy* includes a more general parameterization scheme than most phylogeny estimation programs. Perhaps the most important difference for the user to recognize is the distinction between *local* and *global* parameters. In the simplest form, a local parameter is one that is specific for a single branch on a tree. In contrast, a global parameter is shared by all branches. To illustrate, consider the output generated by the batch file *localglobal.bf* when run using *four.seq*:

```
Original (Local) HKY85 Analysis

Log Likelihood = -608.201788537279;
Tree myTree=((a:0.0143364,b:0.061677)Node1:0.0108616,
c:0.0716517,d:0.0526854);


Global HKY85 Analysis

Log Likelihood = -608.703204177757;
Shared Parameters: S=3.08185

Tree myTree=((a:0.0130548,b:0.0618834)Node1:0.0126785,
c:0.0717394,d:0.052028);
```

In *localglobal.bf*, we have moved beyond the default settings of *HyPhy*, and the details of the batch file will be discussed below. For now, concentrate on the results. *localglobal.bf* performs two analyses of the data in *four.seq*, each using the HKY85 model of sequence evolution. The first, labeled "Original (Local) HKY85 Analysis," is the same analysis that was performed in the previous example (*models.bf*). In this analysis, each branch in the tree was allowed to have its own transition/transversion ratio.

The second analysis performed in *localglobal.bf* is an example of a *global* analysis. In contrast with the previous analysis, the "Global HKY85 Analysis" invokes a *global* transition/transversion ratio, S. In other words, all branches share the same value of S. The estimated global value of S (3.08185) is shown under the heading of Shared Parameters.

The local and global analysis use different numbers of parameters. The local analysis uses a transition and transversion rate for each of the five branches, along with three base frequencies, for a total of 13 parameters. The global analysis includes a transversion rate for each branch, three base frequencies, and a single transition/transversion ratio, for a total of nine parameters. The global analysis is a special case of the local analysis; therefore, the log-likelihood value for the global analysis (-608.703) is lower than that of the local analysis (-608.202). The fact that the addition of four parameters results in such a small difference in model fit suggests that the data harbor little support for the hypothesis that the transition/transversion rate varies among these lineages.

The code for *localglobal.bf* is the following:

```
SetDialogPrompt ("Please specify a nucleotide data file:");

DataSet myData    = ReadDataFile(PROMPT_FOR_FILE);
DataSetFilter myFilter = CreateFilter(myData,1);
HarvestFrequencies(obsFreqs,myFilter,1,1,1);

fprintf(stdout,"\n\n Original (Local) HKY85 Analysis  \n\n");
HKY85RateMatrix = {{*,b,a,b}{b,*,b,a}{a,b,*,b}{b,a,b,*}};
Model HKY85 = (HKY85RateMatrix, obsFreqs);
Tree myTree = ((a,b),c,d);
LikelihoodFunction theLikFun = (myFilter, myTree);
Optimize(results,theLikFun);
fprintf(stdout,theLikFun);

fprintf(stdout,"\n\n Global HKY85 Analysis  \n\n");
global S=2.0;
GlobalHKY85Matrix = {{*,b,b*S,b}{b,*,b,b*S}
                     {b*S,b,*,b}{b,b*S,b,*}};
Model GlobalHKY85 = (GlobalHKY85Matrix, obsFreqs);
Tree myTree = ((a,b),c,d);
LikelihoodFunction theLikFun = (myFilter, myTree);
Optimize(results,theLikFun);
fprintf(stdout,theLikFun);
```

The code for the first analysis is identical to that from *models.bf*. The global analysis introduces a new statement:

```
global S=2.0;
```

This statement declares S to be a global variable. By default, the description of a model (and variables within that model) is used as a template that is copied for every branch on the tree. An important fact is that we cannot later redefine S as a local variable. The scope of a variable is determined at the time of its creation and cannot be altered. In the statement defining *GlobalHKY85Matrix*, one observes that b is used as the transversion rate, while transitions occur at rate $b * S$.

More complex models

HyPhy has support for an infinite number of substitution models. Any Markov chain model using any finite sequence alphabet can be used. Models for codon and amino acid sequences are available through the Standard Analyses menu selection. We refer users who are interested in writing code for such alphabets to the files in the *Examples* subdirectory.

Imposing constraints on variables

Simple constraints: relrate.bf

The primary reason for developing *HyPhy* was to provide a system for performing likelihood analyses on molecular evolutionary data sets. In particular, we wanted to be able to describe and perform likelihood ratio tests (LRTs) easily. In order to perform an LRT, it is first necessary to describe a constraint, or series of constraints, among parameters in the probability model. To illustrate the syntax of parameter constraints in *HyPhy*, examine the code in *relrate.bf*:

```
SetDialogPrompt("Select a nucleotide data file:");
DataSet myData = ReadDataFile (PROMPT_FOR_FILE);
DataSetFilter myFilter = CreateFilter (myData,1);
HarvestFrequencies (obsFreqs, myFilter, 1, 1, 1);
F81RateMatrix = {{* ,mu,mu,mu}{mu,* ,mu,mu}
{mu,mu,* ,mu}{mu,mu,mu,* }};
Model F81 = (F81RateMatrix, obsFreqs);
Tree myTree = (a,b,og);

fprintf(stdout,"\n Unconstrained analysis:\n\n");
LikelihoodFunction theLikFun = (myFilter, myTree, obsFreqs);
Optimize (paramValues, theLikFun);
fprintf (stdout, theLikFun);
lnLA=paramValues[1][0];
dfA=paramValues[1][1];

fprintf(stdout,"\n\n\n Constrained analysis:\n\n");
```

```
myTree.a.mu := myTree.b.mu;
Optimize (paramValues, theLikFun);
fprintf  (stdout, theLikFun);
lnL0=paramValues[1][0];
df0=paramValues[1][1];

LRT=-2*(lnL0-lnLA);
Pvalue=1-CChi2(LRT,dfA-df0);
fprintf(stdout,"\n\nThe statistic ",LRT," has P-value ",
Pvalue,"\n\n");
```

The unconstrained analysis is of the simple type we have discussed previously. In the constrained analysis, however, we impose the constraint of equal substitution rates between lineages a and b with the command

```
myTree.a.mu := myTree.b.mu;
```

The results from this batch file when applied to *three.seq* are:

```
 Unconstrained analysis:

Log Likelihood = -523.374642786834;
Tree myTree=(a:0.0313488,b:0.00634291,og:0.11779);


 Constrained analysis:

Log Likelihood = -525.013303516343;
Tree myTree=(a:0.018846,b:0.018846,og:0.116881);

The statistic 3.27732 has P-value 0.0702435
```

Since these models are nested, we can consider the likelihood ratio statistic, $-2(\ln L_0 - \ln L_A)$, to have an asymptotic chi-squared distribution. In this case, the test statistic has a value of 3.27732. Note in the batch file how the likelihood values and parameter counts are returned by *Optimize* and stored in *paramValues*. The built-in function *CChi2* is the cumulative distribution function of the chi-squared distribution.

Molecular clocks

Perhaps the most common molecular evolutionary hypothesis tested is that a set of sequences has evolved according to a molecular clock. It now seems quite clear that a global molecular clock exists for few, if any, gene sequences. In contrast, the existence of local molecular clocks among more closely related species is more probable. *HyPhy* allows for both types of constraints, including the possibility of testing for multiple local clocks for different user-defined clades in the same tree.

Global clocks. molclock.bf

The batch file *molclock.bf* is a simple example of testing for a global molecular clock. The code should be familiar, except for the new `MolecularClock` statement, which declares that the values of the parameter *mu* should follow a molecular clock on the entire tree *myTree*. An important difference in this batch file is that the `Tree` statement defines a rooted tree. Had an unrooted tree been used, it would have been treated as a rooted tree with a multifurcation at the root. When using time-reversible models, which can't resolve the exact placement of the root on the internal rooting branch, a global molecular clock applied to a rooted tree can be interpreted as: *locate the root on the root branch as to enforce a global molecular clock on the specified rates.* The section of code imposing the molecular clock constraint is:

```
fprintf(stdout,"\n\n Molecular Clock Analysis: \n");
MolecularClock(myTree,m);
LikelihoodFunction theLikFun = (myFilter, myTree);
Optimize(results,theLikFun);
```

Local clocks: localclocks.bf

Particularly when studying data sets consisting of many species spanning a wide level of taxonomic diversity, it may be of interest to assign local molecular clocks to some clades. For instance, in a study of mammalian molecular evolution, one might specify that each genus evolves in a clocklike manner but that different genera evolve at different rates. To allow such analyses, the *MolecularClock* command can be applied to any node on a tree. Unlike the global clock of the previous case, it is not necessary for the *MolecularClock* command to be applied to a rooted tree; the placement of the *MolecularClock* command "roots" the tree, at least locally. To illustrate this feature, we use *localclocks.bf* in conjunction with the file *six.seq*. The relevant new sections of the code are the tree topology definition

```
myTopology = "(((a,b)n1,(c,(d,e))n2),f)";
```

and the declaration of two local molecular clocks:

```
fprintf(stdout,"\n\n Local Molecular Clock Analysis: \n");
ClearConstraints(myTree);
MolecularClock(myTree.n1,m);
MolecularClock(myTree.n2,m);
LikelihoodFunction theLikFun = (myFilter, myTree);
Optimize(results,theLikFun);
```

The topology string used in *localclocks.bf* takes advantage of *HyPhy*'s extended syntax. Notice how we have named two of the internal nodes *n1* and *n2*. Those names override *HyPhy*'s default (and rather cryptic) node-naming convention and allow us to call functions—in this case, *MolecularClock*—on the clades they tag. The syntax of the *MolecularClock* statements is rather C-like. MolecularClock(myTree.n1,m); imposes a local clock on the clade rooted at node *n1* in tree *myTree*. The parameter with clocklike behavior is *m*, the only option for the F81 model being used. The results using the data file *six.seq* are:

```
UNCONSTRAINED ANALYSIS:
Log Likelihood = -685.473598259084;
Tree myTree=((a:0.0296674,b:0.00831723)n1:0.040811,
(c:0.0147138,(d:0.0142457,e:0.0328603)
Node7:0.0309969)n2:0.0130927,f:0.0517146);


GLOBAL MOLECULAR CLOCK ANALYSIS:
Log Likelihood = -690.857603506283;
Tree myTree=((a:0.0181613,b:0.0181613)n1:0.0350919,
(c:0.0385465,(d:0.0195944,e:0.0195944)
Node7:0.0189521)n2:0.0147067,f:0.053838);


P-value for Global Molecular Clock Test: 0.0292988


LOCAL MOLECULAR CLOCK ANALYSIS:
Log Likelihood = -690.761234081996;
Tree myTree=((a:0.0190659,b:0.0190659)n1:0.0386549,
(c:0.0370133,(d:0.0189116,e:0.0189116)
Node7:0.0181017)n2:0.0128865,f:0.0537045);


P-value for Local Molecular Clock Test: 0.0142589
```

By examining the output, one finds that under the local clock model the two subtrees do indeed have clocklike branch lengths, yet the tree as a whole is not clocklike. However, the likelihood ratio test suggests that neither the global nor local clock assumption is correct.

Simulation tools

The use of simulation in molecular evolutionary analysis has always been important. Simulation allows us to test statistical properties of methods, to assess the validity of theoretical asymptotic distributions of statistics, and to study the robustness of procedures to underlying model assumptions. More recently, methods invoking simulation have seen increased use. These techniques include numerical resampling methods for estimating variances or for computing confidence intervals, as well as parametric bootstrap procedures for estimating p-values of test statistics. *HyPhy* provides both parametric and nonparametric simulation tools, and examples of both are illustrated in the following sections.

The bootstrap: bootstrap.bf

The bootstrap provides, among other things, a simple nonparametric approach for estimating variances of parameter estimates. Consider *bootstrap.bf*. The relevant commands from the batch file are as follows. (Some lines of code have been deleted for clarity.)

```
Model F81 = (F81RateMatrix, obsFreqs);
Tree myTree = (a,b,og);
LikelihoodFunction theLikFun = (myFilter, myTree);
Optimize (paramValues, theLikFun);

reps = 100;

for (bsCounter = 1; bsCounter<=reps; bsCounter = bsCounter+1) {
    DataSetFilter bsFilter = Bootstrap(myFilter,1);
    HarvestFrequencies (bsFreqs, bsFilter, 1, 1, 1);
    Model bsModel = (F81RateMatrix, bsFreqs);
    Tree bsTree = (a,b,og);
    LikelihoodFunction bsLik = (bsFilter, bsTree);
    Optimize (bsParamValues, bsLik);
}
```

The first section of code is simply the completion of a typical data analysis, storing and printing results from the analysis of data in *myFilter*. The *for* loop is the heart of the batch file. For each of the *reps* replicates, we generate a new *DataSetFilter* named *bsFilter*. We do this by creating a bootstrap replicate from the existing *DataSetFilter* named *bsFilter*, which was created in the normal fashion. *bsFilter* will contain the same number of columns as *myFilter*. Once the new filter has been created, we recreate a *Model* named *bsModel* and a *Tree* named *bsTree*, which are then used in an appropriate *LikelihoodFunction* command. *Optimize* is used to find MLEs of the parameters. The end result of this batch file is a table consisting of 100 sets of MLEs, each from a bootstrap sample from the original data. Notice in the complete batch file (not shown in the code above) how we use the matrix variable *BSRes* to tabulate and report the average of all bootstrap replicates. More complex analyses, such as bootstrap confidence intervals, based on the bootstrap estimates, can be programmed within the batch file, or the results can be saved and imported into a spreadsheet for statistical analyses.

The *Permute* function, with syntax identical to *Bootstrap*, exists for applications where the columns in the existing *DataSetFilter* must appear exactly once in each of the simulated data sets. This feature may be useful for comparison of the three codon positions or for studies investigating spatial correlations or spatial heterogeneity.

The parametric bootstrap: parboot.bf

Another useful simulation tool is the parametric bootstrap. *HyPhy* provides the *SimulateDataSet* command to provide the type of model-based simulation required. In *parboot.bf*, we find the following lines of code. Again, some lines have been deleted for clarity.

```
for (bsCounter = 1; bsCounter<=reps; bsCounter = bsCounter+1) {
    DataSet bsData = SimulateDataSet(theLikFun);
    DataSetFilter bsFilter = CreateFilter (bsData,1);
    HarvestFrequencies (bsFreqs, bsFilter, 1, 1, 1);
    Model bsModel = (F81RateMatrix, bsFreqs);
    Tree bsTree = (a,b,og);
    LikelihoodFunction bsLik = (bsFilter, bsTree);
    Optimize (bsParamValues, bsLik);
}
```

The end result is analogous to that of *bootstrap.bf*: we simulate *reps* data sets, find MLEs, and tabulate results. The fundamental difference is that the data sets are formed by simulation using the tree structure, evolutionary model, and parameters in *theLikFun* via the function *SimulateDataSet*. An important technical difference is that *SimulateDataSet* generates a *DataSet* as opposed to the *DataSetFilter* created by *Bootstrap*. Thus, we must use the *CreateFilter* command to create an appropriate filter.

Again note the use of *BSRes* for tabulating results and also the use of *fscanf* for acquiring input from the user (see the Batch Language Command Reference for details).

Putting it all together: *positions.bf*

As an example of the type of analysis *HyPhy* was designed to implement, we now describe the batch file *positions.bf*. This file illustrates some of the features of the *CreateFilter* command by ignoring species *C* in *four.seq* and by creating separate filters for each of the three codon positions. The HKY85 model is used as the basic substitution model. A *global* transition:transversion ratio, R, is created; its value is allowed to be shared by all three positions. In the "Combined Analysis," the entire data set is analyzed in the normal way, treating all sites identically. A second *LikelihoodFunction* is then created, in which the data are split into three partitions according to codon position. Each of the three partitions is allowed to evolve with a separate rate. However, the transition/transversion ratio is constrained to be the same for all three codon positions as well as for all lineages. The likelihood ratio test statistic comparing these two models is computed, and the statistical significance of the test is reported using both the chi-squared approximation and nonparametric bootstrapping.

The file *positions.bf* is rather complicated, so we will focus only on some of its key features.

Read and filter the data

It is often the case that molecular data sets have some repeating underlying structure that we would like to exploit or study. For instance, coding regions might be described with the repeating structure 123123123 . . . . In *positions.bf* we create separate *DataSetFilters* for first, second, and third codon positions. The command

```
DataSetFilter myFilter1 =
    CreateFilter (myData,1,"<100>","0,1,3");
```

creates a *DataSetFilter* named *MyData1* that includes only the first nucleotide of each triplet. Likewise, the statement

```
DataSetFilter myFilter3 =
    CreateFilter (myData,1,"<001>","0,1,3");
```

creates a *DataSetFilter* named *MyData3* that includes only the third nucleotide of every triplet. Had we wished to create a filter consisting of both first and second positions, we would have used a statement such as

```
DataSetFilter myFilter12 =
    CreateFilter (myData,1,"<110>","0,1,3");
```

Define a substitution model for each position

The next portion of *positions.bf* creates a vector of observed frequencies for each of the filters using standard syntax.

```
HarvestFrequencies (obsFreqs, myFilter, 1, 1, 1);
HarvestFrequencies (obsFreqs1, myFilter1, 1, 1, 1);
HarvestFrequencies (obsFreqs2, myFilter2, 1, 1, 1);
HarvestFrequencies (obsFreqs3, myFilter3, 1, 1, 1);
```

Next, the basic substitution model is defined. We use the HKY85 model with transversion parameter b and global transition:transversion ratio R. A separate *Model* is created for each partition since each uses different frequencies:

```
global R;
HKY85RateMatrix =
      {{*,b,R*b,b}{b,*,b,R*b}{R*b,b,*,b}{b,R*b,b,*}};
Model HKY85 = (HKY85RateMatrix, obsFreqs);
Tree myTree = (a,b,d);
Model HKY851 = (HKY85RateMatrix, obsFreqs1);
```

```
Tree myTree1 = (a,b,d);
Model HKY852 = (HKY85RateMatrix, obsFreqs2);
Tree myTree2 = (a,b,d);
Model HKY853 = (HKY85RateMatrix, obsFreqs3);
Tree myTree3 = (a,b,d);
```

Define two likelihood functions

We are now ready to set up *LikelihoodFunctions* and *Optimize* them. The analysis of the combined data set is routine:

```
LikelihoodFunction theLikFun = (myFilter,myTree);
Optimize (paramValues, theLikFun);
```

We also store some results for later use:

```
lnLik0 = paramValues[1][0];
npar0 = paramValues[1][1]+3;
fprintf  (stdout, theLikFun, "\n\n");
```

The statement `npar0 = paramValues[1][1]+3;` requires some explanation. The *Optimize* function always returns the number of parameters that were optimized as the `[1][1]` element of its returned matrix of results. Typically, we do not optimize over base frequency values, electing instead to simply use observed frequencies, which are usually very close to the maximum likelihood estimates. Since the frequencies are, in fact, estimated from the data, they need to be included in the parameter count. The value of *npar0* therefore includes the count of independent substitution parameters in the model (the number of which is returned by *Optimize*) along with the three independent base frequencies estimated from the data.

The *LikelihoodFunction* for the "partitioned" analysis simply uses the extended form of the *LikelihoodFunction* command:

```
LikelihoodFunction theSplitLikFun = (myFilter1,myTree1,
                                     myFilter2,myTree2,
                                     myFilter3,myTree3);
Optimize (paramValues, theSplitLikFun);
lnLik1 = paramValues[1][0];
npar1 = paramValues[1][1]+9;
```

Note the addition of the nine estimated frequencies to the model's parameter count, three for each partition.

Find p-values for hypothesis tests

Finally, we compute the *p*-value for the test of the combined analysis (null hypothesis) against the split model (alternative hypothesis). Two approaches are used. First is the normal chi-squared approximation to the LRT statistic:

```
LRT = 2*(lnLik1-lnLik0);
pValueChi2 = 1-CChi2 (LRT, npar1-npar0).
```

One can also estimate the P-value using the parametric bootstrap. The statement for simulating a random data set based on *theLikFun* is

```
DataSet simData = SimulateDataSet(theLikFun);
```

The remaining part of the loop is basically a copy of the original analysis, with variable names adjusted to indicate that they are coming from simulated data. For each simulated data set, we compute the LRT, named *simLRT*, and compare it with the observed LRT. The estimate of the *p*-value is the proportion of simulated datasets with an LRT larger than that of the observed data. We keep track of the number of such events using the variable *count*:

```
simLRT = 2*(simlnLik1-simlnLik0);
if (simLRT > LRT)
{
        count = count+1;
}
```

and report the results:

```
fprintf(stdout,
        "\n\n*** P-value (Parametric BS)= ",count/reps,"\n");
```

The batch file *positions.bf* provides a good example of the flexibility of *HyPhy*, and many of the same ideas could be used to develop analyses of multiple genes. Of particular importance for multilocus analysis is the ability to mix local and global variables. To our knowledge, the type of modeling and testing flexibility demonstrated in *positions.bf* is unique.

Site-to-site rate heterogeneity

One of the most important additions to recent models of sequence evolution is the incorporation of site-to-site rate heterogeneity, which allows the highly desirable property of some positions evolving quickly and some slowly, with others having intermediate rates. In the first portion of this chapter, we demonstrated some of *HyPhy*'s basic functionality with regard to rate heterogeneity. We now continue this discussion, demonstrating the "traditional" approaches to modeling rate heterogeneity as well as some novel features unique to *HyPhy*. We feel that the flexibility in modeling site-to-site rate heterogeneity is one of the strongest aspects of the software package.

The fundamental elements of incorporating site-to-site rate heterogeneity are demonstrated in the file *ratehet.bf*. There one will find an analysis labeled "Variable Rates Model 1," which simply uses the F81 nucleotide model with sites falling into one of four rate classes. The first rate class is an invariant class (i.e., rate 0), while rates of the remaining three categories have relative rates of 1, 2, and 4. The frequencies of the four categories are assumed to be equal for illustration. The key section of code is the following:

```
category rateCat = (4, EQUAL, MEAN, , {{0}{1}{2}{4}}, 0, 4);
```

```
F81VarRateMatrix = {{*,rateCat*m,rateCat*m,rateCat*m}
                     {rateCat*m,*,rateCat*m,rateCat*m}
                     {rateCat*m,rateCat*m,*,rateCat*m}
                     {rateCat*m,rateCat*m,rateCat*m,*}};
```

```
Model F81Var = (F81VarRateMatrix, obsFreqs);
```

The "category" statement defines a discrete probability distribution for the rates. In this case, there are four possible (relative) rates, 0, 1, 2, and 4, and the categories occur with equal frequencies. (See the *HyPhy* documentation and the examples below for further information on the *category* statement.) The second and third statements define a variant of the F81 model of nucleotide evolution. Had we left out the "rateCat" multiplier in the rate matrix, the model would be the standard F81 model. With the inclusion of "rateCat," which is defined in the first statement to be a category variable, we have a model declaring that each site evolves according to the F81 model but that the rates vary from site to site in accordance with the distribution described in the *category* statement. Note that in this case the relative rates are specified by the user, so there is no rate heterogeneity parameter to be estimated from the data.

In the "Variable Rates Model 2" analysis, we find an implementation of the slightly more complex (but more well-known) discrete gamma model first described in [12]. The key element in this analysis is simply a different *category* statement:

```
category rateCat = (4, EQUAL, MEAN,
    GammaDist(_x_,alpha,alpha), CGammaDist(_x_,alpha,alpha),
    0,1e25,CGammaDist(_x_,alpha+1,alpha));
```

We again introduce a discrete distribution with four equiprobable classes, but this time the relative rates of those classes are provided by the gamma distribution. In turn, the arguments in the *category* statement declare

1. Use four rate categories.
2. Assign equal frequencies to the four categories.
3. Use the mean of each discretized interval to represent the rate for the corresponding class.
4. The density function for the rates is the gamma density (which is a built-in function. Alternatively, the formula for any desired density could be entered.)
5. The cumulative density function is provided by the gamma distribution function. (Again, this is a predefined function, and the cdf for any chosen density could be substituted.)

6. The relative rates are limited to the range 0 to 1×10^{25} (to make numerical work simpler).
7. The final argument is optional and specifies a formula for the mean of each interval. If this argument were not provided, the mean would be evaluated numerically.

With this model, *HyPhy* would estimate the branch lengths for each branch in the tree along with the shape parameter α that is specified in the *category* statement.

The third and final example in *ratehet.bf* allows rates to vary according to an exponential distribution. The *category* statement in this case is essentially the same as for the gamma distribution, but with the density and distribution functions for the exponential distribution used instead:

```
category rateCat = (4, EQUAL, MEAN,
    alpha*Exp(-alpha*_x_), 1-Exp(-alpha*_x_), 0, 1e25,
    -_x_*Exp(-alpha*_x) + (1-Exp(-alpha*_x_))/alpha);
```

This fundamental approach can be used to fit any discretized density to data by simply writing an appropriate *category* statement and combining it with any desired substitution matrix. A number of examples are provided in the sample files in the *HyPhy* distribution.

In the file *twocats.bf*, we demonstrate a new idea in modeling rate heterogeneity, the possibility of moving beyond the simple idea of each site having its own rate. For illustration, we show that it is simple to define a model that allows each site to have its own transition and transversion rate, but sites with high transition rates need not also have high transversion rates. We demonstrated an application of this approach to codon-based models based on synonymous and nonsynonymous rates in the first half of the chapter. The basic approach is the same as for the previous examples: we will use the *category* statement to define distributions of rate heterogeneity. However, in this case we will use two *category* statements, one for transitions and one for transversions.

The first analysis in *twocats* is essentially the discrete gamma model found in *ratehet.bf* but with 16 categories rather than four. The second analysis introduces separate distributions for transitions and transversions. Each type of rate is assumed to come from a (discrete) gamma distribution with four categories, but each distribution has its own parameters. This model leads to a model with $4 \times 4 = 16$ rate categories and thus has computational complexity equal to the 16-category discrete gamma in the first analysis. The *category* statements have the same basic format as the previous examples:

```
category catTS = (4, EQUAL, MEAN,
  GammaDist(_x_,alphaS,alphaS), CGammaDist(_x_,alphaS,alphaS),
  0,1e25, CGammaDist(_x_,alphaS+1,alphaS));
```

```
category catTV = (4, EQUAL, MEAN,
  GammaDist(_x_,alphaV,beta), CGammaDist(_x_,alphaV,beta),
  0,1e25, CGammaDist(_x_,alphaV+1,beta)*alphaV/beta);
```

An important mathematical fact arises at this point. Traditionally, the gamma distribution in rate analyses has been described only by its "shape" parameter. The gamma distribution in general is described by a shape parameter and a scale parameter. The confounding of rates and times allows for the (arbitrary) determination of one of the two parameters, and for simplicity the two parameters have simply been assumed to be equal. When we move to the case of two gamma distributions, we still have this level of freedom to arbitrarily assign one parameter. In this example, we have maintained the "traditional" style for the transition rates (see the *category* statement for *catTS*), but we must use both the shape and scale parameters for the second distribution. Thus, we end up with three parameters that govern the distributional form for the transition and transversion rates: *alphaS*, the shape parameter for the transition rate distribution, and *alphaV* and *beta*, the shape and scale parameters for the gamma distribution describing transversion rates.

We must still introduce these category variables into the substitution matrix, and examining the definition of *HKY85TwoVarRateMatrix*, we see that transition rates are multiplied by *catTS*, while transversion rates are multiplied by *catTV*.

Analyzing codon data

So far, we have considered only nucleotide alignments and evolutionary models as examples. Using the example included in the file *codon.bf*, we will discuss how to read and filter codon data and define substitution models that operate at the level of codons.

Defining codon data filters

Codon data sets are nucleotide sequences where the unit of evolution is a triplet of nucleotides, and some states (stop codons) are disallowed. The task of making *HyPhy* interpret a nucleotide alignment as codons is handled by supplying a few additional parameters in a call to *CreateFilter*. Consider the following line in *codons.bf*:

```
DataSetFilter codonFilter =
    CreateFilter(myData,3,"","","TAA,TAG,TGA");
```

The second argument of 3 instructs *HyPhy* to consider triplets of characters in the data set *myData* as units of evolution. If it had been 2, then the filter would consist of dinucleotides. The empty third and fourth arguments include all sequences and sites in the filter. The fifth argument is the comma-separated list of *exclusions* (i.e., character states that are not allowed). One can easily recognize that the list includes the three stop codons for the universal genetic code. All sites in the original nucleotide alignment that contained at least one of the excluded states would be omitted from the filter, and a message would be written to *messages.log*, located in the main *HyPhy* directory.

The filter *myFilter* consists of data for $4^3 - 3 = 61$ states (i.e., all sense codons in the universal genetic code); therefore, any substitution model compatible with this filter must describe a process with 61 states and use a 61×61 rate matrix. Before we proceed with the definition of this matrix, a crucial question must be answered: How does *HyPhy* index codons? For example, which entry in the rate matrix will describe the change from codon ATC to codon TTC? *HyPhy* uses a uniform indexing scheme, which is rather straightforward. The default nucleotide alphabet is ordered as ACGT, and each character is assigned an index in that order: A=0, C=1, G=2, T=3 (note that all indexing starts at 0, as in the programming language C). In previous examples, we used this mapping to define nucleotide rate matrices. For example, the entry in row 2 and column 3 would define the rate of G→T substitutions. Analogously, all sense codons are ordered alphabetically: AAA, AAC, AAG, AAT, ACA, ..., TTG, TTT, excluding stop codons, with the corresponding indexing from 0 to 60. It is easy to check that ATC will have the index of 13, whereas TTC is assigned the index of 58. Consequently, the rate of ATC to TTC substitutions should be placed in row 13 and column 58 of the rate matrix.

A 61×61 rate matrix has 3721 entries, and defining them one by one would be a daunting task. We need a way to avoid an explicit enumeration. Consider the MG94×HKY85 model (6.2) explained in Section 6.2.4. Each substitution rate can be classified by determining the following four attributes: (i) is the change one-step or multistep? (ii) Is the change synonymous or nonsynonymous? (iii) Is the change a transition or a transversion? (iv) What is the equilibrium frequency of the target nucleotide? A compact way to define the model is to loop through all 3721 possible pairs of codons, answer the four questions above, and assign the appropriate rate to the matrix cell. *HyPhy* has no intrinsic knowledge of how codons are translated to amino acids, and this information is needed to decide whether a nucleotide substitution is synonymous or nonsynonymous. *codons.bf* contains such a map for the universal genetic code in the matrix *UniversalGeneticCode*. The 64 codons have 21 possible translations (20 amino acids and a "stop"). Each of the 64 cells of *UniversalGeneticCode* contains an amino acid (or stop) code from 0 to 20, whose meaning is explained in the comments in *codons.bf*. We refer the reader to the code and comments in *codons.bf* for implementation details. The implementation is straightforward but somewhat obtuse. Once the reader becomes comfortable with referencing codons by their indices and interpreting them, the code should be clear. The reason for not having a built-in genetic code translation device is to allow the use of arbitrary (nonuniversal) genetic codes.

The file *codons.bf* illustrates several other useful concepts:

- How to define and call user functions. Function *BuildCodonFrequencies* is employed to compute codon equilibrium frequencies based on observed nucleotide proportions, defined in (6.3).
- The use of a built-in variable to reference the tree string present in the data file (*DATAFILE_TREE*).
- The use of the double underscore operator to substitute numerical values of arguments into formula definitions and avoid unwanted dependencies.

Lastly, *codons.bf* writes out data for further processing with a standard file from the *HyPhy* distribution to perform posterior Bayesian analysis, as discussed in Section 6.2.4.

6.4 Conclusion

This chapter has provided an overview of the basic features and use of the *HyPhy* system. With a programming language at its core, users may elect to write their own likelihood-based molecular evolutionary analyses. A graphical user interface offers much of the power of the batch language, allowing users to fit complex, customizable models to sequence alignments. The user interface also provides access to the parametric bootstrap features of *HyPhy* for carrying out tests of both nested and nonnested hypotheses. Many features of the package, of course, could not be described in this chapter. For instance, *HyPhy* includes a model editor for describing new stochastic models to be used in analyses, and the graphical user interface provides a mechanism to define arbitrary constraints among parameters for construction of likelihood ratio tests. Its authors continue to develop *HyPhy*, with a goal of providing a flexible, portable, and powerful system for carrying out cutting-edge molecular evolutionary analyses.

References

[1] H. Akaike. A new look at the statistical model identification. *IEEE Transactions on Automatic Control*, 119:716–723, 1974.

[2] D. R. Cox. Tests of separate families of hypotheses. In *Proceedings of the 4th Berkeley Symposium*, volume 1, pages 105–123. University of California Press, Los Angeles, CA, 1961.

[3] J. Felsenstein. Evolutionary trees from DNA-sequences — a maximum-likelihood approach. *Journal of Molecular Evolution*, 17:368–376, 1981.

[4] N. Goldman. Statistical tests of models of DNA substitution. *Journal of Molecular Evolution.*, 36:182–198, 1993.

[5] M. Hasegawa, H. Kishino, and T. Yano. Dating of the human-ape splitting by a molecular clock of mitochondrial dna. *Molecular Biology and Evolution*, 21:160–174, 1985.

[6] S. L. Kosakovsky Pond and S. V Muse. Column sorting: Rapid calculation of the phylogenetic likelihood function. *To appear in Systematic Biology*, 2004.

[7] S. V. Muse and B. S. Gaut. A likelihood approach for comparing synonymous and nonsynonymous nucleotide substitution rates, with application to the chloroplast genome. *Molecular Biology and Evolution*, 11:715–724, 1994.

[8] S. V. Muse and B. S. Gaut. Comparing patterns of nucleotide substitution rates among chloroplast loci using the relative ratio test. *Genetics*, 146:393–399, 1997.

[9] R. Nielsen and Z. H. Yang. Likelihood models for detecting positively selected amino acid sites and applications to the HIV-1 envelope gene. *Genetics*, 148:929–936, 1998.

[10] Y. Suzuki and T. Gojobori. A method for detecting positive selection at single amino acid sites. *Molecular Biology and Evolution*, 16:1315–1328, 1999.

[11] Z. Yang. Maximum-likelihood estimation of phylogeny from DNA sequences when substitution rates differ over sites. *Molecular Biology and Evolution*, 10:1396–1401, 1993.

[12] Z. Yang. Maximum likelihood phylogenetic estimation from DNA sequences with variable rates over sites: Approximate methods. *Journal of Molecular Evolution.*, 39:105–111, 1994.

[13] Z. H. Yang. Among-site rate variation and its impact on phylogenetic analyses. *Trends in Ecology and Evolution*, 11:367–372, 1996.

[14] Z. H. Yang, R. Nielsen, N. Goldman, and A. M. K. Pedersen. Codon-substitution models for heterogeneous selection pressure at amino acid sites. *Genetics*, 155:431–449, 2000.

Bayesian Analysis of Molecular Evolution Using MrBayes

John P. Huelsenbeck[1] and Fredrik Ronquist[2]

[1] Division of Biological Sciences, University of California at San Diego, La Jolla, CA 92093, USA, johnh@biomail.ucsd.edu
[2] School of Computational Science and Information Technology, Florida State University, Tallahassee, FL 32306-4120, USA, ronquist@csit.fsu.edu

7.1 Introduction

Stochastic models of evolution play a prominent role in the field of molecular evolution; they are used in applications as far-ranging as phylogeny estimation, uncovering the pattern of DNA substitution, identifying amino acids under directional selection, and in inferring the history of a population using models such as the coalescence. The models used in molecular evolution have become quite sophisticated over time. In the late 1960s one of the first stochastic models applied to molecular evolution was introduced by Jukes and Cantor [38] to describe how substitutions might occur in a DNA sequence. This model was quite simple, really having only one parameter—the amount of change between two sequences—and assumed that all of the different substitution types had an equal probability of occurring. A familiar story, and one of the greatest successes of molecular evolution, has been the gradual improvement of models to describe new observations as they were made. For example, the observation that transitions (substitutions between the nucleotides $A \leftrightarrow G$ and $C \leftrightarrow T$) occur more frequently than transversions (changes between the nucleotides $A \leftrightarrow C$, $A \leftrightarrow T$, $C \leftrightarrow G$, $G \leftrightarrow T$) spurred the development of DNA substitution models that allow the transition rate to differ from the transversion rate [40, 24, 23]. Similarly, the identification of widespread variation in rates across sites led to the development of models of rate variation [72] and also to more sophisticated models that incorporate constraints on amino acid replacement [21, 50]. More recently, rates have been allowed to change on the tree (the covarion-like models of Tuffley and Steel [70]) and can explain patterns such as many substitutions at a site in one clade and few if any substitutions at the same position in another clade of roughly the same age.

The fundamental importance of stochastic models in molecular evolution is this: they contain parameters, and if specific values can be assigned to these parameters based on observations, such as an alignment of DNA sequences, then biologists can learn something about how molecular evolution has occurred. This point is very basic but important. It implies that in addition to careful consideration of the development of models, one needs to be able to efficiently *estimate* the parameters of the model. By efficient we mean the ability to accurately estimate the parameters of an evolutionary model based on as little data as possible. There are only a handful of methods that have been used to estimate parameters of evolutionary models. These include the parsimony, distance, maximum likelihood, and Bayesian methods. In this chapter, we will concentrate on Bayesian estimation of evolutionary parameters. More specifically, we will show how the program MrBayes [35, 59] can be used to investigate several important questions in molecular evolution in a Bayesian framework.

7.2 Maximum Likelihood and Bayesian Estimation

Unlike the parsimony and distance methods, maximum likelihood and Bayesian inference take full advantage of the information contained in an alignment of DNA sequences when estimating parameters of an evolutionary model. Both maximum likelihood and Bayesian estimation rely on the likelihood function. The likelihood is proportional to the probability of observing the data, conditioned on the parameters of the model

$$\ell(\text{Parameter}) = \text{Constant} \times \text{Prob}[\text{Data}|\text{Parameter}],$$

where the constant is arbitrary. The probability of observing the data conditioned on specific parameter values is calculated using stochastic models. Details about how the likelihood can be calculated for an alignment of DNA or protein sequences can be found elsewhere [14]. Here, we have written the likelihood function with only one parameter. However, for the models typically used in molecular evolution, there are many parameters. We make the notational change in what follows by denoting parameters with the Greek symbol θ and the data as $\mathbf{X}$ so that the likelihood function for multiple-parameter models is

$$\ell(\theta_1, \theta_2, \dots, \theta_n) = K \times f(\mathbf{X}|\theta_1, \theta_2, \dots, \theta_n),$$

where K is the constant.

In a maximum likelihood analysis, the combination of parameters that maximizes the likelihood function is the best estimate, called the maximum likelihood estimate. In a Bayesian analysis, on the other hand, the object is to calculate the joint posterior probability distribution of the parameters. This is calculated using Bayes' theorem as

$$f(\theta_1, \theta_2, \dots, \theta_n|\mathbf{X}) = \frac{\ell(\theta_1, \theta_2, \dots, \theta_n) \times f(\theta_1, \theta_2, \dots, \theta_n)}{f(\mathbf{X})},$$

where $f(\theta_1, \theta_2, \ldots, \theta_n | \mathbf{X})$ is the posterior probability distribution, $\ell(\theta_1, \theta_2, \ldots, \theta_n)$ is the likelihood function, and $f(\theta_1, \theta_2, \ldots, \theta_n)$ is the prior probability distribution for the parameters. The posterior probability distribution of parameters can then be used to make inferences.

Although both maximum likelihood and Bayesian analyses are based upon the likelihood function, there are fundamental differences in how the two methods treat parameters. Many of the parameters of an evolutionary model are not of direct interest to the biologist. For example, for someone interested in detecting adaptive evolution at the molecular level, the details of the phylogenetic history of the sequences sampled is not of immediate interest; the focus is on other aspects of the model. The parameters that are not of direct interest but that are needed to complete the model are called nuisance parameters (see [20], for a more thorough discussion of nuisance parameters in phylogenetic inference). There are a few standard ways of dealing with nuisance parameters. One is to maximize the likelihood with respect to them. It is understood, then, that inferences about the parameters of interest depend upon the nuisance parameters taking fixed values. This is the approach usually taken in maximum likelihood analyses and also in empirical Bayes analyses. The other approach assigns a prior probability distribution to the nuisance parameters. The maximum likelihood or posterior probabilities are calculated by integrating over all possible values of the nuisance parameters, weighting each by its (prior) probability. This approach is rarely taken in maximum likelihood analyses (where it is called the integrated likelihood approach [6]) but is the standard method of accounting for nuisance parameters in a Bayesian analysis, where all of the parameters of the model are assigned a prior probability distribution. The advantage of marginalization is that inferences about the parameters of interest do not depend upon any particular value for the nuisance parameters. The disadvantage, of course, is that it may be difficult to specify a reasonable prior model for the parameters.

Maximum likelihood and Bayesian analyses also differ in how they interpret parameters of the model. Maximum likelihood does not treat the parameters of the model as random variables (variables that can take their value by chance), whereas in a Bayesian analysis, everything—the data and the parameters—is treated as random variables. This is not to say that a Bayesian does not think that there is only one actual value for a parameter (such as a phylogenetic tree) but rather that his or her uncertainty about the parameter is described by the posterior probability distribution. In some ways, the treatment of all of the variables as random quantities simplifies a Bayesian analysis. First, one is always dealing with probability distributions. If one is interested in only the phylogeny of a group of organisms, say, then one would base inferences on the marginal posterior probability distribution of phylogeny. The marginal posterior probability of a parameter is calculated by integrating over all possible values of the other parameters, weighting each by its probability. This means that an inference of phylogeny does not critically depend upon another parameter taking a specific value. Another simplification in a Bayesian analysis is that uncertainty in a parameter can be easily described. After all, the probability distribution of the parameter is available, so specifics about the mean, variance, and a range that contains most of the posterior probability for the parameter can be directly calculated from the marginal posterior probability distribution for that parameter. In a maximum likelihood analysis, on the other hand, the parameters of the model are not treated as random variables, so probabilities cannot be directly assigned to the parameters. If one wants to describe the uncertainty in an estimate obtained using maximum likelihood, one has to go through the thought experiment of collecting many data sets of the same size as the original, with parameters set to the maximum likelihood values. One then asks what the range of maximum likelihood estimates would be for the parameter of interest on the imaginary data.

In practice, many studies in molecular evolution apply a hybrid approach that combines ideas from maximum likelihood and Bayesian analysis. For example, in what is now a classic study, Nielsen and Yang [54] identified amino acid positions in a protein-coding DNA sequence under the influence of positive selection using Bayesian methods; the posterior probability that each amino acid position is under directional selection was calculated. However, they used maximum likelihood to estimate all of the parameters of the model. This approach can be called an empirical Bayes approach because of its reliance on Bayesian reasoning for the parameter of interest (the probability a site is under positive selection) and maximum likelihood for the nuisance parameters.

In the following section, we describe three uses of Bayesian methods in molecular evolution: phylogeny estimation, analysis of complex data, and estimating divergence times. We hope to show the ease with which parameters can be estimated, the uncertainty in the parameters can be described, and uncertainty about important parameters can be incorporated into a study in a Bayesian framework.

7.3 Applications of Bayesian Estimation in Molecular Evolution

7.3.1 A Brief Introduction to Models of Molecular Evolution

Before delving into specific examples of the application of Bayesian inference in molecular evolution, the reader needs some background on the modeling assumptions made in a Bayesian analysis. Many of these assumptions are shared by maximum likelihood and distance-based methods. Typically, the models used in molecular evolution have three components. First, they assume a tree relating the samples. Here, the samples might be DNA sequences collected from different species or different individuals within a population. In either case, a basic assumption is that the samples are related to one another through an (unknown) tree. This would be a species tree for sequences sampled from different species, or perhaps a coalescence tree for sequences sampled from individuals from within a population. Second, they assume that the branches of the tree have an (unknown) length. Ideally, the length of a branch on a tree is in terms of time. However, in practice it is difficult to determine the duration of a branch on a tree in terms of time. Instead, the lengths of the branches on the tree are in terms of expected change per character. Figure 7.1 shows some examples of trees with branch lengths. The main points the reader should remember are: (1) Trees can be rooted or unrooted. Rooted trees have a time direction, whereas unrooted trees do not. Most methods of phylogenetic inference, including most implementations of maximum likelihood and Bayesian analysis, are based on time-reversible models of evolution that produce unrooted trees, which must be rooted using some other criterion, such as the outgroup criterion (using distantly related reference sequences to locate the root). (2) The space of possible trees is huge. The number of possible unrooted trees for n species is $B(n) = \frac{(2n-5)!}{2^{n-3}(n-3)!}$ [61]. This means that for a relatively small problem of only $n = 50$ species, there are about $B(50) = 2.838 \times 10^{74}$ possible unrooted trees that can explain the phylogenetic relationships of the species.

The third component of a model of molecular evolution is a process that describes how the characters change on the phylogeny. All model-based methods of phylogenetic inference, including maximum likelihood and Bayesian estimation of phylogeny, currently assume that character change occurs according to a continuous-time Markov chain. At the heart of any continuous-time Markov chain is a matrix of rates specifying the rate of change from one state to another. For example, the instantaneous rate of change under the model described by Hasegawa et al. ([24, 23]; hereafter called the HKY85 model) is

$$\mathbf{Q} = \{q_{ij}\} = \begin{pmatrix} - & \pi_C & \kappa\pi_G & \pi_T \\ \pi_A & - & \pi_G & \kappa\pi_T \\ \kappa\pi_A & \pi_C & - & \pi_T \\ \pi_A & \kappa\pi_C & \pi_G & - \end{pmatrix} \mu.$$

This matrix specifies the rate of change from one nucleotide to another; the rows and columns of the matrix are ordered A, C, G, T, so that the rate of change $C \to G$ is $q_{CG} = \pi_G$. Similarly, the rates of change $C \to T$, $G \to A$, and $T \to C$ are $q_{CT} = \kappa\pi_T$, $q_{GA} = \kappa\pi_A$, and $q_{TG} = \pi_G$, respectively. The diagonals of the rate matrix, denoted with the dash, are specified such that each row sums to zero. Finally, the rate matrix is rescaled such that the mean rate of substitution is one. This can be accomplished by setting $\mu = -1/\sum_{i \in \{A,C,G,T\}} \pi_i q_{ii}$. This rescaling of the rate matrix such that the mean rate is one allows the branch lengths on the phylogenetic tree to be interpreted as the expected number of nucleotide substitutions per site.

We will make a few important points about the rate matrix. First, the rate matrix may have free parameters. For example, the HKY85 model has

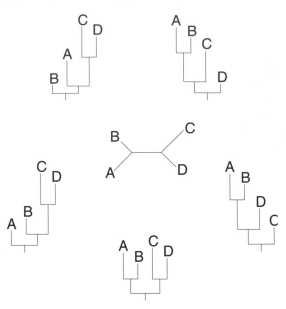

Fig. 7.1. Example of unrooted and rooted trees. An unrooted tree of four species (center) with the branch lengths drawn proportional to their length in terms of expected number of substitutions per site. The five trees surrounding the central, unrooted tree show the five possible rooted trees that result from the unrooted tree.

the parameters κ, π_A, π_C, π_G, and π_T. The parameter κ is the transition/transversion rate bias when $\kappa = 1$ transitions occur at the same rate as transversions. Typically, the transition/transversion rate ratio, estimated using maximum likelihood or Bayesian inference, is greater than one and transitions occur at a higher rate than transversions. The other parameters—π_A, π_C, π_G, and π_T—are the base frequencies and have a biological interpretation as the frequency of the different nucleotides and are also, incidentally, the stationary probabilities of the process (more on stationary probabilities later). Second, the rate matrix, **Q**, can be used to calculate the transition probabilities and the stationary distribution of the substitution process. The transition probabilities and stationary distribution play a key role in calculating the likelihood, and we will spend more time here developing an intuitive understanding of these concepts.

Transition probabilities

Let us consider a specific example of a rate matrix with all of the parameters of the model taking specific values. For example, if we use the HKY85 model and fix the parameters to $\kappa = 5$, $\pi_A = 0.4$, $\pi_C = 0.3$, $\pi_G = 0.2$, and $\pi_T = 0.1$, we get the following matrix of instantaneous rates

$$\mathbf{Q} = \{q_{ij}\} = \begin{pmatrix} -0.886 & 0.190 & 0.633 & 0.063 \\ 0.253 & -0.696 & 0.127 & 0.316 \\ 1.266 & 0.190 & -1.519 & 0.063 \\ 0.253 & 0.949 & 0.127 & -1.329 \end{pmatrix}.$$

Note that these numbers are not special in any particular way. That is to say, they are not based upon any observations from a real data set but are rather arbitrarily picked to illustrate a point. The point is that one can interpret the rate matrix in the physical sense of specifying how changes occur on a phylogenetic tree. Consider the very simple case of a single branch on a phylogenetic tree. Let's assume that the branch is $v = 0.5$ in length and that the ancestor of the branch is the nucleotide G. The situation we have is something like that shown in Figure 7.2(a). How can we simulate the evolution of the site starting from the G at the ancestor? The rate matrix tells us how to do this. First of all, because the current state of the process is G, the only relevant row of the rate matrix is the third one:

$$\mathbf{Q} = \{q_{ij}\} = \begin{pmatrix} \cdot & \cdot & \cdot & \cdot \\ \cdot & \cdot & \cdot & \cdot \\ 1.266 & 0.190 & -1.519 & 0.063 \\ \cdot & \cdot & \cdot & \cdot \end{pmatrix}.$$

The overall rate of change away from nucleotide G is $q_{GA}+q_{GC}+q_{GT} = 1.266+0.190 + 0.063 = 1.519$. Equivalently, the rate of change away from nucleotide G is simply $-q_{GG} = 1.519$. In a continuous-time Markov model, the waiting time between substitutions is exponentially distributed. The exact shape of the exponential distribution is determined by its rate, which is the same as the rate of the corresponding process in the $\mathbf{Q}$ matrix. For instance, if we are in state G, we wait an exponentially distributed amount of time with rate 1.519 until the next substitution occurs. One can easily construct exponential random variables from uniform random variables using the equation

$$t = -\frac{1}{\lambda} \log_e(u),$$

where λ is the rate and u is a uniform(0,1) random number. For example, our calculator has a uniform(0,1) random number generator. The first number it generated is $u = 0.794$. This means that the next time at which a substitution occurs is 0.152 up from the root of the tree (using $\lambda = 1.519$; Figure 7.2(b)). The rate matrix also specifies the probabilities of a change from G to the nucleotides A, C, and T. These probabilities are

$$G \to A : \tfrac{1.266}{1.519} = 0.833, \; G \to C : \tfrac{0.190}{1.519} = 0.125, \; G \to T : \tfrac{0.063}{1.519} = 0.042.$$

To determine the nucleotide to which the process changes, we would generate another uniform(0,1) random number (again called u). If u is between 0 and 0.833, we will say that we had a change from G to A. If the random number

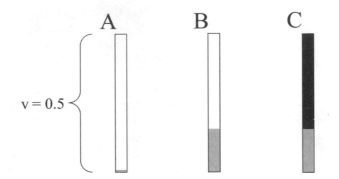

Fig. 7.2. Simulation under the HKY85 substitution process. A single realization of the substitution process under the HKY85 model when $\kappa = 5$, $\pi_A = 0.4$, $\pi_C = 0.3$, $\pi_G = 0.2$, and $\pi_T = 0.1$. The length of the branch is $v = 0.5$ and the starting nucleotide is G (light gray). (a) The process starts in nucleotide G. (b) The first change is 0.152 units up the branch. (c) The change is from G to A (dark gray). The time at which the next change occurs exceeds the total branch length, so the process ends in state C.

is between 0.833 and 0.958, we will say that we had a change from G to C. Finally, if the random number u is between 0.958 and 1.000, we will say we had a change from G to T. The next number generated on our calculator was $u = 0.102$, which means the change was from G to A. The process is now in a different state (the nucleotide A), and the relevant row of the rate matrix is

$$\mathbf{Q} = \{q_{ij}\} = \begin{pmatrix} -0.886 \ 0.190 \ 0.633 \ 0.063 \\ \cdot \quad \cdot \quad \cdot \quad \cdot \\ \cdot \quad \cdot \quad \cdot \quad \cdot \\ \cdot \quad \cdot \quad \cdot \quad \cdot \end{pmatrix}.$$

We wait an exponentially distributed amount of time with parameter $\lambda = 0.886$ until the next substitution occurs. When the substitution occurs, it is to a C, G, or T with probabilities $\frac{0.190}{0.886} = 0.214$, $\frac{0.633}{0.886} = 0.714$, and $\frac{0.063}{0.886} = 0.072$, respectively. This process of generating random and exponentially distributed times until the next substitution occurs and then determining (randomly) which nucleotide has changed is repeated until the process exceeds the length of the branch. The state of the process when it passes the end of the branch is recorded. In the example of Figure 7.2, the process started in state G and ended in state A. (The next uniform random variable generated on our calculator was $u = 0.371$, which means that the next substitution would occur 1.119 units above the substitution $G \to A$. The process was in the state A when it passed the end of the branch.) The only nonrandom part of the entire procedure was the initial decision to start the process in state G. All other aspects of the simulation used a uniform random number generator and our knowledge of the rate matrix to simulate a single realization of the HKY85 process of DNA substitution.

This Monte Carlo procedure for simulating the HKY85 process of DNA substitution can be repeated. The following table summarizes the results of 100 simulations, each of which started with the nucleotide G:

Starting Nucleotide	Ending Nucleotide	Number of Replicates
G	A	27
G	C	10
G	G	59
G	T	4

This table can be interpreted as a Monte Carlo approximation of the *transition probabilities* from nucleotide G to nucleotide $i \in (A, C, G, T)$. Specifically, the Monte Carlo approximations are $p_{GA}(0.5) \approx 0.27$, $p_{GC}(0.5) \approx 0.10$, $p_{GG}(0.5) \approx 0.59$, and $p_{GT}(0.5) \approx 0.04$. These approximate probabilities are all conditioned on the starting nucleotide being G and the branch length being $v = 0.5$. We performed additional simulations in which the starting nucleotide was A, C, or T. Together with the earlier Monte Carlo simulation that started with the nucleotide G, these additional simulations allow us to fill out the following table with the approximate transition probabilities:

		Ending Nucleotide		
		A	C	G	T
	A	0.67	0.13	0.20	0.00
Starting	C	0.13	0.70	0.07	0.10
Nucleotide	G	0.27	0.10	0.59	0.04
	T	0.12	0.30	0.08	0.50

Clearly, these numbers are only crude approximations to the true transition probabilities; after all, each row in the table is based on only 100 Monte Carlo simulations. However, they do illustrate the meaning of the transition probabilities; the transition probability $p_{ij}(v)$ is the probability that the substitution process ends in nucleotide j conditioned on it having started in nucleotide i after an evolutionary amount of time v. The table of approximate transition probabilities above can be interpreted as a matrix of probabilities, usually denoted $\mathbf{P}(v)$. Fortunately, we do not need to rely on Monte Carlo simulation to approximate the transition probability matrix. Instead, we can calculate the transition probability matrix exactly using matrix exponentiation:

$$\mathbf{P}(v) = e^{\mathbf{Q}v}.$$

For the case we have been simulating, the exact transition probabilities (to four decimal places) are

$$\mathbf{P}(0.5) = \{p_{ij}(0.5)\} = \begin{pmatrix} 0.7079 & 0.0813 & 0.1835 & 0.0271 \\ 0.1085 & 0.7377 & 0.0542 & 0.0995 \\ 0.3670 & 0.0813 & 0.5244 & 0.0271 \\ 0.1085 & 0.2985 & 0.0542 & 0.5387 \end{pmatrix}.$$

The transition probability matrix accounts for all the possible ways the process could end up in nucleotide j after starting in nucleotide i. In fact, each of the infinite possibilities is weighted by its probability under the substitution model.

Stationary distribution

The transition probabilities provide the probability of ending in a particular nucleotide after some specific amount of time (or opportunity for substitution, v). These transition probabilities are conditioned on starting in a particular nucleotide. What do the transition probability matrices look like as v increases? The following transition probability matrices show the effect of increasing branch length:

$$\mathbf{P}(0.00) = \begin{pmatrix} 1.000 & 0.000 & 0.000 & 0.000 \\ 0.000 & 1.000 & 0.000 & 0.000 \\ 0.000 & 0.000 & 1.000 & 0.000 \\ 0.000 & 0.000 & 0.000 & 1.000 \end{pmatrix}, \mathbf{P}(0.01) = \begin{pmatrix} 0.991 & 0.002 & 0.006 & 0.001 \\ 0.003 & 0.993 & 0.001 & 0.003 \\ 0.013 & 0.002 & 0.985 & 0.001 \\ 0.003 & 0.009 & 0.001 & 0.987 \end{pmatrix},$$

$$\mathbf{P}(0.10) = \begin{pmatrix} 0.919 & 0.018 & 0.056 & 0.006 \\ 0.024 & 0.934 & 0.012 & 0.029 \\ 0.113 & 0.018 & 0.863 & 0.006 \\ 0.025 & 0.086 & 0.012 & 0.877 \end{pmatrix}, \mathbf{P}(0.50) = \begin{pmatrix} 0.708 & 0.081 & 0.184 & 0.027 \\ 0.106 & 0.738 & 0.054 & 0.100 \\ 0.367 & 0.081 & 0.524 & 0.027 \\ 0.109 & 0.299 & 0.054 & 0.539 \end{pmatrix},$$

$$\mathbf{P}(1.00) = \begin{pmatrix} 0.580 & 0.141 & 0.232 & 0.047 \\ 0.188 & 0.587 & 0.094 & 0.131 \\ 0.464 & 0.141 & 0.348 & 0.047 \\ 0.188 & 0.394 & 0.094 & 0.324 \end{pmatrix}, \mathbf{P}(5.00) = \begin{pmatrix} 0.411 & 0.287 & 0.206 & 0.096 \\ 0.383 & 0.319 & 0.192 & 0.106 \\ 0.411 & 0.287 & 0.206 & 0.096 \\ 0.383 & 0.319 & 0.192 & 0.107 \end{pmatrix},$$

$$\mathbf{P}(10.0) = \begin{pmatrix} 0.401 & 0.299 & 0.200 & 0.099 \\ 0.399 & 0.301 & 0.199 & 0.100 \\ 0.401 & 0.299 & 0.200 & 0.099 \\ 0.399 & 0.301 & 0.199 & 0.100 \end{pmatrix}, \mathbf{P}(100) = \begin{pmatrix} 0.400 & 0.300 & 0.200 & 0.100 \\ 0.400 & 0.300 & 0.200 & 0.100 \\ 0.400 & 0.300 & 0.200 & 0.100 \\ 0.400 & 0.300 & 0.200 & 0.100 \end{pmatrix}.$$

(Each matrix was calculated under the HKY85 model with $\kappa = 5$, $\pi_A = 0.4$, $\pi_C = 0.3$, $\pi_G = 0.2$, and $\pi_T = 0.1$.) Note that as the length of a branch, v, increases, the probability of ending up in a particular nucleotide converges to a single number, regardless of the starting state. For example, the probability of ending up in C is about 0.300 when the branch length is $v = 100$. This is true regardless of whether the process starts in A, C, G, or T. The substitution process has in a sense "forgotten" its starting state.

The stationary distribution is the probability of observing a particular state when the branch length increases without limit ($v \to \infty$). The stationary probabilities of the four nucleotides are $\pi_A = 0.4$, $\pi_C = 0.3$, $\pi_G = 0.2$, and $\pi_T = 0.1$ for the example discussed above. The models typically used in phylogenetic analyses have the stationary probabilities built into the rate matrix, $\mathbf{Q}$. You will notice that the rate matrix for the HKY85 model has parameters

π_A, π_C, π_G, and π_T and that the stationary frequencies of the four nucleotides for our example match the input values for our simulations. Building the stationary frequency of the process into the rate matrix, while somewhat unusual, makes calculating the likelihood function easier. For one, specifying the stationary distribution saves the time of identifying the stationary distribution (which involves solving the equation $\pi\mathbf{Q} = \mathbf{0}$, which simply says that if we start with the nucleotide frequencies reflecting the stationary distribution, the process will have no effect on the nucleotide frequencies). For another, it allows one to more easily specify a time-reversible substitution model. (A time-reversible substitution model has the property that $\pi_i q_{ij} = \pi_j q_{ji}$ for all $i, j \in (A, C, G, T)$, $i \neq j$.) Practically speaking, time reversibility means that we can work with unrooted trees instead of rooted trees (assuming that the molecular clock is not enforced).

Calculating the likelihood

The transition probabilities and stationary distribution are used when calculating the likelihood. For example, consider the following alignment of sequences for five species[1]:

```
Species 1   TAACTGTAAAGGACAACACTAGCAGGCCAGACGCACACGCACAGCGCACC
Species 2   TGACTTTAAAGGACGACCCTACCAGGGCGGACACAAACGGACAGCGCAGC
Species 3   CAAGTTTAGAAAACGGCACCAACACAACAGACGTATGCAACTGACGCACC
Species 4   CGAGTTCAGAAGACGGCACCAACACAGCGGACGTATGCAGACGACGCACC
Species 5   TGCCCTTAGGAGGCGGCACTAACACCGCGGACGAGTGCGGACAACGTACC
```

This is clearly a rather small alignment of sequences to use for estimating phylogeny, but it will illustrate how likelihoods are calculated. The likelihood is the probability of the alignment of sequences, conditioned on a tree with branch lengths. The basic procedure is to calculate the probability of each site (column) in the matrix. Assuming that the substitutions are independent across sites, the probability of the entire alignment is simply the product of the probabilities of the individual sites.

How is the likelihood at a single site calculated? Figure 7.3 shows the observations at the first site (T, T, C, C, and T) at the tips of one of the possible phylogenetic trees for five species. The tree in Figure 7.3 is unusual in that we will assume that the nucleotide states at the interior nodes of the tree are also known. This is clearly a bad assumption because we cannot directly observe the nucleotides that occurred at any point on the tree in the distant past. For now, however, ignore this fact and bear with us. The probability of observing the configuration of nucleotides at the tips and interior nodes of the tree in Figure 7.3 is

---

[1]This alignment was simulated on the tree of Figure 7.3 under the HKY85 model of DNA substitution. Parameter values for the simulation can be found in the caption of Table 7.1.

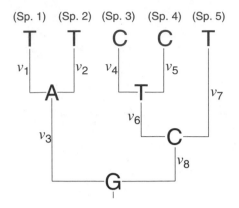

Fig. 7.3. A tree with states assigned to the tips. One of the possible (rooted) trees describing the evolutionary history of the five species. The states at the first site in the alignment of the text are shown at the tips of the tree. The states at the interior nodes of the tree are also shown, though in reality these states are not observed. The length of the ith branch is denoted v_i.

$$\Pr(TTCCT, ATCG | \tau, \mathbf{v}, \theta) =$$
$$\pi_G \, p_{GA}(v_3) \, p_{AT}(v_1) \, p_{AT}(v_2) \, p_{GC}(v_8) \, p_{CT}(v_6) \, p_{CT}(v_7) \, p_{TC}(v_4) \, p_{TC}(v_5).$$

Here we show the probability of the observations (TTCCT) and the states at the interior nodes of the tree (ATCG) conditioned on the tree (τ), branch lengths ($\mathbf{v}$), and other model parameters (θ). Note that to calculate the probability of the states at the tips of the tree, we used the stationary probability of the process (π) and also the transition probabilities [$p_{ij}(v)$]. The stationary probability of the substitution process was used to calculate the probability of the nucleotide at the root of the tree. In this case, we are assuming that the substitution process was running for a very long time before it reached the root of our five-species tree. We then use the transition probabilities to calculate the probabilities of observing the states at each end of the branches. When taking the product of the transition probabilities, we are making the additional assumption that the substitutions on each branch of the tree are independent of one another. This is probably a reasonable assumption for real data sets.

The probability of observing the states at the tips of the tree, described above, was conditioned on the interior nodes of the tree taking specific values (in this case $ATCG$). To calculate the unconditional probability of the observed states at the tips of the tree, we sum over all possible combinations of nucleotide states that can be assigned to the interior nodes of the tree,

$$\Pr(TTCCT | \tau, \mathbf{v}, \theta) = \sum_{w} \sum_{x} \sum_{y} \sum_{z} \Pr(TTCCT, wxyz | \tau, \mathbf{v}, \theta),$$

where $w, x, y, z \in (A, C, G, T)$. Averaging the probabilities over all combinations of states at the interior nodes of the tree accomplishes two things. First, we remove the assumption that the states at the interior nodes take specific values. Second, because the transition probabilities account for all of the possible ways we could have state i at one end of a branch and state j at the other, the probability of the site is also averaged over all possible character histories. Here, we think of a character history as one realization of changes on the tree that is consistent with the observations at the tips of the tree. For example, the parsimony method, besides calculating the minimum number of changes on the tree, also provides a character history; the character history favored by parsimony is the one that minimizes the number of changes required to explain the data. In the case of likelihood-based methods, the likelihood accounts for all possible character histories, with each history weighted by its probability under the substitution model. Nielsen [53] described a method for sampling character histories in proportion to their probability that relies on the interpretation of the rate matrix as specifying waiting times between substitutions. His method provides a means to reconstruct the history of a character that does not inherit the flaws of the parsimony method. Namely, Nielsen's method allows multiple changes on a single branch and also allows for nonparsimonious reconstructions of a character's history. In Chapter 16, Bollback describes how character histories can be mapped onto trees under continuous-time Markov models using the program SIMMAP.

Before moving on to some applications of Bayesian estimation in molecular evolution, we will make two final points. First, in practice, no computer program actually evaluates all combinations of nucleotides that can be assigned to the interior nodes of a tree when calculating the probability of observing the data at a site. There are simply too many combinations for trees of even small size. For example, for a tree of 100 species, there are 99 interior nodes

Table 7.1. Probabilities of individual sites. The probabilities of the 50 sites for the example alignment from the text. The likelihoods are calculated assuming the tree of Figure 7.3 with the branch lengths being $v_1 = 0.1$, $v_2 = 0.1$, $v_3 = 0.2$, $v_4 = 0.1$, $v_5 = 0.1$, $v_6 = 0.1$, $v_7 = 0.2$, and $v_8 = 0.1$. The substitution model parameters were also fixed, with $\kappa = 5$, $\pi_A = 0.4$, $\pi_C = 0.3$, $\pi_G = 0.2$, and $\pi_T = 0.1$.

Site	Prob.	Site	Prob.	Site	Prob.	Site	Prob.	Site	Prob.
1	0.004025	11	0.029483	21	0.179392	31	0.179392	41	0.003755
2	0.001171	12	0.006853	22	0.001003	32	0.154924	42	0.005373
3	0.008008	13	0.024885	23	0.154924	33	0.007647	43	0.016449
4	0.002041	14	0.154924	24	0.179392	34	0.000936	44	0.029483
5	0.005885	15	0.007647	25	0.005719	35	0.024885	45	0.154924
6	0.000397	16	0.024124	26	0.001676	36	0.000403	46	0.047678
7	0.002802	17	0.154924	27	0.000161	37	0.024124	47	0.010442
8	0.179392	18	0.004000	28	0.154924	38	0.154924	48	0.179392
9	0.024124	19	0.154924	29	0.001171	39	0.011088	49	0.002186
10	0.024885	20	0.004025	30	0.047678	40	0.000161	50	0.154924

and 4.02×10^{59} combinations of nucleotides at the ancestral nodes on the tree. Instead, Felsenstein's [14] pruning algorithm is used to calculate the likelihood at a site. Felsenstein's method is mathematically equivalent to the summation shown above but can evaluate the likelihood at a site in a fraction of the time it would take to plow through all combinations of ancestral states. Second, the overall likelihood of a character matrix is the product of the site likelihoods. If we assume that the tree of Figure 7.3 is correct (with all of the parameters taking the values specified in the caption of Table 7.1), then the probability of observing the data is

$$0.004025 \times 0.001171 \times 0.008008 \times \ldots \times 0.154924 = 1.2316 \times 10^{-94},$$

where there are fifty factors, each factor representing the probability of an individual site (column) in the alignment. Table 7.1 shows the probabilities of all fifty sites for the tree of Figure 7.3. Note that the overall probability of observing the data is a very small number ($\approx 10^{-94}$). This is typical of phylogenetic problems and results from the simple fact that many numbers between 0 and 1 are multiplied together. Computers cannot accurately hold very small numbers in memory. Programmers avoid this problem of computer "underflow" by using the log probability of observing the data. The log probability of observing the sample alignment of sequences presented earlier is $\log_e \ell = \log_e(1.2316 \times 10^{-94}) = -216.234734$. The log-likelihood can be accurately stored in computer memory.

7.3.2 Phylogeny Estimation

Frequentist and Bayesian perspectives on phylogeny estimation

The phylogenetic model described in the preceding section has numerous parameters. Minimally, the parameters include the topology of the tree and the lengths of the branches on the tree. In the following, we imagine that every possible tree is labeled: $\tau_1, \tau_2, \ldots, \tau_{B(n)}$. Each tree has its own set of branches, and each branch has a length in terms of expected number of substitutions per site. The lengths of the branches on the ith tree are denoted $\mathbf{v}_i = (v_1, v_2, \ldots, v_{2n-3})$. In addition, there may be parameters associated with the substitution model. The parameters of the substitution model will be denoted θ. For the HKY85 model, the parameters are $\theta = (\kappa, \pi_A, \pi_C, \pi_G, \pi_T)$, but other substitution models may have more or fewer parameters than the HKY85 model. When all of the parameters are specified, one can calculate the likelihood function using the general ideas described in the previous section. The likelihood will be denoted $\ell(\tau_i, \mathbf{v}_i, \theta)$ and is proportional to the probability of observing the data conditioned on the model parameters taking specific values ($\ell(\tau_i, \mathbf{v}_i, \theta) \propto \Pr[\mathbf{X}|\tau_i, \mathbf{v}_i, \theta]$; the alignment of sequences is $\mathbf{X}$).

   Which of the possible trees best explains the alignment of DNA sequences? This is among the most basic questions asked in many molecular evolution studies. In a maximum likelihood analysis, the answer is straightforward: the best estimate of phylogeny is the tree that maximizes the likelihood. This is equivalent to finding the tree that makes the observations most probable. For the toy alignment of sequences given in the previous section, the likelihood is maximized when the tree of Figure 7.3 is used. The 14 other possible trees had a lower likelihood. (This is not surprising because the sequences were simulated on the tree of Figure 7.3.) How was the maximum likelihood tree found? In this case, the program PAUP* [64] visited each of the 15 possible trees. For each tree, it found the combination of parameters that maximized the likelihood. In this analysis, we assumed the HKY85 model, so the parameters included the transition/transversion rate ratio and the nucleotide frequencies. After maximizing the likelihood for each tree, the program picked that tree with the largest likelihood as the best estimate of phylogeny. The approach was described earlier in this chapter; the nuisance parameters (here all of the parameters except for the topology of the tree) are dealt with by maximizing the likelihood with respect to them. The tree of Figure 7.3 has a maximum likelihood score of -211.25187. The parameter estimates on this tree are: $\hat{v}_1 = 0.182$, $\hat{v}_2 = 0.124$, $\hat{v}_{3+8} = 0.226$, $\hat{v}_4 = 0.162$, $\hat{v}_5 = 0.018$, $\hat{v}_6 = 0.159$, $\hat{v}_7 = 0.199$, $\hat{\kappa} = 5.73$, $\hat{\pi}_A = 0.329$, $\hat{\pi}_C = 0.329$, $\hat{\pi}_G = 0.253$, and $\hat{\pi}_T = 0.089$. The method of maximum likelihood is described in more detail in Chapter 2. Importantly, there are many computational shortcuts that can be taken to speed up calculation of the maximum likelihood tree.

In a Bayesian analysis, inferences are based upon the posterior probability distribution of the parameters. The joint posterior probability of all the parameters is calculated using Bayes' theorem as

$$\Pr[\tau_i, \mathbf{v}_i, \theta | \mathbf{X}] = \frac{\Pr[\mathbf{X} | \tau_i, \mathbf{v}_i, \theta] \times \Pr[\tau_i, \mathbf{v}_i, \theta]}{\Pr[\mathbf{X}]}$$

and was only recently applied to the phylogeny problem [44, 45, 57, 46, 74, 41, 47, 52]. The posterior probability is equal to the likelihood ($\Pr[\mathbf{X} | \tau_i, \mathbf{v}_i, \theta]$) times the prior probability of the parameters ($\Pr[\tau_i, \mathbf{v}_i, \theta]$) divided by a normalizing constant ($\Pr[\mathbf{X}]$). The normalizing constant involves a summation over all possible trees and, for each tree, integration over all possible combinations of branch lengths and parameter values. Clearly, the Bayesian method is similar to the method of maximum likelihood; after all, both methods make the same assumptions about the evolutionary process and use the same likelihood function. However, the Bayesian method treats all of the parameters as random variables (note that the posterior probability is the probability of the parameters), and the method also incorporates any prior information the biologist might have about the parameters through their prior probability distribution.

Unfortunately, one cannot calculate the posterior probability distribution of trees analytically. Instead, one resorts to a heuristic algorithm to approximate posterior probabilities of trees. The program MrBayes [35, 59] uses Markov chain Monte Carlo (MCMC; [48, 25]) to approximate posterior probabilities of phylogenetic trees (and the posterior probability density of the model parameters). Briefly, a Markov chain is constructed that has as its state space the parameter values of the model and a stationary distribution that is the posterior probability of the parameters. Samples drawn from this Markov chain while at stationarity are valid, albeit dependent, samples from the posterior probability distribution of the parameters [69]. If one is interested in the posterior probability of a particular phylogenetic tree, one simply notes the fraction of the time the Markov chain visited that tree; the proportion of the time the chain visits the tree is an approximation of that tree's posterior probability. A thorough discussion of MCMC is beyond the scope of this chapter. However, an excellent description of MCMC and its applications in molecular evolution can be found in Chapter 3. We will make only one comment on MCMC as applied to phylogenetics: although MCMC is a wonderful technology that can in many instances practically solve problems that cannot be solved any other way, it is dangerous to apply the method uncritically. It is important when running programs that implement MCMC, such as MrBayes, to critically examine the output from several independent chains for convergence.

We performed a Bayesian analysis on the simulated data set discussed above under the HKY85 model. (We describe how to do the Bayesian analyses performed in this chapter in Appendix 2.) This is an ideal situation because the example data were simulated on the tree of Figure 7.3 under the HKY85 model; the model assumed in the Bayesian analysis is not misspecified. We ran a Markov chain for 1,000,000 cycles using the program MrBayes. The Markov chain visited the tree shown in Figure 7.3 about 99% of the time; the MCMC approximation of the posterior probability of the tree in Figure 7.3 then is about 0.99. This can be considered strong evidence in favor of that tree. The posterior probabilities of phylogenetic trees were calculated by integrating over uncertainty in the other model parameters (such as branch lengths, the transition/tranversion rate ratio, and base frequencies). However, we can turn the study around and ask questions about the parameters of the substitution model. Table 7.2 shows information on the posterior probability density distribution of the substitution model parameters. The table shows the mean, median, and variance of the marginal posterior probability distribution for the tree length (V), transition/transversion rate ratio (κ), and base frequencies $(\pi_A, \pi_C, \pi_G, \pi_T)$. The table also shows the upper and lower limits of an interval that contains 95% of the posterior probability for each parameter. The table shows, for example, that with probability 0.95 the transition/transversion rate ratio is in the interval (2.611, 10.635). In reality, the transition/transversion rate ratio was in that interval. (The data matrix was simulated with $\kappa = 5$.) The mean of the posterior probability distribution for κ was 5.576 (which is fairly close to the true value). The interval we constructed that contains the true value of the parameter with 0.95 probability is called a 95% credible interval. One can construct a credible set of trees in a similar manner; simply order the trees from highest to lowest posterior probability and put the trees into a set (starting from the tree with highest probability) until the cumulative probability of trees in the set is 0.95 [13].

One of the great strengths of the Bayesian approach is the ease with which the results of an analysis can be summarized and interpreted. The posterior probability of a tree has a very simple and direct interpretation: the posterior probability of a tree is the probability that the tree is correct, assuming that the substitution model is correct. It is worth considering how uncertainty in parameter estimates is evaluated in a more traditional phylogenetic approach. Because the tree is not considered a random quantity in other types of analyses, such as a maximum likelihood phylogenetic analysis, one cannot directly assign a probability to the tree. Instead, one has to resort to a rather complicated thought experiment. The thought experiment goes something like this. Assuming that the phylogenetic model is correct and that the parameter estimates take the maximum likelihood values (or better yet, their true values), what would the parameter estimates look like on simulated data sets of the same size as the original data matrix? The distribution of parameter estimates that would be generated in such a study represents the sampling distribution of the parameter. One could construct an interval from the sampling distribution that contains 95% of the parameter estimates from the simulated replicates, and this would be called a confidence interval. A 95% confidence interval is a random interval containing the true value of the parameter with probability 0.95. Very few people have constructed confidence intervals/sets of phylogenetic trees using simulation. The simulation approach we just described is referred to as the parametric bootstrap. A related approach, called the nonparametric bootstrap, generates data matrices of the same size as the original by randomly sampling columns (sites) of the original data matrix with replacement. Each matrix generated using the bootstrap procedure is then analyzed using maximum likelihood under the same model as in the original analysis. The nonparametric bootstrap [16] is widely used in phylogenetic analysis.

Table 7.2. Summary statistics for the marginal posterior probability density distributions of the substitution parameters. The mean, median, variance, and 95% credible interval of the marginal posterior probability density distribution of the substitution parameters of the HKY85 model. The parameters are discussed in the text.

Parameter	Mean	Variance	95% Cred. Interval Lower	Upper	Median
V	0.990	0.025	0.711	1.333	0.980
κ	5.576	4.326	2.611	10.635	5.219
π_A	0.323	0.002	0.235	0.418	0.323
π_C	0.331	0.002	0.238	0.433	0.329
π_G	0.252	0.002	0.176	0.340	0.250
π_T	0.092	0.001	0.047	0.152	0.090

Interpreting posterior probabilities on trees

Trees are rather complex parameters, and it is common to break them into smaller components and analyze these separately. Any tree can be divided into a set of statements about the grouping of taxa. For instance, a rooted tree for four taxa—A, B, C, and D—might contain the groupings (AB) and (ABC). These groupings are called clades, or sometimes taxon bipartitions. In a Bayesian analysis, we can summarize a sample from the posterior distribution of trees in terms of the frequency (posterior probability) of individual clades. This provides an efficient summary of the common characteristics of a possibly large sample of different trees. One of the concerns in Bayesian phylogenetic analysis is the interpretation of the posterior probabilities on trees, or the probabilities of individual clades on trees. The posterior probabilities are usually compared with the nonparametric bootstrap proportions, and many workers have reached the conclusion that the posterior probabilities on clades are too high or that the posterior probabilities do not have an easy interpretation [63]. We find this concern somewhat frustrating, mostly because the implicit assumption is that the nonparametric bootstrap proportions are in some way the correct number that should be assigned to a tree and that any method that gives a different number is in some way suspect. However, it is not clear that the nonparametric bootstrap values on phylogenetic trees should be the gold standard. Indeed, it has been known for at least a decade now that the interpretation of nonparametric bootstrap values on phylogenetic trees is problematic [27]; the bootstrap proportions on trees are better interpreted as a measure of robustness rather than as a confidence interval [28].

What does the posterior probability of a phylogenetic tree represent? Huelsenbeck and Rannala [34] performed a small simulation study that did two things. First, it pointed out that the technique many people used to evaluate the meaning of posterior probabilities was incorrect if the intention was to investigate the best-case scenario for the method (i.e., the situation in which the Bayesian method does not misspecify the model). Second, it pointed out that the common interpretation of the posterior probability of a phylogenetic tree is correct; the posterior probability of a phylogenetic tree is the probability that the tree is correct. The catch is that this is true only when the assumptions of the analysis are correct. Figure 7.4 summarizes the salient points of the Huelsenbeck and Rannala [34] study. The experimental design was as follows. They first randomly sampled a tree, branch lengths, and substitution model parameters from the prior probability distribution of the parameters. (The tree was a small one, with only six species.) This is the main difference between their analysis and all others; they treated the prior model seriously and generated samples from it instead of considering the parameters of the model as fixed when doing the simulations. For each sample from the prior distribution they simulated a data matrix of 100 sites. They then analyzed the simulated data matrix under the correct analysis. Figure 7.4 summarizes the results of 10,000 such simulations for each model. They simulated data

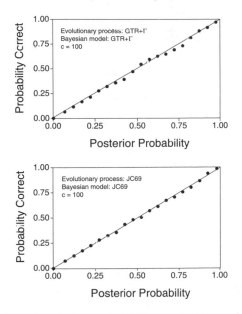

Fig. 7.4. The meaning of posterior probabilities under the model. The relationship between the posterior probability of a phylogenetic tree and the probability that the tree is correct when all of the assumptions of the analysis are satisfied.

under a very simple model (the JC69 model, in which the base frequencies are all equal and the rates of substitution between states are the same) and a complicated model (the GTR+Γ model, in which the nucleotide frequencies are free to vary, the rates of substitution between states are allowed to differ, and the rates across sites are Gamma-distributed). In both cases, the relationship between posterior probabilities and the probability that the tree is correct is linear; the posterior probability of a tree is the probability that the tree is correct, at least when the assumptions of the phylogenetic analysis are satisfied. Importantly, to our knowledge, posterior probabilities are the only measure of support that have this simple interpretation.

Of course, to some extent the simulation results shown in Figure 7.4 are superfluous; the posterior probabilities have always been known to have this interpretation, and the simulations merely confirm the analytical expectation (and incidentally are additional evidence that the program MrBayes is generating valid draws from the posterior probability distribution of trees, at least for simple problems). The more interesting case is when the assumptions of the analysis are incorrect. Suzuki et al. [63] attempted to do such an analysis. Unfortunately, they violated the assumptions of the analysis in a very peculiar way; they simulated data sets in which the underlying phylogeny differed from one gene region to another. This scenario is not a universal concern in phylogenetic analysis (though it can be a problem in the analysis of closely related species, in bacterial phylogenetics, or in population studies).

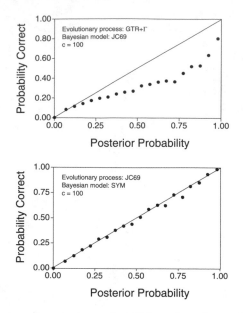

Fig. 7.5. The meaning of posterior probabilities when the model is incorrect. The relationship between the posterior probability of a phylogenetic tree and the probability that the tree is correct when all of the assumptions of the analysis are not met.

The common worry is that the substitution model is incorrect. Huelsenbeck and Rannala [34] performed a few simulations when the assumptions of the analysis are incorrect (Figure 7.5). The top panel in Figure 7.5 shows the case when the evolutionary model is not incorporating some important parameters (the model is underspecified). In this case, the relationship between posterior probabilities and the probability that the tree is correct is not linear. Instead, the method places too much posterior probability on incorrect trees. The situation is not so dire when the evolutionary model has unnecessary parameters (bottom panel in Figure 7.5). These simulation results are consistent with empirical observations of decreasing clade probabilities when the same data are analyzed under increasingly complex models [55].

Bayesian model choice

It appears that Bayesian analysis can be sensitive to model misspecification. It is important to note that the best tree selected under the Bayesian criterion is unlikely to differ significantly from the maximum likelihood tree, mostly because the prior should have a small effect on phylogeny choice when the data set is reasonably large. It is also important to note that it is not really a problem with the Bayesian method but rather with the models used to analyze the data. In a sense, biologists have a method in hand that, in principle, has some very desirable properties: it is fast, allows analysis of complex models in a timely way, and has a correct and simple interpretation when the assumptions of the analysis are satisfied.

The simulation studies summarized in the previous section, along with many simulation studies that examine the performance of phylogenetic methods [29, 30], suggest that it is important to analyze sequence data under as realistic a model as possible. Unfortunately, even the most complicated models currently used in phylogenetic analysis are quite simple and fail to capture important evolutionary processes that generated the sequence data. Phylogenetic models need to be improved to capture evolutionary processes most likely to influence phylogeny estimation. It is impossible to know with certainty what advances will be made in improving phylogenetic models, but we can speculate on what the future might hold. For one thing, it seems important to relax the assumption that the substitution process is homogeneous over the entire phylogenetic history of the organisms under study. This assumption might be relaxed in a number of ways. For example, Foster [17] has relaxed the assumption that nucleotide frequencies are constant over time, and Galtier and Gouy [18] and Galtier et al. [19] relaxed the assumption that the GC content is a constant over a phylogenetic tree. Other such improvements are undoubtedly in store, and Bayesian methods are likely to play an important role in evaluating such models. We can also imagine upper bounds on how many parameters can be added to a phylogenetic model while still maintaining the ability to estimate them from sequence data. It is not clear how close we currently are to that situation. We know that maximum likelihood is consistent for the models typically used in phylogenetic analysis [9, 58], but we do not know whether consistency will be maintained for nonhomogeneous models or other models that account for other evolutionary processes.

We can be certain that analysis of more parameter-rich models will be quite complicated and may require a different perspective on model choice than the one that is widespread in phylogenetics today. Currently, selecting the best model for a particular alignment of DNA sequences is a straightforward affair. For example, the substitution models implemented in the program PAUP* are all a special case of the general time-reversible (GTR) model. The GTR model has instantaneous rate matrix

$$
\mathbf{Q} = \{q_{ij}\} = \begin{pmatrix} - & r_{AC}\pi_C & r_{AG}\pi_G & r_{AT}\pi_T \\ r_{AC}\pi_A & - & r_{CG}\pi_G & r_{CT}\pi_T \\ r_{AG}\pi_A & r_{CG}\pi_C & - & r_{GT}\pi_T \\ r_{AT}\pi_A & r_{CT}\pi_C & r_{GT}\pi_G & - \end{pmatrix} \mu
$$

[67]. Other commonly used models of phylogenetic analysis are all special cases of the GTR model with constraints on its parameters. For example, the HKY85 model constrains the transitions to be one rate ($r_{AG} = r_{CT}$) and the transversions to have another, potentially different rate ($r_{AC} = r_{AT} = r_{CG} = r_{GT}$). The Felsenstein (F81, [14]) model further constrains the transitions and transversions to have the same rate ($r_{AC} = r_{AG} = r_{AT} = r_{CG} = r_{CT=}r_{GT}$).

These models are nested one within the other. The F81 model is a special case of the HKY85 model, and the HKY85 model is a special case of the GTR model. In the programs PAUP* and MrBayes, these different models are set using the "nst" option: nst can be set to 1, 2, or 6 for the F81, HKY85, or GTR models, respectively. Because the models are nested, one can choose an appropriate model using likelihood ratio tests. The likelihood ratio for a comparison of the F81 and HKY85 models is

$$\Lambda = \frac{\max[\ell(\text{F81})]}{\max[\ell(\text{HKY85})]}.$$

Because the models are nested, $\Lambda \leq 1$ and $-2 \log_e \Lambda$ asymptotically follows a χ^2 distribution with one degree of freedom under the null hypothesis. This type of test can be applied to a number of nested models in order to choose the best of them. This approach is easy to perform by hand using a program such as PAUP* but has also been automated in the program Modeltest [56].

The current machinery for model choice appears to work quite well when the universe of candidate models is limited (as is the current case in phylogenetics). But what happens when we reach that happy situation in which the universe of candidate models (pool of models to choose among) is large and the relationship among the models is not nested? There are a number of alternative ways model choice can be performed in this situation. One could use information criteria, such as the Akaike information criterion (AIC), to choose among a pool of candidate models [3]. One could also use the Cox test [10], which uses the likelihood ratio as the test statistic but simulates the null distribution. One might also use Bayes factors to choose among models. Here we will describe how Bayes factors, calculated using MCMC, can be used to choose among a potentially large set of candidate models.

The Bayes factor for a comparison of two models, M_1 and M_2, is

$$BF_{12} = \frac{\Pr[X|M_1]}{\Pr[X|M_2]}.$$

A Bayes factor greater than one is support for M_1, whereas the opposite is true for Bayes factors less than one. Note that the Bayes factor is simply the ratio of the marginal likelihoods of the two models. The Bayes factor integrates over uncertainty in the parameters. The likelihood ratio, on the other hand, maximizes the likelihood with respect to the parameters of the model. Jeffreys [36] provided a table for the interpretation of Bayes factors. In general, the Bayes factor describes the degree by which you change your opinion about rival hypotheses after observing data.

Here we will describe how Bayes factors can be used to choose among substitution models ([32]; also see [62]). First, we will note that the universe of possible time-reversible substitution models is much larger than typically implemented in phylogenetic programs. Appendix 1 shows all of the possible time-reversible substitution models. There are 203 of them, though only a few of them have been named (formally described in a paper). (For the reader interested in the combinatorics, the number of substitution models is given by the Bell [5] numbers.) We use a special notation to describe each of these models. We assign index values to each of the six substitution rates in the order AC, AG, AT, CG, CT, GT. If a model has the constraint that $r_i = r_j$, then the index value for those two rates is the same. Moreover, the index number for the first rate is always 1, and indices are labeled sequentially. So, for example, "111111" denotes the Jukes and Cantor [38] or Felsenstein [14] model and "121121" denotes the Kimura [40], Hasegawa et al. [24, 23], or Felsenstein [15] model. The simplest model is "111111" and the most complex is the GTR model, "123456." The program PAUP* can implement all of these models through a little-used option. (The command "lset nst=6 rmatrix=estimate rclass=(abbcba)" implements one of the unnamed models, constraining $r_{AC} = r_{GT}$ and $r_{AG} = r_{AT} = r_{CT}$, with r_{CG} having another independent rate.) The interested reader can contact J.P.H. for a file that instructs the program PAUP* to maximize the likelihood for each of the 203 possible substitution models. This would allow one to choose among substitution models using AIC or related information criteria.

To calculate the Bayes factors for the different substitution models, we first need to calculate the posterior probability for each of the possible models. We do this using MCMC. Here, the goal is to construct a Markov chain that visits substitution models in proportion to their posterior probability. We could not use the normal theory for constructing a Markov chain for MCMC analysis because the dimensionality of the problem changes from model to model; the 203 models often differ in the number of substitution rates. Instead, we constructed a Markov chain using reversible jump to visit candidate substitution models [22]. Reversible jump MCMC is described in more detail by Larget (Chapter 3). The program we wrote uses two proposal mechanisms to move among models. One proposal mechanism takes a group of substitution rates that are constrained to be the same and splits them into two groups with potentially different rates. The other mechanism takes two groups of substitution rates, each of which has substitutions constrained to be the same, and merges the two groups into one.

To begin, let's examine the simple data matrix that we have been using throughout this chapter: the five-species matrix of 50 sites simulated under the HKY85 model on the tree of Figure 7.3. Up to now, we have been performing all of our analyses—maximum likelihood and Bayesian—under the HKY85 model of DNA substitution (the true model) for this alignment. However, which model is selected as best using the Bayesian reversible jump MCMC approach? Is the true model, or at least one similar to the true model, chosen as the best? We ran the reversible jump MCMC program for a total of 10,000,000 cycles on the small simulated data set. The true model (M_{15}, 121121) was visited with the highest frequency; this model was visited 14.2% of the time, which means the posterior probability of this model is about 0.142. What is the Bayes factor for a comparison of M_{15} with all of the other models (M_{15}^C)?

As described above, the Bayes factor is the ratio of the marginal likelihoods. It also can be calculated, however, as the ratio of the posterior odds to the prior odds of the two hypotheses of interest:

$$BF_{12} = \frac{\Pr[X|M_1]}{\Pr[X|M_2]} = \frac{\frac{\Pr[M_1|X]}{\Pr[M_2|X]}}{\frac{\Pr[M_1]}{\Pr[M_2]}}.$$

The posterior probability of M_{15} is $\Pr[M_{15}|X] = 0.142$, and the posterior probability of all of the other models against which we are comparing M_{15} is just $\Pr[M_{15}^C|X] = 1 - \Pr[M_{15}|X] = 1 - 0.142 = 0.858$. We also know the prior probabilities of the hypotheses. We assumed a uniform prior distribution on all of the possible models, so the prior probability of any specific model is $1/203 = 0.0049$. The Bayes factor for a comparison of M_{15} with the other models is then

$$BF_{12} = \frac{\frac{\Pr[M_{15}|X]}{\Pr[M_{15}^C|X]}}{\frac{\Pr[M_{15}]}{\Pr[M_{15}^C]}} = \frac{\frac{0.142}{0.858}}{\frac{1/203}{202/203}} = 33.4.$$

This means that we change our mind about the relative tenability of the two hypotheses by a factor of about 33 after observing the small data matrix. A Bayes factor of 33 would be considered strong evidence in favor of the model [36]. We can also construct a 95% credible set of models. This is a set of models that has a cumulative posterior probability of 0.95. The 95% credible set included 41 models, which in order were 121121, 121131, 123123, 121321, 121341, 123143, 121323, 123321, 121343, 123121, 123341, 121123, 123323, 123141, 121134, 123343, 121331, 121345, 123423, 123421, 123451, 123453, 123145, 121324, 123124, 123324, 123424, 123454, 123345, 123456, 121133, 123441, 121334, 121333, 123443, 123425, 123313, 121111, 123131, 121344, and 123331. Note that the best of these models (the first 16, in fact, which have a cumulative posterior probability of 0.72) do not constrain a transition to have the same rate as a transversion. One can see that the second-best model (M_{40}, 121131) has this property. The second best-model also happens to be a named one (it is the model described by Tamura and Nei, [66]). The third-best model, however, is not a named one.

Huelsenbeck et al. [32] examined 16 data sets using the approach described here. The details about the data sets can be found in that paper. Table 7.3 summarizes the results. In most cases, the posterior probability was spread across a handful of models. The Bayes factors ranged from 52.3 to about 500, suggesting that all of the alignments contained considerable information about which models are preferred. Also, one can see that for 14 of the 16 data matrices, the 95% credible set contains models that do not constrain transitions to have the same rate as transversions. The best models are usually variants of the model first proposed by Kimura [40]. The exceptions are the HIV-*env* and vertebrate β-globin alignments. The Bayesian approach helped us find these unusual models, which would not usually be considered in a more traditional approach to model choice.

Practicing biologists already favor "automated" approaches to choosing among models. The program Modeltest [56] is very popular for this reason; even though the universe of models of interest to the biologist (i.e., implemented in a computer program) is of only moderate size, it is convenient to have a program that automatically considers each of these models and returns the best of them. The program Modeltest, for example, typically looks at seven of the 203 possible time-reversible substitution models, considering only nested models that are implemented in most phylogeny packages. One could reasonably argue that the number of models currently implemented is small enough that one could perform model choice by hand, with the corresponding advantage that it promotes a more intimate exploration of the data by the biologist, promotes understanding of the models, and keeps the basic scientific responsibility of choosing which hypotheses to investigate in the biologist's hands. However, as models become more complicated and the number of possible models increases, it becomes more difficult to perform model choice by hand. In such cases, an approach like the one described here might be useful.

Table 7.3. The best models for 16 data sets using Bayes factors. PP, the model with the highest posterior probability, with its corresponding probability; BF, the Bayes factor for the best model.

Name	PP	BF	95% Credible Set of Models
Angiosperms	189 (0.41)	142.7	(189, 193, 125, 147, 203)
Archaea	198 (0.70)	472.1	(198, 168, 203)
Bats	112 (0.32)	95.0	(112, 50, 162, 147, 125, 152, 90, 183, 157, 122, 15, 189)
Butterflies	136 (0.32)	93.7	(136, 162, 112, 90, 168, 40, 125, 191, 201, 183, 198, 152, 189)
Crocodiles	40 (0.27)	74.2	(40, 125, 166, 134, 168, 189, 191, 162, 193)
Gophers	112 (0.28)	77.5	(112 ,162, 15, 50, 40, 189, 125, 147, 95, 90, 138, 201, 183, 136, 117, 152, 122, 191)
HIV-1 (*env*)	25 (0.29)	83.0	(25, 60, 50, 64, 100, 125, 102, 97, 164, 169, 152, 159, 173, 157, 175, 147, 171, 191, 193, 189, 140, 117)
HIV-1 (*pol*)	50 (0.62)	335.2	(50, 125, 157, 152, 147, 193)
Lice	15 (0.56)	260.0	(15, 40, 117, 90, 50, 122, 136, 95, 166, 112, 125)
Lizards	193 (0.70)	481.1	(193, 138, 200, 203)
Mammals	193 (0.64)	364.3	(193, 203)
Parrotfish	162 (0.56)	258.0	(162, 189, 201)
Primates	15 (0.31)	91.0	(15, 40, 112, 95, 138, 162, 90, 136, 50, 125, 168, 122, 166, 117, 134)
Vertebrates	125 (0.21)	52.3	(125, 40, 168, 64, 134, 189, 166, 193, 191, 162, 136, 171, 198, 138, 50, 175, 173)
Water snakes	166 (0.55)	242.9	(166, 191, 117, 152, 134, 200, 198, 177)
Whales	15 (0.60)	300.1	(15, 40, 117, 95, 85, 122, 112, 90, 134, 50, 166)

7.3.3 Inferring Phylogeny under Complex Models

Alignments that contain multiple genes, or data of different types, are becoming much more common. It is now relatively easy to sequence multiple genes for any particular phylogenetic analysis, leading to data sets that were uncommon just a few years ago. For example, consider the data set collected by Kim et al. [39], which is fairly typical of those that are now collected for phylogenetic problems. They looked at sequences from three different genes sampled from 27 leaf beetles: the second variable region (D2) of the nuclear rRNA large subunit (28S) and partial sequences from a nuclear gene (EF-1α) and a mitochondrial gene (COI). They also had information from 49 morphological characters. (Although the program MrBayes can analyze morphological data in combination with molecular data, using the approach described by Lewis [43], we do not examine the morphological characters of the Kim et al. study in this chapter. This is a book on molecular evolution, after all. The reader interested in Bayesian analysis of combined morphological and molecular data is referred to the paper by Nylander et al. [55].) The molecular characters of the Kim et al. [39] study were carefully aligned; the ribosomal sequences were aligned using the secondary structure as a guide, and the protein-coding genes were aligned first by the translated amino acid sequence. For illustrative purposes, we are going to consider the amino acid sequences from the COI gene and not the complete DNA sequence. This is probably not the best approach because there is information in the DNA sequence that is being lost when only the amino acid sequence of the gene is considered. However, we want to show how data of different types can be analyzed in MrBayes.

The data from the Kim et al. [39] study that we examine, then, consists of three parts: the nucleotide sequences from the 28S rRNA gene, the nucleotide sequences from the EF-1α gene, and the amino acid sequences from the COI gene. Each of these partitions of the data requires careful consideration. To begin with, it is clear that the same sort of continuous-time Markov chain model is not going to be appropriate for each of these gene regions. After all, the nucleotide part of the alignment has only four states whereas the amino acid part of the alignment (the COI gene) has 20 potential states. We could resort to a very simple partitioned analysis, treating all of the nucleotide sequences with one model and the amino acid sequences with another. However, this approach, too, has problems. Is it really reasonable to treat the protein-coding DNA sequences in the same way as the ribosomal sequences? Moreover, in this case we have information on the secondary structure of the ribosomal gene; we know which nucleotides probably form Watson-Crick pairs in the stem regions of the ribosomal gene. It seems sensible that this information should be accommodated in the analysis of the sequences.

One of the strengths of likelihood-based approaches in general, and the program MrBayes in particular, is that heterogeneous data of the type collected by Kim et al. [39] can be included in a single analysis, with the peculiarities of the substitution process in each partition accounted for. Here are the special considerations we think each data partition of the Kim et al. [39] study raise:

Stem regions of the 28S rRNA nucleotide sequences. Although the assumption of independence across sites (invoked when one multiplies the probabilities of columns in the alignment to get the likelihood) is not necessarily a good one for any data set, it seems especially bad for the stem regions of ribosomal genes. The secondary structure in ribosomal genes plays an important functional role. The functional importance of secondary structure in ribosomal genes causes nonindependence of substitutions in sites participating in a Watson-Crick pair: specifically, if a mutation occurs in one member of a base pair in a functionally important stem, natural selection causes the rate of substitution to be higher for compensatory changes. That is, individuals with a mutation that restores the base pairing have a higher fitness than individuals that do not carry the mutation, and the mutation may eventually become fixed in the population. The end result of natural selection acting on maintenance of stems is a signature of covariation between paired nucleotides.

Schöniger and von Haeseler [60] described a model that accounts for the nonindependence of substitutions in stem regions of ribosomal genes. They suggest that instead of modeling the substitution process on a site-by-site basis using the models described earlier in this chapter, as was then common, substitutions should be modeled on both of the nucleotides participating in the stem pair bond—the doublet. Instead of four states, the doublet model of Schöniger and von Haeseler [60] has 16 states (all possible doublets: AA, AC, AG, AU,..., UA, UC, UG, UU). The instantaneous rate matrix instead of being 4×4 is now 16×16. Each element of the rate matrix, $\mathbf{Q}$, can be specified as follows:

$$q_{ij} = \begin{cases} \kappa \pi_j & : & \text{transition} \\ \pi_j & : & \text{transversion} \\ 0 & : & i \text{ and } j \text{ differ at two positions} \end{cases}.$$

Note that this model only allows a single substitution in an instant of time; substitutions between doublets like $AA \to CG$ have an instantaneous rate of zero. This is not to say that transitions between such doublets are not allowed, only that a minimum of two substitutions is required. Just as there are different parameterizations of the 4×4 models, one can have different parameterizations of the doublet model. The one described here allows a transition/transversion rate bias. However, one could construct a doublet model under any of the models shown in Appendix 1.

Loop regions of the 28S rRNA nucleotide sequences. We will use a more traditional 4×4 model for the loop regions of the ribosomal genes. Nucleotides in the loop regions presumably do not participate in any strong interactions with other sites (at least that we can identify beforehand).

EF-1α nucleotide sequences. Special attention should be paid to the choice of model for protein-coding genes, where the structure of the codon causes heterogeneity at the different codon positions, along with potential nonindependence of substitutions within the codon. The rate of substitution is the most obvious difference at different codon positions. Because of the redundancy of the genetic code, typically second positions are the most conservative and third codon positions are the least conservative. Often people approach this problem of rate variation by grouping the nucleotides at the first, second, and third codon positions into different partitions and allow the overall rate of substitution to differ at the different positions. Another approach, and the one we take here, is to stretch the model of DNA substitution around the codon [21, 50]. We now have 64 possible states (the triplets AAA, AAC, AAG, AAT, ACA,..., TTT), and instead of a 4×4—or even a 16×16—rate matrix, we have a 64×64 instantaneous rate matrix describing the continuous-time Markov chain. Usually, the stop codons are excluded from the state space, and the rate matrix, now 61×61 for the universal code, is

$$q_{ij} = \begin{cases} \omega\kappa\pi_j & : & \text{nonsynonymous transition} \\ \omega\pi_j & : & \text{nonsynonymous transversion} \\ \kappa\pi_j & : & \text{synonymous transition} \\ \pi_j & : & \text{synonymous transversion} \\ 0 & : & i \text{ and } j \text{ differ at more than one position,} \end{cases}$$

where ω is the nonsynonymous/synonymous rate ratio, κ is the transition/transversion rate ratio, and π_j is the stationary frequency of codon j [21, 50]. This matrix specifies the rate of change from codon i to codon j. This rate matrix, like the 4×4 and 16×16 rate matrices, only allows one substitution at a time.

The traditional codon model, described here, does not allow the nonsynonymous/synonymous rate to vary across sites. This assumption has been relaxed. Nielsen and Yang [54] allowed the ω at a site to be a random variable. Their method allows ω to vary across the sequence and also the identification of amino acid positions under directional, or positive, selection. The program PAML [73] implements an empirical Bayes approach to identifying amino acid positions under positive selection. MrBayes uses the same general idea to identify positive selection but implements a fully Bayesian approach, integrating over uncertainty in model parameters [31]. Here, we will not allow the nonsynonymous/synonymous rate to vary across sites.

COI amino acid sequences. In some ways, modeling the amino acid sequences is more complicated for the nucleotide sequences. Some sort of continuous-time Markov chain with 20 states seems appropriate. The most general time-reversible substitution model for amino acids is

$$\mathbf{Q} = \{q_{ij}\} = \begin{pmatrix} - & r_{AR}\pi_R & r_{AN}\pi_N & \cdots & r_{AW}\pi_W & r_{AY}\pi_Y & r_{AV}\pi_V \\ r_{AR}\pi_A & - & r_{RN}\pi_N & \cdots & r_{RW}\pi_W & r_{RY}\pi_Y & r_{RV}\pi_V \\ r_{AN}\pi_A & r_{RN}\pi_R & - & \cdots & r_{NW}\pi_W & r_{NY}\pi_Y & r_{NV}\pi_V \\ \vdots & \vdots & \vdots & \ddots & \vdots & \vdots & \vdots \\ r_{AW}\pi_A & r_{RW}\pi_R & r_{NW}\pi_N & \cdots & - & r_{WY}\pi_Y & r_{WV}\pi_V \\ r_{AY}\pi_A & r_{RY}\pi_R & r_{NY}\pi_N & \cdots & r_{YW}\pi_W & - & r_{YV}\pi_V \\ r_{AV}\pi_A & r_{RV}\pi_R & r_{NV}\pi_N & \cdots & r_{WV}\pi_W & r_{YV}\pi_Y & - \end{pmatrix} \mu.$$

(The dots represent rows and columns that are not shown. The entire matrix is too large to be printed nicely on the page.) There are a total of 208 free parameters; 19 of them involve the stationary frequencies of the amino acids. Knowing 19 of the amino acid frequencies allows you to calculate the frequency of the 20th, so there are a total of 19 free parameters. Similarly, there are a total of $20 \times 19/2 - 1 = 189$ rate parameters. Contrast this with the codon model. The size of the rate matrix for the codon model is much larger than the size of the amino acid rate matrix ($61 \times 61 = 3721$ versus $20 \times 20 = 400$). However, there are fewer free parameters for even the most general time-reversible codon model (given that it is formulated as specified above) than there are for the most general time-reversible amino acid model (66 and 208 for the codon and amino acid matrix, respectively). Of course, the reason the codon model has so few parameters for its size is that many of the entries in the matrix are zero.

Molecular evolutionists have come up with a unique solution to the problem of the large number of potential free parameters in the amino acid matrices. They fix them all to specific values. The parameters are estimated once on large databases of amino acid sequence alignments. The details of how to do this are beyond the scope of this chapter. But, the end result is that we have a number of amino acid rate matrices, each with no free parameters (nothing to estimate), that are designed for specific types of data. These matrices go by different names: Poisson [7], Jones [37], Dayhoff [11], Mtrev [1], Mtmam [8], WAG [71], Rtrev [12], Cprev [2], Blossum [26], and Vt [49]. The amino acid models are designed for use with different types of data. For example, WAG was estimated on nuclear genes, Cprev on chloroplast genes, and Rtrev on viral genes. Which of these models is the appropriate one for the mitochondrial COI gene sequences for leaf beetles? It is not clear which one we should use; nobody has ever designed a mitochondrial amino acid model for insects, much less leaf beetles. It might make sense to use one of the mitochondrial matrices, such as the Mtrev or Mtmam models. However, we can do better than this. Instead of assuming a specific model for the analyses, we can let the amino acid model be a random variable. We will assume that the ten amino acid models listed above all have equal prior probability. We will use MCMC to sum over the uncertainty in the models. This is the same approach described in the previous section, where we used reversible jump MCMC to choose among all possible time-reversible nucleotide substitution models. Fortunately, we do not need to resort to reversible jump MCMC here because all of the parameters of the models are fixed. We do not change dimensions when going from one amino acid model to another.

There are only a few other caveats to consider before we can actually start our analysis of the leaf beetle data with the complex substitution model. Many of the parameters of the model for the individual partitions are shared across partitions. These parameters include the tree, branch lengths, and the rates of substitution under the GTR model for the nucleotide data. Because we are mostly interested in estimating phylogeny here, we will assume that the same tree underlies each of the partitions. That is, we will not allow one tree for the EF-1α gene and another for the loop regions of the 28S ribosomal gene. This seems like a reasonable choice as we have no a priori reason to expect the trees for each partition to differ. However, we might expect the rates of substitution to differ systematically across genes (some might be more evolutionarily constrained) and also for rates to vary from site to site within a gene. We do the following to account for rate variation across and within partitions. Across partitions, we apply a site-specific model by introducing a single parameter for each partition that increases or decreases the rate of substitution for all of the sites within the gene. For example, if the rate multipliers were $m_1 = 0.1$, $m_2 = 1.0$, $m_3 = 2.0$, and $m_4 = 0.9$, then the first and fourth partitions would have, on average, a rate of substitution lower than the mean rate, and the third partition would have a rate greater than the mean rate. In this hypothetical example, the second partition has a rate exactly equal to the mean rate of substitution. Site-specific models are often denoted in the literature by SS; the GTR model with site-specific rate variation is denoted GTR+SS. The site-specific model, although it allows rates to vary systematically from one partition to another, does not account for rate variation among site within a partition. Here we assume that the rate at a site is a random variable drawn from a Gamma distribution. This is commonly assumed in the literature, and Gamma rate variation models are often denoted with a Γ. We are assuming a mixture of rate variation models, so our models could be denoted something like GTR+SS+Γ. The modeling assumptions we are making can be summarized in a table:

Partition	# States	Model	Substitution Rate Variation
Stem	16	GTR	Gamma
Loop	4	GTR	Gamma
EF-1α	61	GTR	Equal
COI	20	Mixture	Gamma

We will also allow parameters that could potentially be constrained to be equal across partitions, such as the shape parameters of the Gamma rate variation model, to be different. The parameters of the model that need to be estimated include:

Parameters	Notes
τ & $\mathbf{v}$	Tree and branch lengths, shared across all of the partitions
$\pi_{AA} \ldots \pi_{UU}$	State frequencies for the stem region partition
$\pi_A \ldots \pi_T$	State frequencies for the loop region partition
$\pi_{AAA} \ldots \pi_{TTT}$	Codon frequencies for the EF-1α gene
$\pi_A \ldots \pi_V$	Amino acid frequencies for the COI gene
$r_{AC}^{(1)} \ldots r_{GT}^{(1)}$	The GTR rate parameters for the loop region partition
$r_{AC}^{(2)} \ldots r_{GT}^{(2)}$	The GTR rate parameters for the stem region partition
$r_{AC}^{(3)} \ldots r_{GT}^{(3)}$	The GTR rate parameters for the EF-1α gene
ω	The nonsynonymous/synonymous rate ratio for the EF-1α gene
α_1	The Gamma shape parameter for the loop region partition
α_2	The Gamma shape parameter for the stem region partition
α_4	The Gamma shape parameter for the COI amino acid data
m_1	The rate multiplier for the loop region partition
m_2	The rate multiplier for the stem region partition
m_3	The rate multiplier for the EF-1α gene
m_4	The rate multiplier for the COI gene
S	The amino acid model for the COI gene

Note that we are allowing most of the parameters to be estimated independently for each gene partition. It is not clear that this is the best strategy. For example, the data might be consistent with some of the parameters being constrained to be the same across partitions. This would allow us to be more parsimonious with our parameters. However, at this time there is no easy way of deciding which pattern of constraints is the best for partitioned data.

We used MrBayes to analyze the data under the complicated substitution model. We ran an MCMC algorithm for 3,000,000 update cycles, sampling the chain every one hundredth cycle. Figure 7.6 shows a majority rule consensus tree of the trees that were visited during the course of the MCMC analysis. (The tree is based on samples taken during the last two million cycles of the chain.) The tree has additional information on it. For one thing, the numbers at the interior nodes represent the posterior probability of that clade being correct (again assuming the model is correct). For another, the branch lengths on the majority rule tree are proportional to the mean of the posterior probability of the branch length.

The Bayesian analysis also provided information on the parameters of the model. Appendix 3 summarizes the marginal posterior probability of each parameter. There are a few points to note here. First, the nonsynonymous/synonymous rate ratio (ω) is estimated to be a very small number. This is consistent with the EF-1α gene being under strong purifying selection. (Substitutions leading to amino acid changes are strongly selected against.) Second, the rate multiplier parameters for the site specific model (m_1, m_2, m_3, m_4) indicate that the rate of substitution is different for the gene regions. The stem partition of the ribosomal gene is the most conservative. Third, the doublet stationary frequency parameters ($\pi_{AA} \ldots \pi_{TT}$) are consistent with a pattern of higher rates for Watson-Crick doublets; note that the stationary frequency is highest for the AT, TA, GC, and CG doublets. Finally, in this analysis, we

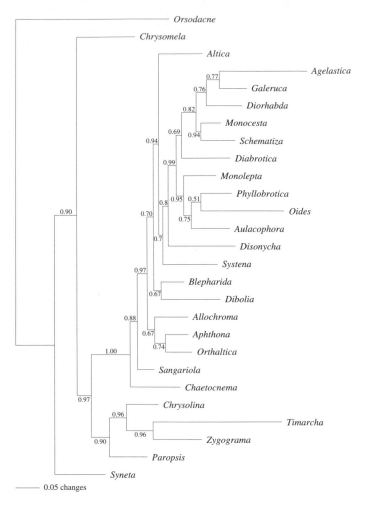

Fig. 7.6. Bayesian phylogenetic tree of leaf beetles. A majority rule tree of the trees sampled during the course of the MCMC analysis. The numbers at the interior nodes are the marginal posterior probability of the clade being correct.

allowed the stationary frequencies of the states to be random variables and integrated over their uncertainty. All of the state frequency parameters were given a flat Dirichlet prior distribution. Although the base frequencies are commonly estimated via maximum likelihood for simple (4 × 4) models, they are rarely estimated for codon models. Instead, they are usually estimated by using the observed frequencies of the nucleotides at the three codon positions to predict the codon frequencies. In the Bayesian analysis, on the other hand, estimating these parameters is not too onerous.

The only parameter not shown in Appendix 3 is the amino acid model, which was treated as unknown in this analysis. The Markov chain proposed moves among the ten different amino acid models listed earlier. The chain visited the Mtrev model almost all of the time, giving it a posterior probability of 1.0. The results of the Bayesian analysis confirm our guess that the Mtrev should be the most reasonable of the amino acid models because it was estimated using a database of mitochondrial sequences. Importantly, we did not need to rely on our guess of what amino acid model to use and could let the data inform us about the fit of the alternative models.

7.3.4 Estimating Divergence Times

The molecular clock hypothesis states that substitutions accumulate at roughly the same rate along different lineages of a phylogenetic tree [75, 76]. Besides being among the earliest ideas in molecular evolution, the molecular clock hypothesis is an immensely useful one. If true, it suggests a way to estimate the divergence times of species with poor fossil records. The idea in its simplest form is shown in Figure 7.7. The figure shows a tree of three species. The numbers on the branches are the branch lengths in terms of expected number of substitutions per site. Note that the branch lengths on the tree satisfy the molecular clock hypothesis; if you sum the lengths of the branches from the root to each of the tips, you get the same number (0.4). One can estimate branch lengths under the molecular clock hypothesis by constraining the branch lengths to have this property. Figure 7.7 shows the second key assumption that must be made to estimate divergence times. We assume that the divergence of at least one of the clades on the tree is known. In this hypothetical example, we assume that species A and B diverged five million years ago. We have calibrated the molecular clock. The calibration is this: if five million years have elapsed since the common ancestor of A and B, then 0.1 substitutions is equal to five million years. Together, the assumptions of a molecular clock and a calibration allow us to infer that the ancestor of the three species must have diverged 20 million years ago.

There are numerous potential problems with the simple picture we presented:

- Substitutions may not accumulate at the same rate along different lineages. In fact, it is easy to test the molecular clock hypothesis using, for example, a likelihood ratio test [14]. The molecular clock hypothesis is usually rejected for real data sets.
- Even if the molecular clock is true, we do not know the lengths of the branches with certainty. In fact, there are potential errors not only in the branch lengths but also in the tree.
- We do not know the divergence times of any of the species on the tree with absolute certainty. This uncertainty should in some way be accommodated.

The first problem—that substitutions may not accumulate at a constant rate along the phylogenetic tree—has received the most attention from biologists. Many statistical tests have been devised to examine whether rates really are

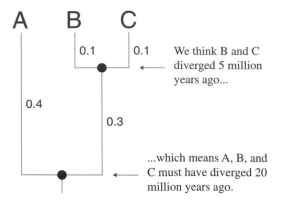

Fig. 7.7. Estimating divergence times using the molecular clock. A tree of three species showing how divergence times can be estimated.

constant over the tree. As already mentioned, applying these tests to real data usually results in the molecular clock being rejected. However, it is still possible that divergence times can be estimated even if the clock is not perfect. Perhaps the tests of the molecular clock are sensitive enough to detect small amounts of rate variation, but the degree of rate variation does not scupper our ability to estimate divergence times. Some biologists have attempted to account for the variation in rates. One approach is to find taxa that are the worst offenders of the clock and either eliminate them [65] or allow a different rate just for those taxa. Another approach specifies a parametric model describing how substitution rates change on the tree. These relaxed clock models still allow estimation of divergence times but may correct for limited degrees of rate variation across lineages. To date, two different models have been proposed for allowing rates to vary across the tree [68, 33] and, in both cases, a Bayesian MCMC approach was taken to estimate parameters.

In the remainder of this section, we will assume that the molecular clock is true or at least that if the molecular clock is violated, we can still meaningfully estimate divergence times. The point of this section is not to provide a definitive answer to the divergence time of any particular group but rather to show how uncertainty in the tree, branch lengths, and calibration times can be accounted for in a Bayesian analysis. We examine two data sets. The first data set included complete mitochondrial protein-coding sequences from 23 mammals [4]. We excluded the platypus (*Ornithorhynchus anatinus*) and the guinea pig (*Cavia porcellus*) from our analysis. We analyzed the alignment of mitochondrial sequences under the GTR+SS model of DNA substitution. The data were partitioned by codon position, and the rates for the first, second, and third positions were estimated. The second data set consists of 104 amino acid sequences sampled from mouse, rat, an artiodactyl, human, and chicken collated by Nei et al. [51]. Nei et al. [51] were mainly interested in estimating the divergence times of the rodents and the rodent-human split and pointed out the importance of taking a multigene approach to divergence time estimation. We analyze their data using the partitioned approach described in the previous section. We partition the data by gene, resulting in 104 divisions in the data. We allow rates to vary systematically across genes using the site-specific model. We allow rates to vary within genes by treating the rate of substitution at an amino acid position as a Gamma-distributed random variable. We allow different Gamma shape parameters for each partition. Moreover, we allow a different amino acid model for each partition, with the actual identity of the amino acid model being unknown. For both data sets, we constrained the branch lengths to obey the molecular clock hypothesis. MrBayes was used to approximate the joint posterior probability of all of the parameters of the evolutionary model. For the mammalian mitochondrial alignment, we ran the MCMC algorithm for a total of one million cycles and based inferences on samples taken during the last 900,000 MCMC cycles. For the amino acid alignments, we ran each of the two independent Markov chains for a total of three million update cycles. We combined the samples taken after the five hundred thousandth cycle.

For the mammalian data set, we had a total of 9000 trees with branch lengths that were sampled from the posterior probability distribution of trees. Each of the trees obeyed the molecular clock, meaning that if one were to take a direct path from each tip of the tree to the root and sum the lengths of the branches on each path, one would obtain the same number. Importantly, the lengths of the branches and the topology of the tree differed from one sample to another. The differences reflect the uncertainty in the data about the tree and branch lengths. The final missing ingredient is a calibration time for some divergence time on the tree. We used the divergence between the cows and the whales as the calibration. Our first analysis of these samples will reflect the typical approach taken when estimating divergence times; we will assume that the divergence between cows and whales was *precisely* 56.5 million years ago. This is a reasonable guess at the divergence time of cows and whales. Figure 7.8 shows the posterior probability distribution of the divergence time at the root of the tree, corresponding to the divergence of marsupial and placental mammals. The top-left panel, marked "Fixed(56.5)", shows the posterior probability of the marsupial-placental split when the cows and whales are assumed to diverge precisely 56.5 million years ago. It shows that even when we assume that the molecular clock is true and the calibration time is known without error, there is considerable uncertainty about the divergence time. The 95% credible interval for the divergence of marsupials from placentals is (115.6, 145.1), a span of about 30 million years in the early Cretaceous period. In fact, it is easy to calculate the probability that the divergence time was in any specific time interval; with (posterior) probabilities 0.0, 0.97, 0.03, and 0.0, the divergence was in the late Cretaceous, early Cretaceous, late Jurassic, and middle Jurassic periods, respectively. These probabilities account for the uncertainty in the topology of the tree, branch lengths on the tree, and

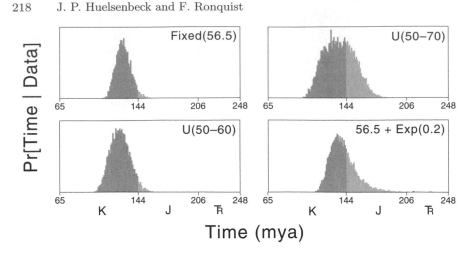

Fig. 7.8. The posterior probability density distribution of the divergence time of placental and marsupial mammals. The distributions were calculated assuming the divergence time between cows and whales was precisely 56.5 million years [Fixed(56.5)], uniformly distributed between two times (U), or no less than 56.5 million years, with an exponentially declining prior distribution into the past [56.5 + Exp(0.2)]. K, J, and Tr are the Cretaceous, Jurassic, and Triassic time periods, respectively.

parameters of the substitution model but do assume that the calibration time was perfectly known.

The three other panels in Figure 7.8 show the posterior probability distribution of the divergence of marsupial and placental mammals when the calibration is not assumed known with certainty. In two of the analyses, we assumed that the cows and whales diverged at some unknown time, constrained to lie in an interval. The probability of the divergence at any time in the interval was uniformly distributed. The last analysis, shown in the lower-right panel of Figure 7.8, assumed that the divergence of cows and whales occurred no more recently than 56.5 million years and was exponentially distributed before then (an offset exponential prior distribution). As expected, the effect of introducing uncertainty in the calibration times is reflected in a posterior probability distribution that is more spread out. The additional uncertainty can be neatly summarized by the 95% credible intervals:

Prior	Credible Interval	Size
Fixed(56.5)	(115.6, 145.1)	29.5
U(50, 60)	(107.8, 145.8)	38.0
U(50, 70)	(110.3, 166.9)	56.6
56.5 + Exp(0.2)	(119.8, 175.6)	55.8

The column marked "Size" shows the duration of the credible interval in millions of years. Clearly, introducing uncertainty in the calibration time is

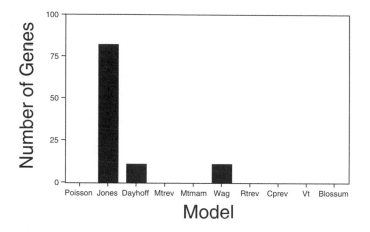

Fig. 7.9. The distribution of best amino acid models for the 104 amino acid alignments. The number of alignments for which each amino acid model was best for the Nei et al. [51] study.

reflected in the posterior probability distribution, and the credible interval becomes larger as more uncertainty is introduced into the calibration time.

The results from the analysis of the 104 concatenated amino acid alignments was similar to that of the mammalian mitochondrial data. However, the model for the amino acid data sets was quite complicated. Besides the tree and branch lengths, there were 104 Gamma shape parameters, 104 rate multipliers for the site-specific model, and 104 unknown amino acid models to estimate. We do not attempt to summarize the information for all of these parameters here. We only show the results for the amino acid models. Figure 7.9 shows which models were chosen as best for the various amino acid alignments. In 82 cases, the model of Jones et al. [37] was chosen as best. The Dayhoff and Wag models [11, 71] were chosen 11 times each. The seven other amino acid models were never chosen as the best one in any of the 104 alignments, though some did receive considerable posterior probability. There was no uncertainty in the topology of the tree chosen using the Bayesian method (Figure 7.10).

As a calibration, Nei et al. [51] assumed that the divergence of birds and mammals occurred exactly 310 million years ago. Table 7.4 summarizes the results of the divergence times for three clades on the tree, assuming the calibration time of Nei et al. [51] as well as three other calibrations that allow for uncertainty in the divergence time of birds and mammals. As might be expected, the uncertainty is greater for the older divergences. Also, having a calibration time that is older than the group of interest makes the posterior probability distribution less vulnerable to errors in the calibration time.

The prior models for the uncertainty in the calibration times we used here are largely arbitrary and chosen mostly to make the point that errors in cali-

Table 7.4. Credible intervals for divergence times of the amino acid data. The 95% credible intervals for the divergence of mouse from rat, human from rodents, and the time at the root of the tree for four different calibrations of the bird-mammal split.

Calibration	Mouse-Rat	Human-Rodent	Root
310	(25.9, 33.4)	(84.5, 97.5)	(448.3, 487.8)
U(288, 310)	(25.0, 33.0)	(80.6, 97.5)	(427.7, 491.8)
288 + Exp(0.1)	(24.6, 32.6)	(79.8, 96.6)	(423.3, 495.1)
288 + Exp(0.05)	(24.9, 34.9)	(80.4, 106.5)	(426.4, 551.6)

bration times can be accounted for in a Bayesian analysis and that these errors can make a difference in the results (at least, these errors can make a difference in how much one believes the results). Experts in the fossils from these groups would place very different prior distributions on the calibration times. For example, Philip Gingerich (pers. comm.) would place a much smaller error on the divergence times between cows and whales than we did here; the fossil record for this group is rich, and it is unlikely that cows and whales diverged as early as 100 million years ago (our offset exponential prior distribution places some weight on this hypothesis along with divergences that are much earlier). Lee [42] pointed out that the widely used bird-mammal calibration of 310 million years is poorly chosen. The earliest synapsids (fossils on the lineage leading to modern-day mammals) are from the upper Pennsylvanian period, about 288 million years ago. This is much more recent than the calibration of 310 million years used by some to calibrate the molecular clock. The Bayesian framework makes it possible to explore how different prior distributions affect the conclusions drawn from a particular data set. When the data are highly informative about the parameters examined, as is commonly the case, the exact choice of prior distribution is likely to have little influence on the results. In dating exercises, however, particularly when only one calibration point is used, the precision of the calibration is likely to affect the dating significantly.

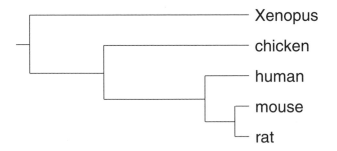

Fig. 7.10. The best tree for the 104 amino acid alignments. This tree had a posterior probability approximated to be 1.0 by the MCMC algorithm. The length of the branch is the mean of the posterior probability distribution.

7.4 Conclusions

In this chapter, we have attempted to demonstrate some of the power and flexibility of the Bayesian approach to the inference of phylogeny and molecular evolution. The most important aspect we want to convey is the efficiency of the Bayesian MCMC methodology in addressing complex models. Current statistical analyses of molecular evolution are based on very simple models inspired by the apparent simplicity of molecular sequences. But beyond the simple sequences of symbols lies tremendous evolutionary complexity. Approaches that ignore this complexity do not utilize the molecular information efficiently and are prone to produce erroneous inferences. Modeling the complexity of molecular evolution more accurately will be critical to future progress in statistical analysis of molecular evolution. The Bayesian MCMC approach provides promising tools for the analysis of these realistic evolutionary models.

References

[1] J. Adachi and M. Hasegawa. MOLPHY version 2.3: Programs for molecular phylogenetics based on maximum likelihood. *Computer Science Monographs of Institute of Statistical Mathematics*, 28:1–150, 1996.

[2] J. Adachi, P. Waddell, W. Martin, and M. Hasegawa. Plastid genome phylogeny and a model of amino acid substitution for proteins encoded by chloroplast DNA. *Journal of Molecular Evolution*, 50:348–358, 2000.

[3] H. Akaike. Information theory as an extension of the maximum likelihood principle. In B. N. Petrov and F. Csaki, editors, *Second International Symposium on Information Theory*, pages 267–281. Akademiai Kiado, Budapest, 1973.

[4] U. Arnason, A. Gullberg, and A. Janke. Phylogenetic analyses of mitochondrial DNA suggest a sister group relationship between Xenartha (Edentata) and Ferungulates. *Molecular Biology and Evolution*, 14:762–768, 1997.

[5] E. T. Bell. Exponential numbers. *American Mathematical Monthly*, 41:411–419, 1934.

[6] J. O. Berger, B. Liseo, and R. L. Wolpert. Integrated likelihood methods for eliminating nuisance parameters. *Statistical Science*, 14:1–28, 1999.

[7] M. J. Bishop and A. E. Friday. Tetrapod relationships: The molecular evidence. In C. Patterson, editor, *Molecules and Morphology in Evolution*, pages 123–139. Cambridge University Press, Cambridge, England, 1987.

[8] Y. Cao, A. Janke, P. J. Waddell, M. Westerman, O. Takenaka, S. Murata, N. Okada, S. Paabo, and M. Hasegawa. Conflict amongst individual mitochondrial proteins in resolving the phylogeny of eutherian orders. *Journal of Molecular Evolution*, 47:307–322, 1998.

[9] J. T. Chang. Full reconstruction of Markov models on evolutionary tree: Identifiability and consistency. *Mathematical Bioscience*, 137:51–73, 1996.

[10] D. R. Cox. Further results on tests of families of alternate hypotheses. *Journal of the Royal Statistical Society B*, 24:406–424, 1962.

[11] M. O. Dayhoff, R. M. Schwartz, and B. C. Orcutt. A model of evolutionary change in proteins. In *Atlas of Protein Sequence and Structure*, volume 5, pages 345–352. National Biomedical Research Foundation, Washington, DC, 1978. Suppl. 3.

[12] M. W. Dimmic, J. S. Rest, D. P. Mindell, and D. Goldstein. rtREV: An amino acid substitution matrix for inference of retrovirus and reverse transcriptase phylogeny. *Journal of Molecular Evolution*, 55:65–73, 2002.

[13] J. Felsenstein. *Statistical inference and the estimation of phylogenies*. PhD thesis, University of Chicago, 1968.

[14] J. Felsenstein. Evolutionary trees from DNA sequences: A maximum likelihood approach. *Journal of Molecular Evolution*, 17:368–376, 1981.

[15] J. Felsenstein. Distance methods for inferring phylogenies: A justification. *Evolution*, 38:16–24, 1984.

[16] J. Felsenstein. Confidence limits on phylogenies: An approach using the bootstrap. *Evolution*, 39:783–791, 1985.

[17] P. G. Foster. Modeling compositional heterogeneity. *Systematic Biology*, 53:485–495, 2004.

[18] N. Galtier and M. Gouy. Inferring pattern and process: Maximum-likelihood implementation of a nonhomogeneous model of DNA sequence evolution for phylogenetic analysis. *Molecular Biology Evolution*, 15:871–879, 1998.

[19] N. Galtier, N. Tourasse, and M. Gouy. A nonhyperthermophilic common ancestor to extant life forms. *Science*, 283:220–221, 1999.

[20] N. Goldman. Maximum likelihood inference of phylogenetic trees with special reference to a Poisson process model of DNA substitution and to parsimony analyses. *Systematic Zoology*, 39:345–361, 1990.

[21] N. Goldman and Z. Yang. A codon-based model of nucleotide substitution for protein-coding DNA sequences. *Molecular Biology and Evolution*, 11:725–736, 1994.

[22] P. J. Green. Reversible jump Markov chain Monte Carlo computation and Bayesian model determination. *Biometrika*, 82:711–732, 1995.

[23] M. Hasegawa, H. Kishino, and T. Yano. Dating the human-ape splitting by a molecular clock of mitochondrial DNA. *Journal of Molecular Evolution*, 22:160–174, 1985.

[24] M. Hasegawa, T. Yano, and H. Kishino. A new molecular clock of mitochondrial DNA and the evolution of Hominoids. *Proceedings of the Japan Academy Series B*, 60:95–98, 1984.

[25] W. K. Hastings. Monte Carlo sampling methods using Markov chains and their applications. *Biometrika*, 57:97–109, 1970.

[26] S. Henikoff and J. G. Henikoff. Amino acid substitution matrices from protein blocks. *Proceedings of the National Academy of Sciences, USA*, 89:10915–10919, 1992.

[27] D. M. Hillis and J. J. Bull. An empirical test of bootstrapping as a method for assessing confidence in phylogenetic analysis. *Systematic Biology*, 42:182–192, 1993.

[28] S. Holmes. Bootstrapping phylogenetic trees: Theory and methods. *Statistical Science*, 18:241–255, 2003.

[29] J. P. Huelsenbeck. Performance of phylogenetic methods in simulation. *Systematic Biology*, 44:17–48, 1995.

[30] J. P. Huelsenbeck. The robustness of two phylogenetic methods: Four taxon simulations reveal a slight superiority of maximum likelihood over neighbor joining. *Molecular Biology and Evolution*, 12:843–849, 1995.

[31] J. P. Huelsenbeck and K. A. Dyer. Bayesian estimation of positively selected sites. *Journal of Molecular Evolution*, 58:661–672, 2004.

[32] J. P. Huelsenbeck, B. Larget, and M. E. Alfaro. Bayesian phylogenetic model selection using reversible jump Markov chain Monte Carlo. *Mol. Biol. Evol*, 21:1123–1133, 2004.

[33] J. P. Huelsenbeck, B. Larget, and D. Swofford. A compound Poisson process for relaxing the molecular clock. *Genetics*, 154:1879–1892, 2000.

[34] J. P. Huelsenbeck and B. Rannala. Frequentist properties of Bayesian posterior probabilities of phylogenetic trees under simple and complex substitution models. *Systematic Biology*. In press.

[35] J. P. Huelsenbeck and F. Ronquist. MrBayes: Bayesian inference of phylogenetic trees. *Bioinformatics*, 17:754–755, 2001.

[36] H. Jeffreys. *Theory of Probability*. Oxford University Press, Oxford, 1961.

[37] D. T. Jones, W. R. Taylor, and J. M. Thornton. The rapid generation of mutation data matrices from protein sequences. *Computer Applications in the Biosciences*, 8:275–282, 1992.

[38] T. H. Jukes and C. R. Cantor. Evolution of protein molecules. In H. N. Munro, editor, *Mammalian Protein Metabolism*, pages 21–123. Academic Press, New York, 1969.

[39] S. Kim, K. M. Kjer, and C. N. Duckett. Comparison between molecular and morphological-based phylogenies of galerucine/alticine leaf beetles (Coleoptera: Chrysomelidae). *Insect Systematic Evolution*, 34:53–64, 2003.

[40] M. Kimura. A simple method for estimating evolutionary rates of base substitutions through comparative studies of nucleotide sequences. *Journal of Molecular Evolution*, 16:111–120, 1980.

[41] B. Larget and D. Simon. Markov chain Monte Carlo algorithms for the Bayesian analysis of phylogenetic trees. *Molecular Biology and Evolution*, 16:750–759, 1999.

[42] M. S. Y. Lee. Molecular clock calibrations and metazoan divergence dates. *Journal of Molecular Evolution*, 49:385–391, 1999.

[43] P. O. Lewis. A likelihood approach to estimating phylogeny from discrete morphological character data. *Systematic Biology*, 50:913–925, 2001.

[44] S. Li. *Phylogenetic tree construction using Markov chain Monte Carlo.* PhD thesis, Ohio State University, Columbus, 1996.

[45] B. Mau. *Bayesian phylogenetic inference via Markov chain Monte Carlo methods.* PhD thesis, University of Wisconsin, Madison, 1996.

[46] B. Mau and M. Newton. Phylogenetic inference for binary data on dendrograms using Markov chain Monte Carlo. *Journal of Computational and Graphical Statistics*, 6:122–131, 1997.

[47] B. Mau, M. Newton, and B. Larget. Bayesian phylogenetic inference via Markov chain Monte Carlo methods. *Biometrics*, 55:1–12, 1999.

[48] N. Metropolis, A. W. Rosenbluth, M. N. Rosenbluth, A. W. Teller, and E. Teller. Equations of state calculations by fast computing machines. *Journal of Chemical Physics*, 21:1087–1091, 1953.

[49] T. Muller and M. Vingron. Modeling amino acid replacement. *Journal of Computational Biology*, 7:761–776, 2000.

[50] S. V. Muse and B. S. Gaut. A likelihood approach for comparing synonymous and nonsynonymous nucleotide substitution rates with application to the chloroplast genome. *Molecular Biology and Evolution*, 11:715–724, 1994.

[51] M. Nei, P. Xu, and G. Glazko. Estimation of divergence times from multiprotein sequences for a few mammalian species and several distantly related organisms. *Proceedings of the National Academy of Sciences, USA*, 98:2497–2502, 2001.

[52] M. Newton, B. Mau, and B. Larget. Markov chain Monte Carlo for the Bayesian analysis of evolutionary trees from aligned molecular sequences. In F. Seillier-Moseiwitch, T. P. Speed, and M. Waterman, editors, *Statistics in Molecular Biology*. Monograph Series of the Institute of Mathematical Statistics, 1999.

[53] R. Nielsen. Mapping mutations on phylogenies. *Systematic Biology*, 51:729–739, 2002.

[54] R. Nielsen and Z. Yang. Likelihood models for detecting positively selected amino acid sites and applications to the HIV-1 envelope gene. *Genetics*, 148:929–936, 1998.

[55] J. A. A. Nylander, F. Ronquist, J. P. Huelsenbeck, and J. L. Nieves-Aldrey. Bayesian phylogenetic analysis of combined data. *Systematic Biology*, 53:47–67, 2004.

[56] D. Posada and K. A. Crandall. Modeltest: Testing the model of DNA substitution. *Bioinformatics*, 14:817–818, 1998.

[57] B. Rannala and Z. Yang. Probability distribution of molecular evolutionary trees: A new method of phylogenetic inference. *Journal of Molecular Evolution*, 43:304–311, 1996.

[58] J. S. Rogers. On the consistency of maximum likelihood estimation of phylogenetic trees from nucleotide sequences. *Systematic Biology*, 46:354–357, 1997.

[59] F. Ronquist and J. P. Huelsenbeck. MrBayes 3: Bayesian phylogenetic inference under mixed models. *Bioinformatics*, 19:1572–1574, 2003.

[60] M. Schöniger and A. von Haeseler. A stochastic model and the evolution of autocorrelated DNA sequences. *Mol. Phyl. Evol.*, 3:240–247, 1994.

[61] E. Schröder. Vier combinatorische probleme. *Z. Math. Phys.*, 15:361–376, 1870.

[62] M. A. Suchard, R. E. Weiss, and J. S. Sinsheimer. Bayesian selection of continuous-time Markov chain evolutionary models. *Molecular Biology and Evolution*, 18:1001–1013, 2001.

[63] Y. Suzuki, G. V. Glazko, and M. Nei. Overcredibility of molecular phylogenies obtained by Bayesian phylogenetics. *Proceedings of the National Academy of Sciences, USA*, 99:15138–16143, 2002.

[64] D. L. Swofford. *PAUP*. Phylogenetic Analysis Using Parsimony (*and Other Methods). Version 4.* Sinauer Associates, Sunderland, MA, 2002.

[65] N. Takezaki, A. Rzhetsky, and M. Nei. Phylogenetic test of molecular clock and linearized trees. *Molecular Biology and Evolution*, 12:823–833, 1995.

[66] K. Tamura and M. Nei. Estimation of the number of nucleotide substitutions in the control region of mitochondrial DNA in humans and chimpanzees. *Molecular Biology and Evolution*, 10:512–526, 1993.

[67] S. Tavaré. Some probabilistic and statistical problems on the analysis of DNA sequences. *Lectures in Mathematics in the Life Sciences*, 17:57–86, 1986.

[68] J. L. Thorne, H. Kishino, and I. S. Painter. Estimating the rate of evolution of the rate of molecular evolution. *Molecular Biology and Evolution*, 15:1647–1657, 1998.

[69] L. Tierney. Markov chains for exploring posterior distributions. *Annals of Statistics*, 22:1701–1762, 1994.

[70] C. Tuffley and M. Steel. Modeling the covarion hypothesis of nucleotide substitution. *Mathematical Biosciences*, 147:63–91, 1998.

[71] S. Whelan and N. Goldman. A general empirical model of protein evolution derived from multiple protein families using a maximum likelihood approach. *Molecular Biology and Evolution*, 18:691–699, 2001.

[72] Z. Yang. Maximum likelihood estimation of phylogeny from DNA sequences when substitution rates differ over sites. *Molecular Biology and Evolution*, 10:1396–1401, 1993.

[73] Z. Yang. PAML: A program package for phylogenetic analysis by maximum likelihood. *Comptuer Applications in Bioscience*, 15:555–556, 1997.

[74] Z. Yang and B. Rannala. Bayesian phylogenetic inference using DNA sequences: A Markov chain Monte Carlo method. *Molecular Biology and Evolution*, 14:717–724, 1997.

[75] E. Zuckerkandl and L. Pauling. Molecular disease, evolution, and genetic heterogeneity. In M. Kasha and B. Pullman, editors, *Horizons in Biochemistry*, pages 189–225. Academic Press, New York, 1962.

[76] E. Zuckerkandl and L. Pauling. Evolutionary divergence and convergence in proteins. In V. Bryson and H. J. Vogel, editors, *Evolving Genes and Proteins*, pages 97–166. Academic Press, New York, 1965.

Appendix 1. All Possible Time-Reversible Models of DNA Substitution

$M_1 = 111111$	$M_{35} = 122322$	$M_{69} = 121322$	$M_{103} = 112132$	$M_{137} = 121314$	$M_{171} = 112343$
$M_2 = 122222$	$M_{36} = 122232$	$M_{70} = 121232$	$M_{104} = 112123$	$M_{138} = 121134$	$M_{172} = 112334$
$M_3 = 121111$	$M_{37} = 122223$	$M_{71} = 121223$	$M_{105} = 111233$	$M_{139} = 112341$	$M_{173} = 112342$
$M_4 = 112111$	$M_{38} = 123111$	$M_{72} = 122312$	$M_{106} = 111232$	$M_{140} = 112314$	$M_{174} = 112324$
$M_5 = 111211$	$M_{39} = 121311$	$M_{73} = 122321$	$M_{107} = 111223$	$M_{141} = 112134$	$M_{175} = 112234$
$M_6 = 111121$	$M_{40} = 121131$	$M_{74} = 122132$	$M_{108} = 112233$	$M_{142} = 111234$	$M_{176} = 123412$
$M_7 = 111112$	$M_{41} = 121113$	$M_{75} = 122123$	$M_{109} = 112323$	$M_{143} = 123344$	$M_{177} = 123421$
$M_8 = 112222$	$M_{42} = 112311$	$M_{76} = 122211$	$M_{110} = 112332$	$M_{144} = 123434$	$M_{178} = 123142$
$M_9 = 121222$	$M_{43} = 112131$	$M_{77} = 122213$	$M_{111} = 121233$	$M_{145} = 123443$	$M_{179} = 123124$
$M_{10} = 122122$	$M_{44} = 112113$	$M_{78} = 123311$	$M_{112} = 121323$	$M_{146} = 123244$	$M_{180} = 123241$
$M_{11} = 122212$	$M_{45} = 111231$	$M_{79} = 123131$	$M_{113} = 121332$	$M_{147} = 123424$	$M_{181} = 123214$
$M_{12} = 122221$	$M_{46} = 111213$	$M_{80} = 123113$	$M_{114} = 122133$	$M_{148} = 123442$	$M_{182} = 121342$
$M_{13} = 122111$	$M_{47} = 111123$	$M_{81} = 121331$	$M_{115} = 122313$	$M_{149} = 122344$	$M_{183} = 121324$
$M_{14} = 121211$	$M_{48} = 122333$	$M_{82} = 121313$	$M_{116} = 122331$	$M_{150} = 122343$	$M_{184} = 121234$
$M_{15} = 121121$	$M_{49} = 123233$	$M_{83} = 121133$	$M_{117} = 123123$	$M_{151} = 122334$	$M_{185} = 122341$
$M_{16} = 121112$	$M_{50} = 123323$	$M_{84} = 123323$	$M_{118} = 123132$	$M_{152} = 123423$	$M_{186} = 122314$
$M_{17} = 112211$	$M_{51} = 123332$	$M_{85} = 123121$	$M_{119} = 123213$	$M_{153} = 123432$	$M_{187} = 122134$
$M_{18} = 112121$	$M_{52} = 123322$	$M_{86} = 123112$	$M_{120} = 123231$	$M_{154} = 123243$	$M_{188} = 123455$
$M_{19} = 112112$	$M_{53} = 123232$	$M_{87} = 122311$	$M_{121} = 123312$	$M_{155} = 123234$	$M_{189} = 123454$
$M_{20} = 111221$	$M_{54} = 123223$	$M_{88} = 122131$	$M_{122} = 123321$	$M_{156} = 123342$	$M_{190} = 123445$
$M_{21} = 111212$	$M_{55} = 122332$	$M_{89} = 122113$	$M_{123} = 123444$	$M_{157} = 123324$	$M_{191} = 123453$
$M_{22} = 111122$	$M_{56} = 122323$	$M_{90} = 121321$	$M_{124} = 123433$	$M_{158} = 123144$	$M_{192} = 123435$
$M_{23} = 111222$	$M_{57} = 122233$	$M_{91} = 121312$	$M_{125} = 123343$	$M_{159} = 123414$	$M_{193} = 123345$
$M_{24} = 112122$	$M_{58} = 121333$	$M_{92} = 121231$	$M_{126} = 123334$	$M_{160} = 123441$	$M_{194} = 123452$
$M_{25} = 112212$	$M_{59} = 123133$	$M_{93} = 121213$	$M_{127} = 123422$	$M_{161} = 121344$	$M_{195} = 123425$
$M_{26} = 112221$	$M_{60} = 123313$	$M_{94} = 121132$	$M_{128} = 123242$	$M_{162} = 121343$	$M_{196} = 123245$
$M_{27} = 121122$	$M_{61} = 123331$	$M_{95} = 121123$	$M_{129} = 123224$	$M_{163} = 121334$	$M_{197} = 122345$
$M_{28} = 121212$	$M_{62} = 112333$	$M_{96} = 112331$	$M_{130} = 122342$	$M_{164} = 123413$	$M_{198} = 123451$
$M_{29} = 121221$	$M_{63} = 112322$	$M_{97} = 112313$	$M_{131} = 122324$	$M_{165} = 123431$	$M_{199} = 123415$
$M_{30} = 122112$	$M_{64} = 112232$	$M_{98} = 112133$	$M_{132} = 122234$	$M_{166} = 123143$	$M_{200} = 123145$
$M_{31} = 122121$	$M_{65} = 112223$	$M_{99} = 112321$	$M_{133} = 123411$	$M_{167} = 123134$	$M_{201} = 121345$
$M_{32} = 122211$	$M_{66} = 123122$	$M_{100} = 112312$	$M_{134} = 123141$	$M_{168} = 123341$	$M_{202} = 112345$
$M_{33} = 123333$	$M_{67} = 123212$	$M_{101} = 112231$	$M_{135} = 123114$	$M_{169} = 123314$	$M_{203} = 123456$
$M_{34} = 123222$	$M_{68} = 123221$	$M_{102} = 112213$	$M_{136} = 121341$	$M_{170} = 112344$	

Appendix 2. Using MrBayes 3.0

MrBayes 3.0 [35, 59] is a program distributed free of charge and can be downloaded from the web at `http://www.mrbayes.net`. The program takes as
input an alignment of DNA, RNA, amino acid, or restriction site data. (Matrices of morphological characters can be input, too.) The program uses Markov
chain Monte Carlo methods to approximate the joint posterior probability
distribution of the phylogenetic tree, branch lengths, and substitution model
parameters. The parameter values sampled by the Markov chain are saved to
two files; one file contains the trees that were sampled, whereas the other file
has the parameter values that were sampled. The program also provides some
commands for summarizing the results. The basic steps (and commands) that
need to be executed to perform a Bayesian analysis of phylogeny using Mr-
Bayes include: (1) reading in the data file ("execute [file name]"); (2) setting
the model (using the "lset" and "prset" commands); (3) running the Markov
chain Monte Carlo algorithm (using the "mcmc" command); and (4) summarizing the results (using the "sumt" and "sump" commands). The program
has extensive online help, which can be reached using the "help" command.
We urge the user to explore the available commands and the extensive amount
we have written about each by exploring the "help" option.

Analyzing the "toy" example of simulated data. The data matrix analyzed in numerous places in the text was simulated on the tree of Figure 7.3
under the HKY85 model of DNA substitution. The specific HKY85 parameter values and the branch lengths used for the simulation can be found in the
text. The input file contained the alignment of sequences and the commands:

```
begin data;
   dimensions ntax=5 nchar=50;
   format datatype=dna;
   matrix
   Species_1   TAACTGTAAAGGACAACACTAGCAGGCCAGACGCACACGCACAGCGCACC
   Species_2   TGACTTTAAAGGACGACCCTACCAGGGCGGACACAAACGGACAGCGCAGC
   Species_3   CAAGTTTAGAAAACGGCACCAACACAACAGACGTATGCAACTGACGCACC
   Species_4   CGAGTTCAGAAGACGGCACCAACACAGCGGACGTATGCAGACGACGCACC
   Species_5   TGCCCTTAGGAGGCGGCACTAACACCGCGGACGAGTGCGGACAACGTACC
   ;
end;

begin mrbayes;
   lset nst=2 rates=equal;
   mcmc ngen=1000000 nchains=1 samplefreq=100 printfreq=100;
   sumt burnin=1001;
   sump burnin=1001;
end;
```

The actual alignment is in a NEXUS file format. More accurately, the input
file format is NEXUS(ish) because we do not implement all of the NEXUS
standards in the program, and have extended the format in some (unlawful)
ways. The data are contained in the "data block", which starts with a "begin
data" command and ends with an "end" command. The next block is specific
to the program and is called a "MrBayes" block. Other programs will simply

skip this block of commands, just as MrBayes skips over foreign blocks it does not understand. All of the commands that can be issued to the program via the command line can also be embedded directly into the file. This facilitates batch processing of data sets.

The first command sets the model to HKY85 with no rate variation across sites. The second command runs the MCMC algorithm, and the third and fourth commands summarize the results of the MCMC analysis, discarding the first 1001 samples taken by the chain. Inferences then are based on the last 9000 samples taken from the posterior probability distribution.

Analyzing the leaf beetle data under a complicated model. The following shows the data and MrBayes block used in the analysis of the Kim et al. [39] alignment of three different genes. We do not show the entire alignment, though we do show the most relevant portions of the data block. Specifically, we show that you need to specify the data type as mixed when you perform a simultaneous Bayesian analysis on different types of data

```
begin data;
    dimensions ntax=27 nchar=1090;
    format datatype=mixed(rna:1-516,dna:517-936,protein:937-1090) gap=- missing=?;
    matrix
    Orsodacne     gGGUAAACCUNAGaA [ 1060 other sites ] DPILYQHLFWFFGHP
    Chrysomela    GGGUAAACCUGAGAA [ 1060 other sites ] DPILYQHLFWFFGHP
    Altica        --------------- [ 1060 other sites ] DPILYQHLFWFFGHP
    Agelastica    GGGUAAACCUGAGAA [ 1060 other sites ] DPILYQHLFWFFGHP
    Monolepta     GGGUAAACCUGAGAA [ 1060 other sites ] DPILYQHLFWFFGHP
    Phyllobrotica ---------UGANAA [ 1060 other sites ] DPILYQHLFWFFGHP
    Allochroma    GGGUAAaCcUGAgAA [ 1060 other sites ] DPILYQHLFWFFGHP
    Chrysolina    GGGUAAACCUGAGAA [ 1060 other sites ] DPILYQHLFWFFGHP
    Aphthona      GGGUAAACCCUGAGAA [ 1060 other sites ] ??????????????
    Chaetocnema   --------------- [ 1060 other sites ] DPILYQHLFWFFGHP
    Systena       ---CCGACCUGAGAA [ 1060 other sites ] DPILYQIILFWFFGHP
    Monocesta     ----------GAGAA [ 1060 other sites ] DPILYQHLFWFFGHP
    Disonycha     -------------AA [ 1060 other sites ] DPILYQHLFWFFGHP
    Blepharida    --------------- [ 1060 other sites ] DPILYQHLFWFFGHP
    Galeruca      GGGUAAACCUGAGAA [ 1060 other sites ] DPILYQHLFWFFGHP
    Orthaltica    GGGUAAACCUGAGAA [ 1060 other sites ] DPILYQHLFWFFGHP
    Paropsis      GGGUAAACCUGAGAA [ 1060 other sites ] DPILYQHLFWFFGHP
    Timarcha      -----AACCUGAGAA [ 1060 other sites ] DPILYQHLFWFFGHP
    Zygograma     GGGUAAACCUGAGAA [ 1060 other sites ] DPILYQHLFWFFGHP
    Syneta        -----GAACUUACAA [ 1060 other sites ] DPILYQHLFWFFGHP
    Dibolia       ggguaaaccugagaa [ 1060 other sites ] DPILYQHLFWFFGHP
    Sangariola    --------------- [ 1060 other sites ] DPILYQHLFWFFGHP
    Aulacophora   ----------AGAA [ 1060 other sites ] DPILYQHLFWFFGHP
    Diabrotica    GGGUAAACcUGAgAA [ 1060 other sites ] DPILYQHLFWFFGHP
    Diorhabda     ----------AGAA [ 1060 other sites ] DPTLYQHLFWFFGHP
    Schematiza    -----????UGAGAA [ 1060 other sites ] DPILYQHLFWFFGHP
    Oides         GGGUAACCCUGAGAA [ 1060 other sites ] DPILYQHLFWFFGHP
    ;
end;

begin mrbayes;
    pairs 22:497,  21:498,  20:499,  19:500,  18:501,  17:502,  16:503,  33:172,
          34:171,  35:170,  36:169,  37:168,  38:167,  45:160,  46:159,  47:158,
          48:157,  49:156,  50:155,  51:154,  53:153,  54:152,  55:151,  59:150,
          60:149,  61:148,  62:147,  63:146,  86:126,  87:125,  88:124,  89:123,
          187:484, 186:485, 185:486, 184:487, 183:488, 182:489, 191:295, 192:294,
          193:293, 194:292, 195:291, 196:290, 197:289, 198:288, 199:287, 200:286,
          201:283, 202:282, 203:281, 204:280, 205:279, 206:278, 213:268, 214:267,
          215:266, 216:265, 217:264, 226:259, 227:258, 228:257, 229:256, 230:255,
```

```
        231:254, 232:253, 233:252, 304:477, 305:476, 306:475, 307:474, 308:473,
        316:335, 317:334, 318:333, 319:332, 336:440, 337:439, 338:438, 339:437,
        340:436, 341:435, 343:422, 344:421, 345:420, 346:419, 347:418, 348:417,
        349:416, 351:414, 352:413, 353:412, 354:411, 355:408, 356:407, 357:406,
        358:405, 359:404, 360:403, 361:402, 369:400, 370:399, 371:398, 372:397,
        373:396, 376:394, 377:393, 379:392, 380:391, 381:390;
charset ambiguously_aligned = 92-103 108-122 234-251 320-327 449-468;
charset stems               = 22 497  21 498  20 499  19 500  18 501  17 502
                              16 503  33 172  34 171  35 170  36 169  37 168
                              38 167  45 160  46 159  47 158  48 157  49 156
                              50 155  51 154  53 153  54 152  55 151  59 150
                              60 149  61 148  62 147  63 146  86 126  87 125
                              88 124  89 123 187 484 186 485 185 486 184 487
                             183 488 182 489 191 295 192 294 193 293 194 292
                             195 291 196 290 197 289 198 288 199 287 200 286
                             201 283 202 282 203 281 204 280 205 279 206 278
                             213 268 214 267 215 266 216 265 217 264 226 259
                             227 258 228 257 229 256 230 255 231 254 232 253
                             233 252 304 477 305 476 306 475 307 474 308 473
                             316 335 317 334 318 333 319 332 336 440 337 439
                             338 438 339 437 340 436 341 435 343 422 344 421
                             345 420 346 419 347 418 348 417 349 416 351 414
                             352 413 353 412 354 411 355 408 356 407 357 406
                             358 405 359 404 360 403 361 402 369 400 370 399
                             371 398 372 397 373 396 376 394 377 393 379 392
                             380 391 381 390;
charset loops               = 1-15 23-32 39-44 52 56-58 64-85 90-122 127-145
                             161-166 173-181 188-190 207-212 218-225 234-251
                             260-263 269-277 284 285 296-303 309-315 320-331
                             342 350 362-368 374 375 378 382-389 395 401 409
                             410 415 423-434 441-472 478-483 490-496 504-516;
charset rna                 = 1-516;
charset dna                 = 517-936;
charset protein             = 937-1090;
charset D2                  = 1-516;
charset EF1a                = 517-936;
charset EF1a1st             = 517-936\3;
charset EF1a2nd             = 518-936\3;
charset EF1a3rd             = 519-936\3;
charset CO1aa               = 937-1090;
partition by_gene_and_pos   = 5:rna,EF1a1st,EF1a2nd,EF1a3rd,CO1aa;
partition by_gene           = 3:rna,EF1a,CO1aa;
partition by_gene_and_struct = 4:stems,loops,EF1a,CO1aa;
exclude ambiguously_aligned;
set partition = by_gene_and_struct;
lset applyto=(1) nucmodel=doublet;
lset applyto=(2) nucmodel=4by4;
lset applyto=(3) nucmodel=codon;
lset applyto=(1,2,4) rates=gamma;
lset nst=6;
prset ratepr=variable aamodelpr=mixed;
unlink shape=(all) revmat=(all);
mcmc ngen=3000000 nchains=1 samplefreq=100 printfreq=100;
sumt burnin=10001;
sump burni=10001;
end;
```

The commands in the MrBayes block show how to specify a very complicated model. First, we specify which nucleotides pair with one another using the pairs command. We then specify a number of character sets using the "charset" command. Specifying character sets saves the hassle of having to type in a long list of character numbers every time you want to refer to some division of the data (such as a gene). We then specify three character partitions. A character partition divides the data into groups of characters. Each character in the matrix must be assigned to one, and only one, group. For example, one of the partitions we define (by_gene) divides the characters into three groups. When a data file is executed, it sets up a default partition of the data that groups characters by data type. We need to tell the program which of the four partitions to use (where the four partitions are default, by_gene_and_pos, by_gene, and by_gene_and_struct). We do this using the set command. Finally, we use lset and prset to specify different models for different groups of characters. In fact, with the applyto option in lset and prset and the link and unlink commands, one can specify a very large number of possible models that currently cannot be implemented with any other phylogeny program. The last three commands will run the MCMC algorithm and then summarize the results.

Analyzing the 104 amino acid alignments. The analysis of the data collated by Nei et al. [51] was conceptually simple, though laborious, to set up. The data block, as usual, has the alignment, this time in interleaved format. The MrBayes block has 104 character set definitions, specifies a partition, grouping positions by gene, sets the partition, and then sets up a model in which the parameters are estimated independently for each gene and that enforces the molecular clock. The "outgroup" command can be used to specify the location of the root in output trees. The trees are simply rooted between the outgroup and the rest of the taxa. By default, MrBayes uses the first taxon in the matrix as the outgroup.

```
begin data;
    dimensions ntax=5 nchar=48092;
    format datatype=protein interleave=yes;
    matrix
    [The data for the 104 alignments were here. We do not include
    them here for obvious reasons (see the nchar command above).]
    ;
end;

begin mrbayes;
    charset M00007  =     1 -   112;
    charset M00008  =   113 -   218;
    charset M00037  =   219 -   671;
    [There were another 98 character set definitions, which we have deleted here.]
    charset N01447  = 45917 - 46694;
    charset N01456  = 46695 - 47285;
    charset N01479  = 47286 - 48092;
    partition by_gene = 104:M00007,M00008,[100 other partitions],N01456,N01479;
    set autoclose=yes nowarn=yes;
    set partition=by_gene;
    outgroup xenopus;
    lset rates=gamma;
    prset ratepr=variable aamodel=mixed brlenspr=clock:uniform;
    unlink shape=(all) aamodel=(all);
    mcmcp ngen=30000000 nchains=1 samplefreq=1000 savebrlens=yes;
end;
```

Appendix 3. Parameter Estimates for the Leaf Beetle Data

The numbers are the mean and 95% credible interval of the posterior probability density distribution for each parameter.

Param.	Mean (CI)	Param.	Mean (CI)	Param.	Mean (CI)
V	3.495 (3.209, 3.828)	π_G	0.222 (0.180, 0.267)	π_{GAC}	0.012 (0.008, 0.016)
$r_{CT}^{(1)}$	0.428 (0.187, 0.850)	π_T	0.285 (0.240, 0.332)	π_{GAG}	0.007 (0.006, 0.009)
$r_{CG}^{(1)}$	0.616 (0.166, 1.616)	π_{AAA}	0.023 (0.020, 0.024)	π_{GAT}	0.018 (0.016, 0.019)
$r_{AT}^{(1)}$	2.130 (0.703, 5.436)	π_{AAC}	0.006 (0.006, 0.008)	π_{GCA}	0.014 (0.012, 0.018)
$r_{AG}^{(1)}$	0.780 (0.340, 1.594)	π_{AAG}	0.019 (0.014, 0.023)	π_{GCC}	0.023 (0.019, 0.027)
$r_{AC}^{(1)}$	0.828 (0.214, 2.240)	π_{AAT}	0.005 (0.004, 0.006)	π_{GCG}	0.005 (0.005, 0.005)
$r_{CT}^{(2)}$	3.200 (2.037, 4.915)	π_{ACA}	0.011 (0.007, 0.013)	π_{GCT}	0.036 (0.034, 0.037)
$r_{CG}^{(2)}$	0.335 (0.116, 0.683)	π_{ACC}	0.021 (0.017, 0.024)	π_{GGA}	0.019 (0.014, 0.022)
$r_{AT}^{(2)}$	0.994 (0.522, 1.699)	π_{ACG}	0.006 (0.004, 0.009)	π_{GGC}	0.013 (0.006, 0.015)
$r_{AG}^{(2)}$	2.805 (1.702, 4.447)	π_{ACT}	0.025 (0.019, 0.027)	π_{GGG}	0.004 (0.004, 0.006)
$r_{AC}^{(2)}$	1.051 (0.541, 1.880)	π_{AGA}	0.020 (0.013, 0.021)	π_{GGT}	0.018 (0.015, 0.019)
$r_{CT}^{(3)}$	2.292 (1.471, 3.555)	π_{AGC}	0.016 (0.014, 0.019)	π_{GTA}	0.022 (0.017, 0.028)
$r_{CG}^{(3)}$	1.021 (0.400, 2.127)	π_{AGG}	0.004 (0.001, 0.007)	π_{GTC}	0.014 (0.008, 0.014)
$r_{AT}^{(3)}$	1.320 (0.766, 2.184)	π_{AGT}	0.001 (0.001, 0.002)	π_{GTG}	0.014 (0.012, 0.016)
$r_{AG}^{(3)}$	2.276 (1.424, 3.621)	π_{ATA}	0.003 (0.003, 0.004)	π_{GTT}	0.020 (0.016, 0.020)
$r_{AC}^{(3)}$	1.041 (0.575, 1.756)	π_{ATC}	0.025 (0.024, 0.029)	π_{TAC}	0.033 (0.030, 0.034)
ω	0.010 (0.010, 0.012)	π_{ATG}	0.014 (0.009, 0.017)	π_{TAT}	0.011 (0.010, 0.016)
π_{AA}	0.001 (0.000, 0.004)	π_{ATT}	0.026 (0.016, 0.029)	π_{TCA}	0.020 (0.017, 0.026)
π_{AC}	0.004 (0.000, 0.008)	π_{CAA}	0.015 (0.011, 0.019)	π_{TCC}	0.026 (0.023, 0.033)
π_{AG}	0.006 (0.003, 0.012)	π_{CAC}	0.010 (0.009, 0.014)	π_{TCG}	0.015 (0.014, 0.016)
π_{AT}	0.122 (0.086, 0.170)	π_{CAG}	0.009 (0.006, 0.011)	π_{TCT}	0.025 (0.024, 0.037)
π_{CA}	0.003 (0.000, 0.008)	π_{CAT}	0.009 (0.005, 0.010)	π_{TGC}	0.003 (0.003, 0.005)
π_{CC}	0.005 (0.001, 0.013)	π_{CCA}	0.022 (0.021, 0.024)	π_{TGG}	0.014 (0.008, 0.016)
π_{CG}	0.257 (0.191, 0.319)	π_{CCC}	0.012 (0.011, 0.014)	π_{TGT}	0.001 (0.001, 0.003)
π_{CT}	0.002 (0.000, 0.005)	π_{CCG}	0.008 (0.003, 0.010)	π_{TTA}	0.020 (0.013, 0.025)
π_{GA}	0.001 (0.000, 0.003)	π_{CCT}	0.008 (0.007, 0.010)	π_{TTC}	0.045 (0.044, 0.049)
π_{GC}	0.284 (0.222, 0.353)	π_{CGA}	0.002 (0.001, 0.004)	π_{TTG}	0.025 (0.025, 0.026)
π_{GG}	0.003 (0.000, 0.008)	π_{CGC}	0.009 (0.009, 0.009)	π_{TTT}	0.011 (0.010, 0.011)
π_{GT}	0.078 (0.057, 0.106)	π_{CGG}	0.001 (0.000, 0.000)	α_1	0.422 (0.308, 0.570)
π_{TA}	0.145 (0.103, 0.190)	π_{CGT}	0.016 (0.014, 0.016)	α_2	0.381 (0.296, 0.484)
π_{TC}	0.004 (0.001, 0.008)	π_{CTA}	0.005 (0.004, 0.010)	α_4	0.226 (0.175, 0.288)
π_{TG}	0.073 (0.056, 0.093)	π_{CTC}	0.016 (0.015, 0.020)	m_1	0.708 (0.553, 0.894)
π_{TT}	0.003 (0.001, 0.008)	π_{CTG}	0.042 (0.036, 0.046)	m_2	0.870 (0.732, 1.027)
π_A	0.252 (0.209, 0.301)	π_{CTT}	0.042 (0.034, 0.048)	m_3	1.274 (1.171, 1.378)
π_C	0.239 (0.199, 0.284)	π_{GAA}	0.034 (0.031, 0.044)	m_4	0.856 (0.651, 1.100)

8

Estimation of Divergence Times from Molecular Sequence Data

Jeffrey L. Thorne[1] and Hirohisa Kishino[2]

[1] Bioinformatics Research Center, Box 7566,
North Carolina State University, Raleigh, NC 27695-7566, USA,
thorne@statgen.ncsu.edu
[2] Laboratory of Biometrics, Graduate School of Agriculture and Life Sciences,
University of Tokyo, Yayoi 1-1-1, Bunkyo-ku, Tokyo 113-8657, Japan
kishino@peach.ab.a.u-tokyo.ac.jp

8.1 Introduction

On page 143 of his 1959 book *The Molecular Basis of Evolution*, Christian
B. Anfinsen listed some of the main components of the then-nascent field of
molecular evolution:

> A comparison of the structures of homologous proteins (i.e., proteins
> with the same kinds of biological activity or function) from different
> species is important, therefore, for two reasons. First, the *similarities*
> found give a measure of the minimum structure for biological func-
> tion. Second, the *differences* found may give us important clues to the
> rate at which successful mutations have occurred throughout evolu-
> tionary time and may also serve as an additional basis for establishing
> phylogenetic relationships.

Three years later, Zuckerkandl and Pauling [52] combined paleontological information with molecular sequence data to fulfill Anfinsen's prediction that
rates of molecular evolution could be estimated. Zuckerkandl and Pauling applied their estimate of the chronological rate of hemoglobin evolution to infer
both times since gene duplication and times since speciation events. They
noted that:

> It is possible to evaluate very roughly and tentatively the time that
> has elapsed since any two of the hemoglobin chains present in given
> species and controlled by non-allelic genes diverged from a common
> chain ancestor. The figures used in this evaluation are the number of
> differences between these chains, the number of differences between
> corresponding chains in different animal species, and the geological
> age at which the common ancestor of the different species in question

may be considered to have lived.

(Zuckerkandl and Pauling, pp. 200-201 in [52]).

Based on their finding that vertebrate hemoglobins change at an approximately constant rate, Zuckerkandl and Pauling ([53], p. 148) later concluded that there "... may thus exist a molecular clock." However, these authors acknowledged that the assumption of a constant chronological rate of molecular evolution was regarded by some investigators as biologically implausible. They wrote:

> Ernst Mayr recalled at this meeting that there are two distinct aspects to phylogeny: the splitting of lines, and what happens to the lines subsequently by divergence. He emphasized that, after splitting, the resulting lines may evolve at very different rates, and, in particular, along different lines different individual systems – say, the central nervous system along one given line – may be modified at a relatively fast rate, so that proteins involved in the function of that system may change considerably, while other types of proteins remain nearly unchanged. How can one then expect a given type of protein to display constant rates of evolutionary modification along different lines of descent?

(Zuckerkandl and Pauling, p. 138 in [53]).

In the decades that have passed since Anfinsen's prescient statement and Zuckerkandl and Pauling's work, what has changed regarding the estimation of evolutionary divergence times from molecular sequence data? The most obvious development is that today there is a much greater availability of DNA and protein data to analyze. Unfortunately, techniques for analyzing data have improved at a slower pace than have those for collecting data.

During most of the period following Zuckerkandl and Pauling's seminal papers, divergence time estimation from molecular sequence data has centered on the hypothesis of a constant chronological rate of sequence change. One set of methods [13] aims to estimate divergence times if this hypothesis of a constant rate of evolution is true. Another group of methods [22, 9, 48, 27, 42] aims to test this null hypothesis versus the alternative that rates change over time.

One extreme perspective is that the developers and users of both groups of methods have been misguided. This view posits that a constant chronological rate of molecular evolution is biologically implausible. Rates of molecular evolution are the outcome of complex interactions between the features of biological systems and their environments. Indisputably, biological systems and their environments change over time. Therefore, rates of molecular evolution must change.

For example, rates of molecular evolution are affected by the mutation rate per generation as well as by the generation length, and both of these depend intricately on biology. As biological systems evolve, generation lengths and mutation frequencies will vary. Moreover, the probability of fixation of mutations that are not selectively neutral depends on population size [28]. Population size itself fluctuates, as do the selective regimes to which genomes are exposed. The extreme perspective is that evolutionary rates are never constant and therefore the molecular clock should never be applied to the estimation of divergence times. Likewise, if the molecular clock hypothesis of constant chronological rates is always false, there is little value in making this a null hypothesis and then testing a hypothesis that is already known to be false.

A more balanced perspective may be that, in many cases, evolutionary rates are approximately—if not exactly— constant. Because evolutionary rates are largely determined by the biological systems that they affect, they may be likely to be nearly identical in two biological systems that are closely related and therefore highly similar. The molecular clock assumption may always be formally incorrect but it may sometimes be almost correct. An unsolved question is: When do the weaknesses of the molecular clock assumption outweigh its convenience?

Until recently, all practical approaches for estimating evolutionary divergence times relied upon the molecular clock hypothesis. In [33], Sanderson introduced a pioneering technique that allowed for the estimation of evolutionary divergence times without the restriction of a constant rate. Soon after, multiple other approaches were formulated for estimating divergence times without relying upon a constant rate [45, 15, 34]. Our goal here is to overview these newer divergence-time estimation methods. In particular, we emphasize the Bayesian framework that we have been developing [45, 20, 44]. Before doing this, we highlight the statistical issues that arise even if a constant chronological rate of molecular evolution could be safely assumed.

8.2 Branch Lengths as Products of Rates and Times

Amounts of evolution are usually expressed as expected numbers of nucleotide substitutions or amino acid replacements per site. To convert between the observed percentage of sites that differ between sequences and the expected number of substitutions or amino acid replacements, probabilistic models of sequence change can be adopted. These models range in complexity from the simple process of nucleotide substitution proposed by Jukes and Cantor [18] to comparatively realistic descriptions. Widely used models of sequence change are well-summarized elsewhere [41, 11]. We make no attempt to overview them here, but we do emphasize two points. The first is that models of sequence evolution are typically phrased in terms of rates, but these are relative rates rather than rates with chronological time units. Second, a poor model may lead to inaccurate divergence times. There may exist no sets of parameter values for which a particular model is a good description of the evolutionary process.

The effects on divergence time estimates of poor model choice are conceivably quite large but are generally difficult to assess. One potential impact of poor model choice is biased branch lengths. Because the bias in a branch-length estimate is likely to be a nonlinear function of the branch lengths, flawed evolutionary models can lead to systematic errors in divergence dating.

When solely DNA or protein sequence data are available, amounts of evolution can be estimated but rates and times are confounded. This confounding is problematic whether or not the molecular clock hypothesis is correct. For instance, imagine two homologous DNA sequences that differ by about 0.065 substitutions per site. One possibility is that their common ancestral sequence existed 6.5 million years ago and that substitutions per site have been accumulating at a rate of approximately 0.01 per million years. One of arbitrarily many alternative possibilities is that the two sequences had a common ancestor 1 million years ago and the substitutions have been accumulating at a rate of 0.065 per million years. The number of substitutions per site is an amount of evolution, and the expected amount of evolution is the rate multiplied by the time duration. In the absence of fossil or other information external to the sequence data, the rates and times cannot be separated, and only their product, the expected amount of evolution, can be estimated. Although infinitely long sequences could conceivably lead to a perfect estimate of the expected amount of evolution per site, the confounding of rates and times means that, even in the ideal situation of infinitely long sequences, neither rates nor times can be estimated solely from molecular data (see Figure 8.1).

When constant evolutionary rates are assumed, paleontological information can disentangle rates and times. Because the expected amount of evolution is the product of the rate and time, a known time leads to an estimated rate. With a molecular clock, the resulting rate can then be employed to infer other divergence times. For a rate of 0.1 substitutions per site per million years, a branch with an expected 0.065 substitutions per site would have a time duration of 0.65 million years (Figure 8.2). By mapping available paleontological information to a phylogeny and then calibrating the chronological rate of molecular change, the molecular clock allows divergence times to be estimated for phylogenetic groups where fossil data are sparse or absent.

Because the fossil record is very incomplete, there has been great interest in supplementing it with molecular data via the constant evolutionary rate assumption. Often, a statistical test of the constant chronological rate assumption is applied prior to divergence time estimation. With the null hypothesis of a constant rate, the expected amount of difference between a common ancestral sequence and its descendants should not vary among descendants. This reasoning underlies the most widely applied approaches that purport to test the constant chronological rate assumption.

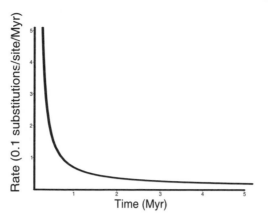

Fig. 8.1. Rates and times that yield an expected amount of evolution equal to 0.065 substitutions per site. Because the expected amount of evolution on a branch is equal to the product of the average evolutionary rate multiplied by the time duration of the branch, a perfectly estimated amount of sequence evolution would not suffice to allow rates and times to be disentangled. In this figure as well as in Figures 8.2 through 8.5, the time units are millions of years (Myr) and the rate units are 0.1 expected substitutions per site per million years.

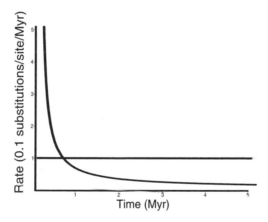

Fig. 8.2. When the expected amount of sequence evolution on a branch is 0.065 substitutions per site, knowledge that the evolutionary rate is 0.1 substitutions per million years implies that the time duration of the branch is 0.65 million years.

However, widely used approaches actually correspond to a more general null hypothesis than a constant chronological rate. When evolutionary rates do not vary over time, the expected amount of evolution is the product of the rate and the amount of time during which evolution occurs. If the evolutionary rate is instead variable over time, the expected amount of evolution b on a branch that begins at time 0 and ends at time T would be

$$b = \int_{t=0}^{T} r(t)dt,$$

where $r(t)$ is the rate at time t. We will refer to the path of $r(t)$ between 0 and T as a rate trajectory. Now consider a second branch that begins at the same node as the branch with length b. If the second branch also has time duration T, then the length of the second branch will be

$$b^* = \int_{t=0}^{T} r^*(t)dt,$$

where $r^*(t)$ represents the rate of evolution at time t on the second branch. There are infinitely many possible trajectories of rates for $r(t)$ and $r^*(t)$ that yield $b^* = b$. Widely applied tests [9, 48, 27, 42] evaluate a more general null hypothesis than a molecular clock. They do not examine whether the rates of evolution on different branches of an evolutionary tree are all identical and invariant. Instead, these tests focus on whether the sum of branch lengths between a root node and its descendant tips is the same for all tips. One biologically plausible situation is that the rates on all branches that are extant at any given instant are identical but that this shared rate itself changes over time. This "shared-rate" scenario could occur when all evolutionary lineages belong to a single species that experiences changes in evolutionary rates over time due to fluctuations in population size or environment [38, 6].

A good fossil record can help to differentiate between the shared-rate scenario and the molecular clock hypothesis, but often it can be challenging to compare these two possibilities when all sequences in a data set are isolated at effectively the same time. For RNA viruses and other quickly evolving organisms, the shared-rate and molecular clock hypotheses can be more easy to distinguish. A retroviral sequence isolated five years ago may be substantially more similar to a common ancestral sequence than is a sequence isolated today. By incorporating dates of sequence isolation into procedures for studying divergence times, the rate of a molecular clock can be inferred even in the absence of fossil information [24, 23, 21, 30].

Noncontemporaneously isolated (i.e., serially sampled) viral sequence data sets can be an especially rich source of potential evolutionary information [7, 37]. With serially sampled data, the "shared-rate" and clock hypotheses can be distinguished because rates of evolution during different time periods can be estimated and statistically compared [38, 6]. By exploiting serially sampled data, Drummond et al. [5] have convincingly demonstrated that rates of viral evolution can change when the treatment regimen of an infected patient is modified. The relatively high information content of serially sampled data is one reason that the study of viral change appears to be among the most promising avenues for understanding the process of molecular evolution.

8.3 Beyond the Classical Molecular Clock

8.3.1 The Overdispersed Molecular Clock

When defining branch lengths as the expected amount of change, a fixed rate trajectory is assumed. The actual number of changes per sequence for a fixed rate trajectory with independently evolving sequence positions has either exactly or very nearly a Poisson distribution, depending on the details of the substitution model [51]. For a Poisson distribution, the mean is equal to the variance. One focus of classic work in population genetics has been to examine whether a Poisson distribution accurately summarizes the process of molecular evolution or whether the process can be characterized as overdispersed because the variance of the number of changes exceeds the mean [12].

Most existing models of molecular evolution assume independent change among sites. In reality, however, a change at one site is apt to affect the rate of change at other sequence positions. Dependent change among positions can produce overdispersion. When sequence positions evolve in a compensatory fashion but independent change is modelled, the variance in branch length estimates may be underestimated. As a result, a hypothesis test based on an independent change model may be prone to rejecting the null hypothesis of a constant rate even when that null hypothesis is true.

Cutler [4] developed an innovative procedure for inferring divergence times. Although this procedure assumes a constant chronological rate of change, it accounts for the overdispersion of branch-length estimates that can arise when sites do not change independently. It does this by inflating the variance of branch-length estimates beyond what is expected with the particular independent change model that is used to infer branch-lengths. It might be preferable to explicitly model the dependence structure among sites, and limited progress is being made toward evolutionary inference when sequence positions change in a dependent fashion due to overlapping reading frames [17, 29] or constraints imposed by protein tertiary structure [32], but a variance adjustment such as suggested by Cutler [4] may be the most practical alternative for the near future.

8.3.2 Local Clocks

With maximum likelihood and a known rooted tree topology, divergence dating via the molecular clock and a single calibration point is relatively simple. Let X represent an aligned data set of molecular sequences and T denote the node times on the tree. With the molecular clock hypothesis, only a single rate R is needed. A probabilistic model of sequence evolution may include other parameters (e.g., the transition-transversion rate, and the stationary frequencies of the nucleotide types), but these are minor complications that will be omitted from our notation and discussion. The likelihood is then $p(X|R,T)$, the probability density of the data given the rate and node times. The values of

R and T fully determine the branch lengths B on the tree. Because rates and times are confounded with molecular sequence data, $p(X|B) = p(X|R,T)$. The branch lengths $\widehat{B}$ that maximize $p(X|B)$ are therefore the maximum likelihood estimates.

Not all sets of branch lengths B are consistent with the molecular clock assumption. Specifically, a constant chronological rate means that the sum of branch lengths from an internal node to a descendant tip will not differ among descendant tips. The branch lengths are therefore constrained by the molecular clock, and the maximization that yields $\widehat{B}$ only considers sets of branch lengths that satisfy these constraints. When the branch lengths $\widehat{B}$ are obtained, a node that serves as the single calibration time immediately yields estimates for both the rate R and the node times T.

Because maximum likelihood estimators have so many appealing statistical properties (see [26]), it is desirable to apply them to the estimation of divergence times without relying upon the unrealistic molecular clock hypothesis. Conceivably, there could be a separate parameter representing the evolutionary rate for each branch and also a parameter representing each node time on the rooted topology. A difficulty with such a parameter-rich model is that the molecular sequence data only have information about branch-lengths, and external evidence is necessary to disentangle rates and times.

Fortunately, maximum likelihood can be salvaged as a divergence-time estimation technique for models that are intermediate between the single-rate clock model and the parameter-rich model. One possibility is to construct a model where each branch is assigned to a category, with the total number of categories being substantially less than the total number of branches and with all branches that belong to the same rate category being forced to share identical rates [50] (see also [19, 46, 31, 49]). This scheme has been referred to as a local clock model [50] because it has clocklike evolution within prespecified local regions of a tree but also allows the clock to tick at different rates among the prespecified regions. With a local clock model, multiple calibration points can be incorporated so that the rates of branches belonging to individual categories can be more accurately determined.

Although assignment of branches to rate categories prior to data analysis is not absolutely necessary for local clock procedures, preassignment is computationally attractive and has been a feature of previously proposed local clock techniques. Often, it may not be clear how many categories of rates should be modelled. Likewise, assignment of branches to individual rate categories can be somewhat arbitrary. Despite these limitations, local clock models appear to yield divergence time estimates that are similar to those produced by the Bayesian methods described below [2, 49].

8.3.3 Penalized Likelihood

A very different strategy for divergence time estimation has been developed by Sanderson [34, 35]. Underlying this strategy is the biologically plausible notion that evolutionary rates will diverge as the lineages that these rates affect diverge. This node-dating technique allows each branch to have its own average rate. Rather than have R represent a single rate, we use R here to denote the set of average rates for the different branches on the tree. Instead of maximizing the likelihood $p(X|B) = p(X|R,T)$, Sanderson [34] finds the combination $(\tilde{R}, \tilde{T})$ of rates R and times T that obeys any constraints on node times and that maximizes

$$\log(p(X|R,T)) - \lambda \Phi(R),$$

where $\Phi(R)$ is a penalty function and where λ determines the contribution of this penalty function. There are many possible forms for the penalty function $\Phi(R)$, but the idea is to discourage, though not prevent, rate change. One form of $\Phi(R)$ explored by Sanderson [34] is the sum of two parts. The first part is the variance among rates for those branches that are directly attached to the root. The second part involves those branches that are not directly attached to the root. For each of these branches, the difference between its average rate and the rate of its ancestral branch is squared, and these squared differences are then summed. Only positive values of λ are permitted. Small λ values allow extensive rate variation over time, whereas large values encourage a clocklike pattern of rates. To select the value of λ, Sanderson [34] has developed a clever cross-validation strategy which will not be detailed here.

For computational expediency, existing implementations of this penalized likelihood technique (see [34, 35]) do not fully evaluate $\log(p(X|R,T))$ when finding $\tilde{R}$ and $\tilde{T}$. Instead, a two-step procedure is followed. The first step yields an integer-valued estimate of the number of nucleotide substitutions or amino acid replacements that have affected the sequence on each branch of the tree. The second step treats the integer-valued estimate for a branch as if it were a directly observed realization from a Poisson distribution. The mean value of the Poisson distribution is determined from R and T by the product of the average rate and time duration of the branch.

The penalized log-likelihood can be connected to Bayesian inference. The posterior density of rates and times is

$$p(R,T|X) = \frac{p(X|R,T)p(R,T)}{p(X)}.$$

Taking the logarithm of both sides,

$$\log(p(R,T|X)) = \log(p(X|R,T)) + \log(p(R,T)) - \log(p(X)).$$

Because $p(X)$ is only a normalization constant, the rates R and times T that maximize the posterior density are those values that maximize $\log(p(X|R,T))$ $+ \log(p(R,T))$. Furthermore, the $-\lambda \Phi(R)$ term is analogous to the logarithm of the prior distribution for rates and times $\log(p(R,T))$, since the penalized likelihood procedure maximizes $\log(p(X|R,T)) - \lambda \Phi(R)$. In other words, the penalized likelihood procedure is akin to finding the rates and times that maximize the posterior density $p(R, T|X)$.

There is a subtle but important point about the uncertainty of divergence times that are inferred by the penalized likelihood procedure. If the branch lengths B were known, maximizing

$\log(p(X|R, T)) - \lambda\Phi(R) = \log(p(X|B)) - \lambda\Phi(R)$ over rates R and times T would be equivalent to maximizing $-\Phi(R)$ over R and T. Although $\tilde{R}$ and $\tilde{T}$ represent the rates and times that optimize the penalized likelihood criterion, any rates and times are possible in this artificial scenario if they are consistent with both the known branch lengths and the constraints on node times. Since the branch lengths are being considered known, the penalized likelihood estimates $\tilde{R}$ and $\tilde{T}$ would not be influenced by the data X. Therefore, this artificial scenario would yield the same rate and time estimates for any nonparametric bootstrap replicate data set that is created by sampling sites with replacement [8, 10]. Because $\tilde{R}$ and $\tilde{T}$ would not be the only possible set of rates and times that yield the branch lengths B, there must be uncertainty about the rates and times that the nonparametric bootstrap does not reflect. The problem is that the nonparametric bootstrap procedure can adequately summarize uncertainty of branch lengths but cannot adequately reflect the uncertainty in rates and times conditional upon the branch lengths. Although this flaw of the nonparametric bootstrap procedure is more apparent when the branch lengths are considered known, it is also present in the more realistic situation where the branch lengths are unknown.

A more satisfactory procedure than solely the nonparametric bootstrap for quantifying rate and time uncertainty with the penalized likelihood approach would be an attractive goal for future research. In a standard maximum likelihood situation, the variances and covariances of parameter estimates can be approximated by the curvature of the log-likelihood surface around its peak ([40], pp. 675–676). This approximation improves with the amount of data because, as the amount of data grows, the shape of the log-likelihood surface near its peak asymptotically approaches (up to a constant of proportionality) a multivariate normal distribution. This asymptotic behavior is subject to certain regularity conditions being satisfied. Lack of parameter identifiability due to times and rates being confounded can lead to variances and covariances of estimated rates and times being poorly approximated by the curvature of the log-likelihood surface. Nevertheless, there may be certain penalty functions $\Phi(R)$ that would lead to variances and covariances of rates and times being well-approximated by the curvature of the penalized log-likelihood surface.

8.4 Bayesian Divergence Time Estimation

Bayesian inference of divergence times can be performed when a chronologically constant rate of molecular evolution is assumed. An advantage of Bayesian inference with the molecular clock is that prior information about rates and times can be better exploited. Consider the time that has elapsed since a common ancestral gene of human and chimp homologues. Before seeing or analyzing the sequence data, an expert might believe that there is a high probability of the elapsed time being within some interval that spans only several million years. Likewise, without inspecting the sequence data, the expert could confidently state that the rate of nucleotide substitution in the human and chimp lineages is likely to be somewhere in a range encompassing only a few orders of magnitude. When biological systems are not as well-studied as the human and chimp cases, the prior distributions of rates and times will be less concentrated, but even very diffuse prior information can be combined with sequence data to separate rates and times to some extent.

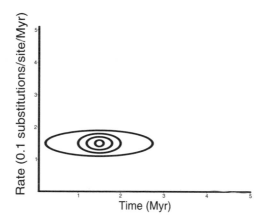

Fig. 8.3. A contour plot of the prior distribution for rates and times. Prior densities are highest for rates and times within the innermost ellipse and are lowest for rates and times outside the largest ellipse.

Figure 8.1 shows that, in the absence of prior information, even an infinite amount of sequence data cannot separate rates and times. In Figure 8.1, neither a time of 10 seconds nor a time of 10 million years can be excluded. Nevertheless, Figure 8.1 reveals that the actual rate and time must be some point on a one-dimensional slice through the two-dimensional parameter space of rates and times. The problem is that sequence data are no help in determining which points on the one-dimensional slice are most reasonable. Figure 8.3 shows a possible prior distribution for rates and times. By combining the prior distribution summarized in Figure 8.3 with the sequence information summarized in Figure 8.1, the posterior distribution of rates and times in Figure 8.4 is generated. As with Figure 8.1, Figure 8.4 again illustrates that the true rate and time must be a point on the one-dimensional slice through the two-dimensional parameter space of rate and time. However, Figure 8.4 indicates that some rates and times have higher posterior densities than others. The

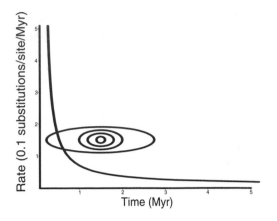

Fig. 8.4. Prior information about rates and times combined with a perfectly estimated branch length of 0.065 substitutions per site leads to a posterior distribution
of rates and times. All points of the posterior distribution that have positive density
are found on the line corresponding to the product of rate and time being 0.065
substitutions per site. The points with the highest posterior density are those on
the line with the highest prior density.

relative posterior densities of the points on the slice are determined by the
prior distribution depicted in Figure 8.3. Constraints on times are especially
helpful because they can concentrate the posterior distribution (Figure 8.5).

When rates are not assumed constant, the divergence time estimation
problem becomes more challenging, but Bayesian inference can still be performed. Here, we focus on our techniques for Bayesian dating [45, 20, 44].
Other Bayesian techniques have been introduced [15, 21, 2], but these are
generally similar to our approach and we highlight only the most important
differences.

To allow rates to change, we adopt a stochastic model [20]. We assume the
average rate on a branch is the mean of the rate at the nodes that begin and
end the branch. This assumption is a weakness of our approach and can be
interpreted as forcing rates to change in a linear fashion from the beginning to
ending nodes of a branch. Our alternative interpretation is that the mean of
the beginning and ending rates on a branch hopefully is a good approximation
of the average rate on the branch. A more satisfactory treatment of rates would
not have the average rate on the branch be a deterministic function of the
rates at the nodes. In this sense, the approach of Huelsenbeck et al. [15] is
superior. These investigators had rates change in discrete jumps that were
statistically modelled so that they could occur at any point on any lineage on
the rooted tree.

Because the average rates on branches for our Bayesian procedures are
completely determined by the rates at nodes, we employ a stochastic model
for the node rates. Given the rate at the beginning of a branch, our simple

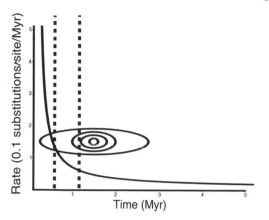

Fig. 8.5. The effect on the posterior distribution of constraints on time. The dashed vertical lines represent the highest and lowest possible values for the time duration of the branch. When the branch length is known to be 0.065 substitutions per site, the constraints represented by the vertical lines force the posterior distribution of rates and times to be confined to relatively small intervals. Due to the prior density and the constraints, the rates and times with the highest posterior densities are those combinations where the time duration slightly exceeds the lower bound on time.

model assigns a normal distribution to the logarithm of the rate at the end. The mean of this normal distribution is set so that the expected rate at the end of the branch (and not the expected logarithm of the rate) is equal to the rate at the beginning. The variance of the normal distribution is the product of a rate variation parameter ν and the time duration of the branch. When $\nu = 0$, there is a constant chronological rate of sequence change. As the value of ν increases, the tendency to deviate from a clocklike pattern of evolution increases.

Because this model describes the probability distribution for the rate at the node that ends a branch in terms of the beginning node rate and because the root node on a tree is the only node that does not end a branch, we assign a prior distribution for the rate at the root node. Letting R now represent the rates at the nodes on a tree, the prior distribution for the rate at the root and the model for rate change together specify $p(R|T, \nu)$, the probability density of the rates R given the node times T and the rate change parameter ν.

Aris-Brosou and Yang [2] explored other diverse models for rate change. They concluded that, while the data may contain enough information for comparing the adequacy of competing rate change models, the details of the rate change model have little impact on divergence time estimates as long as rate change is permitted. The conclusions of Aris-Brosou and Yang [2] coincide with our experience that divergence time estimates do not seem to be very sensitive to either the value of ν or its prior distribution.

In contrast with their comparisons of different rate change models, Aris-Brosou and Yang [2] found that divergence time estimates were sensitive to whether rates were allowed to change or whether they were forced to be constant. By comparing analyses performed with a molecular clock ($\nu = 0$) and without ($\nu > 0$), we find that the similarity between clock and nonclock divergence time estimates varies widely among data sets (pers. obs.). However, because they account for stochasticity of evolutionary rates, credibility intervals for node times tend to be wider when produced by nonclock analyses than they do when clocks are enforced [20, 2, 3].

Bayesian procedures also require a prior distribution $p(T)$ for the node times T. Unfortunately, it is difficult to appropriately quantify this prior distribution $p(T)$. Our implementation specifies $p(T)$ in two parts. The first component of $p(T)$ is a gamma distribution for the prior distribution of the time since the root node. For a given root node time, every other interior node time can be specified by the time between the node and its descendant tips divided by the time separating the root and the tips. To handle these proportional times, our implementation invokes a prior distribution that is a generalization to rooted tree structures of the Dirichlet distribution [20]. The primary advantages of the generalized Dirichlet are its simplicity and its relative flexibility. A disadvantage is the lack of a biological rationale for its form.

Prior distributions for node times can also be generated via a stochastic process with explicit descriptions of speciation, extinction, and taxon sampling [2]. Although this latter class of prior distributions has the strong advantage of being biologically interpretable, it is unclear to us whether these prior distributions are realistic enough to improve rather than hamper divergence time estimation. Aris-Brosou and Yang [2] showed that the value of a parameter related to the intensity of taxon sampling in their prior distribution for divergence times had a substantial influence on the posterior distribution. Both the prior mean and prior variance of a parameter can affect the posterior distribution. In the absence of strong biological justification for having divergence-time prior distributions with small variances, we believe a large prior variance is preferable. It would be interesting to investigate the relationship between the prior variance for divergence times and the value of the taxon-sampling parameter of Aris-Brosou and Yang [2].

To incorporate fossil and other evidence about divergence dates that is external to the sequence data, we follow Sanderson [33] by adopting constraints on node times. We find by simulation that evolutionary dating is greatly aided if at least one interior node time is bounded by a constraint that forces it to exceed some age and if at least one node time is bounded by a constraint that forces it to be less than some age (data not shown). The credibility intervals on node times in a tree should become narrower as the times become more severely constrained. The improvements in divergence time estimation that stem from incorporation of constraints are so large as to suggest to us that research effort invested in gathering and interpreting fossil data will often be more productive for evolutionary dating than is effort invested in collecting additional sequence data.

We denote the constraints on node times by C. The prior distribution for divergence times has density $p(T)$ and is often much different from the distribution conditional upon the constraints, which has density $p(T|C)$. Even when the unconditional divergence time distribution can be analytically studied, formulas for means and variances of node times may be complicated to express for the distribution with density $p(T|C)$. For example, the prior distribution for the time of the root node might be a gamma distribution with mean 80 million years and standard deviation 10 million years. If a constraint that some interior node time must exceed 100 million years is then added, the root node time will also have to exceed 100 million years and cannot continue to have a prior mean of 80 million years.

Rather than deriving explicit formulas for the effects of constraints on the means and variances of our generalized Dirichlet prior distributions, we usually summarize the distribution with density $p(T|C)$ in a less elegant manner. Specifically, we randomly sample times T from the distribution with density $p(T|C)$. If the sample yields a biologically implausible distribution of node times, we modify the mean and variance of the distribution with density $p(T)$ until the distribution with density $p(T|C)$ more adequately reflects our beliefs. Only after finding a suitable distribution with density $p(T|C)$ will we analyze the sequence data X [20].

We employ the Metropolis-Hastings algorithm [25, 14] to approximate $p(T, R, \nu | X, C)$, the distribution of the times T, rates R, and rate change parameter ν conditional upon the sequence data X and the node time constraints C. We can do this because $p(T, R, \nu | X, C)$ can be written as a ratio with a denominator that is not a function of the parameters T, R, and ν and with a numerator that is a product of terms that we can calculate,

$$p(T, R, \nu | X, C) = \frac{p(X, T, R, \nu | C)}{p(X|C)}$$

$$= \frac{p(X|T, R, \nu, C)p(T, R, \nu | C)}{p(X|C)}$$

$$= \frac{p(X|T, R, \nu, C)p(R|T, \nu, C)p(T|\nu, C)p(\nu | C)}{p(X|C)}.$$

The structure of our model permits further simplification:

$$p(T, R, \nu | X, C) = \frac{p(X|T, R)p(R|T, \nu)p(T|C)p(\nu)}{p(X|C)}$$

$$= \frac{p(X|B)p(R|T, \nu)p(T|C)p(\nu)}{p(X|C)}.$$

In the numerator of the final expression, $p(R|T, \nu)$ is defined by our model for rate evolution and $p(\nu)$ is the prior distribution for the rate change parameter. The density $p(T|C)$ is identical to the density $p(T)$ up to a proportionality constant to. The proportionality constant adjusts for the fact that many possible node times T are inconsistent with the constraints C. Fortunately, this proportionality constant need not be calculated to apply the Metropolis-Hastings algorithm.

The most problematic term in the numerator of the final expression is the likelihood $p(X|B)$. For widely used models of sequence evolution that have independent changes among positions, $p(X|B)$ can be calculated via the pruning algorithm [9]. Unfortunately, the pruning algorithm is computationally demanding, and the Metropolis-Hastings procedure can require evaluations of $p(X|B)$ for many different sets of branch lengths B. As computing speeds inevitably improve, the requirement to repeatedly evaluate the likelihood $p(X|B)$ will become less onerous. In fact, the pruning algorithm has been applied to Bayesian data with and without a clock in studies done by Aris-Brosou and Yang [2, 3]. However, full evaluation of $p(X|B)$ can still be so onerous as to make statistical inference with some data sets impractical.

Because multivariate normal densities are quick to calculate, we have approximated $p(X|B)$ up to a constant of proportionality with a multivariate normal distribution centered about the maximum likelihood estimates of branch lengths $\widehat{B}$ [45]. The covariance matrix of this multivariate normal distribution is estimated from the curvature of the log-likelihood surface (i.e., the inverse of the information matrix—[40], pp. 675–676). A defect of this multivariate normal approximation is that it is likely to be inaccurate when maximum likelihood branch lengths are zero or are very short. We have addressed this inaccuracy with ad hoc procedures, but Korber et al. [21] improved upon the approximation by using the gradient of the log-likelihood surface for branch lengths with maximum likelihood estimates of zero. The Poisson approximation of $p(X|B)$ by Sanderson [33, 34] may be superior to the multivariate normal when branch lengths are short. In contrast, when branch lengths are long, the multivariate normal may be a better approximation than the Poisson distribution. One unexplored option would be to construct a hybrid approximation that uses the Poisson distribution for some short branch lengths and the multivariate normal distribution for the other branch lengths.

8.5 Discussion

8.5.1 Uncertainties in the Estimated Divergence Times

Model choice is not the only consideration when inferring divergence times from molecular sequence data. Here, we discuss several other potential sources of inaccuracy [47]. The effects of some of these sources are straightforward to quantify, while the effects of others are more difficult to assess.

With only occasional exceptions, the process of evolution is not amenable to direct observation. Therefore, the changes that have transformed DNA sequences over time can only be inferred. This inference can be relatively accurate when the sequences being analyzed are closely related. When the sequences are fairly diverged, the changes may be so numerous as to have altered individual positions multiple times. In this situation, the estimated number of changes that occurred on a branch could be substantially different from the actual number. Some implementations of divergence-dating techniques involve estimating the number of changes on a branch but then treating the estimates as if they were actual observations [22, 33, 4, 34]. One concern might be whether these implementations are prone to underestimating the uncertainty in divergence dates because they treat estimated numbers of changes as if they were actual numbers of changes. Fortunately, this source of uncertainty is not an important practical problem because it can be reflected with a nonparametric bootstrap approach that involves sampling sites with replacement.

Given only a trajectory of rates over time, the expected number of changes can be determined, but the actual number is a random variable. This randomness increases the uncertainty of divergence dating. Provided the model is correct, this increased uncertainty is properly reflected with dating techniques that rely upon a probabilistic model of sequence change.

Stochasticity in Rates of Evolution Over Time

One source of overdispersion is dependent change among sequence positions, but overdispersion can even arise with independently evolving positions. Earlier, we gave an example where two sister branches existed for the same amount of time but experienced different rate trajectories that led to differing branch lengths b and b^*. With independently evolving sequence positions, the actual number of changes occurring on one branch would be exactly or very nearly a realization from a Poisson distribution with mean b, whereas the actual number of changes occurring on the other branch would be exactly or very nearly a sample from a Poisson distribution with mean b^*.

However, rates of evolution can themselves be considered random variables, and expectations over possible rate trajectories can be taken. Because branch lengths vary among rate trajectories, the evolutionary process is overdispersed when there is stochasticity of rate trajectories. Therefore, the stochasticity of evolutionary rates is yet another factor that can increase the uncertainty of node timing on a phylogeny. This explains why credibility intervals for node times are wider when a clock is not assumed than when it is [20].

Fossil Uncertainty

Conventionally, fossil information has been incorporated into molecular clock analyses in the form of calibration points. Adoption of a calibration point implies that fossil evidence can precisely pinpoint the time of an internal node. Unfortunately, fossil data are not usually so directly informative. Geological dating of fossils has associated uncertainty. Moreover, the probability is almost zero that a particular fossil is a remnant of the exact organism that harbored the ancestral gene corresponding to a specific node on a phylogeny.

Rather than specify calibration points, Sanderson [33] translated fossil evidence into constraints on node times. This treatment has subsequently been adopted by others [20, 4]. Typically, it is more straightforward to determine from fossils that a particular node must exceed some age than it is to determine that a node must be less than some age. Although an improvement over calibration points, constraints on node ages are still a primitive summary of fossil data. More sophisticated treatments of fossil evidence have been proposed [16, 43] and could be combined with molecular information in future divergence time analyses.

Topological Uncertainty

By treating the topology as known, methods for divergence time estimation that require a prespecified topology are prone to underestimating uncertainties in divergence times. In practice, we believe this topological source of uncertainty in divergence times should often be small because the parts of a topology that are the most prone to error are likely to involve branches with short time duration [50]. Cases where topological error is most relevant to divergence time estimation may be at two extremes. One extreme is so much evolution that some branches are saturated for change. With saturated branches, topological errors may involve branches with long length, and the wrong topology may substantially impact divergence time estimates. At the other extreme, some data sets may consist almost exclusively of very closely related sequences. With these data sets, the topology may be difficult to determine, and each of the few nucleotide substitutions that occur can have a relatively significant impact on divergence dating.

8.5.2 Multigene Analyses

Variances of branch length estimates that are based upon a single gene may be substantial. When these variances are large, accurate determination of divergence dates becomes unlikely. One way to reduce the variance of branch-length estimates is to collect and then concatenate multiple gene sequences from each taxon of interest. The concatenated data can then be analyzed as though they represented a single gene [39].

There is also a less general potential benefit of concatenation than variance reduction. By virtue of its bigger size, a concatenated multigene data set should yield more evolutionary changes per sequence than would a data set of only a single gene. The multivariate normal approximation of the likelihood surface is apt to be particularly poor when branches are represented by zero or no more than a few changes per sequence. This means that shortcomings of the multivariate normal approximation may be less problematic for concatenated data sets than for data sets with only a single gene.

Advantages of concatenation depend on the extent of correlation of rate change over time between genes. Concatenation can reduce variance in divergence time estimates that is generated by branch-length uncertainty, but it does not reduce the variance in divergence time estimates that stems from the stochasticity of evolutionary rates. If rate change is mainly gene-specific rather than lineage-specific, it may be preferable to allow each gene in a multigene data set to have its own rate trajectory but to assume that all genes share a common set of divergence times [44, 49]. For determining node times, the benefit of allowing rate trajectories to differ among genes is potentially large. However, the process of molecular evolution is poorly characterized, and the relative contributions of gene-specific versus lineage-specific factors toward rate variation over time are unclear. To some extent, the relative contributions of these factors can be assessed by studying the posterior distribution of rates and times when the prior distribution has genes change rates independently of one another [44]. Still, it is clear that concatenation would be preferred if all rate variation over time were exclusively lineage-specific. For multigene analyses with independent rate trajectories among genes, poorly chosen prior distributions for rates and rate change can have a substantial and misleading impact on posterior distributions for rates and times [44]. Future models of rate change could incorporate both lineage-specific and gene-specific components.

The assumption that all loci share a common set of divergence times should be satisfactory when all taxa are rather distantly related. For closely related taxa, variations in history among genes should not be ignored and the assumption of a common set of divergence times among loci is unwarranted. Therefore, we caution against making this assumption for closely related taxa.

8.5.3 Correlated Rate Change

Although the focus of this chapter is divergence times, better tools for inferring node dates are likely to be better tools for characterizing rate variation over time and could lead to much-needed improvements in the understanding of molecular evolution. These tools could help to identify groups of genes that have had strongly correlated patterns of rate change over time. Genes with correlated patterns of rate evolution may have a shared involvement in generating some phenotype that has been subjected to natural selection. Identification of genes with correlated patterns of rate change would thereby be a potential means of functional annotation. A crude test for detecting correlated rate change among genes has been proposed [44].

8.5.4 Synonymous and Nonsynonymous Rate Change

Rate variations over time could have different patterns at different sites within a gene or for different kinds of changes within a gene. For protein-coding genes, it is natural to distinguish between the synonymous nucleotide substitutions that do not alter the amino acid sequences of a protein and the nonsynonymous nucleotide substitutions that do change the protein sequence. Variations in synonymous substitution rates over time may be largely explained by changes in generation time or mutation rate whereas variation in nonsynonymous rates could also be attributable to changes in natural selection or population size. By comparing patterns of synonymous and nonsynonymous rate variation over time, the forces affecting sequence change can be better understood.

With conventional approaches, only amounts of synonymous and nonsynonymous change and their ratio are examined. By using codon-based models where both synonymous and nonsynonymous rates can change over time, Bayesian techniques for studying protein-coding gene evolution permit the estimation of chronological rates of both synonymous and nonsynonymous change [36]. In addition to providing improved tools for studying the process of protein-coding gene evolution, a codon-based Bayesian approach potentially leads to improved estimates of divergence times.

8.6 Conclusion

For more than thirty years following the ground-breaking molecular clock work by Zuckerkandl and Pauling [52, 53], divergence time estimation from molecular sequence data was dominated by the constant chronological rate assumption. A recent burst of activity has led to improved techniques for separating rates of molecular evolution from divergence times. Just as the interest in divergence time estimation seems to have grown, we expect that the ability of newly introduced methods to better determine chronological rates of sequence change will soon become better appreciated by those who study the process of molecular evolution. We anticipate that methods for characterizing evolutionary rates and times will continue to advance at a rapid rate.

Acknowledgments

We thank Stéphane Aris-Brosou, David Cutler, Michael Sanderson, Eric Schuettpelz, and Tae-Kun Seo for helpful discussions. We also thank an anonymous reviewer for comments. This work was supported by the Institute for Bioinformatics Research and Development of the Japanese Science and Technology Corporation, the Japanese Society for the Promotion of Science, and the National Science Foundation (grants INT-9909348, DEB-0089745 and DEB-0120635).

References

[1] C. B. Anfinsen. *The Molecular Basis of Evolution*. John Wiley and Sons, Inc, New York, 1959.

[2] S. Aris-Brosou and Z. Yang. Effects of models of rate evolution on estimation of divergence dates with special reference to the metazoan 18S ribosomal RNA phylogeny. *Syst. Biol.*, 51:703–714, 2002.

[3] S. Aris-Brosou and Z. Yang. Bayesian models of episodic evolution support a late Precambrian explosive diversification of the Metazoa. *Mol. Biol. Evol.*, 20:1947–1954, 2003.

[4] D. J. Cutler. Estimating divergence times in the presence of an overdispersed molecular clock. *Mol. Biol. Evol.*, 17:1647–1660, 2002.

[5] A. Drummond, R. Forsberg, and A. G. Rodrigo. The inference of stepwise changes in substitution rates using serial sequence samples. *Mol. Biol. Evol.*, 18:1365–1371, 2001.

[6] A. Drummond and A. G. Rodrigo. Reconstructing genealogies of serial samples under the assumption of a molecular clock using serial—sample UPGMA. *Mol. Biol. Evol.*, 17:1807–1815, 2000.

[7] A. J. Drummond, G. K. Nicholls, A. G. Rodrigo, and W. Solomon. Estimating mutation parameters, population history and genealogy simultaneously from temporally spaced sequence data. *Genetics*, 161:1307–1320, 2002.

[8] B. Efron and R. J. Tibshirani. *An Introduction to the Bootstrap*. Chapman and Hall, London, 1993.

[9] J. Felsenstein. Evolutionary trees from DNA sequences: A maximum likelihood approach. *J. Mol. Evol.*, 17:368–376, 1981.

[10] J. Felsenstein. Confidence limits on phylogenies: An approach using the bootstrap. *Evolution*, 39:783–791, 1985.

[11] J. Felsenstein. *Inferring Phylogenies*. Sinauer Associates, Sunderland, MA, 2004.

[12] J. H. Gillespie. *The Causes of Molecular Evolution*. Oxford University Press, New York, 1991.

[13] M. Hasegawa, H. Kishino, and T. Yano. Dating of the human-ape splitting by a molecular clock of mitochondrial DNA. *J. Mol. Evol.*, 22:160–174, 1985.

[14] W. K. Hastings. Monte Carlo sampling methods using Markov chains and their applications. *Biometrika*, 57:97–109, 1970.

[15] J. P. Huelsenbeck, B. Larget, and D. L. Swofford. A compound Poisson process for relaxing the molecular clock. *Genetics*, 154:1879–1892, 2000.

[16] J. P. Huelsenbeck and B. Rannala. Maximum likelihood estimation of phylogeny using stratigraphic data. *Paleobiology*, 23:174–180, 1997.

[17] J. L. Jensen and A.-M. K. Pedersen. Probabilistic models of DNA sequence evolution with context dependent rates of substitution. *Adv. Appl. Prob.*, 32:499–517, 2000.

[18] T. H. Jukes and C. R. Cantor. Evolution of protein molecules. In H. N. Munro, editor, *Mammalian Protein Metabolism*. Academic Press, New York, 1969.

[19] H. Kishino and M. Hasegawa. Converting distance to time: An application to human evolution. *Methods Enzymol.*, 183:550–570, 1990.

[20] H. Kishino, J. L. Thorne, and W. J. Bruno. Performance of a divergence time estimation method under a probabilistic model of rate evolution. *Mol. Biol. Evol.*, 18:352–361, 2001.

[21] B. Korber, M. Muldoon, J. Theiler, F. Gao, R. Gupta, A. Lapedes, B. H. Hahn, S. Wolinsky, and T. Bhattacharya. Timing the ancestor of the HIV-1 pandemic strains. *Science*, 288:1789–1796, 2000.

[22] C. H. Langley and W. M. Fitch. An examination of the constancy of the rate of molecular evolution. *J. Mol. Evol.*, 3:161–177, 1974.

[23] T. Leitner and J. Albert. The molecular clock of HIV-1 unveiled through analysis of a known transmission history. *Proc. Natl. Acad. Sci. USA*, 96:10752–10757, 1999.

[24] W.-H. Li, M. Tanimura, and P. M. Sharpi. Rates and dates of divergence between AIDS virus nucleotide sequences. *Mol. Biol. Evol.*, 5:313–330, 1988.

[25] N. Metropolis, A. W. Rosenbluth, M. N. Rosenbluth, A. H. Teller, and E. Teller. Equations of state calculations by fast computing machines. *J. Chem. Phys.*, 21:1087–1092, 1953.

[26] A. M. Mood, F. A. Graybill, and D. C. Boes. *Introduction to the theory of statistics*. McGraw-Hill, New York, 3rd edition, 1974.

[27] S. V. Muse and B. S. Weir. Testing for equality of evolutionary rates. *Genetics*, 132:269–276, 1992.

[28] T. Ohta. Population size and rate of evolution. *J. Mol. Evol.*, 1:305–314, 1972.

[29] A.-M. K. Pedersen and J. L. Jensen. A dependent-rates model and an MCMC-based methodology for the maximum likelihood analysis of sequences with overlapping reading frames. *Mol. Biol. Evol.*, 18:763–776, 2001.

[30] A. Rambaut. Estimating the rate of molecular evolution: Incorporating non—contemporaneous sequences into maximum likelihood phylogenies. *Bioinformatics*, 16:395–399, 2000.

[31] A. Rambaut and L. Bromham. Estimating divergence dates from molecular sequences. *Mol. Biol. Evol.*, 15:442–448, 1998.

[32] D. M. Robinson, D. T. Jones, H. Kishino, N. Goldman, and J. L. Thorne. Protein evolution with dependence among codons due to tertiary structure. *Mol. Biol. Evol.*, 20:1692–1704, 2003.

[33] M. J. Sanderson. A nonparametric approach to estimating divergence times in the absence of rate constancy. *Mol. Biol. Evol.*, 14:1218–1232, 1997.

[34] M. J. Sanderson. Estimating absolute rates of molecular evolution and divergence times: A penalized likelihood approach. *Mol. Biol. Evol.*, 19:101–109, 2002.

[35] M .J. Sanderson. R8s: Inferring absolute rates of molecular evolution and divergence times in the absence of a molecular clock. *Bioinformatics*, 19:301–302, 2003.

[36] T.-K. Seo, H. Kishino, and J .L. Thorne. Estimating absolute rates of synonymous and nonsynonymous nucleotide substitution in order to characterize natural selection and date species divergences. *Mol. Biol. Evol.*, 21(7):1201–1213, 2004.

[37] T.-K. Seo, J. L. Thorne, M. Hasegawa, and H. Kishino. Estimation of effective population size of HIV-1 within a host: A pseudomaximum–likelihood approach. *Genetics*, 160:1283–1293, 2002.

[38] T.-K. Seo, J. L. Thorne, M. Hasegawa, and H. Kishino. A viral sampling design for estimating evolutionary rates and divergence times. *Bioinformatics*, 18:115–123, 2002.

[39] M. S. Springer, W. J. Murphy, E. Eizirik, and S. J. O'Brien. Placental mammal diversification and the Cretaceous-Tertiary boundary. *Proc. Natl. Acad. Sci. USA*, 100:1056–1061, 2003.

[40] A. Stuart and J. K. Ord. *Kendall's Advanced Theory of Statistics*, volume 2. Oxford University Press, Oxford, 5th edition, 1991.

[41] D. L. Swofford, G. J. Olsen, P. J. Waddell, and D. M. Hillis. Phylogenetic inference. In D. M. Hillis, C. Moritz, and B. K. Mable, editors, *Molecular Systematics*. Sinauer Associates, Sunderland, MA, 2nd edition, 1996.

[42] F. Tajima. Simple methods for testing the molecular evolutionary clock hypothesis. *Genetics*, 135:599–607, 1993.

[43] S. Tavaré, C. R. Marshall, O. Will, C. Soligo, and R. D. Martin. Using the fossil record to estimate the age of the last common ancestor of extant primates. *Nature*, 416:726–729, 2002.

[44] J. L. Thorne and H. Kishino. Divergence time and evolutionary rate estimation with multilocus data. *Syst. Biol.*, 51:689–702.

[45] J. L. Thorne, H. Kishino, and I. S. Painter. Estimating the rate of evolution of the rate of molecular evolution. *Mol. Biol. Evol.*, 15:1647–1657, 1998.

[46] M. K. Uyenoyama. A generalized least—squares estimate for the origin of sporophytic self-incompatibility. *Genetics*, 139:975–992, 1995.

[47] P. J. Waddell and D. Penny. Evolutionary trees of apes and humans from DNA sequences. In *Handbook of Human Symbolic Evolution*. Oxford University Press, Oxford, Englan, 1996.

[48] C.-I. Wu and W.-H. Li. Evidence for higher rates of nucleotide substitution in rodents than in man. *Proc. Natl. Acad. Sci. USA*, 82:1741–1745, 1985.

[49] Z. H. Yang and A. D. Yoder. Comparison of likelihood and bayesian methods for estimating divergence times using multiple gene loci and

calibration points, with application to a radiation of cute-looking mouse lemur species. *Syst. Biol.*, 52:705–716, 2003.

[50] A. D. Yoder and Z. H. Yang. Estimation of primate speciation dates using local molecular clocks. *Mol. Biol. Evol.*, 17:1081–1090, 2000.

[51] Q. Zheng. On the dispersion index of a Markovian molecular clock. *Math. Biosc.*, 172:115–128, 2001.

[52] E. Zuckerkandl and L. Pauling. Molecular disease, evolution, and genic heterogeneity. In M. Kasha and B. Pullman, editors, *Horizons in Biochemistry: Albert Szent-Györgyi Dedicatory Volume*. Academic Press, New York, 1962.

[53] E. Zuckerkandl and L. Pauling. Evolutionary divergence and convergence in proteins. In V. Bryson and H. J. Vogel, editors, *Evolving Genes and Proteins*. Academic Press, New York, 1965.

Part III

Models of Molecular Evolution

9

Markov Models of Protein Sequence Evolution

Matthew W. Dimmic

Department of Biological Statistics and Computational Biology,
Cornell University, Ithaca, NY 14853, USA, mwd8@cornell.edu

9.1 Introduction

Proteins play a vital role in almost every process of life. There are over 2100 known protein families; many are crucially involved in biochemical metabolism, cellular signaling and transport, reaction catalysis, cytoskeletal structure, immune recognition, and sensory input. Because single mutations in the amino acid sequences can have a drastic effect on the protein's function– and potentially on the fitness of the individual–proteins can be considered the primary unit of phenotypic expression. The complex relationship among protein chemistry, structure, function, and evolution is therefore a significant piece of the evolutionary puzzle, and models of protein evolution are used to test hypotheses about these relationships.

There are several applications where protein evolutionary models have had particular success. For example:

1. detecting and aligning remote homologs,
2. measuring divergence times between sequences and species,
3. inferring the evolutionary history of related proteins (the phylogenetic tree), and
4. determining the physicochemical factors that have been important to the function and evolution of a protein family.

This chapter focuses on models that can be used for the last two applications, specifically those that treat evolution as a Markov chain with transitions between amino acid states. When combined with the statistical toolbox of likelihood methods (Chapter 2), Markov models have proven to be a powerful tool for phylogenetic inference and hypothesis testing. Rather than attempting to provide an exhaustive description of all available models, this chapter will highlight a few that illustrate the important distinguishing features of protein sequence evolution.

9.2 Basic Features of Protein Sequences

The factors that are important to the evolution of a protein can be complex and subtle, and the study of protein evolution is a very active field. A detailed discussion of protein structure and evolution is beyond the scope of this chapter (see, e.g., references [10, 60]), but a few of the most relevant features from an evolutionary perspective follow.

- **Most proteins function natively with the amino acid chain folded into a stable three-dimensional structure.** This structure is called the *tertiary structure* and is thought to be determined primarily by the amino acid chain's interaction with a solvent and with itself (a proposition known as the Thermodynamic Hypothesis [4]). While the protein-folding pathway is still not well-understood for most sequences, some general principles of protein folding are known. For example, in aqueous solution, a combination of entropic and enthalpic factors combine to cause hydrophobic (oily) amino acids to fold into the interior of a protein, exposing charged and polar residues to solvent. These factors also give rise to stable substructures that occur ubiquitously in protein families (called *secondary structure*), such as alpha helices and beta strands. One consequence of this folded protein structure is that each residue in a protein sequence is exposed to a different local environment, potentially resulting in different evolutionary constraints at each site.

- **In general, protein structure is more conserved during evolution than protein sequence.** In the SCOP database of protein structure classification [49], there are over 2100 protein structural families of homologous sequences, while there are nearly 170,000 amino acid sequences in Pfam [8], a protein families database. A good rule of thumb is that two sequences with more than 30% sequence identity are likely to be homologous and fold into the same tertiary structure or domain, although homology can sometimes be inferred at a lower identity. The conservation of function is less clear-cut; sometimes proteins can perform the same function with less than 15% identity, while in other cases the function can be completely altered by the mutation of a few key amino acids.

- **Protein sequences are generally subject to greater selective constraint than noncoding DNA sequences.** For a protein to function properly in an organism, it must be transcribed and transported, fold, interact with its binding partners or substrate, perform its function efficiently (catalysis, recognition, transport, etc.), and be properly disposed of when no longer needed. If a mutation in the amino acid sequence affects any of these steps, it can potentially affect the function of the protein and therefore the fitness of the organism. The combination of these factors constrains the evolution of protein sequences much more than the evolution of most noncoding DNA sequences, an effect that can be observed as a low nonsynonymous/synonymous substitution rate ratio. This allows protein sequences to be used to infer homology among more distant evolutionary

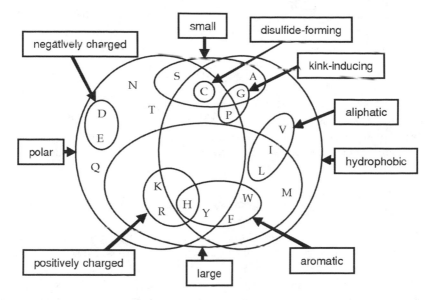

Fig. 9.1. Some commonly used amino acid properties. For other properties, a variety of empirically determined scales have been collected in the AAindex database [39].

relatives, but it also complicates phylogenetic inference because the effect of these constraints can be difficult to predict.

- **Amino acids each have unique properties that are utilized in different contexts within the protein.** Figure 9.1 shows several different sets of physicochemical properties; the importance of each property depends upon the local environment of the amino acid. For example, tyrosine is a bulky, aromatic amino acid that is often found in the same context as other large aromatics such as tryptophan and phenylalanine. But tyrosine also has a polar hydroxyl group like serine and threonine, and like those amino acids it can be involved in hydrogen-bonding interactions as well. In modeling protein evolution, it is therefore important to take into account not only the properties of the amino acids but also the context in which they are used.

9.3 The REV Model

A Markov model of protein evolution must at a minimum provide a substitution probability matrix $\mathbf{P}(t)$, where $P_{ij}(t)$ is the probability that an amino acid substitution $i \to j$ will occur in evolutionary time t. In the likelihood function (see Chapter 2), a different $\mathbf{P}$-matrix is needed for each branch of the tree. Rather than estimating each matrix separately, typically a single *instantaneous transition rate matrix*[1] $\mathbf{Q}$ is used for the entire tree, and the different $\mathbf{P}$-matrices can be calculated using the standard equation for a continuous Markov process:

$$P_{ij}(t) = \left[e^{\mathbf{Q}t}\right]_{ij} .\tag{9.1}$$

The challenge for any model of protein evolution is to determine the best estimate of the transition rates in the $\mathbf{Q}$-matrix given the available data. In general these rates are not equal; some amino acid substitutions are more likely than others, and this is directly related to the physical and chemical characteristics of the amino acids in protein structures.

9.3.1 Counting Methods for Model Estimation

Early models of protein evolution were limited by the computational issues associated with likelihood inference on phylogenetic trees. To overcome these limitations, empirical methods were devised to approximate the transition matrix by counting the number of inferred substitutions. The first widely used model of protein evolution was developed by Margaret Dayhoff and co-workers in the *Atlas of Protein Sequence and Structure* [17]. Using sets of closely related protein sequences (more than 85% similar), they used an assumption of parsimony to infer the ancestral amino acids at each site in the protein. Once the amino acid at each node has been determined, the internodal substitutions can be counted, resulting in a 20×20 symmetric matrix of amino acid replacement counts (the $\mathbf{A}$-matrix).

These counts depend on the exposure of each amino acid during the evolutionary process; a rare but mutable amino acid can be indistinguishable from a common amino acid that rarely changes. To discriminate between these possibilities, Dayhoff defined the mutability of an amino acid m_i as the number of inferred changes for each amino acid divided by its total number of nodal appearances on the tree N_i:

$$m_i = \frac{\sum_{k \neq i} A_{ik}}{N_i} .\tag{9.2}$$

The mutabilities and the count matrix were then multiplied to calculate $\mathbf{M}$, the "mutation probability matrix"[2],

$$M_{ij} = \lambda m_i \frac{A_{ij}}{\sum_{k \neq i} A_{ik}} \quad \text{for} \quad j \neq i ,\tag{9.3}$$

---

[1] Because "transition" has a formal meaning when dealing with sequence evolution, the Markov transition rate matrix $\mathbf{Q}$ is sometimes called a mutation rate matrix or substitution rate matrix in the biological literature.

[2] For consistency with standard Markov model notation, the i, j indices have been reversed from Dayhoff's original notation.

where λ is a scaling factor that determines the average probability of a substitution in a specified unit of evolutionary time. Because this is a probability matrix, each row must sum to 1, and therefore the diagonals are

$$M_{ii} = 1 - \lambda m_i \ . \tag{9.4}$$

When λ is set so that the mean probability of substitution is 0.01, on average one substitution will be observed per 100 amino acid sites. This is called 1 PAM, or "point accepted mutation," a commonly used measure of distance between protein sequences.

The Dayhoff model was originally developed to aid in alignment of distant homologs and to help determine evolutionary distances between sequences. To convert it into a continuous-time Markov chain model that can be used for statistical inference and likelihood calculations [22, 41, 1], the matrix is commonly converted into a slightly different form, the symmetric "relative rate matrix" $\mathbf{R}$:

$$R_{ij} = \frac{A_{ij}}{N_i N_j} \ . \tag{9.5}$$

Typically this is estimated empirically from the sequence alignment since in a stationary process with sufficient data $\pi_j^{obs} \rightarrow \pi_j$. The instantaneous transition matrix $\mathbf{Q}$ is then defined as

$$Q_{ij} = \delta R_{ij} \pi_j / s \quad \text{for} \quad j \neq i \tag{9.6}$$

and

$$Q_{ii} = -\sum_{k \neq i} Q_{ik} \ , \tag{9.7}$$

where π_j is the stationary frequency of amino acid j, δ is the number of expected substitutions per site in a unit of evolutionary time (typically 0.01), and s is a normalization constant,

$$s = \sum_{i,j,i \neq j} \pi_i \pi_j R_{ij} \ . \tag{9.8}$$

This $\mathbf{R}$-matrix parameterization is generally called the REV model.[3] As the most general reversible amino acid model, it is analogous in form to the GTR model for nucleotide substitution. One important difference is that in contrast with the GTR nucleotide model, the REV matrix values are fixed and not estimated for each new dataset of interest. Similar empirical counting methods have been used to update the REV model parameters as more data have become available. Dayhoff's matrix represents data from 1572 counted substitutions; in 1992, Jones et al. updated the matrix parameters using 59,190

---

[3]The values of the Dayhoff $\mathbf{R}$-matrix are different from the Dayhoff log-odds PAM matrix used for sequence alignment (although they use the same underlying $\mathbf{A}$ count matrix), so care should be taken not to confuse the two.

substitutions from 16,130 protein sequences in what is now commonly called the JTT model [37]. These sequences were generally globular proteins in an aqueous solvent; a model has also been tallied for transmembrane proteins and is called the tmREV matrix [38]. Each of these REV models–Dayhoff, JTT, and tmREV–differs only in the particular (fixed) values of their **R**-matrices.

Counting methods are relatively rapid, but they can only utilize the information from closely related sequence comparisons. To test the effects of this assumption, Benner, Cohen, and Gonnet [9] computed a set of matrices from proteins related by different PAM distances. When comparing these with extrapolated Dayhoff matrices, they found that the Dayhoff parameter values reflected the structure of the genetic code, while over long timescales the more accurate matrix values were better correlated with physicochemical properties of the amino acids. Even with methods that allow more divergent sequences to be used, the parsimony assumption inherent in the counting methods can cause bias in parameter estimation [16].

9.3.2 Likelihood Methods for Model Estimation

Maximum likelihood (ML) methods are a natural choice for optimizing models over divergent datasets [22, 76], as they can account for the probability of multiple substitutions over long branches and can be tested using a rigorous statistical framework. The likelihood equation for a phylogenetic tree utilizes a continuous-time Markov chain, which determines the probability of substitution over the evolutionary time t of each branch by exponentiating the **Q**-matrix as shown in equation (9.1). This equation can be approximated as

$$\mathbf{P}(t) = e^{\mathbf{Q}t} \approx \mathbf{1} + \mathbf{Q}t + (\mathbf{Q}t)^2/2 + \dots . \qquad (9.9)$$

The higher-order terms in this expansion account for the nonzero probability of multiple substitutions over long branches. As t increases and/or the off-diagonal terms in **Q** increase, the higher-order terms become more significant and the assumptions of parsimony no longer hold true.

The first use of ML estimation (MLE) methods to optimize REV model parameters was by Adachi and Hasegawa [1], who were interested in modeling the evolution of mitochondrial proteins. The Dayhoff and JTT matrices were developed as an average over many protein families from the nuclear genome, but mitochondrial proteins evolve under different selective constraints. Translated using a different genetic code with a different nucleotide compositional bias, most mitochondrial proteins also function in a lipid membrane rather than in aqueous solution. To account for these differences, the mtREV model [1] was developed on a tree of mitochondrial proteins from a diverse set of vertebrate species. Instead of using a counting method to infer the values of the **R**-matrix, each of the R_{ij} values was treated as a free parameter of the model and estimated using maximum likelihood. The MLE REV model was therefore estimated with 210 parameters: the 190 values of the symmetric **R**-matrix and the 20 amino acid frequencies. (The model has 208 degrees of freedom, because the **R**-matrix values are relative and the amino acid frequencies must sum to 1.) The resulting mtREV matrix had a significantly higher likelihood on mitochondrial proteins than the JTT matrix, and some of the known differences between mitochondrial and nuclear proteins are evident from mtREV's parameter values. For example, the Arg↔Lys substitution rate is much lower in mtREV than in JTT, a difference that is attributed to the fact that it requires two nucleotide mutations in the mitochondrial genetic code while only requiring one in the universal code. Other MLE REV models developed for specific datasets include the cpREV model for chloroplast proteins [2] and the rtREV model for retroviral polymerase proteins [19]. As with the Dayhoff model, all of these ML-estimated REV models have their **R**-matrix values fixed for further analyses; they are not adjusted for each new dataset.

For a general model applicable to many different protein families, the MLE equivalent of the Dayhoff and JTT matrices is the WAG matrix [74]. To create this matrix, 3905 protein sequences were divided into 182 protein families. A neighbor-joining tree was inferred for each family, and then the combined likelihood was maximized by adjusting the values of the **R**-matrix. Using the likelihood ratio test (LRT), the increase in likelihood over the former models was found to be statistically significant for all families in the analyzed dataset. In fact, the increase in likelihood from the JTT matrix to the WAG matrix is even greater than the increase from Dayhoff to JTT, despite the fact that WAG was optimized using fewer protein sequences than JTT, an indication of the power of the ML estimation method. Because ML estimation can be computationally expensive, approximate methods have been developed as a compromise between accuracy and speed [54].

The selective constraints acting on the amino acid level are reflected in the parameter values for these REV models. For example, in the universal genetic code, Ala is fourfold degenerate–represented by the codons GC*–while Trp is only represented by one codon (UGG). Therefore, there are six possible nucleotide mutations away from Ala and nine mutations away from Trp. If there were no selection on the amino acid level, one would predict from entropic principles that Trp would show a greater propensity for substitution than Ala since there are more "escape routes." But according to the mutabilities calculated for example by Jones et al. [37], tryptophan has the *lowest* mutability, while alanine has a mutability four times larger, an effect due in part to tryptophan's unique chemical characteristics. In the mtREV model, Cys is more mutable than it is in matrices optimized on proteins that function in an intracellular environment [1], probably because Cys-Cys disulfide bonds are not thought to be as important to membrane proteins as they are to aqueous proteins. The importance of such factors can be tested by comparison with the codon Poisson model [41], which disallows all single-step amino acid substitutions requiring more than one nucleotide mutation ($R_{ij} = 0$), while all other R_{ij} values are set to 1. This simple model is almost always statistically rejected in favor of models such as the Dayhoff model, an indication that simple nucleotide models are inadequate for modeling protein sequences because they do not account for the different properties of amino acid residues.

The simplicity and generality of the REV model are attractive, and its similarity in form to nucleotide models has made its implementation in phylogenetic software a fairly straightforward task. But this simplicity can be a limitation when attempting to understand the complex determinants of evolution at the protein level. Alternative models have concentrated on correcting some of these limitations, focusing on four features of amino acid sequence evolution: site heterogeneity, time heterogeneity, site dependence, and the physicochemical properties of amino acids. The rest of this chapter will be devoted to a discussion of these features and the models that address them.

9.4 Modeling Heterogeneity Across Sites

The parameter values in the REV models are typically an average over many amino acid sites from many different proteins. The implicit assumption is that every site in the protein is subject to the same evolutionary constraints, or at least that the constraints are evenly distributed about some mean value. But most proteins fold into an intricate three-dimensional structure, creating a different chemical environment for each amino acid residue. This heterogeneity in environments leads to heterogeneity in evolutionary constraints, which can have a dramatic effect on protein evolution and inference [56] (see Figure 9.2). The concept of a single transition matrix that can describe the evolutionary process at every site becomes difficult to justify.

9.4.1 Rate Heterogeneity Across Sites (RHAS)

One useful approximation for modeling site heterogeneity is the use of a distribution of evolutionary rates, or rate heterogeneity across sites (RHAS). According to the Neutral Theory [40], functionally important sites are under more stringent evolutionary constraints and will therefore exhibit a lower overall substitution rate. One of the simplest methods for adding rate heterogeneity to a phylogenetic model is to use a Gamma distribution of rates [77]. This is done exactly as with the nucleotide models (see Chapter 1), where each site is assigned an equal prior probability ϕ_k of evolving at rate λ_k. The possible values for λ_k are drawn from a discretized Gamma distribution with a specified number of categories K. The likelihood function in each column in the alignment D_n is determined by summing the conditional likelihood for each possible rate:

$$L\left(D_n|\theta'\right) \equiv \sum_{k=1}^{K} f\left(D_n|\lambda_k, \theta'\right) \phi_k . \tag{9.10}$$

In this case, θ' represents the other parameters in the model; for example, the **R**-matrix and amino acid frequencies. The shape of the Gamma distribution

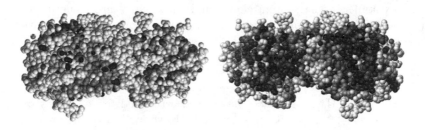

Fig. 9.2. An empirical Bayesian mapping of rates onto sites in the trypsin family. The darker the residue, the slower the rate of evolution at that site. Residues on the surface of the protein (*left*) tend to be less conserved than those seen in a cutaway view of the interior (*right*). The posteriors were calculated using CONSURF [63] and mapped onto the structure using MOLMOL [43].

can be adjusted with a single parameter, while the relative rates of substitution at any particular site are determined by a REV model. The result is a set of K rate categories, each related by a common REV model but multiplied by a different rate constant to determine the **P**-matrices:

$$P_{ij}^k(t) = \left[e^{\mathbf{Q}\lambda_k t}\right]_{ij} . \qquad (9.11)$$

The improvement in likelihood with a rate distribution is almost always significant relative to a site-homogeneous model even with just a few rate categories, making the Gamma distribution a commonly used approximation. This difference is not just a statistical nuance; failure to account for rate heterogeneity can also cause errors when inferring phylogenies and divergence times [13]. For this reason, the inclusion of some form of site heterogeneity–at least a REV+Γ model–is almost always recommended for phylogenetic analysis.

9.4.2 Pattern Heterogeneity Across Sites (PHAS)

The fact that rate heterogeneity is ubiquitous among protein sequences is evidence of the diversity of selective constraints operating on the amino acid level. Still, a rate distribution cannot account for variability in the *pattern* of evolutionary constraints. It allows a site to evolve more slowly or quickly, but a simple rate distribution does not, for example, allow a Gly→Ala substitution rate to be higher than a Gly→Pro substitution at one site but lower at another. Due to the diversity of amino acid environments in a folded protein, such differences can be pronounced. For example, glycine and alanine are the smallest amino acids, while proline (although still small) is bulkier. In the folded core of the protein, where steric constraints might preclude a bulky amino acid, the Gly→Pro substitution may be less favorable than the more conservative Gly→Ala substitution. But glycine and proline are also known to induce kinks in the protein chain, and these kinks are sometimes necessary for terminating alpha helices and creating turns. In these locations, a Gly→Pro substitution becomes the "conservative" one, accepted more often than a Gly→Ala substitution. This can only be modeled by a change in the *relative* rates of substitutions, not just the overall rate.

To account for this type of heterogeneity (called here pattern heterogeneity across sites, or PHAS), matrices have been estimated for specific structural classes of sites [59, 73, 44, 53]. For example, Koshi and Goldstein [44] divided sites into four different structural categories–helix (H), turn (T), strand/sheet (E), and coil (C)–and subdivided those into two accessibility categories: buried (b) and exposed (e). Then ML estimation was used along with an evolutionary tree to estimate structure-specific substitution matrices for each category.

The Koshi-Goldstein structure-based matrices were log-odds matrices designed for sequence alignment and structural prediction rather than Markov substitution matrices. To optimize matrices for phylogenetic use, Goldman and co-workers used an across-sites hidden Markov model (HMM) called the PASSML model [29]. PASSML begins with the assumption that any sequence site is in one of the eight structural categories mentioned above. These categories are further divided into a total of 38 possible classes by position along the sequence. For example, there are six each of the possible buried and exposed sheet classes $[Eb_i, Ee_i \ (i \in \{1, 2, ..., 6\})]$, ten each of the buried and exposed helix classes, two buried and two exposed turn classes, and one buried and one exposed coil class. This seemingly complicated division has an empirical basis: if each sequence site were completely independent (with no $i, i+1$ dependence) and transitions between structural categories were random, the length of each structure in a protein would be geometrically distributed with a mean of 1. This is physically unrealistic; helices and sheets by definition involve more than one amino acid. By adding in site dependence with an HMM, the PASSML model's mean structure length better resembles the empirically observed distribution.

A large training database of over 200 globular protein families with known structure was used to estimate PASSML's parameters, with the **R**-matrix for each site category estimated using a technique similar to the Dayhoff counting method. The result was a set of eight **Q**-matrices and their associated equilibrium amino acid frequencies (one set for each combination of secondary structure type and solvent accessibility):

$$\theta_{\text{passml}} = \{\mathbf{Q}_k, \boldsymbol{\pi}_k\} \text{ for } k \in \{1, 2, ..., 8\} \ . \tag{9.12}$$

To model site dependence, the PASSML model also includes a set of ρ_{kl} parameters, the transition probabilities between the hidden classes that were estimated by empirical fit to the data. Once the model was estimated on the training set, all parameter values were then fixed for further analysis.

To apply PASSML to nontraining datasets, it is not necessary to know the protein's structure. Because this is an HMM, the true state of a site is

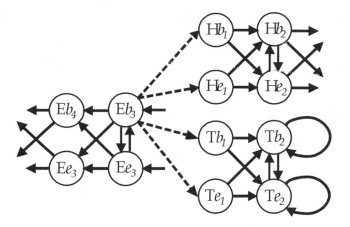

Fig. 9.3. An example of some allowed transitions within categories (solid lines) and between categories (dashed lines) in the PASSML model. The structural categories shown are helix, sheet, and turn (H, E, and T); the accessibility categories are buried and exposed (*b* and *e*). Not all transitions or states are shown. For more details, see Goldman et al. [29]

considered to be "hidden", and the likelihood is a sum over the conditional likelihood of each site class at each site,

$$L(S_n|T) = \sum_{k_n} f(S_n, k_n|T) , \qquad (9.13)$$

where

$$f(S_n, k_n|T) = \sum_{k_{n-1}} f(D_n|k_n, T)P(S_{n-1}, k_{n-1}|T)\rho_{k_{n-1}k_n} . \qquad (9.14)$$

Here D_n denotes the nth column in the N-column alignment, S_n denotes the set of columns $\{D_x\}, x \in \{1, 2, ..., n\}$, k_n is the structural category for site n, $\rho_{k_{n-1}k_n}$ is the transition probability from a category at column $n-1$ to a category at column n, and T is the tree topology. The difference between this equation and an alignment-based HMM approach (see Chapter 14) is in the likelihood function; in this case $f(D_n|k_n, T)$ is the phylogenetic likelihood function for site n.

Using likelihood ratio tests on several different protein families, the authors found that simple models that did not include structural categories were always rejected in favor of those that did. Even HMMs with just eight site classes with no site-to-site dependence (all ρ_{kl} equal) yielded a much higher likelihood. (Each site class's mean rate is also variable, so rate heterogeneity is an implicit feature of the model.) The inclusion of additional solvent accessibility categories was also found to be significant, but the dependence among adjacent sites was a less important feature of the model. Using the same technique but different category designations, the PASSML-TM model [47] and

MT126 model [48] have been optimized for transmembrane and mitochondrial proteins, respectively.

An interesting twist on structure-based evolutionary modeling uses simulated evolution on a known protein structure to create substitution matrices. The IS-SCPE method [25] (for Independent Sites–Structurally Constrained Protein Evolution) requires a representative structure and sequence for the protein family of interest. The sequence is repeatedly mutated, and the mean field energies at each site in the structure are computed using prespecified energy potentials. The structural perturbation of the new sequence from the reference structure is then calculated, and mutations causing a perturbation smaller than a specified cutoff are accepted. Finally, the accepted mutations are tallied in a set of replacement count matrices that are categorized by structural class (for example, position in an alpha helix).

Matrices created using this method were found to have a significantly higher likelihood than the JTT+Γ model, another indication that rate heterogeneity alone can sometimes be insufficient for modeling and that pattern heterogeneity also plays a large role. The IS-SCPE method is promising for proteins of known structure, although its assumptions that energetic stability and local structural integrity are important evolutionary constraints should be kept in mind when this method is applied.

9.5 Mechanistic Models

The models discussed in Section 9.4 emphasize structural features that are often easily observable: alpha helices, beta sheets, solvent accessibility, structural stability, etc. The ubiquity and strong conservation of such features favor this assumption, but protein evolution can also be affected by subtleties that are difficult to ascertain a priori. Transient recognition binding patches, allosteric regulatory networks, and dynamic hinge regions are just a few examples of evolutionary constraints that may be crucial but not obvious, even when the protein structure is available. In fact, there is evidence indicating that many functional regions of proteins are disordered and do not exist in a single stable structural state [20].

When little is known about the structural determinants of evolution in a protein family, ideally one would prefer to estimate the model's parameters for each dataset of interest. This is typical when applying nucleotide models such as the Jukes-Cantor or GTR models, where the ML estimates of the parameters are jointly estimated while searching for the ML tree topology. Contrast this with all the amino acid models mentioned thus far, which have been trained on a set of reference sequences or a reference structure and then the parameters are fixed for further analysis. The reasons for this are both theoretical and practical. The GTR model has only six parameters (ten if the nucleotide frequencies are also estimated), while the full REV model has 190 parameters (or 210 with estimated frequencies). When using a PHAS model, this number is multiplied by the number of site classes. A model with eight site classes, each represented by a REV matrix, can have over 1600 parameters! Precise estimation with so many degrees of freedom requires an extremely large dataset. Even when hundreds of sequences are used, the time required for model estimation can be prohibitive, especially when simultaneously estimating the tree topology and branch lengths.

One way to reduce the number of parameters is to use a mechanistic model. The parameters of the models discussed previously can all be considered to be somewhat empirical: substitutions are tallied in parameters that have more statistical convenience than physical meaning. In reality, these relative rates are an aggregate measure of the physicochemical characteristics of the amino acids and their interactions with their local environment. In contrast, mechanistic models explicitly utilize these physicochemical characteristics, facilitating the testing of hypotheses related to these properties. This reparameterization reduces the degrees of freedom, allowing the use of a realistic number of site classes while estimating mechanistic model parameters for each dataset of interest. Mechanistic models are frequently used in combination with multiple site classes; in these cases they are a type of PHAS model.

Several examples of mechanistic models can be summarized as physicochemical amino acid fitness models [45, 79]. In these types of models the **Q**-matrix for each site class k is divided into a mutation rate λ and an amino acid substitution function Ω_{ij}^k:

$$Q_{ij}^k = \lambda \Omega_{ij}^k (F_i^k, F_j^k) . \tag{9.15}$$

F_i^k and F_j^k are amino acid fitnesses,[4] parameters that are explicitly dependent upon the physicochemical properties of the amino acids. For example, in the model of Koshi and Goldstein [45] (called the FIT-PC model here), these fitnesses are determined as quadratic functions of the amino acid's hydrophobicity (h) and volume (v):

$$F_i^k = a_k \left(h_i - h_o^k \right)^2 + b_k \left(v_i - v_o^k \right)^2 . \tag{9.16}$$

In this model, a_k, b_k, h_o^k, and v_o^k are all parameters of the model and estimated from the data using maximum likelihood. The first two parameters (a_k and b_k) determine the strength of the selective pressure from each chemical characteristic, while h_o^k and v_o^k determine the optimal value of that characteristic in the site class. The substitution rate for nonsynonymous changes in the FIT-PC model is determined by a fitness function,

---

[4]These parameters have been called fitnesses as an analogy to fitness functions on an energy landscape rather than as fitnesses in the genetic sense of the term. Nevertheless, they could be made mathematically equivalent with the proper choice of fitness function. Also, in a reckless abuse of notation, a superscript k will indicate that the parameter is particular to that site class, not that k is a numerical exponent.

$$\Omega_{ij}^k = \begin{cases} \omega_k e^{\left(F_j^k - F_i^k\right)}, & \text{if } F_j^k < F_i^k, \\ \omega_k, & \text{if } F_j^k \geq F_i^k. \end{cases} \tag{9.17}$$

The value of the parameter ω_k is estimated using ML, and it can be regarded as a general selective disadvantage for making any nonsynonymous change (or as an adaptive advantage if $\omega_k > 1$). This type of fitness function is also known as a Metropolis-Hastings function; its form is chosen for its mathematical convenience, as it allows the straightforward derivation of the equilibrium frequencies for each site class as a function of the fitnesses:

$$\pi_i^k = \frac{e^{F_i^k}}{\sum_{i'} e^{F_{i'}^k}} . \tag{9.18}$$

Qualitatively, favorable mutations to "more fit" amino acid are all accepted at the same rate, while unfavorable mutations are tolerated depending on the difference between the amino acid properties. The larger the difference, the lower the substitution probability.

There are several alternatives for calculating λ in equation (9.15). One possibility is to set λ as an estimated parameter. This can only be done if ω_k is fixed, as they are indistinguishable on the amino acid level (one is actually estimating $\{\lambda\omega\}_k$). Another possibility is to use a Gamma rate distribution to subdivide each site class into r rate categories,

$$Q_{ij}^{kr} = \Omega_{ij}^k \lambda^{kr} , \tag{9.19}$$

where each λ^{kr} is determined from category r of a discretized $\Gamma(\alpha, \omega_k \alpha)$ distribution that has a mean ω_k (see Chapter 5).

A third possibility, suggested by Yang et al. [79], is to specify λ as a weighted sum of the mutation rates on the codon level, independent of site class k but dependent on the set of codons $u \in i$ and $v \in j$ coding for amino acids i and j, respectively:

$$\lambda_{ij} = \sum_{u \in i} \sum_{v \in j} \lambda_{uv} \left(\frac{\pi_u}{\sum_{u' \in i} \pi_{u'}} \right) . \tag{9.20}$$

The frequency of codon u, π_u, can be estimated empirically from the data. λ_{uv} can itself be set as a function of mechanistic parameters on the nucleotide level such as the transition/transversion rate ratio κ:

$$\lambda_{uv} = \begin{cases} 0, & \text{if the two codons differ at more than one position,} \\ \pi_v, & \text{for transversion,} \\ \kappa\pi_v, & \text{for transition.} \end{cases} \tag{9.21}$$

Because these fitness models do not make prior assumptions about which fitness characteristics best describe each specific site, they must deal with the issue of how to assign the site classes. For example, a large amino acid may be appropriate for some sites in the protein (represented by high values of v_o^k and b_k), while a small amino acid might be important at another site. To avoid any prior assumptions about which sites are represented by which selective constraints, *hidden site classes* are used. These are analogous to the hidden rate classes used by the Gamma distribution and to the hidden Markov classes of the PASSML model, albeit without site dependence. Rather than explicitly assigning classes to sites, all sites instead have a prior probability ϕ_k of being modeled by each site class k. The likelihood at each column in the alignment D_n is the sum of the conditional probabilities weighted by these prior distributions:

$$L\left(D_n | \theta\right) \equiv \sum_k f(D_n | \theta_k, k)\phi_k .$$ (9.22)

Typically the prior distribution are set equal for all classes and all sites (flat priors), although they can be specified on a site-by-site basis if prior information about each site is to be included in the model.

Under the simplifying assumption that $\lambda = 1$, the FIT-PC model has just five parameters per site class (a_k, b_k, h_o, v_o, and ω_k) compared with over 200 for a REV model. This allows the values of the parameters to be ML-estimated for each dataset of interest rather than estimated on a training set and then fixed as with the REV models. The FIT-PC model generally yields higher likelihoods than site-homogeneous REV models once a moderate number of site classes is specified (generally five or more) [45, 18]. The fact that higher likelihoods can be achieved even with such a simplified model is further evidence of the importance of site heterogeneity in protein modeling.

Other parameterizations of the fitnesses and fitness function have also been explored [79, 78]. For example, in the DIST-PC model [79], the fitnesses can more accurately be called distances, where the distance d_{ij} between amino acids is (in their example) based on polarity (p) and volume (v) [52]:

$$d_{ij} = \sqrt{(p_i - p_j)^2/\sigma_{\Delta p}^2 + (v_i - v_j)^2/\sigma_{\Delta v}^2} .$$ (9.23)

Here $\sigma_{\Delta p}^2$ and $\sigma_{\Delta v}^2$ are the standard deviations of $|p_i - p_j|$ and $|v_i - v_j|$, respectively. The substitution rate is an exponential function of this distance:

$$\Omega_{ij}^k = \omega_k e^{-(b_k d_{ij}/d_{\max})} .$$ (9.24)

The parameter b_k is a measure of the strength of selection upon the particular physical properties of the amino acid; a larger value of b_k indicates that more radical amino acid changes are less likely to be accepted as substitutions. The λ parameter can be specified as described above in the FIT-PC model. The differences between the Ω functions of the two models reflect two distinct philosophies about the manner in which evolution proceeds. The DIST-PC function can be thought of as a neutral walk through the fitness landscape; what matters most is not the direction of changes but whether or not they are conservative or radical. Given enough mutational steps, an amino acid position could change from small to large with little penalty. As a result, the equilibrium amino acid frequencies for the Ω function are all equal in the DIST-PC model if reversibility is assumed. By contrast, the FIT-PC model assumes that the protein site has an optimal set of physicochemical properties, and the favorability of an amino acid change is measured relative to both the former amino acid and this ideal value. Favorable mutations are all accepted at the same rate, while unfavorable ones are tolerated depending on their distance from the former amino acid's properties. In reality, the sites in a protein probably evolve in a mixture of these two regimes, and so a mixture of site classes or fitness functions can be appropriate [68].

The FIT-PC and DIST-PC models relax assumptions about which protein structural types are important, but they still require the specification of particular amino acid characteristics. Hydrophobicity, bulk, and polarity have been shown to be three of the most dominant [52, 44, 70], but other characteristics are sometimes crucial, such as the turn-inducing properties of glycine and proline or the delocalized electrons of the aromatic amino acids. To avoid any assumptions about which characteristics are important, a general fitness model can be used [18]. FIT-GEN is nested with the FIT-PC model, instead setting each F_i in equation (9.17) as a free parameter rather than as a function of physicochemical properties. This yields 21 parameters per site class: the 20 amino acid fitnesses F_i^k and the nonsynonymous rate ω_k (there are 20 free parameters because the fitnesses are relative). With adequate data, FIT-GEN is better able to capture the nuances of evolution than FIT-PC at the cost of some simplicity, while still using 188 fewer parameters per site class than a REV model. FIT-GEN can be used in an iterative manner with FIT-PC; general fitnesses can first be determined, and these can then be correlated with physicochemical characteristics. The dominant characteristics can then be utilized in a FIT-PC or DIST-PC model for later analysis on the same protein family.

The FIT-GEN model still assumes a specific number of site classes; the most general approach would be to assign one site class per location in the protein. Bruno [11] used an EM algorithm to obtain site-specific amino acid frequencies, with one frequency vector per site. The obstacle then becomes a lack of data; at short evolutionary distances, the inferred substitutions at each site may be just a fraction of the allowable substitutions, so a large, diverse sequence set is required. Although the parameters from this method are not directly applicable to phylogenetic analysis, they can provide a starting point for further site classifications, such as by principle component analysis [45] or as initial groupings in a FIT-GEN model.

Part of the power of these types of mechanistic hidden site class models is that they lend themselves well to empirical Bayesian mapping [58]. In this technique, the posterior probability of each site class k at each site n can easily be calculated using the likelihood and the prior distributions:

$$\Pr(k|D_n, \theta) = \frac{f(D_n|\theta_k, k)\phi_k}{\sum_{k'} f(D_n|\theta_k, k')\phi_{k'}} . \tag{9.25}$$

These posteriors probabilities can then be mapped onto the sequence alignment or protein structure of interest to determine which sites are more likely to be evolving under the different selective constraints [78, 5, 68, 69]. For example, site class 1 could model a fitness function based on polarity, site class 2 on bulk, and so on. When the posterior probabilities are mapped onto the sequence alignment, sites where bulk has been more important to evolution than polarity will have a higher posterior probability for site class 2. When mapped onto the protein structure, these posterior probabilities can reveal important evolutionary features such as transmembrane regions and dimerization interfaces [69]. Empirical Bayesian mapping is not limited to PHAS models; it has also been applied to RHAS models to map rate heterogeneity onto protein structures [63] (see Figure 9.2).

9.6 Modeling Heterogeneity over Time

The phylogenetic models discussed above assume that the rate and pattern of evolution have remained constant over the entire evolutionary tree, an assumption called *homotachy* [50]. This assumption can be violated when a protein is adapting to a new function or structure; according to the Neutral Theory, sites that are involved in the change in function will appear to evolve at a different rate. By developing models that allow a change in rate over time, these types of functional shifts can be detected.

The concept of an explicitly heterotachous model (or RHAT model, for Rate Heterogeneity Across Time) was first outlined in a maximum parsimony framework as the covarion model [23]. With this model, only a fraction of the sites in a protein-coding gene are "on" and can accept mutations: the concomitantly variable codons. All others are "off", and no substitutions are observed; these sites are assumed to be completely functionally constrained. A site may switch from "on" to "off" (and vice versa) with a certain persistence time, indicating that the site has acquired (or lost) functional significance.

More recently, covarion-like models for proteins have been cast into a likelihood framework, allowing the application of likelihood ratio tests for hypothesis testing. In 1999, Gu developed a time-heterogeneous ML method and applied it to the detection of functional divergence between gene duplicates [34]. Gene duplication is thought to be a factory for evolutionary diversification; one copy of the gene can continue to perform its native function, while the other can be adapted for a distinct task [15]. This adaptation results in different rates of substitution for each of the two paralogous protein families, a phenomenon dubbed type I divergence.

Consider the subfamilies in Figure 9.4, with two possible states: S_0 and S_1. S_0 is the null hypothesis that there are no altered functional constraints

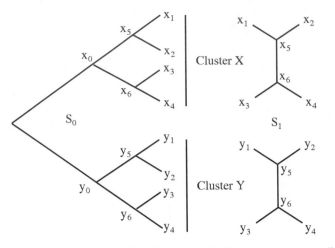

Fig. 9.4. In state S_0, both subfamilies are scaled by the same overall rate λ. In state S_1, each family subtree may have a different overall rate. Adapted from Gu [35].

in either subfamily. In this state, the substitution rates for each subfamily are completely correlated: $\lambda_X = \lambda_Y = \lambda$. The other possibility, S_1, is that the function of either or both subfamilies has diverged since the common ancestor, and therefore λ_X and λ_Y are treated as independent. The other parameter requiring estimation is θ_{12}, the probability that the site is in state S_1 (also called the coefficient of type I divergence).

To calculate the likelihoods, it is assumed that the subtrees are statistically independent, so that $f(X_n|\lambda_X)$ and $f(Y_n|\lambda_Y)$ are the likelihoods at site n for the unrooted subfamilies X and Y, respectively, conditional upon the rates for each subfamily. Since it is a difference in rate that is important and not the absolute rates, the values of λ_X and λ_Y are not explicitly specified but integrated out by using a Gamma rate distribution [77],

$$p(X_n) = E[f(X_n|\lambda)] = \sum_{\lambda'} f(X|\lambda')\phi(\lambda') , \qquad (9.26)$$

where $\phi(\lambda')$ is the probability of each λ' partition from the Gamma distribution. The joint probabilities conditional on being in either state S_0 or S_1 can then be written as

$$f^*(X_n, Y_n|S_0) = \sum_{\lambda'} f(X_n|\lambda')f(Y_n|\lambda')\phi(\lambda')$$
$$= E[f(X_n|\lambda)f(Y_n|\lambda)] ,$$
$$f^*(X_n, Y_n|S_1) = p(X_n)p(Y_n)$$
$$= E[f(X_n|\lambda_X)] \times E[f(Y|\lambda_Y)] . \qquad (9.27)$$

Finally, the full joint probability for the two subtrees at this site is

$$p^*(X_n, Y_n) = (1 - \theta_{12}) f^* (X_n, Y_n | S_0) + \theta_{12} f^* (X_n, Y_n | S_1) , \qquad (9.28)$$

where θ_{12} is a parameter called the coefficient of divergence. Over the whole tree, the likelihood is

$$L(X, Y | \text{data}) = \prod_n p^*(X_n, Y_n) . \qquad (9.29)$$

The null hypothesis is that $\theta_{12} = 0$, while the alternate hypothesis of functional divergence is that $\theta_{12} > 0$; these can be compared with a likelihood ratio test.[5] As the support for a rate change in the data increases, so should the θ_{12} parameter.

This RHAT model has been shown to successfully detect functional divergence on a variety of protein families, including COX enzymes [35] and tyrosine kinases [33]. It has been extended to the comparison of multiple clusters [35] and for the detection of type II divergence, where the evolutionary rate immediately after duplication is different from that in either subfamily. Empirical Bayesian mapping has been applied to detect the specific sites most likely to be involved in the functional change [35, 42, 28], and a faster approximate method has been devised that uses the ML estimates of the substitution counts in each subfamily to test for significance [34].

Note that the RHAT model, like the RHAS model, specifies only the rate parameter in conjunction with a REV matrix such as JTT and does not address differences in the pattern of mutations. This could be important if, for example, a particular site may evolve at the same rate in two subfamilies but with positively charged residues selected in one subfamily and negatively charged residues selected in the other. These types of questions have been addressed for nucleotide models [72, 27, 36] and in qualitative fashion for proteins [69], but they have not yet been applied in a rigorously testable "PHAT" context for proteins. As an example, one could set S_0 as the null hypothesis that sites in the subtrees evolve with the same mechanistic site class ($k_{X_n} = k_{Y_n} = k$) and evaluate S_1 as the alternative hypothesis that k_{X_n} and k_{Y_n} are independent.

The existence of rate heterogencity between widely divergent sequences and between paralogous protein subfamilies is generally well-accepted. But there is mounting evidence indicating that heterotachy, like site heterogeneity, may be quite common even within protein families where function is largely maintained [28]. For example, Lopez and co-workers [50] performed a thorough analysis of heterotachy on over 3000 sequences of vertebrate cytochrome b, a protein whose function in the electron transport pathway is generally conserved throughout vertebrates. They used a modified RHAT model to examine several evolutionary groupings of cytochrome b, finding significant evidence of protein heterotachy among birds, mammals, and fish, as well as among

---

[5]Since the θ_{12} parameter is at the boundary of its state space in the null model, the corrected χ^2 test should be used for significance testing [30].

four different groupings of murids. This significance was not caused by just a few extremely adaptive locations but was instead seen at a large percentage of sites. The fact that rates could vary significantly through time within a protein family with ostensibly conserved function is an indication that heterotachy may be an important component of protein models.

9.7 Modeling Correlated Evolution Between Sites

Most models of protein evolution treat sites independently, but this is mainly a mathematical convenience that helps to keep the likelihood equations tractable. In reality, a protein sequence does not generally function as an extended floppy chain, but as a globular structure where the amino acids pack tightly against one another. Since it is these interactions that determine the structure and function of a protein, there is significant interest in modeling correlated evolution between sites.

Correlation between sites can be classified as indirect or direct. Indirect correlation occurs when sites are in the same structural category and therefore subject to the same selective constraints. Their rates and patterns may be correlated, but a substitution at one site does not necessarily affect the rate of substitution at another. This type of correlation is the basis for models such as the PASSML models discussed previously. To measure the strength of correlation between adjacent sites, Gonnet and co-workers used a 400×400 dipeptide matrix [31]. This matrix was created using a parsimony-counting method similar to the Dayhoff method, but in this case there are 400 character states, one representing each two-residue pair. The resulting matrix was significantly different from matrices created by assuming site independence, indicating that nearby sites can undergo correlated evolution. For example, on average, conservation at the first position was likely to be correlated with conservation in the second, a reflection of the fact that nearby residues tend to be in the same types of environments.

Direct correlation, or coevolution, occurs when a substitution at one site alters the fitness landscape at other sites, potentially creating an adaptive evolutionary regime.[6] For example, the salt bridge is one type of stabilizing interaction in proteins, potentially formed when a positively charged amino acid residue is in the proximity of a negatively charged residue. If one member of the salt bridge mutates into an oppositely charged residue, it can destabilize the protein structure or disrupt its function. Assuming the mutation is accepted as a substitution, the salt bridge can be reestablished by a compensatory mutation at the other site, and the substitution rate can temporarily increase as a result [24]. Other possibilities for compensatory coevolution include small-large amino acid pairs and a polar-polar to nonpolar-nonpolar compensation.

---

[6]This is sometimes called "covariation," but that term is avoided here to minimize confusion with the covarion model mentioned in Section 9.6.

The degree to which such coevolution occurs is still debated. Most evidence seems to indicate that compensatory substitution does occur but that the prevalence is low [57, 66, 7, 46, 51, 32]. Some possible explanations for this weak signal are: (a) the first mutation is generally so deleterious that no chance for compensatory change is allowed, (b) an unfavorable substitution can be effectively compensated by subtle shifts in the protein structure, and/or (c) compensatory substitutions are important but occur at just a few sites in a particular protein, making them hard to detect among the many comparisons that must be made. Because of the potential for predicting protein structures and interactions, there has been significant interest in developing methods to detect the sites that may be strongly coevolving. Most of these methods have been primarily based upon detecting mutually informative sites in the alignment [67, 46, 6, 3, 21], but these types of methods can be misled unless proper correction due to evolutionary relationships is taken into account [61, 71, 75]. Even in methods that do explicitly utilize the phylogenetic tree, the tests are generally not based on a particular model of evolution [14, 26].

As an example of a coevolutionary Markov model, the site-independent fitness models in Section 9.5 are readily applicable to a coevolutionary framework by adding a correlation term:

$$F^{AB}(a, b) = F^A_{\text{ind}}(a) + F^B_{\text{ind}}(b) + F^{AB}_{\text{dep}}(a, b) \ . \tag{9.30}$$

In this equation, $F^A_{\text{ind}}(a)$ is the fitness for amino acid a if site A evolved independently of site B; for example, the fitness in equation (9.16) can be used. $F^{AB}_{\text{dep}}(a, b)$ is the coevolution term, an increase or decrease in the fitness of amino acid a due to the presence of amino acid b at site B. This dependent fitness function can itself be made mechanistic:

$$F^{AB}_{\text{dep}}(a, b) = \rho_{AB}\psi_{ab} \ . \tag{9.31}$$

ρ_{AB} is the strength of interaction between the site pairs, and $\boldsymbol{\Psi}$ is a symmetric interaction matrix, where ψ_{ab} describes the interaction between amino acids a and b. In the salt-bridge example given above, $\psi_{ab} > 0$ when a and b are of opposite charge, and $\psi_{ab} < 0$ when their charge has the same sign. Assuming $\rho_{AB} > 0$, the overall fitness F^{AB} will be increased when $\psi_{ab} > 0$–when the interaction between a and b is favorable. If a mutation is attempted at site A to amino acid j, when $\psi_{jb} > \psi_{ib}$, the result will potentially be an increase in Ω_{aj} and therefore Q_{aj}, the transition rate (equations (9.17) and (9.15)). When specifying $\boldsymbol{\Psi}$ using for example empirically determined contact energies, this model can be nested with the FIT-PC model; the two are equivalent when $\rho_{AB} = 0$ for a site pair. This model is similar to a codon-based model described in [64].

While such a coevolutionary fitness model is theoretically attractive, it is computationally impractical when performing full-likelihood calculations. One of the barriers to the development of any coevolutionary Markov model is the size of the state space. Instead of the 20 amino acid states in the site-independent model, there are $20 \times 20 = 400$ possible pairs of amino acids, leading to a 400×400 **Q**-matrix. Even if the data were available to estimate the ρ_{AB} parameter for each site pair, it is computationally expensive to exponentiate such a matrix and to use it in the phylogenetic likelihood function. (RNA coevolutionary models, with only $4 \times 4 = 16$ possible states, have had more success because they do not suffer from this limitation [55, 65].)

To simplify the state space, Pollock and co-workers [62] created a coevolutionary Markov model by reducing amino acids to two states (designated A and a or B and b, depending on their position in the pair). For example, all large amino acids might be designated A and small residues called a, or the split could be based on amino acid charge (positive or negative). There are then four possible states at an amino acid site pair—AB, Ab, aB, and ab—and the transition matrix is

$$\mathbf{Q} = \begin{matrix} AB \\ Ab \\ aB \\ ab \end{matrix} \left\{ \begin{matrix} -\sum_{AB} & \lambda_B \pi_{Ab}/\pi_A & \lambda_A \pi_{aB}/\pi_B & 0 \\ \lambda_B \pi_{AB}/\pi_A & -\sum_{Ab} & 0 & \lambda_A \pi_{ab}/\pi_B \\ \lambda_B \pi_{AB}/\pi_B & 0 & -\sum_{aB} & \lambda_B \pi_{ab}/\pi_a \\ 0 & \lambda_A \pi_{Ab}/\pi_b & \lambda_B \pi_{aB}/\pi_a & -\sum_{ab} \end{matrix} \right\}, \quad (9.32)$$

where λ_A and λ_B are the rates at each site and the π's are the stationary frequencies for each possible pair. The independent frequencies π_A and π_B are constrained by the pairwise frequencies:

$$\pi_A = \pi_{AB} + \pi_{Ab}. \quad (9.33)$$

The coevolutionary model has six parameters per site pair:

$$\theta_{\text{coev}} = \{\lambda_A, \lambda_B, \pi_{AB}, \pi_{Ab}, \pi_{aB}, \pi_{ab}\}. \quad (9.34)$$

(Since the π values must sum to 1, one of them is constrained by the others, and there are five free parameters). This yields one degree of freedom in comparison with the site-independent model, which assumes that $\pi_{xy} = \pi_x \pi_y$. The degree of correlation can be examined as a residue disequilibrium value, $RD = \pi_{AB}\pi_{ab} - \pi_{Ab}\pi_{aB}$, where a higher RD value indicates greater correlation. Pollock et al. found that the likelihood ratios did not fit the usual chi-squared distribution, so they used simulation to determine significance levels.

When applied to myoglobin, their model indicated the presence of coevolution, especially among neighboring sites, but as with most studies, the signal is weak. For example, they tested 2259 site pairs for coevolution using a charge metric to determine the character states. Due to Type I error, at the 5% significance level one would expect to erroneously report 113 pairs as false positives where no coevolution actually occurred. Pollock et al. found 158 significant pairs, indicating that 43 truly coevolving site pairs are probably mixed in with those 113. This is an example of the multiple testing problem that arises when testing all site pairs in a protein: the number of comparisons increases as the square of the number of sites, threatening to swamp the small number of true positives with false positives. Therefore, it is often important to reduce the number of comparisons either by making some prior assumptions about which sites are to be tested, by combining the data from groups of sites using total likelihood, or by making only relative comparisons.

9.8 Final Notes

Although protein evolutionary models have taken great strides since the formation of the Dayhoff matrix, their development and implementation are still nascent when compared with nucleotide models. For example, there is no commonly accepted hierarchy of nested protein models, and most of the more complex models detailed here have not been incorporated into popular tree-searching software packages. Therefore, to perform ML tree estimation on amino acid sequences, REV matrices such as JTT and WAG are generally the only options in commonly used software. At the very least, it is important to include rate heterogeneity among sites, such as with the $+\Gamma$ option, as failure to do so can cause errors in topology and divergence estimation [13]. Studies seem to tentatively indicate that the tree topology is fairly robust to model misspecification, as long as some site heterogeneity is included [12] in either RHAS or PHAS form. Therefore, it may be an adequate approximation to choose a credible set of trees using a REV+Γ model and then test more detailed hypotheses with the specialized models. Nevertheless, the full potential of recent advances in protein modeling will not be realized until these models are better integrated with tree-searching methods.

Another practical decision is whether to use amino acid or codon models. Codons contain information about the underlying mutation rate, and this information can be valuable for detecting selection at a particular site or along a particular lineage (see Chapter 5). But with this increase in information comes a decrease in computational speed. The Felsenstein pruning algorithm for likelihood calculation [22] is $\mathcal{O}(N^3)$; computational time increases as the cube of the number of states. Since codon state space is over three times larger than amino acid state space, computations with amino acid models are generally about 27 times faster than with codon models. For larger datasets and/or longer divergence times, amino acid models are often a more pragmatic choice and may provide more information about the origin of evolutionary constraints such as protein structure and amino acid characteristics. For smaller, more closely related sets of sequences, codon models offer higher fidelity and may provide more information about the different "directions" of Darwinian selection (purifying or adaptive) that act upon the evolution of the protein.

One practical constraint on the development of protein phylogenetic models has been the computational time involved in ML estimation and significance testing. Bayesian phylogenetic methods hold great promise for alleviating these concerns. Bayesian methods can provide estimates of the variance on the parameters of interest and integrate over the uncertainty in other parameters, allowing models that are more complex than those estimated using ML methods. With recent computational strides in this field, it is possible that Bayesian methods may facilitate the combination of site dependence with rate, pattern, and time heterogeneity into a unified framework.

References

[1] J. Adachi and M. Hasegawa. Model of amino acid substitution in proteins encoded by mitochondrial DNA. *J Mol Evol*, 42(4):459–468, Apr 1996.

[2] J. Adachi, P. J. Waddell, W. Martin, and M. Hasegawa. Plastid genome phylogeny and a model of amino acid substitution for proteins encoded by chloroplast DNA. *J Mol Evol*, 50(4):348–358, Apr 2000.

[3] D. A. Afonnikov, D. Y. Oshchepkov, and N. A. Kolchanov. Detection of conserved physico-chemical characteristics of proteins by analyzing clusters of positions with co-ordinated substitutions. *Bioinformatics*, 17(11):1035–1046, Nov 2001.

[4] C. B. Anfinsen. Principles that govern the folding of protein chains. *Science*, 181(96):223–230, Jul 1973.

[5] M. Anisimova, J. P. Bielawski, and Z. Yang. Accuracy and power of Bayes prediction of amino acid sites under positive selection. *Mol Biol Evol*, 19(6):950–958, Jun 2002.

[6] W. R. Atchley, K. R. Wollenberg, W. M. Fitch, W. Terhalle, and A. W. Dress. Correlations among amino acid sites in bHLH protein domains: An information theoretic analysis. *Mol Biol Evol*, 17(1):164–178, Jan 2000.

[7] E. Azarya-Sprinzak, D. Naor, H. J. Wolfson, and R. Nussinov. Interchanges of spatially neighbouring residues in structurally conserved environments. *Protein Eng*, 10(10):1109–1122, Oct 1997.

[8] A. Bateman, E. Birney, L. Cerruti, R. Durbin, L. Etwiller, S. R. Eddy, S. Griffiths-Jones, K. L. Howe, M. Marshall, and E. L. L. Sonnhammer. The Pfam protein families database. *Nucleic Acids Res*, 30(1):276–280, Jan 2002.

[9] S. A. Benner, M. A. Cohen, and G. H. Gonnet. Amino acid substitution during functionally constrained divergent evolution of protein sequences. *Protein Eng*, 7(11):1323–1332, Nov 1994.

[10] C. Branden and J. Tooze. *Introduction to Protein Structure*. Garland Publishing, New York, 1999.

[11] W. J. Bruno. Modeling residue usage in aligned protein sequences via maximum likelihood. *Mol Biol Evol*, 13(10):1368–1374, Dec 1996.

[12] T. R. Buckley. Model misspecification and probabilistic tests of topology: Evidence from empirical data sets. *Syst Biol*, 51(3):509–523, Jun 2002.

[13] T. R. Buckley, C. Simon, and G. K. Chambers. Exploring among-site rate variation models in a maximum likelihood framework using empirical data: Effects of model assumptions on estimates of topology, branch lengths, and bootstrap support. *Syst Biol*, 50(1):67–86, Feb 2001.

[14] G. Chelvanayagam, A. Eggenschwiler, L. Knecht, G. H. Gonnet, and S. A. Benner. An analysis of simultaneous variation in protein structures. *Protein Eng*, 10(4):307–316, Apr 1997.

[15] C. Chothia, J. Gough, C. Vogel, and S. A. Teichmann. Evolution of the protein repertoire. *Science*, 300(5626):1701–1703, Jun 2003.

[16] T. M. Collins, P. H. Wimberger, and G. J. P. Naylor. Compositional bias, character-state bias, and character-state reconstruction using parsimony. *Sys Biol*, 43:482–496, 1994.

[17] M. O. Dayhoff, R. M. Schwartz, and B. C. Orcutt. A model of evolutionary change in proteins. In M. O. Dayhoff, editor, *Atlas of Protein Sequence and Structure*, volume 5, chapter 22, pages 345–352. National Biomedical Research Foundation, Washington, DC, 1978.

[18] M. W. Dimmic, D. P. Mindell, and R. A. Goldstein. Modeling evolution at the protein level using an adjustable amino acid fitness model. In *Pacific Symposium on Biocomputing*, pages 18–29. World Scientific, Singapore, 2000.

[19] M. W. Dimmic, J. S. Rest, D. P. Mindell, and R. A. Goldstein. rtREV: an amino acid substitution matrix for inference of retrovirus and reverse transcriptase phylogeny. *J Mol Evol*, 55(1):65–73, Jul 2002.

[20] A. K. Dunker, C. J. Brown, J. D. Lawson, L. M. Iakoucheva, and Z. Obradovic. Intrinsic disorder and protein function. *Biochemistry*, 41(21):6573–6582, May 2002.

[21] P. Fariselli, O. Olmca, A. Valencia, and R. Casadio. Progress in predicting inter-residue contacts of proteins with neural networks and correlated mutations. *Proteins*, Suppl 5:157–162, 2001. Evaluation Studies.

[22] J. Felsenstein. Evolutionary trees from DNA sequences: a maximum likelihood approach. *J Mol Evol*, 17(6):368–376, 1981.

[23] W. M. Fitch and E. Markowitz. An improved method for determining codon variability in a gene and its application to the rate of fixation of mutations in evolution. *Biochem Genet*, 4(5):579–593, Oct 1970.

[24] K. M. Flaherty, D. B. McKay, W. Kabsch, and K. C. Holmes. Similarity of the three-dimensional structures of actin and the ATPase fragment of a 70-kDa heat shock cognate protein. *Proc Natl Acad Sci USA*, 88(11):5041–5045, Jun 1991.

[25] M. S. Fornasari, G. Parisi, and J. Echave. Site-specific amino acid replacement matrices from structurally constrained protein evolution simulations. *Mol Biol Evol*, 19(3):352–356, Mar 2002, letter.

[26] K. Fukami-Kobayashi, D. R. Schreiber, and S. A. Benner. Detecting compensatory covariation signals in protein evolution using reconstructed ancestral sequences. *J Mol Biol*, 319(3):729–743, Jun 2002.

[27] N. Galtier. Maximum-likelihood phylogenetic analysis under a covarion-like model. *Mol Biol Evol*, 18(5):866–873, May 2001.

[28] E. A. Gaucher, X. Gu, M. M. Miyamoto, and S. A. Benner. Predicting functional divergence in protein evolution by site-specific rate shifts. *Trends Biochem Sci*, 27(6):315–321, Jun 2002.

[29] N. Goldman, J. L. Thorne, and D. T. Jones. Assessing the impact of secondary structure and solvent accessibility on protein evolution. *Genetics*, 149(1):445–458, May 1998.

[30] N. Goldman and S. Whelan. Statistical tests of gamma-distributed rate heterogeneity in models of sequence evolution in phylogenetics. *Mol Biol Evol*, 17(6):975–978, Jun 2000, letter.

[31] G. H. Gonnet, M. A. Cohen, and S. A. Benner. Analysis of amino acid substitution during divergent evolution: The 400 by 400 dipeptide substitution matrix. *Biochem Biophys Res Commun*, 199(2):489–496, Mar 1994.

[32] S. Govindarajan, J. E. Ness, S. Kim, E. C. Mundorff, J. Minshull, and C. Gustafsson. Systematic variation of amino acid substitutions for stringent assessment of pairwise covariation. *J Mol Biol*, 328(5):1061–1069, May 2003.

[33] J. Gu, Y. Wang, and X. Gu. Evolutionary analysis for functional divergence of Jak protein kinase domains and tissue-specific genes. *J Mol Evol*, 54(6):725–733, Jun 2002.

[34] X. Gu. Statistical methods for testing functional divergence after gene duplication. *Mol Biol Evol*, 16(12):1664–1674, Dec 1999.

[35] X. Gu. Mathematical modeling for functional divergence after gene duplication. *J Comput Biol*, 8(3):221–234, 2001.

[36] J. P. Huelsenbeck. Testing a covariotide model of DNA substitution. *Mol Biol Evol*, 19(5):698–707, May 2002.

[37] D. T. Jones, W. R. Taylor, and J. M. Thornton. The rapid generation of mutation data matrices from protein sequences. *Comput Appl Biosci*, 8(3):275–282, Jun 1992.

[38] D. T. Jones, W. R. Taylor, and J. M. Thornton. A mutation data matrix for transmembrane proteins. *FEBS Lett*, 339(3):269–275, Feb 1994.

[39] S. Kawashima and M. Kanehisa. AAindex: Amino acid index database. *Nucleic Acids Res*, 28(1):374, Jan 2000.

[40] M. Kimura. *Population Genetics, Molecular Evolution, and the Neutral Theory: Selected Papers*. University of Chicago Press, Chicago, 1994.

[41] H. Kishino, T. Miyata, and M. Hasegawa. Maximum likelihood inference of protein phylogeny and the origin of chloroplasts. *J Mol Evol*, 30:151–160, 1990.

[42] B. Knudsen and M. M. Miyamoto. A likelihood ratio test for evolutionary rate shifts and functional divergence among proteins. *Proc Natl Acad Sci USA*, 98(25):14512–14517, Dec 2001.

[43] R. Koradi, M. Billeter, and K. Wuthrich. MOLMOL: A program for display and analysis of macromolecular structures. *J Mol Graph*, 14(1):51–55, Feb 1996.

[44] J. M. Koshi and R. A. Goldstein. Context-dependent optimal substitution matrices. *Protein Eng*, 8(7):641–645, Jul 1995.

[45] J. M. Koshi and R. A. Goldstein. Models of natural mutations including site heterogeneity. *Proteins*, 32(3):289–295, Aug 1998.

[46] S. M. Larson, A. A. Di Nardo, and A. R. Davidson. Analysis of covariation in an SH3 domain sequence alignment: Applications in tertiary contact prediction and the design of compensating hydrophobic core substitutions. *J Mol Biol*, 303(3):433–446, Oct 2000.

[47] P. Lió and N. Goldman. Using protein structural information in evolutionary inference: Transmembrane proteins. *Mol Biol Evol*, 16(12):1696–1710, Dec 1999.

[48] P. Lió and N. Goldman. Modeling mitochondrial protein evolution using structural information. *J Mol Evol*, 54(4):519–529, Apr 2002.

[49] L. Lo Conte, B. Ailey, T. J. Hubbard, S. E. Brenner, A. G. Murzin, and C. Chothia. SCOP: A structural classification of proteins database. *Nucleic Acids Res*, 28(1):257–259, Jan 2000.

[50] P. Lopez, D. Casane, and H. Philippe. Heterotachy, an important process of protein evolution. *Mol Biol Evol*, 19(1):1–7, Jan 2002.

[51] Y. Mandel-Gutfreund, S. M. Zaremba, and L. M. Gregoret. Contributions of residue pairing to beta-sheet formation: Conservation and covariation of amino acid residue pairs on antiparallel beta-strands. *J Mol Biol*, 305(5):1145–1159, Feb 2001.

[52] T. Miyata, S. Miyazawa, and T. Yasunaga. Two types of amino acid substitutions in protein evolution. *J Mol Evol*, 12(3):219–236, Mar 1979.

[53] K. Mizuguchi and T. Blundell. Analysis of conservation and substitutions of secondary structure elements within protein superfamilies. *Bioinformatics*, 16(12):1111–1119, Dec 2000.

[54] T. Muller, R. Spang, and M. Vingron. Estimating amino acid substitution models: A comparison of Dayhoff's estimator, the resolvent approach and a maximum likelihood method. *Mol Biol Evol*, 19(1):8–13, Jan 2002.

[55] S. V. Muse. Evolutionary analyses of DNA sequences subject to constraints of secondary structure. *Genetics*, 139(3):1429–1439, Mar 1995.

[56] G. J. Naylor and W. M. Brown. Structural biology and phylogenetic estimation. *Nature*, 388(6642):527–528, Aug 1997, letter.

[57] E. Neher. How frequent are correlated changes in families of protein sequences? *Proc Natl Acad Sci USA*, 91(1):98–102, Jan 1994.

[58] R. Nielsen and Z. Yang. Likelihood models for detecting positively selected amino acid sites and applications to the HIV-1 envelope gene. *Genetics*, 148(3):929–936, Mar 1998.

[59] J. Overington, D. Donnelly, M. S. Johnson, A. Sali, and T. L. Blundell. Environment-specific amino acid substitution tables: Tertiary templates and prediction of protein folds. *Protein Sci*, 1(2):216–226, Feb 1992.

[60] L. Patthy. *Protein Evolution*. Blackwell Science, London, 1999.

[61] D. D. Pollock and W. R. Taylor. Effectiveness of correlation analysis in identifying protein residues undergoing correlated evolution. *Protein Eng*, 10(6):647–657, Jun 1997.

[62] D. D. Pollock, W. R. Taylor, and N. Goldman. Coevolving protein residues: Maximum likelihood identification and relationship to structure. *J Mol Biol*, 287(1):187–198, Mar 1999.

[63] T. Pupko, R. E. Bell, I. Mayrose, F. Glaser, and N. Ben-Tal. Rate4Site: An algorithmic tool for the identification of functional regions in proteins by surface mapping of evolutionary determinants within their homologues. *Bioinformatics*, 18 (Suppl 1):71–77, Jul 2002.

[64] D. M. Robinson, D. T. Jones, H. Kishino, N. Goldman, and J. L. Thorne. Protein evolution with dependence among codons due to tertiary structure. *Mol Biol Evol*, 20(10):1692–1704, Oct 2003.

[65] M. Schoeniger and A. von Haeseler. Toward assigning helical regions in alignments of ribosomal RNA and testing the appropriateness of evolutionary models. *J Mol Evol*, 49(5):691–698, Nov 1999.

[66] O. Schueler and H. Margalit. Conservation of salt bridges in protein families. *J Mol Biol*, 248(1):125–135, Apr 1995.

[67] I. N. Shindyalov, N. A. Kolchanov, and C. Sander. Can three-dimensional contacts in protein structures be predicted by analysis of correlated mutations. *Protein Eng*, 7(3):349–358, Mar 1994.

[68] O. Soyer, M. W. Dimmic, R. R. Neubig, and R. A. Goldstein. Using evolutionary methods to study G-protein coupled receptors. In *Pacific Symposium on Biocomputing*, pages 625–636. World Scientific, Singapore, 2002.

[69] O. Soyer, M. W. Dimmic, R. R. Neubig, and R. A. Goldstein. Dimerization in aminergic G-protein coupled receptors: Application of a hidden site-class model of evolution. *Biochemistry*, 42(49):14522–14531, Dec 2003.

[70] K. Tomii and M. Kanehisa. Analysis of amino acid indices and mutation matrices for sequence comparison and structure prediction of proteins. *Protein Eng*, 9(1):27–36, Jan 1996.

[71] P. Tufféry and P. Darlu. Exploring a phylogenetic approach for the detection of correlated substitutions in proteins. *Mol Biol Evol*, 17(11):1753–1759, Nov 2000.

[72] C. Tuffley and M. Steel. Modeling the covarion hypothesis of nucleotide substitution. *Math Biosci*, 147(1):63–91, Jan 1998.

[73] H. Wako and T. L. Blundell. Use of amino acid environment-dependent substitution tables and conformational propensities in structure prediction from aligned sequences of homologous proteins. I. Solvent accessibility classes. *J Mol Biol*, 238(5):682–692, May 1994.

[74] S. Whelan and N. Goldman. A general empirical model of protein evolution derived from multiple protein families using a maximum-likelihood approach. *Mol Biol Evol*, 18(5):691–699, May 2001.

[75] K. R. Wollenberg and W. R. Atchley. Separation of phylogenetic and functional associations in biological sequences by using the parametric bootstrap. *Proc Natl Acad Sci USA*, 97(7):3288–3291, Mar 2000.

[76] Z. Yang. Estimating the pattern of nucleotide substitution. *J Mol Evol*, 39(1):105–111, Jul 1994.

[77] Z. Yang. Maximum likelihood phylogenetic estimation from DNA sequences with variable rates over sites: Approximate methods. *J Mol Evol*, 39(3):306–314, Sep 1994.

[78] Z. Yang. Relating physicochemical properties of amino acids to variable nucleotide substitution patterns among sites. In *Pacific Symposium on Biocomputing*, pages 81–92. World Scientific, Singapore, 2000.

[79] Z. Yang, R. Nielsen, and M. Hasegawa. Models of amino acid substitution and applications to mitochondrial protein evolution. *Mol Biol Evol*, 15(12):1600–1611, Dec 1998.

Models of Microsatellite Evolution

Peter Calabrese[1] and Raazesh Sainudiin[2]

[1] University of Southern California, Los Angeles, CA 90089-2532, USA,
petercal@usc.edu
[2] Cornell University, Ithaca, NY 14853, USA, rs228@cornell.edu

10.1 Introduction

Microsatellites are simple sequence repeats in DNA; for example, the motif AT repeated twenty-five times in a row. Microsatellites mutate by changing the number of their repeats; for example, the $(AT)_{25}$ mentioned in the previous sentence might become an $(AT)_{24}$ or $(AT)_{26}$ in that individual's offspring. These length-changing mutations occur at rates several orders of magnitude higher than point mutations. The reason for microsatellites' popularity as genetic markers is that their high mutation rates make them highly polymorphic, and they are relatively dense in the genomes of many organisms. For a review, see the article by Ellegren [16], and for a collection of articles, see the book edited by Goldstein and Schlötterer [23]. Ellegren [16] succinctly wrote, "simple repeats do not evolve simply." In this chapter, then, we will discuss many different models for microsatellite evolution.

Researchers have exploited microsatellites for many purposes. They are commonly used in the construction of genome-wide maps, in the search for disease-causing genes, and in identification for both forensics applications and paternity tests. As an example, the controversy over whether Thomas Jefferson fathered a child with his slave Sally Hemings was rekindled by a microsatellite study [19]. In these applications, two individuals are considered closely related if a large percentage of the microsatellite markers studied have the same number of repeats. However, microsatellites in different individuals are not just the same or different; they can differ by just a few repeat units or by many repeat units. Because pedigree experiments have shown that most mutations are a change in one repeat unit (85% in [54], 78% in [50]), some researchers have used microsatellites as molecular clocks. By studying the average number of repeat differences in many microsatellite loci, one can infer the time to the most recent common ancestor of two individuals. Microsatellites have been used to estimate the age of nonmicrosatellite mutations; for example, the $CCR5 - \Delta 32$ AIDS-resistance allele [46]. In cancer research, hypermutable microsatellites with deficient DNA mismatch repair systems have been used to reconstruct tumor progression [48]. Microsatellites have been used to infer selective sweeps (for a recent review, see, e.g., [41]), demographic history ([22], [10], [37], [2]), and population structure ([38], [17]).

The vast majority of microsatellites in higher organisms are believed to evolve neutrally (i.e., there is no selection pressure on the number of repeats). Nonetheless, some microsatellites exist in promoter regions and may be sites for protein binding or be near such sites. In this case, the number of repeats in these microsatellites has an effect on transcription and the degree of protein binding [29]. Furthermore, other microsatellites play a direct role in such human diseases as Fragile X syndrome, myotonic dystrophy, and Huntington's disease; these diseases are caused by trinucleotide microsatellites at specific locations expanding beyond a disease-specific threshold [39].

The predominant mechanism by which microsatellites mutate is believed to be replication slippage [15]. When DNA replicates, the two strands sometimes disassociate. In nonrepetitive DNA, the strands reassociate the same way they were before the slippage event, with matching base pairs on the opposing strands. But in repetitive DNA, since there are so many possible matching base pair alignments, sometimes the strands realign differently, forming an unmatched loop on one of the strands. Then, when the two strands completely disassociate and begin replication anew, the strand that had the loop will contain a longer microsatellite than the opposing strand. The microsatellite on the template strand will always have the same length before and after the slippage event. If the loop is on the template strand, then the microsatellite on the replicating strand will be shorter, and if the loop is on the replicating strand, then the microsatellite on its side will be longer. For a figure of this process, see [15], p. 38. These primary mutations, which depend exclusively on the DNA replication machinery, occur at rates several orders of magnitude higher than the observed mutation rate and are countered by the DNA mismatch repair machinery (for a recent review, see [42]). Thus the observed mutations are those replication slippage events that have escaped repair.

Since longer microsatellites present more opportunity for slippage, we would expect mutation rates to increase as a function of microsatellite length; this prediction is experimentally supported [53]. Some microsatellites are interrupted; for example, $(AT)_{12}GT(AT)_7$. Since these interruptions allow fewer realignments after a possible slippage event, we would expect interrupted microsatellites to have lower mutation rates, and this is also experimentally supported [36].

Several other factors are also known to be associated with the heterogeneity in mutation rates across microsatellite loci. Dinucleotides have a lower mutation rate than tetranucleotides (see Table 1 of [16]). Moreover, different dinucleotide motifs have strikingly different length distributions in the human genome [7], possibly due to motif-specific differences in the efficacy of mismatch repair [25]. A significant number of uninterrupted compound repeats (> 30000) such as $(TG)_m$-$(TA)_n$, with both m and $n \geq 5$ repeat units, occur in the human genome (Sainudiin and Durrett, unpublished results); their evolutionary dynamics are complex [5] and not well-understood. A further complication to measuring microsatellite variability is that insertions/deletions in the flanking regions can also affect the PCR fragment length (see, e.g., [24]).

10.2 Models

The oldest model for microsatellite evolution is the stepwise mutation model originally proposed by Ohta and Kimura [35] for electrophoretic alleles. In this model, the number of repeat units is equally likely to increase or decrease by one at a rate independent of the microsatellite's length, subject to the constraint that the number of repeat units cannot become smaller than one. Let X be the length of the microsatellite; then

$$X \to X + 1 \text{ at rate } \gamma \text{ and}$$
$$X \to X - 1 \text{ at rate } \gamma. \tag{10.1}$$

This is a symmetric random walk with a lower boundary condition. Numerous complications to this basic model have been introduced.

The first complication we will discuss is allowing the mutation rate to depend on the microsatellite's length. For example, Kruglyak et al. [30] proposed a proportional slippage model where the mutation rate increases linearly with the microsatellite's length

$$X \to X + 1 \text{ at rate } b(X - 1) \text{ and}$$
$$X \to X - 1 \text{ at rate } b(X - 1). \tag{10.2}$$

Sibly, Whittaker, and Talbot [45] proposed a model with an additional constant term

$$X \to X + 1 \text{ at rate } b_0 + b_1(X - 1) \text{ and}$$
$$X \to X - 1 \text{ at rate } b_0 + b_1(X - 1). \tag{10.3}$$

This constant term is analogous to the "indel slippage" term in [12]. Calabrese, Durrett, and Aquadro [8] further extended this model to prevent microsatellites shorter than a threshold κ from mutating:

$$X \to X + 1 \text{ at rate } b(X - \kappa)^+ \text{ and}$$
$$X \to X - 1 \text{ at rate } b(X - \kappa)^+ \tag{10.4}$$

(where $(X - \kappa)^+ = \max(X - \kappa, 0)$). The symmetric random walk models do not have a stationary distribution on their countable state space [32]. Nauta and Weissing [33] proposed a finite alleles version of the stepwise mutation model by imposing range constraints with reflecting boundaries to assure a uniform stationary distribution (also see [18]).

Most, but not all, observed microsatellite mutations are by one repeat unit. Therefore, Di Rienzo et al. [11] proposed a model that allows for larger mutations. With probability p, a mutation is one repeat unit, and with probability $1 - p$, the mutation could be larger. In their formulation, the one-step mutations followed the stepwise mutation model, and the larger mutations were equally likely to be contractions or expansions, with the magnitude of these mutations following a truncated geometric distribution.

Another complication is to allow the mutation rates to be asymmetric. Walsh [49] proposed a linear birth-death chain (i.e., a proportional slippage model where the mutation rate increases linearly with the microsatellite's length in the presence of biased contractions,

$$
\begin{aligned}
X &\to X + 1 \text{ at rate } bX, \\
X &\to X - 1 \text{ at rate } dX,
\end{aligned}
\tag{10.5}
$$

for $X \in \{2, 3, \ldots, \infty\}$ and $1 \to 2$ at a much smaller birth rate ν). He showed that a stationary distribution exists for this model when $d/b > 1$ (see also [47]). Fu and Chakraborty [20] proposed a model that allows larger mutations according to a geometric distribution in the presence of a constant mutational bias. Calabrese and Durrett [7] generalized the models described earlier to asymmetric linear and quadratic models: for the linear model

$$
\begin{aligned}
X &\to X + 1 \text{ at rate } u_0 + u_1(X - \kappa)^+ \text{ and} \\
X &\to X - 1 \text{ at rate } d_0 + d_1(X - \kappa)^+,
\end{aligned}
\tag{10.6}
$$

and for the quadratic model

$$
\begin{aligned}
X &\to X + 1 \text{ at rate } u_0 + u_1(X - \kappa)^+ + [u_2(X - \kappa)^+]^2 \text{ and} \\
X &\to X - 1 \text{ at rate } d_0 + d_1(X - \kappa)^+ + [d_2(X - \kappa)^+]^2.
\end{aligned}
\tag{10.7}
$$

The expansion and contraction rates can take the same parametric form with distinct parameters as above or take different parametric forms as well. Xu et al. [54] suggested that the expansion rate be independent of microsatellite length while the contraction rate increases exponentially with microsatellite length. Using an approximation to the Ornstein-Uhlenbeck process, Garza, Slatkin, and Freimer [21] proposed that microsatellites have a focal length in the sense that microsatellites shorter than this length have a bias upwards, whereas longer microsatellites have a bias downwards (also see [58]). These models can also allow larger mutations by specifying the expectation and variance of the size of mutations and thus nest the stepwise mutation model and the model due to Di Rienzo et al. [11] within them. While most of the previously described asymmetric models do not stipulate this focal property a priori, when these models are fit to data, generally their parameter estimates do satisfy this property.

Two recent studies attempt to capture several features of microsatellite evolution just described. Whittaker et al. [52] proposed a class of models with the following transition rates from microsatellite length $X = i$ to length j:

$$
q_{ij} = \begin{cases} \gamma_u e^{\alpha_u i} e^{-\lambda_u(j-i)}, & i < j, \\ \gamma_d e^{\alpha_d i} e^{-\lambda_d(i-j)}, & i > j. \end{cases}
\tag{10.8}
$$

Sainudiin [40] proposed another class of models,

$$q_{ij} = \begin{cases} \mu(1 + (i - \kappa)s)[u - v(i - \kappa)]_0^1 m(1 - m)^{|i-j|-1}, i < j, \\ \mu(1 + (i - \kappa)s)\{1 - [u - v(i - \kappa)]_0^1\}\frac{m(1-m)^{|i-j|-1}}{1-(1-m)^{i-\kappa}}, i > j. \end{cases} \quad (10.9)$$

In equation (10.9), the notation means

$$[\alpha]_0^1 = \begin{cases} 1 \text{ if } \alpha > 1, \\ 0 \text{ if } \alpha < 0, \text{ and} \\ \alpha \text{ otherwise}. \end{cases} \quad (10.10)$$

Both classes of models allow the mutation rates to increase with microsatellite length, the bias to change as a function of microsatellite length, and larger mutations to have a geometrical distribution. However, the parametric forms of these models differ.

The final model complication we will consider is point mutations. Point mutations can interrupt a microsatellite, for example transforming $(AT)_{20}$ into $(AT)_{12}GT(AT)_7$. Since most researchers measure the length of microsatellites without sequencing, they would not detect this transformation. Bell and Jurka [3] proposed that such point mutations constrain the growth of microsatellites. Kruglyak et al. [30] proposed a model with two processes

1. single-step proportional slippage (described above): $X \to X + 1$ at rate $b(X - 1)$ and $X \to X - 1$ at rate $b(X - 1)$, and
2. point mutations: $X \to j$, where $j < X$ at rate a.

Thus a point mutation is uniformly likely to affect any of the repeat units. Durrett and Kruglyak [14] showed that this model has a stationary distribution. Sibly, Whittaker, and Talbot [45] and Calabrese and Durrett [7] followed this paradigm of considering a slippage process and a point mutation process, but they considered more general slippage processes. (Now that we are considering interrupted microsatellites, the state space is more complicated. The studies referenced above chose counting schemes to limit the state space to one dimension, but they all chose to do this in different ways, and this has been the source of some confusion. For every point in the genome, Kruglyak et al. [30] and Durrett and Kruglyak [14] counted all uninterrupted repeats to the left. The other studies did not consider every point in the genome: Sibly, Whittaker, and Talbot [45] counted only the left half of an interrupted repeat, and Calabrese and Durrett [7] counted only uninterrupted repeats.)

One final caveat is in order. Many microsatellite models have been proposed, and we believe this summary captures most of the important concepts, but we do not claim to be exhaustive.

10.3 Experiments and Analysis

One of the statistical tools used in this section is the Akaike information criterion (AIC) [1]. The formula for computing the AIC score for a model is simple:

$$\text{AIC} = -2\log(\text{maximum likelihood}) + 2(\text{number parameters}). \qquad (10.11)$$

Given a list of models, we compute the AIC score for each model and choose the models with the lowest scores. This scheme has the advantage over the likelihood ratio test and parametric bootstrap (see, e.g., [4] and [28]) of being able to select from a large set of models without considering all pairwise comparisons. The AIC score is intuitive because the best models should have high likelihoods, and models are penalized for having a large number of parameters. But this scoring system also has a firm statistical foundation. The book by Burnham and Anderson [6] discusses the model selection problem in general and also presents a heuristic justification for the AIC scoring system; see pp. 239–247.

There are several types of data sets to consider when studying microsatellites. The first is pedigree data. Two of the largest such data sets (both in humans) were by Xu et al. [54] and Huang et al. [26]. Xu et al. [54] observed that the rate of expansion is independent of microsatellite "length" but that the rate of contraction increases exponentially as a function of microsatellite "length." Huang et al. [26] found a statistically significant negative relationship between the magnitude and direction of mutation and "length." In the two preceding sentences, we have put the word length in quotation marks because both groups of researchers did not measure the actual length of a microsatellite but rather the total length of the polymerase chain reaction (PCR) product that consists of the microsatellite and a variable amount of flanking sequence. They then applied the inverse of the distribution function of the observed lengths to obtain a number in $[0, 1]$ that they called the "length." This near universal practice of measuring the PCR product length rather than the actual number of repeat units has complicated modeling efforts.

In another large human pedigree study, Whittaker et al. [52] have taken the further step of using the human genome sequence and the primer sequence to infer the number of repeat units from the PCR product length. While this method cannot tell whether an individual microsatellite has been interrupted by point mutations, it is an important advance over simply using PCR product length. They measured 118,866 parent-offspring transmissions of AC repeats and observed 53 mutations, for a mutation rate of 4.5×10^{-4} per generation. Approximately 72% of the mutations were of one repeat unit. The mutation rate clearly increased superlinearly with the repeat length (see Figure 2 in [52]). They used the AIC scoring system to compare models of the class in equation (10.8). The cases where mutation rate increases with microsatellite length ($\alpha > 0$) were significant improvements over the cases where mutation rate was independent of microsatellite length ($\alpha = 0$). The best model in this class had asymmetric γ and α terms ($\gamma_u \neq \gamma_d$ and $\alpha_u \neq \alpha_d$), implying a mutation rate bias, but a symmetric λ term, implying the distribution of the size of the larger mutations was symmetric. The estimated parameters implied that microsatellites shorter than 20 repeat units had a bias towards expansions and longer microsatellites had a bias towards contractions. However, there were large confidence intervals for all of the parameter estimates.

Another type of data set is in vitro experiments. During PCR, microsatellites are duplicated and there are opportunities for slippage just as in in vivo cell division. In single-molecule PCR experiments, Shinde et al. [43] found that slippage rates increase linearly with microsatellite length, and there is a threshold of eight base pairs, below which microsatellites do not appear to slip. For all microsatellite lengths, they found a higher rate for contractions than expansions (14 times greater for AC microsatellites and five times greater for poly-A microsatellites). There are thermodynamic reasons to expect this asymmetry in vitro (see references in [43]). Clearly there are differences, however, between these in vitro experiments with *Taq* DNA polymerase and no mismatch repair system and in vivo cell division. Another set of related experiments studies microsatellite mutations in vivo but in organisms whose mismatch repair system has been knocked out. For an example in mice, see [55], and in yeast see [36]. There are many more microsatellite mutations in individuals with deficient mismatch repair systems, and this is informative for studying microsatellite models, but in addition to the rate, the pattern of mutations also appears to be different in these individuals.

In population data, many unrelated individuals are typed at numerous microsatellite loci. Nielsen [34] suggested using such data sets and likelihood ideas for model selection. The problem with this approach is that assumptions must be made about the genealogies of individuals. These assumptions will in turn affect the evaluation of the models.

Another type of data set is genome data. There are now complete (or nearly complete) genome sequences available for many species. For each such species, we thus have the length distribution of all microsatellites in one idealized individual. If we assume this distribution is at equilibrium and we consider models that have a stationary distribution, then we can fit these models to genome data. All the references we will discuss assume that microsatellites are "born" at some minimum length. Kruglyak et al. [30] fit the proportional slippage (equation (10.2)) with point mutation model to the then-available genome sequence of humans, mice, fruit flies, and yeast. They later fit this model to the complete genome sequence of yeast [31]. Assuming different microsatellites evolve independently, Sibly, Whittaker, and Talbot [45] then used likelihood ideas to compare different models of microsatellite evolution. They considered symmetric slippage models of the form in equation (10.3) with a point mutation process and found support for the parameters $b_0 \neq 0$ and $b_1 \neq 0$.

Calabrese and Durrett [7] used genomic data and the AIC scoring system to consider many different slippage models, including most of those then in the literature. They considered general slippage processes with a uniform point mutation process as described in the previous section. The data they considered were moderately spaced dinucleotide microsatellites uninterrupted by point mutations. They found the asymmetric models explained the genome data significantly better than the symmetric models. One of the best models had asymmetric quadratic slippage (equation (10.7)), where the parameters were such that dinucleotide microsatellites with length longer than 25 repeat units had a bias towards contractions. Moreover, for humans (but not *Drosophila*), they found that the different dinucleotide motifs had strikingly different distributions, as shown in Figure 10.1.

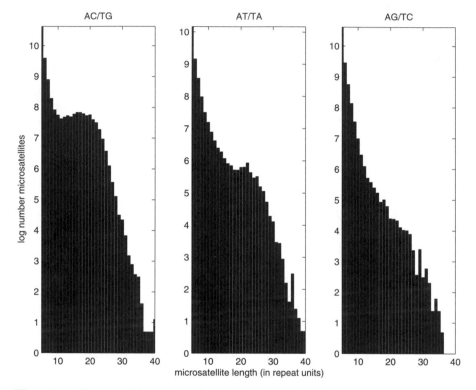

Fig. 10.1. Separated by motif, the natural logarithm of one plus the number of dinucleotide microsatellites of different lengths in the human genome.

Calabrese and Durrett [7] exploited a connection with queueing theory in order to calculate the stationary distribution. In the language of continuous-time Markov chains, each model was specified by a set of exponential holding times $\mu(j)$ for microsatellites of length j and the probabilities $p(j,i)$ a microsatellite of length j will next mutate to length i. The total number of microsatellites in the genome was modeled as a network of queues (specifically $M/M/\infty$ queues, where in the usual queueing theory terminology microsatellites correspond to customers, microsatellite lengths correspond to stations, and at each length or station there are an infinite number of servers); when a microsatellite is interrupted by a point mutation, it exits the network. Since all microsatellites have a positive probability of leaving the network, there exists a stationary distribution. Define arrival rates

$$r(\kappa) = \lambda + \sum_i r(i)p(i, \kappa) , \tag{10.12}$$

$$r(j) = \sum_i r(i)p(i, j), j > \kappa , \tag{10.13}$$

where λ is a scaling parameter that is the rate at which microsatellites are born at the minimum considered length κ. Then the stationary distribution is for all j the number of microsatellites with length j that are independent and Poisson distributed with mean $r(j)/\mu(j)$ (see e.g., [13], p. 192). Let $l(j)$ be the number of microsatellites of length j in the genome, and define the normalizing constant $Z = \sum_j r(j)/\mu(j)$. Since Calabrese and Durrett assumed moderately spaced microsatellites evolve independently, conditioning on the number of microsatellites, the likelihood of the data is

$$\prod_j \left(\frac{r(j)}{Z\mu(j)} \right)^{l(j)} . \tag{10.14}$$

For each model, Calabrese and Durrett numerically solved the linear system of equations (10.12), (10.13) to determine the arrival rates $r(j)$, numerically maximized the likelihood of the genome (10.14) over the space of model parameters, and computed the AIC score.

The final type of data set we will consider collects microsatellites from two closely related populations or species. Since longer microsatellites are more mutable and hence more useful to experimentalists, when microsatellites are selected in one species and then compared in another species, there is an ascertainment bias. Two studies that avoid this problem are Cooper, Rubinsztein, and Amos [9] and Webster, Smith, and Ellegren [51]. Despite accounting for this ascertainment bias, Cooper, Rubinsztein, and Amos [9] found that human dinucleotide microsatellites are significantly longer than their chimpanzee orthologs. Webster, Smith, and Ellegren [51] considered many more microsatellites and concurred with the findings of Cooper et al. They also found that human mononucleotide microsatellites are more likely to be *shorter* than their chimpanzee counterparts. Webster, Smith, and Ellegren [51] selected an unbiased sample of AC dinucleotide microsatellites from a region of genomic DNA and compared the orthologs in humans and chimpanzees.

Sainudiin [40] followed this strategy and used the AIC scoring system to compare slippage models of the form in equation (10.9). Sainudiin initially assumed that the same microsatellite model and parameters applied both to the human and chimpanzee lineages, and concluded $s \not\models 0$, so longer microsatellites are more mutable, and $v \neq 0$, so there is a bias term that depends linearly on the microsatellite's length. The estimated parameters imply microsatellites shorter than 18 repeat units have a bias towards expansion, while longer microsatellites have a bias towards contraction. When Sainudiin relaxed the assumption that the same model parameters applied in both the human and chimpanzee lineages, it was found that this focal length increased to 21 repeats in humans while remaining at 18 repeats in chimpanzees, further confirming the findings in [9] and [51].

Sainudiin [40] considered three Markov chains, one each on the ancestral, human, and chimpanzee lineages. These three Markov chains had rate matrices $\mathbf{Q}^{(a)}$, $\mathbf{Q}^{(h)}$, and $\mathbf{Q}^{(c)}$, specified by equation (10.9), possibly with different parameters. Let λ_h and λ_c be the branch lengths of the human and chimpanzee lineages, respectively. Then the transition probability matrices $\mathbf{P}^{(h)}(\lambda_h)$ and $\mathbf{P}^{(c)}(\lambda_c)$ were obtained by matrix exponentiation of the product of the corresponding rate matrix and branch length (e.g., $\mathbf{P}^{(h)}(\lambda_h) = \exp\{\mathbf{Q}^{(h)}\lambda_h\}$). The stationary distribution of the ancestral Markov chain $\mathbf{X}^{(a)}$ was denoted by $\boldsymbol{\pi}^{(a)}$. The data considered were N homologous microsatellite lengths (H_i, C_i) in the human and chimpanzee genomes. Then the likelihood of the data is

$$\prod_i \sum_j \pi_j^{(a)} \, P_{j,C_i}^{(c)}(\lambda_c) \, P_{j,H_i}^{(h)}(\lambda_h). \tag{10.15}$$

Since the ancestral state is unknown, the likelihood can be thought of as a weighted sum over all possible ancestral states, where the weights come from the stationary distribution of the ancestral chain. The product term comes from the assumption of independence among the N loci. For each model, Sainudiin [40] numerically optimized the likelihood (10.15) over the space of model parameters, and computed the AIC score.

10.4 Discussion

We have discussed numerous microsatellite models. There is evidence from a number of different sources that the best models have the following properties

1. long microsatellites are more likely to mutate, and
2. long microsatellites have a bias towards contraction, while
3. short microsatellites have a bias towards expansion.

For dinucleotide repeats in humans, this focal length appears to be around 20 repeat units. Moreover, all the model parameters depend on both the length and composition of the repeat motif. In our opinion, the parametric form of the "best" model is still unclear.

We believe that the best type of data set to determine this model is pedigree data, where the actual number of repeat units has been inferred (rather than just using the PCR fragment length) as in Whittaker et al. [52]. It would be interesting to infer the number of repeat units and reanalyze the data sets in Xu et al. [54] and Huang et al. [26]. The main advantage of using genome data is that it is plentiful. The disadvantage is that we do not observe mutations but rather the stationary distribution; and distinct models may have very similar stationary distributions. For example, this makes it difficult to determine the percentage of large mutations using genome data. Further, Sibly et al. [44] specifically investigated the distribution of interrupted repeats and found the existing slippage/point mutation models inadequate to explain the data.

One question is whether the choice of model matters. For example, let us consider using the statistic $(\delta\mu)^2$ to measure the time to the most recent common ancestor of two individuals. Let X_i and Y_i be the microsatellite lengths at the ith locus in the two individuals; define the statistic

$$(\delta\mu)^2 = \sum_{i=1}^{I}(X_i - Y_i)^2/I$$

as the average over I loci. Goldstein et al. [22] showed that for the stepwise mutation model $E(\delta\mu)^2(t) = 2\gamma t$, where t is the time to the most recent common ancestor and γ is the mutation rate. For the more complicated models that we have discussed, it is unlikely that there is such a simple formula, but we can simulate these models.

Let us compare the mutation model in equation (10.8) and the stepwise mutation model. One difficulty with length-dependent mutation models that we do not encounter with the stepwise mutation model is that we have to make additional assumptions about the length of the ancestor. Since the model in equation (10.8) has a stationary distribution, let us assume that the common ancestor has a length chosen randomly from this stationary distribution. Furthermore we can use this stationary distribution to find the average mutation rate for a microsatellite chosen randomly from this distribution. In order to fairly compare models, let us set the mutation parameter in the stepwise mutation model equal to this average. For the model parameters estimated in [52], this average mutation rate is $\gamma_1 = 1 \times 10^{-4}$. This rate is smaller than both the observed rate in [52] (4.5×10^{-4}) and that in other dinucleotide studies (e.g. 5.6×10^{-4} in [22]). Consequently, we also consider the model in equation (10.8) where the γ parameters have been increased to match the average mutation rate $\gamma_2 = 5 \times 10^{-4}$. When we increase the γ parameters, we preserve the estimated ratio γ_u/γ_d and the α and λ parameters; these new γ parameters are still well within the confidence intervals estimated from the data. Likewise, we consider the stepwise mutation model with elevated rate $\gamma_2 = 5 \times 10^{-4}$.

Figure 10.2 shows the mean of $(\delta\mu)^2$ for these two models as a function of time. The left-hand plot has average mutation rate $\gamma_1 = 1 \times 10^{-4}$, and the right-hand plot is a rescaling with average mutation rate $\gamma_2 = 5 \times 10^{-4}$. For the left-hand plot at times less than 25000 generations, the two models are in good agreement. For greater times, the two models diverge; this is because under the stepwise mutation model, the $(\delta\mu)^2$ statistic continues to grow linearly, while for the models with a stationary distribution, this statistic eventually plateaus. For the right-hand plot, since the mutation rate is five times greater, the two models start to diverge about five times earlier, at around 5000 generations. If we are interested in short divergence times, then the stepwise mutation model seems a reasonable approximation. If we are interested in divergence times that are too long and we believe the true microsatellite mutation model has a stationary distribution, then microsatellites will not be useful because any divergence statistic will eventually plateau due to this stationarity.

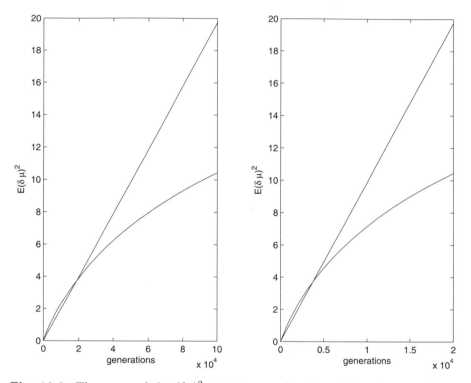

Fig. 10.2. The mean of the $(\delta\mu)^2$ statistic as a function of time in generations. The linear curve is for the stepwise mutation model; the nonlinear curve is for the slippage model in equation (10.8). The left-hand plot has average mutation rate $\gamma_1 = 1 \times 10^{-4}$, and the right-hand plot is a rescaling with average mutation rate $\gamma_2 = 5 \times 10^{-4}$. Since the right-hand plot has a mutation rate five times greater, the two models start to diverge five times earlier.

The times for which the models diverge are germane to the study of human populations. If we model the genealogy of unrelated individuals with the neutral coalescent (see, e.g., [27]) and assume the commonly used estimate of 10000 for the effective population size of humans, then the average time to the most recent common ancestor of two individuals is 20000 generations, and the average time to the most recent common ancestor of a large sample is 40000 generations. Assuming the stepwise mutation model, various statistics of microsatellite lengths have been used to infer aspects of demographic history (e.g., [57], [10], [37], [56]). We have simulated the coalescent process with effective population size 10000 and sample size 50, and alternately used the stepwise mutation model and the model in equation (10.8) with the two parameter sets discussed in the previous paragraphs. In Table 10.1, we show the median (5% quantile, 95% quantile) for several summary statistics. We can see that such statistics and related tests will be dependent on the mutation model used.

Table 10.1. The median (5% quantile, 95% quantile) of the sample variance, homozygosity, and number of alleles for the stepwise mutation model (SMM) and the model in equation (10.8) (WHB) at two average mutation rates. These values were simulated using the coalescent with effective population size 10000, and sample size 50.

Model	$\gamma_1 = 1 \times 10^{-4}$			$\gamma_2 = 5 \times 10^{-4}$		
	sam.var.	homo.	num.all.	sam.var.	homo.	num.all.
SMM	1.2 (0.3, 5.9)	0.3 (0.2, 0.6)	5 (3, 7)	6.2 (1.8, 29)	0.2 (0.1, 0.3)	9 (6,13)
WHB	1.9 (0.2, 11)	0.4 (0.2, 1.0)	5 (2, 10)	7.0 (1.0, 20)	0.2 (0.1, 0.5)	10 (4, 14)

References

[1] H. Akaike. A new look at the statistical model identification. *IEEE Trans. Autom. Control*, 19:716–723, 1974.

[2] M. Beaumont. Detecting population expansion and decline using microsatellites. *Genetics*, 153:2013–2029, 1999.

[3] G. I. Bell and J. Jurka. The length distribution of perfect dimer repetitive DNA is consistent with its evolution by an unbiased single-step mutation process. *J. Mol. Evol.*, 44:414–421, 1997.

[4] P. J. Bickel and K. A. Doksum. *Mathematical Statistics*. Prentice-Hall, Englewood Cliffs, NJ, 1977.

[5] L. N. Bull, Pabón-Peña C. R., and N. B. Freimer. Compound microsatellite repeats: Practical and theoretical features. *Genome Res.*, 9:830–838, 1999.

[6] K. P. Burnham and D. R. Anderson. *Model Selection and Inference*. Springer, New York, 1998.

[7] P. Calabrese and R. Durrett. Dinucleotide repeats in the *Drosophila* and human genomes have complex, length-dependent mutation processes. *Mol. Biol. Evol.*, 20:715–725, 2003.

[8] P. P. Calabrese, R. T. Durrett, and C. F. Aquadro. Dynamics of microsatellite divergence and proportional slippage/point mutation models. *Genetics*, 159:839–852, 2001.

[9] G. Cooper, D. C. Rubinsztein, and W. Amos. Ascertainment bias cannot entirely account for human microsatellites being longer than their chimpanzee homologues. *Hum. Mol. Gen.*, 7:1425–1429, 1998.

[10] A. Di Rienzo, P. Donnelly, C. Toomajian, B. Sisk, A. Hill, M. L. Petzl-Erler, G. K. Haines, and D. H. Barch. Heterogeneity of microsatellite mutations within and between loci, and implications for human demographic histories. *Genetics*, 148:1269–1284, 1998.

[11] A. Di Rienzo, A. C. Peterson, J. C. Garza, A. M. Valdes, and M. Slatkin et al. Mutational processes of simple-sequence repeat loci in human populations. *Proc. Natl. Acad. Sci. USA*, 91:3166–3170, 1994.

[12] D. Dieringer and C. Schlötterer. Two distinct modes of microsatellite mutation processes: Evidence from the complete genomic sequences of nine species. *Genome Res.*, 13:2242–2250, 2003.

[13] R. Durrett. *Essentials of Stochastic Processes*. Springer, New York, 1999.

[14] R. Durrett and S. Kruglyak. A new stochastic model of microsatellite evolution. *J. Appl. Probab.*, 36:621–631, 1999.

[15] J. A. Eisen. Mechanistic basis for microsatellite instability. In D.B. Goldstein and C. Schlötterer, editors, *Microsatellites: Evolution and Applications*. Oxford University Press, Oxford, 1999.

[16] H. Ellegren. Microsatellite mutations in the germ line: Implications for evolutionary inference. *Trends Genet.*, 16:551–558, 2000.

[17] D. Falush, M. Stephens, and J. K. Pritchard. Inference of population structure using multilocus genotype data: Linked loci and correlated allele frequencies. *Genetics*, 164:1567–1587, 2003.

[18] M. W. Feldman, A. Bergman, D. D. Pollock, and D. B. Goldstein. Microsatellite genetic distances with range constraints: Analytic description and problems of estimation. *Genetics*, 145:207–216, 1997.

[19] E. A. Foster, M. A. Jobling, P. G. Taylor, P. Donnelly, P. de Knijff, R. Mieremet, T. Zerjal, and C. Tyler-Smith. Jefferson fathered slave's last child. *Nature*, 396:27–28, 1998.

[20] Y. Fu and R. Chakraborty. Simultaneous estimation of all the parameters of a step-wise mutation model. *Genetics*, 150:487–497, 1998.

[21] J. C. Garza, M. Slatkin, and N. B. Freimer. Microsatellite allele frequencies in humans and chimpanzees, with implications for constraints on allele size. *Mol. Biol. Evol.*, 12:594–603, 1995.

[22] D. B. Goldstein, A. Ruiz-Linares, L. L. Cavalli-Sforza, and M. W. Feldman. Genetic absolute dating based on microsatellites and modern human origins. *Proc. Natl. Acad. Sci. USA*, 92:6723–6727, 1995.

[23] D. B. Goldstein and C. Schlötterer. *Microsatellites: Evolution and Applications*. Oxford University Press, Oxford, 1999.

[24] M. C. Grimaldi and B. Crouau-Roy. Microsatellite allelic homoplasy due to variable flanking sequences. *J. Mol. Evol.*, 44(3):336–340, 1997.

[25] B. Harr, J. Todorova, and C. Schlötterer. Mismatch repair-driven mutational bias in *D. melanogaster*. *Mol. Cell*, 10:199–205, 2002.

[26] Q-Y. Huang, F-H. Xu, H. Shen, H-Y. Deng, Y-J. Liu, Y-Z. Liu, J-L. Li, R. R. Recker, and H-W. Deng. Mutational patterns at dinucleotide microsatellite loci in humans. *Am. J. Hum. Genet.*, 70:625–634, 2002.

[27] R. R. Hudson. Gene genealogies and the coalescent process. *Oxford Surveys Evol. Biol.*, 7:1–44, 1990.

[28] J. P. Huelsenbeck and B. Rannala. Phylogenetic methods come of age: Testing hypotheses in an evolutionary context. *Science*, 276:227 – 232, 1997.

[29] Y. Kashi and M. Soller. Functional roles of microsatellites and minisatellites. In C. Schlötterer and D. B. Goldstein, editors, *Microsatellites: Evolution and Applications*. Oxford University Press, Oxford, 1999.

[30] S. Kruglyak, R. Durrett, M. D. Schug, and C. F. Aquadro. Equilibrium distributions of microsatellite repeat length resulting from a balance between slippage events and point mutations. *Proc. Natl. Acad. Sci. USA*, 95:10774–10778, 1998.

[31] S. Kruglyak, R. Durrett, M. D. Schug, and C. F. Aquadro. Distribution and abundance of microsatellites in the yeast genome can be explained by a balance between slippage events and point mutations. *Mol. Biol. Evol.*, 17:1210–1219, 2000.

[32] P. A. P. Moran. Wandering distributions and the electrophoretic profile. *Theor. Pop. Bio.*, 8:318–330, 1975.

[33] M. J. Nauta and F. J. Weissing. Constraints on allele size at microsatellite loci: Implications for genetic differentiation. *Genetics*, 143:1021–1032, 1996.

[34] R. Nielsen. A likelihood approach to population samples of microsatellite alleles. *Genetics*, 146:711–716, 1997.

[35] T. Ohta and M. Kimura. A model of mutation appropriate to estimate the number of electrophoretic detectable alleles in a finite population. *Genet. Res.*, 22:201–204, 1973.

[36] T. D. Petes, P. W. Greenwell, and M. Dominska. Stabilization of microsatellite sequences by variant repeats in the yeast *Saccharomyces cerevisiae*. *Genetics*, 146:491–498, 1997.

[37] D. E. Reich and D. B. Goldstein. Genetic evidence for a paleolithic human population expansion in africa. *Proc. Natl. Acad. Sci. USA*, 95:8119–8123, 1998.

[38] N. A. Rosenberg, J. K. Pritchard, J. L. Weber, H. W. Cann, K. K. Kidd, L. A. Zhivotovsky, and M. W. Feldman. Genetic structure of human populations. *Science*, 298:2381–2385, 2002.

[39] D. C. Rubinsztein. Trinucleotide expansion mutations cause disease which do not conform to classical Mendelian expectations. In C. Schlötterer and D. B. Goldstein, editors, *Microsatellites: Evolution and Applications*. Oxford University Press, Oxford, 1999.

[40] R. Sainudiin. Statistical inference of microsatellite models: An application to humans and chimpanzees, 2003. M.S. Thesis, Cornell University, NY.

[41] C. Schlötterer. Hitchhiking mapping–functional genomics from the population genetics perspectve. *Trends Genet.*, 19:32–38, 2003.

[42] M. J. Schofield and P. Hsieh. DNA mismatch repair: Molecular mechanisms and biological function. *Annu. Rev. Microbiol.*, 57:579–608, 2003.

[43] D. Shinde, Y. Lai, F. Sun, and N. Arnheim. *Taq* DNA polymerase slippage mutation rates measured by PCR and quasi-likelihood analysis: $(ca/gt)_n$ and $(a/t)_n$ microsatellites. *Nucleic Acids Research*, 31:974–980, 2003.

[44] R. M. Sibly, A. Meade, N. Boxall, M. J. Wilkinson, D. W. Corne, and J. C. Whittaker. The structure of interrupted human ac microsatellites. *Mol. Biol. Evol.*, 20:453–459, 2003.

[45] R. M. Sibly, J. C. Whittaker, and M. Talbot. A maximum-likelihood approach to fitting equilibrium models of microsatellite evolution. *Mol. Biol. Evol.*, 18:413–417, 2001.

[46] J. C. Stephens, D. E. Reich, D. B. Goldstein, H. D. Shin, and M. W. Smith et al. Dating the origin of the $ccr5 - \delta32$ AIDS-resistance allele by the coalescence of haplotypes. *Am. J. Hum. Genet.*, 62:1507–1515, 1998.

[47] H. Tachida and M. Iizuka. Persistence of repeated sequences that evolve by replication slippage. *Genetics*, 131:471–478, 1992.

[48] J-L. Tsao, Y. Yatabe, R. Salovaara, H. J. Järvinen, J-P. Mecklin, L. A. Aaltonen, S. Tavaré, and D. Shibata. Genetic reconstruction of individual colorectal tumor histories. *Proc. Natl. Acad. Sci. USA*, 97:1236–1241, 2000.

[49] J. B. Walsh. Persistence of tandem arrays: Implications for satellite and simple-sequence DNAs. *Genetics*, 115:553–567, 1997.

[50] J. L. Weber and C. Wong. Mutation of human short tandem repeats. *Hum. Mol. Genet.*, 2:1123–1128, 1993.

[51] M. T. Webster, N. G. C. Smith, and H. Ellegren. Microsatellite evolution inferred from human-chimpanzee genomic sequence alignments. *Proc. Natl. Acad. Sci. USA*, 99:8748–8753, 2002.

[52] J. C. Whittaker, R. M. Harbord, N. Boxall, I. Mackay, G. Dawson, and R. M. Sibly. Likelihood-based estimation of microsatellite mutation rates. *Genetics*, 164:781–787, 2003.

[53] M. Wierdl, M. Dominska, and T. D. Petes. Microsatellite instability in yeast: Dependence on the length of the microsatellite. *Genetics*, 146:769–779, 1997.

[54] X. Xu, M. Peng, Z. Fang, and X. Xu. The direction of microsatellite mutation is dependent upon allele length. *Nature Genet.*, 24:396–399, 2000.

[55] X. Yao, A. B. Buermeyer, L. Narayanan, D. Tran, S. M. Baker, T. A. Prolla, P. M. Glazer, R. M. Liskay, and N. Arnheim. Different mutator phenotypes in *Mlh1*- versus *Pms2*-deficient mice. *Proc. Natl. Acad. Sci. USA*, 96:6850–6855, 1999.

[56] L. A. Zhivotovsky, L. Bennett, A. M. Bowcock, and M. W. Feldman. Human population expansion and microsatellite variation. *Mol. Biol. Evol.*, 17:757–767, 2000.

[57] L. A. Zhivotovsky and M. W. Feldman. Microsatellite variability and genetic distances. *Proc. Natl. Acad. Sci. USA*, 92:11549–11552, 1995.

[58] L. A. Zhivotovsky, M. W. Feldman, and S. A. Grishechkin. Biased mutations and microsatellite variation. *Mol. Biol. Evol.*, 14:926–933, 1997.

11

Genome Rearrangement

Rick Durrett[1]

Dept. of Math., Cornell University Department of Mathematics, Ithaca, NY 14853, USA, rtd1@cornell.edu

Genomes evolve by chromosomal fissions and fusions, reciprocal translocations between chromosomes, and inversions that change gene order within chromosomes. For more than a decade, biologists and computer scientists have studied these processes by parsimony methods, asking what is the minimum number of events needed to turn one genome into another. We have recently begun to develop a stochastic approach to this and related questions that has the advantage of producing confidence intervals for estimates and allowing tests of hypotheses concerning mechanisms.

11.1 Inversions

We begin with the simplest problem of the comparison of two chromosomes where the genetic material differs only due to a number of inversions that have reversed the order of chromosomal segments. This occurs for mitochondrial DNA, mammalian X chromosomes, and chromosome arms in some insect species (e.g., *Drosophila* and *Anopheles*). To explain the problem, we begin with an example. The relationship between the human and mouse X chromosomes may be given by a signed permutation (see Figure 2 in [19])

$$1 \quad -7 \quad 6 \quad -10 \quad 9 \quad -8 \quad 2 \quad -11 \quad -3 \quad 5 \quad 4$$

In words, if we look at the positions of genes, then in the first segment of each chromosome, the genes appear in the same order. The genes in the second segment of the mouse X chromosome are the same as those in the seventh segment of the human X chromosome but the order is reversed, and so on.

Hannenhalli and Pevzner [11] developed a polynomial algorithm for computing the inversion distance between chromosomes (i.e., the smallest number of inversions needed to transform one chromosome into another). The first step in preparing to use the HP algorithm is to double the markers. When segment i is doubled, we replace it by two consecutive numbers $2i - 1$ and

$2i$ (e.g., 6 becomes 11 and 12). A reversed segment $-i$ is replaced by $2i$ and $2i - 1$ (e.g., -7 is replaced by 14 and 13). The doubled markers use up the integers 1 to 22. To these we add a 0 at the front and a 23 at the end. Using commas to separate the ends of the markers, we can write the two genomes as follows:

$$\text{mouse} \quad 0, 1\,2, 14\,13, 11\,12, 20\,19, 17\,18, 16\,15,$$
$$3\,4, 22\,21, 6\,5, 9\,10, 7\,8, 23$$
$$\text{human} \quad 0, 1\,2, 3\,4, 5\,6, 7\,8, 9\,10, 11\,12, 13\,14,$$
$$15\,16, 17\,18, 19\,20, 21\,22, 23$$

The next step is to construct the breakpoint graph that results when the commas are replaced by edges that connect vertices with the corresponding numbers. In Figure 11.1, we write the vertices in their order in the mouse genome. Commas in the mouse order become thick lines (black edges), while those in the human genome are thin lines (gray edges).

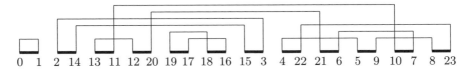

Fig. 11.1. Breakpoint graph for human-mouse X chromosome comparison.

Each vertex has one black and one gray edge, so its connected components are easy to find: start with a vertex and followed the connections in either direction until you come back to where you started. In this example, there are five cycles:

$$0 - 1 - 0 \qquad 2 - 14 - 15 - 3 - 2 \qquad 4 - 22 - 23 - 8 - 9 - 5 - 4$$
$$19 - 17 - 16 - 18 - 19 \qquad 13 - 11 - 10 - 7 - 6 - 21 - 20 - 12 - 13$$

To compute a lower bound for the distance, we first count the number of commas seen when we write out one genome. In this example, that is 1 plus the number of segments ($n = 11$). We then subtract the number of connected components, $c(n)$, in the breakpoint graph. This is a lower bound on the distance since any inversion can at most reduce this quantity by 1, and it is 0 when the two genomes are the same. In symbols,

$$d(\pi) \geq n + 1 - c(\pi) = 12 - 5 = 7 \ .$$

In general, the distance between genomes can be larger than the lower bound from the breakpoint graph. There can be obstructions, called *hurdles*, that can prevent us from decreasing the distance, and hurdles can be intertwined in a *fortress of hurdles* that takes an extra move to break (see [11]). If π is the signed permutation that represents the relative order and orientation of segments in the two genomes, then

$$d(\pi) = n + 1 - c(\pi) + h(\pi) + f(\pi),$$

where $h(\pi)$ is the number of hurdles, $f(\pi)$ is the indicator of the event, and π is a fortress of hurdles.

Fortunately, the complexities associated with hurdles rarely arise in biological data sets. Bafna and Pevzner [1] considered the inversion distance problem for 11 chloroplast and mitochondrial data sets, and in all cases they found that the distance was equal to the lower bound. We can verify that 7 is the minimum distance for the human-mouse comparison by constructing a sequence of seven moves that transforms the mouse X chromosome into the human order. There are thousands of solutions, so we leave this as an exercise for the reader. Here are some hints: (i) To do this, it suffices to choose at each step an inversion that increases the number of cycles by 1. (ii) This never occurs if the two chosen black edges are in different cycles. (iii) If the two black edges are in the same cycle and are (a, b) and (c, d) as we read from left to right, this will occur unless in the cycle minus these two edges a is connected to d and b to c, in which case the number of cycles will not change. For example, in Figure 11.1, an inversion that breaks black edges 19-17 and 18-16 will increase the number of cycles, but the one that breaks 2-14 and 15-3 will not. See Section 5.2 of [7] or Chapter 10 of [18] for more details.

Ranz, Segarra, and Ruiz [21] did a comparative study of chromosome 2 of *Drosophila repleta* and chromosome arm 3R of *D. melanogaster*. If we number the 26 genes that they studied according to their order on the *D. repleta* chromosome, then their order on *D. melanogaster* is given by

12 7 4 *2 3 21 20* 18 1 13 9 16 6 14 *26 25 24* 15 *10 11* 8 5 *23 22* 19 17

where we have used italics to indicate adjacencies that have been preserved. Since the divergence of these two species, this chromosome region has been subjected to many inversions. Our first question is: How many inversions have occurred? To answer this question, we need to formulate and analyze a model. Before we do this, the reader should note that in contrast with the human-mouse comparison, here we do not have enough markers to determine the relative orientation of the segments, so we have an unsigned permutation.

11.1.1 *n*-inversion Chain

Consider n markers on a chromosome, which we label with $1, 2, \ldots n$, and that can be in any of the $n!$ possible orders. To these markers we add two others: one called 0 at the beginning and one called $n + 1$ at the end. Finally, for convenience of description, we connect adjacent markers by edges. For example, when $n = 7$, the state of the chromosome might be

$$0 - 5 - 3 - 4 - 1 - 7 - 2 - 6 - 8$$

In biological applications, the probability of an inversion in a given generation is small, so we will formulate the dynamics in continuous time. The labels 0 and $n + 1$ never move. To shuffle the others, at times of a rate one Poisson process, we pick two of the $n+1$ edges at random and invert the order of the markers in between. For example, if we pick the edges $5 - 3$ and $7 - 2$, the result is

$$0 - 5 - 7 - 1 - 4 - 3 - 2 - 6 - 8$$

If we pick $3 - 4$ and $4 - 1$ in the first arrangement, there is no visible change. However, allowing this move will simplify the mathematical analysis and only amounts to a small time change of the dynamics in which one picks two markers $1 \leq i < j \leq n$ at random and reverses the segment with those endpoints.

It is clear that if the chromosome is shuffled repeatedly, then in the limit all of the $n!$ orders for the interior markers will have equal probability. The first question is how long it takes for the marker order to be randomized. To explain the answer, we recall that the total variation distance between two distributions μ and ν is $\sup_A |\mu(A) - \nu(A)|$.

Theorem 11.1. *Consider the state of the system at time $t = cn \ln n$ starting with all markers in order. If $c < 1/2$, then the total variation distance to the uniform distribution ν goes to 1 as $n \to \infty$. If $c > 2$, then the total variation distance goes to 0.*

For a proof, see [8]. There is a gap between the upper bound and the lower bound, but on the basis of other results it is natural to guess that the lower bound is right (i.e., convergence to equilibrium takes about $(n \ln n)/2$ shuffles). When $n = 26$, this is 42.3. Consequently, when the number of inversions is large (in the example, more than 40), the final arrangement is almost independent of the initial one and we do not expect to be able to accurately estimate the actual number of inversions.

While Theorem 11.1 may be interesting for card-shuffling algorithms, its conclusion does not tell us much about the number of inversions that occurred in our data set. To begin to investigate this question, we note that there are six conserved adjacencies. This means that at least $27 - 6 = 21$ edges have been disturbed, so at least 11 inversions have occurred. Biologists often use this easy-to-compute estimate, which is called the *breakpoint distance*. However, this lower bound is usually not sharp. In this example, it can be shown that at least 14 inversions are needed to put the markers in order.

The maximum parsimony solution is 14, but there is no guarantee that nature took the shortest path between the two genomes. York, Durrett, and

Nielsen [22] have introduced a Bayesian approach to the problem of inferring the history of inversions separating two chromosomes. They assume that the differences between the gene arrangements in two species come from running the n-inversion chain for some unknown time λ. Given a number of inversions ℓ, let $\pi_0, \pi_1, \ldots, \pi_\ell$ be the proposed evolutionary sequence that connects the two genomes, with each π_k differing from the previous one by one inversion. Let Ω be the set of all such sequences (of any length) and X be a generic member of Ω.

Let D (for data) be the marker order in the two sampled genomes. The Markov chain Monte Carlo method of York, Durrett, and Nielsen [22] consists of defining a Markov chain on $\Omega \times [0, \infty)$ with stationary density $P(X, \lambda | D)$. They alternate updating λ and X. First, a new λ is chosen according to $P(\lambda | X, D)$, then a new path is produced by choosing a segment to cut out of the current path, and then the two endpoints are reconnected. In generating the new path, they use the graph distance $n + 1 - c(\pi)$ as a guide and prefer steps that reduce the distance. We refer the reader to the cited paper for more details. Figure 11.2 shows a picture of the posterior distribution of the number of inversions for the Ranz, Segarra, and Ruiz [21] data set. Note that this density assigns a small probability to the shortest path (with length 14) and has a mode at 19.

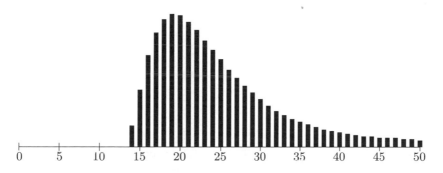

Fig. 11.2. Posterior distribution of inversions for *Drosophila* data.

An alternative and simpler approach to our question comes from considering $\phi(\eta) =$ the number of conserved edges minus 2. Subtracting 2 makes ϕ orthogonal to the constant eigenfunction. A simple calculation shows that ϕ is an eigenfunction of the chain with eigenvalue $(n - 1)/(n + 1)$. In our case, $n = 26$ and $\phi = 4$, so solving

$$27 \left(\frac{25}{27} \right)^m = 4 \quad \text{gives} \quad m = \frac{\ln(4/27)}{\ln(25/27)} = 24.8 \,,$$

which gives a moment estimate of the number of inversions that seems consistent with the distribution in Figure 11.2.

Ranz, Casals, and, Ruiz [20] enriched the comparative map so that 79 markers can be located in both species. Again numbering the markers on the *D. repleta* chromosome by their order on *D. melanogaster*, we have:

36	*37*	17	40	*16*	*15*	*14*	63	*10*	*9*	55	28
13	51	22	79	39	70	66	5	6	7	35	64
33	*32*	60	*61*	18	65	62	12	1	11	23	20
4	52	68	29	48	3	21	53	8	43	72	*58*
57	*56*	19	49	34	59	30	77	31	67	44	2
27	38	50	*26*	*25*	76	69	41	24	75	71	78
73	47	54	45	74	42	46					

The number of conserved adjacencies (again indicated with italics) is 11, so our moment estimate is

$$m = \frac{\ln(9/80)}{\ln(78/80)} = 86.3 \ .$$

This agrees with the Bayesian analysis in [22], where the mode of the posterior distribution is 87. However, these two numbers differ drastically from the parsimony analyses. The breakpoint distance is $(80 - 11)/2 = 35$, while the parsimony distance is 54. This lies outside the 95% credible interval of $[65, 120]$ that comes from the Bayesian estimate. Indeed the posterior probability of 54 is so small that this value was never seen in the 258 million MCMC updates in the simulation run.

11.2 Distances

In the last two examples, we saw that the breakpoint distance was likely to be an underestimate of the true distance. This brings up the question of when the parsimony estimate is reliable. Bourque and Pevzner [3] have approached this question by taking 100 markers in order, performing k randomly chosen inversions, computing D_k, the minimum number of inversions needed to return to the identity, and then plotting the average value of $D_k - k \leq 0$ (the circles in Figure 11.3). They concluded based on this and other simulations that the parsimony distance based on n markers was good as long as the number of inversions was at most $0.4n$. The smooth curve, which we will describe in Theorem 11.2 below, gives the limiting behavior of $(D_{cn} - cn)/n$.

The first step is to consider the analogous but simpler problem for . In that case, the distance from the identity can be easily computed: it is the number of markers n minus the number of cycles in the permutation. For an example, consider the following permutation of 14 objects written in its cyclic decomposition:

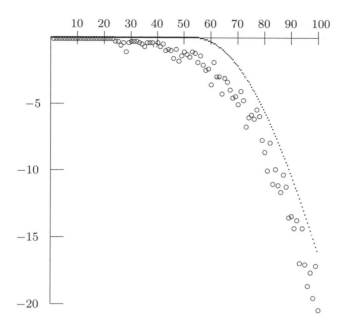

Fig. 11.3. Bourque-Pevzner simulation results vs. Theorem 11.2.

$$(1\,7\,4)\ (2)\ (3\,12)\ (5\,13\,9\,11\,6)\ (8\,10\,14)$$

which indicates that $1 \to 7$, $7 \to 4$, $4 \to 1$, $2 \to 2$, $3 \to 12$, $12 \to 3$, and so on. There are five cycles, so the distance from the identity is 9. If we perform a transposition that includes markers from two different cycles (e.g., 7 and 9), the two cycles merge into one, while if we pick two in the same cycle (e.g., 13 and 11), it splits into two.

The situation is similar but slightly more complicated for inversions. There, if we ignore the complexity of hurdles, the distance is $n + 1$ minus the number of components in the breakpoint graph. An inversion that involves edges in two different components merges them into one, but an inversion that involves two edges of the same cycle may or may not increase the number of cycles. To have a cleaner mathematical problem, we will consider the biologically less relevant case of random transpositions and ask a question that in terms of the rate 1 continuous-time random walk on the permutation group is: How far from the identity are we at time cn?

The first step in attacking this problem is to notice that by our description the cycle structure evolves according to a *coagulation-fragmentation process*. Suppose that for the moment we ignore fragmentation and draw an edge from i to j whenever we transpose i and j. In this case, the cycles are the components of the resulting random graph. There are $n(n-1)/2$ potential edges, so results of Erdös and Renyi imply that when $c < 1/2$ there are no very large components and we can ignore fragmentations. In this phase, the distance will typically increase by 1 on each step, or in the notation of Bourque and Pevzner [3], $D_k - k \approx 0$. When $n = 100$, this phase lasts until there have been about 50 inversions.

When $c > 1/2$, a giant component emerges in the percolation model, and its behavior is much different from the large cycles in the permutation that experience a number of fragmentations and coagulations. The dynamics of the large components are quite complicated but (i) there can never be more than $\sqrt{n}$ of size $\sqrt{n}$ or larger and (ii) an easy argument shows that the number of fragmentations occurring to clusters of size $\leq \sqrt{n}$ is $O(\sqrt{n})$. These two observations plus results from the theory of random graphs (see Theorem 12 in Section V.2 of [2]) imply the following theorem.

Theorem 11.2. *The number of cycles at time $cn/2$ is $g(c)n + O(\sqrt{n})$, where*

$$g(c) = \sum_{k=1}^{\infty} \frac{1}{k} p_k(c) \quad \text{and} \quad p_k(c) = \frac{1}{c} \frac{k^{k-1}}{k!} (ce^{-c})^k .$$

Using Stirling's formula $k! \sim k^k e^{-k} \sqrt{2\pi k}$, it is easy to see that g' is continuous but $g''(1)$ does not exist. It is somewhat remarkable that $g(c) = 1 - c/2$ for $c < 1$. Thus there is a phase transition in the behavior of the distance of the random transposition random walk from the identity at time $n/2$.

As stated, the result only applies to transpositions. However, the same exact conclusion applies to inversions. To show this, we note that the only difference between the two systems is that picking the same cycle twice may or may not increase the number of cycles in the breakpoint graph, and our proof has shown that fragmentations can be ignored.

To explain the strange function $g(c)$ that appears in the answer, we begin with Cayley's result that there are k^{k-2} trees with k labeled vertices. At time cn, each edge is present with probability $\approx (cn/2)/\binom{n}{2} \approx c/n$, so the expected number of trees present of size k is

$$\binom{n}{k} k^{k-2} \left(\frac{c}{n}\right)^{k-1} \left(1 - \frac{c}{n}\right)^{k(n-k)+\binom{k}{2}-(k-1)}$$

since each of the $k - 1$ edges need to be present and there can be no edges connecting the k point set to its complement ($k(n - k)$ edges) or any other edges connecting the k points ($\binom{k}{2} - (k-1)$ edges). For fixed k, $\binom{n}{k} \approx n^k/k!$, so the equation above is

$$\approx n \frac{k^{k-2}}{k!} (2c)^{k-1} \left(1 - \frac{2c}{n}\right)^{kn} ,$$

from which the result follows easily. We have written the conclusion in the form given above so that $p_k(c)$ is the probability in an Erdös-Renyi graph with edge occupancy probability c/n that 1 belongs to a component of size k.

Having found laws of large numbers for the distance, it is natural to ask about fluctuations. This project is being carried out as part of the Ph.D. thesis of Nathaniel Berestycki. Since these results are only exact for transpositions and are merely a lower bound for inversions, we will only state the first two results. The subcritical regime ($cn/2$ with $c < 1$) is easy. Let F_t be the number of fragmentations at time t in a system in which transpositions occur at rate one. The continuous time setting is more convenient since it leads to a random graph with independent edges. If N_t is the number of transpositions at time t, then $D_t - N_t = -2F_t$, so we study the latter quantity.

Theorem 11.3 *Suppose* $0 \leq c < 1$. *As* $n \to \infty$, $F_{cn/2}$ *converges in distribution to a Poisson random variable with mean* $(-\ln(1-c) - c)/2$.

Since a Poisson random variable with large mean rescales to approximate a normal, it should not be surprising that if we change the time to make the variance linear, the result is a Brownian motion.

Theorem 11.4. *Let* $c_n(r) = 1 - n^{-r/3}$ *for* $0 \leq r \leq 1$. *As* $n \to \infty$,

$$X_n(r) = (F_{c_n(r)n/2} - (r/6)\log n)/((1/6)\log n)^{1/2}$$

converges to a standard Brownian motion.

Expected value estimates (see [15]) imply that the number of fragmentations in $[1 - n^{-1/3}, 1]$ is $O(1)$ and hence can be ignored. It follows from this that

$$(F_{n/2} - (1/6)\log n)/((1/6)\log n)^{1/2}$$

has an approximately normal distribution. To connect this with the simulations of Bourque and Pevnzer, we note that this implies $EF_{50} \approx (1/6)\log 50 = 0.767$, which seems consistent with the data in Figure 11.3, even though all we know from the comparison is that this is an upper bound on the difference between N_t and the distance.

11.3 Genomic Distance

In general, genomes evolve not only by inversions within chromosomes but also due to translocations between chromosomes and due to fissions and fusions that change the number of chromosomes. To reduce the number of events considered from four to two, we note that a translocation splits two chromosomes (into, say, $a - b$ and $c - d$) and then recombines the pieces (to make $a - d$ and $b - c$, say). A fission is the special case in which the segments c and d are empty, and a fusion is when b and c are empty. To illustrate the problem, we will consider part of the data of Doganlar et al. [6], who constructed a comparative genetic linkage map of eggplant (*Solanum melongena*) with 233

markers based on tomato cDNA, genomic DNA, and ESTs. Using the first
letter of the common name to denote the species, they found that the marker
order on T1 and E1 and on T8 and E8 were identical, while in four other cases
(T2 vs. E2, T6 vs. E6, T7 vs. E7, T9 vs. E9) the collections of markers were
the same and the order became the same after a small number of inversions
were performed (3, 1, 2, and 1, respectively).

In our example, we will compare the remaining six chromosomes from the
two species. The first step is to divide the chromosomes into *conserved segments* where the adjacency of markers has been preserved between the two
species, allowing for the possibility of the overall order being reversed. When
such segments have two or more markers, we can determine the relative orientation. However, as the HP algorithm assumes one knows the relative orientation of segments, we will have to assign orientations to conserved segments
consisting of single markers in order to minimize the distance. In the case of
the tomato-eggplant comparison, there are only five singleton segments, so
one can easily consider all $2^5 = 32$ possibilities. The next table shows the two
genomes with an assignment of signs to the singleton markers that minimizes
the distance.

Eggplant	Tomato
1 2 3 4 5 6	1 −5 2 6
7 8	21 −22 −20 8
9 10	−4 14 11 −15 3 9
11 12 13 14 15 16 17 18	7 16 −18 17
19 20 21 22	−19 24 −26 27 25
23 24 25 26 27	−12 23 13 10

As in the inversion distance problem, our first step is to double the markers. The second step is to add ends to the chromosomes and enough empty
chromosomes to make the number of chromosomes equal. In this example,
no empty chromosomes are needed. We have labeled the ends in the first
genome by 1000 to 1011 and in the second genome by 2000 to 2011. The next
table shows the result of the first two preparatory steps. Commas indicate
separations between two segments or between a segment and an end.

Eggplant
1000, 1 2 , 3 4 , 5 6 , 7 8 , 9 10 , 11 12 , 1001
1002, 13 14 , 15 16 , 1003
1004, 17 18 , 19 20 , 1005
1006, 21 22 , 23 24 , 25 26 , 27 28 , 29 30 , 31 32 , 33 34 , 35 36 , 1007
1008, 37 38 , 39 40 , 41 42 , 43 44 , 1009
1010, 45 46 , 47 48 , 49 50 , 51 52 , 53 54 , 1011

Tomato
2000, 1 2 , 10 9 , 3 4 , 11 12 , 2001
2002, 41 42 , 44 43 , 40 39 , 15 16 , 2003

2004, 8 7 , 27 28 , 21 22 , 30 29 , 5 6 , 17 18 , 2005
2006, 13 14 , 31 32 , 36 35 , 33 34 , 2007
2008, 38 37 , 47 48 , 52 51 , 53 54 , 49 50 , 2009
2010, 24 23 , 45 46 , 25 26 , 19 20 , 2011

As before, the next step is to construct the breakpoint graph that results when the commas are replaced by edges that connect vertices with the corresponding numbers. We did not draw the graph since to compute the distance we only need to know the connected components of the graph. Since each vertex has degree two, these are easy to find: start with a vertex and follow the connections. The resulting component will either be a path that connects two ends or a cycle that consists of markers and no ends. In our example there are five paths of length three: $1000 - 1 - 2000$, $1001 - 12 - 2001$, $1002 - 13 - 2006$, $1003 - 16 - 2003$, and $1005 - 20 - 2011$. These paths tell us that end 1000 in genome 1 corresponds to end 2000 in genome 2, and so forth. The other correspondences between ends will be determined after we compute the distance. The remaining components in the breakpoint graph are listed below.

1004 17 6 7 27 26 19 18 2005
1006 21 28 29 5 4 11 10 2 3 9 8 2004
1007 36 32 33 35 34 2007
1008 37 47 46 25 24 2010
1009 44 42 43 40 41 2002
1010 45 23 22 30 31 14 15 39 38 2008
1011 54 49 48 52 53 51 50 2009

To compute a lower bound for the distance, we start with the number of commas seen when we write out one genome. In this example, that is 33. We subtract the number of connected components in the breakpoint graph. In this example, that is $5 + 7 = 12$, and then we add the number of paths that begin and end in the same genome, which in this case is 0. The result, which is 21 in this case, is a lower bound on the distance since any inversion or translocation can at most reduce this quantity by 1, and it is 0 when the two genomes are the same. As before, this is only a lower bound. For the genomic distance problem, the full answer is quite complicated and involves seven quantities associated with the genome. (For more details, see [12] or [18].)

At least in this example, nature is simpler than the mathematically worst possible case. It is easy to produce a path of length 21 to show that the lower bound is achieved. For a solution, see [9]. That paper extends the methods of [22] to develop a Bayesian estimate of the number of inversions and translocations separating the two genomes. As we have just calculated, the parsimony solution for the comparison of all 12 chromosomes is $21 + 7 = 28$. The Bayesian analysis produces 95% credible intervals of $[5, 7]$, $[21, 31]$, and $[28, 37]$ for the number of translocations and inversions and the total number of events (respectively) separating tomato and eggplant. The mode of the posterior joint distribution of the number of translocations and inversions occurs at $(6.6, 25.9)$. Thus even in the case of these two closely related genomes, the most likely numbers of inversions and translocations are somewhat higher than their parsimony estimates.

When distances between the markers are known in one genome, there is another method due to Nadeau and Taylor [17], that can be used to estimate the number of inversions and translocations that have occurred. The basic data for the process is the set of lengths of conserved segments (i.e., two or more consecutive markers in one genome that are adjacent, possibly in the reverse order, in the other). The actual conserved segment in the genome is larger than the distance r between the two markers at the ends of the conserved interval. Thinking about what happens when we put n points at random in the unit interval, which produces $n + 1$ segments with $n - 1$ between the leftmost and the rightmost points, we estimate the length of the conserved segment containing these markers by $\hat{r} = r(n + 1)/(n - 1)$, where n is the number of markers in the segment.

Let D be the density of markers (i.e., the total number divided by the size of the genome). If the average length of conserved segments is L and we assume that their lengths are exponentially distributed, then since we only detect segments with two markers, the distribution of their lengths is

$$(1 - e^{-Dx} - Dxe^{-Dx})\frac{1}{L}e^{-x/L}$$

normalized to be a probability density. A little calculus shows that the mean of this distribution is $(L^2D + 3L)/(LD + 1)$.

Historically, the first application of this technique was to a human-mouse comparative map with a total of 56 markers. Based on this limited amount of data, it was estimated that there were 178 ± 39 conserved segments. For more than fifteen years, this estimate held up remarkably well as the density of the comparative map increased, see [16]. However, the completion of the sequencing of the mouse genome ([5]; see Figure 11.3) has revealed 342 conserved segments of size > 300 kb (kilobases).

To illustrate the Nadeau and Taylor computation, we will use a comparative map of the human and cattle autosomes (nonsex chromosomes) (see Figure 11.4) constructed by Band et al. [10]. Using resources on the NCBI home page, we were able to determine the location in the human genome of 422 genes in the map. These defined 125 conserved segments of actual average length 7.188 Mb (megabases), giving rise to an adjusted average length 14.501 Mb. Assuming 3200 Mb for the size of the human genome, the marker density was $D = 1.32 \times 10^{-4}$ or one every 7.582 Mb. Setting $14.501 = (L^2D + 3L)/(LD + 1)$ and solving the quadratic equation for L gives an estimate $\hat{L} = 7.144$ Mb, which translates into approximately 448 segments. Subtracting 22 chromosome ends, we infer that there were 424 breakpoints, which leads to an estimate of 212 inversions and translocations. As a check on the assumptions of the Nadeau and Taylor computation, we note that if

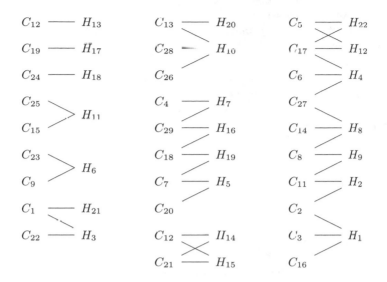

Fig. 11.4. Comparison of cattle and human autosomes.

markers and segment endpoints are distributed randomly, then the number of markers in a conserved segment would have a geometric distribution. The next table compares the observed counts with what was expected.

markers	observed	expected
0	–	222.9
1	85	108.1
2	76	52.5
3	29	25.4
4	10	12.3
5	5	6.0
6	3	2.9
7	1	1.4
8	1	0.7

To get an idea of the number of translocations that have occurred, we will look at the human-cattle correspondence through the eyes of FISH (fluorescent in situ hybridization) data of Hayes [14] and Chowdhary et al. [4]. In this technique, one takes individual human chromosomes, labels them with fluorescent chemicals, and then determines where they hybridize to cattle chromosomes. To visualize the relationship between the genomes it is useful to draw the bipartite graph with vertices being the chromosome numbers in the two genomes and an edge from C_i to H_j if part of cattle chromosome i is homologous to part of human chromosome j. We call this the *Oxford graph* since the adjacency matrix of this graph is what biologists would call an *Oxford grid*.

Parsimony analysis reveals that a minimum of 155 moves (20 translocations and 135 inversions) are needed to rearrange the cattle genome to match the chromosomes of the human genome. Durrett, Nielsen, and York [9] have applied their Bayesian methods to this example but experienced convergence problems. Figures 11.5 and 11.6 of their paper give posterior distributions from four runs. In the case of inversions, the modes are 20, 21, 21, and 25, with the overall shape of the fourth posterior distribution being considerably different. The modes for translocations are all in the range 185–191, but the variance differs considerably from run to run.

11.4 Nonuniformity of Inversions

Define a *syntenic segment* to be a segment of chromosome where all of the markers come from the same chromosome in the other species but not necessarily in the same order. A remarkable aspect of the cattle data is that although our estimates suggest that there have been roughly 20 translocations and 190 inversions, each chromosome consists of only a few syntenic segments. If inversions were uniformly distributed on the chromosome, we would expect that inversions that occur after a translocation would mingle the two segments.

A second piece of evidence that not all inversions are equally likely comes from the 79 marker *Drosophila* data. The estimated number of inversions is large, but there is still a strong correlation between the marker orders in the two genomes. Spearman's rank correlation $\rho = 0.326$, which is significant at the $p = 0.001$ level. From the point of view of Theorem 11.1 this is not surprising: our lower bound on the mixing time predicts that $39.5 \ln 75 = 173$ inversions are needed to completely randomize the data. However, simulations in [8] show that the rank correlation is randomized well before that time. In 10,000 runs, the average rank correlation is only 0.0423 after 40 inversions, and only 4.3% of the runs had a rank correlation larger than 0.325.

To seek a biological explanation of the nonuniformity, we note that the gene-to-gene pairing of homologous chromosomes implies that if one chromosome of the pair contains an inversion that the other does not, a loop will form in the region in which the gene order is inverted. (See, e.g., page 367 of [13].) If a recombination occurs in the inverted region, then the recombined chromosomes will contain two copies of some regions and zero copies of others, which can have unpleasant consequences. A simple way to take this into account is with the θ-inversion model.

11.4.1 θ-inversion model

Inversions that reverse markers i to $i+j$ occur at rate $\theta^{j-1}/n(1-\theta)$. The reasoning here is that the probability of no recombination decreases exponentially with the length of the segment reversed.

We expect that the likelihood methods of Durrett, Nielsen, and York can be extended to the θ-inversion model in order to estimate inversion tract lengths. A second way to approach the problem is to see how estimates of the number of inversions depend on the density of markers in the map. If n markers (blue balls) are randomly distributed and we pick two inversion endpoints (red balls) at random, then the relative positions of the $n+2$ balls are all equally likely. The inversion will not be detected by the set of markers if there are 0 or 1 blue balls between the two red ones, an event of probability

$$\frac{n+1+n}{\binom{n+2}{2}} = \frac{4n+2}{(n+2)(n+1)} \approx \frac{4}{n+2} .$$

This means that the 26 markers in the first *Drosophila* data set should have missed only 1/7 of the inversions, in sharp contrast to the fact that our estimate jumped from 24.8 with 26 markers to 86.3 with 79.

Suppose now that markers are distributed according to a Poisson process with mean spacing M while inversion tract lengths have an exponential distribution with mean L. If we place one inversion endpoint at random on the chromosome and then move to the right to locate the second one, then the probability that a marker comes before the other inversion endpoint is

$$\frac{1/M}{1/M + 1/L} = \frac{L}{L+M}$$

so the fraction detected is $L^2/(L+M)^2$. If we take 30 Mb as an estimate for the size of the chromosome arm studied, we see that the marker spacings in the two studies are $M_1 = 30/27 = 1.11$ Mb and $M_2 = 30/80 = .375$ Mb, respectively. Taking ratios, we can estimate L by

$$\frac{86.3}{24.8} = \frac{(L+1.1)^2}{(L+0.375)^2} .$$

Taking square roots of each side and solving, we have $1.865L+0.375 = L+1.1$ or $L = 0.725/0.765 = 0.948$ Mb. If this is accurate, then the larger data set only detects

$$\left(\frac{0.948}{1.273}\right)^2 = 0.554 ,$$

or 55.4% of the inversions that have occurred. Our best guess is that the chromosome arm has experienced $86.3/0.554 = 157$ inversions. This simple calculation is only meant to illustrate the possibilities of the method, which needs to be developed further and tested on other examples.

References

[1] V. Bafna and P. Pevzner. Sorting by reversals: Genome rearrangement in plant organelles and evolutionary history of X chromosome. *Mol. Biol. Evol*, 12:239–246, 1995.

[2] B. Bollobás. *Random Graphs*. Academic Press, New York, 1985.

[3] G. Bourque and P. A. Pevzner. Genome-scale evolution: Reconstructing gene orders in ancestral species. *Genome Res.*, 12:26–36, 2002.

[4] B. P. Chowdhary, L. Fronicke, I. Gustavsson, and H. Scherthan. Comparative analysis of cattle and human genomes: Detection of ZOO-FISH and gene mapping-based chromosomal homologies. *Mammalian Genome*, 7:297–302, 1996.

[5] Mouse Genome Sequencing Consortium. Initial sequencing and comparative analysis of the mouse genome. *Nature*, 420:520–562, 2002.

[6] S. Doganlar, A. Frary, M. C. Daunay, R. N. Lester, and S. D. Tanksley. A comparative genetic linkage map of eggplant (*Solanum melongea*) and its implications for genome evolution in the Solanaceae. *Genetics*, 162:1697–1711, 2002.

[7] R. Durrett. *Probability Models for DNA Sequence Evolution*. Springer-Verlag, New York, 2002.

[8] R. Durrett. Shuffling chromosomes. *J. Theor. Prob.*, 16:725–750, 2003.

[9] R. Durrett, R. Nielsen, and T. L. York. Bayesian estimation of genomic distance. 2004.

[10] M. J. Band et al. An ordered comparative map of the cattle and human genomes. *Genome Res.*, 10:1359–1368, 2000.

[11] S. Hannenhalli and P. A. Pevzner. Transforming cabbage into turnip (polynomial algorithm for sorting signed permutations by reversals). In *Proceedings of the 27th Annual ACM Symposium on the Theory of Computing*, pages 178–189. Association for Computing Machinery, New York, 1995.

[12] S. Hannenhalli and P. A. Pevzner. Transforming men into mice (polynomial algorithm for the genomic distance problem). In *Proceedings of the 36th Annual IEEE Symposium on Foundations of Computer Science*, pages 581–592. IEEE, New York, 1995.

[13] D. L. Hartl and E. W. Jones. *Genetics: The Analysis of Genes and Genomes*. Jones and Bartlett, Sudbury, MA, 2000.

[14] H. Hayes. Chromosome painting with human chromosome-specific DNA libraries reveals the extent and distribution of conserved segments in bovine chromosomes. *Cytogenet. Cell. Genetics*, 71:168–174, 1995.

[15] T. Luczak, B. Pittel, and J. C. Weirman. The structure of the random graph at the point of phase transition. *Trans. Am. Math. Soc.*, 341:721–748, 1994.

[16] J. H. Nadeau and D. Sankoff. The lengths of undiscovered conserved segments in comparative maps. *Mammalian Genome*, 9:491–495, 1998.

[17] J. H. Nadeau and B. A. Taylor. Lengths of chromosomal segments conserved since divergence of man and mouse. *Proc. Natl. Acad. Sci. USA*, 81:814–818, 1984.

[18] P. A. Pevzner, *Computational Molecular Biology: An Algorithmic Approach*. MIT Press, Cambridge, 2000.

[19] P. A. Pevzner and G. Tesler. Genome rearrangement in mammalian evolution: Lessons from human and mouse genomes. *Genome Research.*, 13:37–45, 2003.

[20] J. M. Ranz, F. Casals, and A. Ruiz. How malleable is the eukaryotic genome? Extreme rate of chromosomal rearrangement in the genus *Drosophila*. *Genome Res.*, 11:230–239, 2001.

[21] J. M. Ranz, S. Segarra, and A. Ruiz. Chromosomal homology and molecular organization of Muller's element *D* and *E* in the *Drosophila repleta* species group. *Genetics*, 145:281–295, 1997.

[22] T. L. York, R. Durrett, and R. Nielsen. Bayesian estimation of the number of inversions in the history of two chromosomes. *J. Comp. Biol.*, 9:805–818, 2002.

Phylogenetic Hidden Markov Models

Adam Siepel[1] and David Haussler[2]

[1] Center for Biomolecular Science and Engineering,
University of California, Santa Cruz, CA 95064, USA, acs@soe.ucsc.edu
[2] Center for Biomolecular Science and Engineering,
University of California, Santa Cruz, CA 95064, USA, haussler@soe.ucsc.edu

Phylogenetic hidden Markov models, or phylo-HMMs, are probabilistic models that consider not only the way substitutions occur through evolutionary history at each site of a genome but also the way this process changes from one site to the next. By treating molecular evolution as a combination of two Markov processes—one that operates in the dimension of *space* (along a genome) and one that operates in the dimension of *time* (along the branches of a phylogenetic tree)—these models allow aspects of both sequence structure and sequence evolution to be captured. Moreover, as we will discuss, they permit key computations to be performed exactly and efficiently. Phylo-HMMs allow evolutionary information to be brought to bear on a wide variety of problems of sequence "segmentation," such as gene prediction and the identification of conserved elements.

Phylo-HMMs were first proposed as a way of improving phylogenetic models that allow for variation among sites in the rate of substitution [9, 52]. Soon afterward, they were adapted for the problem of secondary structure prediction [11, 47], and some time later for the detection of recombination events [20]. Recently there has been a revival of interest in these models [41, 42, 43, 44, 33], in connection with an explosion in the availability of comparative sequence data, and an accompanying surge of interest in comparative methods for the detection of functional elements [5, 3, 24, 46, 6]. There has been particular interest in applying phylo-HMMs to a multispecies version of the ab initio gene prediction problem [41, 43, 33].

In this chapter, phylo-HMMs are introduced, and examples are presented illustrating how they can be used both to identify regions of interest in multiply aligned sequences and to improve the goodness of fit of ordinary phylogenetic models. In addition, we discuss how hidden Markov models (HMMs), phylogenetic models, and phylo-HMMs all can be considered special cases of general "graphical models" and how the algorithms that are used with these models can be considered special cases of more general algorithms. This chapter is written at a tutorial level, suitable for readers who are familiar with phylogenetic models but have had limited exposure to other kinds of graphical models.

12.1 Background

A phylo-HMM can be thought of as a machine that probabilistically generates
a multiple alignment, column by column, such that each column is defined by
a phylogenetic model. As with the single-sequence HMMs ordinarily used in
biological sequence analysis [7], this machine probabilistically proceeds from
one state to another[1], and at each time step it "emits" an observable object, which is drawn from the distribution associated with the current state
(Figure 12.1). With phylo-HMMs, however, the distributions associated with
states are no longer multinomial distributions over a set of characters (e.g.,
{A,C,G,T}) but are more complex distributions defined by phylogenetic models.

Phylogenetic models, as considered here, define a stochastic process of substitution that operates independently at each site in a genome. (The question
of independence will be revisited below.) In the assumed process, a character
is first drawn at random from the background distribution and assigned to the
root of the tree; character substitutions then occur randomly along the tree's
branches from root to leaves. The characters that remain at the leaves when
the process has been completed define an alignment column. Thus, a phylogenetic model induces a distribution over alignment columns having a correlation structure that reflects the phylogeny and substitution process (see [11]).
The different phylogenetic models associated with the states of a phylo-HMM
may reflect different overall rates of substitution (as in conserved and noncon-
served regions), different patterns of substitution or background distributions
(as in coding and noncoding regions), or even different tree topologies (as with
recombination [20]).

Typically with HMMs, a sequence of observations (here denoted $\mathbf{X}$) is
available to be analyzed, but the sequence of states (called the "path") by
which the observations were generated is "hidden" (hence the name "hidden
Markov model"). Efficient algorithms are available to compute the maximum-
likelihood path, the posterior probability that any given state generated any
given element of $\mathbf{X}$, and the total probability of $\mathbf{X}$ considering all possible
paths (the likelihood of the model). The usefulness of HMMs in general, and
phylo-HMMs in particular, is in large part a consequence of the fact that
these computations can be performed exactly and efficiently. In this chapter,
three examples of applications of phylo-HMMs will be presented that parallel these three types of computation—prediction based on the maximum-
likelihood path (Example 12.1), prediction based on posterior probabilities
(Example 12.2), and improved goodness of fit, as evidenced by model likelihood (Example 12.3). Finally, it will be shown how these algorithms may be
considered special cases of more general algorithms by regarding phylo-HMMs
as graphical models.

---

[1]Throughout this chapter, it is assumed that the Markov chain for state transitions is discrete, first-order, and homogeneous.

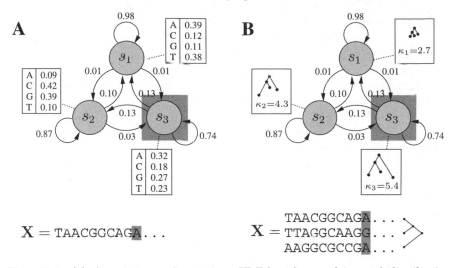

Fig. 12.1. (a) A 3-state single-sequence HMM with a multinomial distribution associated with each state (boxed tables). A new state is visited at each time step according to the indicated transition probabilities (numbers on arcs), and a new character is emitted according to the probability distribution for that state. The shaded boxes indicate the current state and a newly emitted character, which is appended to the sequence **X**. In this example, one state has an A+T-rich distribution (s_1), one has a G+C-rich distribution (s_2), and one favors purines (s_3). (b) An analogous phylo-HMM. In this case, the multinomial distributions are replaced by phylogenetic models, and at each time step a new column in a multiple alignment **X** is emitted. The phylogenetic models include parameters describing the overall shape and size of the tree as well as the background distribution for characters and the pattern of substitution. For simplicity, the tree parameters are represented graphically, and only one auxiliary parameter is shown.

12.2 Formal Definition of a Phylo-HMM

Formally, we define phylo-HMM $\boldsymbol{\theta} = (S, \boldsymbol{\psi}, \mathbf{A}, \mathbf{b})$ to be a four-tuple, consisting of a set of states, $S = \{s_1, \ldots, s_M\}$, a set of associated phylogenetic models, $\boldsymbol{\psi} = \{\boldsymbol{\psi}_1, \ldots, \boldsymbol{\psi}_M\}$, a matrix of state-transition probabilities, $\mathbf{A} = \{a_{j,k}\}$ ($1 \leq j, k \leq M$), and a vector of initial-state probabilities, $\mathbf{b} = (b_1, \ldots, b_M)$. In particular, $\boldsymbol{\psi}_j$ is the phylogenetic model associated with state s_j ($1 \leq j \leq M$), $a_{j,k}$ ($1 \leq j, k \leq M$) is the conditional probability of visiting state k at some site i given that state j is visited at site $i-1$, and b_j ($1 \leq j \leq M$) is the probability that state j is visited first (thus, $\sum_k a_{j,k} = 1$ for all j, and $\sum_j b_j = 1$). Let **X** be the given alignment, consisting of L columns (sites) and n rows (one for each of n species), with the ith column denoted $\mathbf{X}_i$ ($1 \leq i \leq L$).

Each phylogenetic model $\boldsymbol{\psi}_j$, in turn, consists of several components. For our purposes, a phylogenetic model $\boldsymbol{\psi}_j = (\mathbf{Q}_j, \boldsymbol{\pi}_j, \boldsymbol{\tau}_j, \boldsymbol{\beta}_j)$ is a four-tuple consisting of a substitution rate matrix $\mathbf{Q}_j$, a vector of background (or equilibrium) frequencies $\boldsymbol{\pi}_j$, a binary tree $\boldsymbol{\tau}_j$, and a set of branch lengths $\boldsymbol{\beta}_j$. The model is defined with respect to an alphabet Σ (e.g., $\Sigma = \{A,C,G,T\}$) whose size is denoted d. Generally, $\mathbf{Q}_j$ has dimension $d \times d$, and $\boldsymbol{\pi}$ has dimension d (but see Example 12.3). The tree τ_j has n leaves, corresponding to n present-day taxa. The elements of $\boldsymbol{\beta}_j$ are associated with the branches (edges) of the tree. It is assumed that all phylogenetic models in $\boldsymbol{\psi}$ are defined with respect to the same alphabet and number of species.

The probability that a column $\mathbf{X}_i$ is emitted by state s_j is simply the probability of $\mathbf{X}_i$ under the corresponding phylogenetic model, $P(\mathbf{X}_i|\boldsymbol{\psi}_j)$. This quantity can be computed efficiently by a recursive dynamic programming algorithm known as Felsenstein's "pruning" algorithm [8]. Felsenstein's algorithm requires conditional probabilities of substitution for all bases $a, b \in \Sigma$ and branch lengths $t \in \boldsymbol{\beta}_j$. The probability of substitution of a base b for a base a along a branch of length t, denoted $P(b|a, t, \boldsymbol{\psi}_j)$, is based on a continuous-time Markov model of substitution, defined by the rate matrix $\mathbf{Q}_j$. In particular, for any given nonnegative value t, the conditional probabilities $P(b|a, t, \boldsymbol{\psi}_j)$ for all $a, b \in \Sigma$ are given by the $d \times d$ matrix $\mathbf{P}_j(t) = \exp(\mathbf{Q}_j t)$, where $\exp(\mathbf{Q}_j t) = \sum_{k=0}^{\infty} \frac{(\mathbf{Q}_j t)^k}{k!}$ [28]. $\mathbf{Q}_j$ can be parameterized in various more or less parsimonious ways [50]. For most of this chapter, we will assume the parameterization corresponding to the "HKY" model [13], which implies that $\mathbf{Q}_j$ has the form

$$
\mathbf{Q}_j = \begin{pmatrix} - & \pi_{Cj} & \kappa_j \pi_{Gj} & \pi_{Tj} \\ \pi_{Aj} & - & \pi_{Gj} & \kappa_j \pi_{Tj} \\ \kappa_j \pi_{Aj} & \pi_{Cj} & - & \pi_{Tj} \\ \pi_{Aj} & \kappa_j \pi_{Cj} & \pi_{Gj} & - \end{pmatrix},
\tag{12.1}
$$

where $\boldsymbol{\pi}_j = (\pi_{Aj}, \pi_{Cj}, \pi_{Gj}, \pi_{Tj})$, κ_j represents the transition/transversion rate ratio for model $\boldsymbol{\psi}_j$, and the $-$ symbols indicate quantities required to make each row sum to zero.

A "path" through the phylo-HMM is a sequence of states, $\boldsymbol{\phi} = (\phi_1, \ldots, \phi_L)$, such that $\phi_i \in \{1, \ldots, M\}$ for $1 \leq i \leq L$. The joint probability of a path and an alignment is[2]

$$
P(\boldsymbol{\phi}, \mathbf{X}|\boldsymbol{\theta}) = b_{\phi_1} P(\mathbf{X}_1|\boldsymbol{\psi}_{\phi_1}) \prod_{i=2}^{L} a_{\phi_{i-1}, \phi_i} P(\mathbf{X}_i|\boldsymbol{\psi}_{\phi_i}).
\tag{12.2}
$$

The likelihood is given by the sum over all paths, $P(\mathbf{X}|\boldsymbol{\theta}) = \sum_{\boldsymbol{\phi}} P(\boldsymbol{\phi}, \mathbf{X}|\boldsymbol{\theta})$, and the maximum-likelihood path is $\hat{\boldsymbol{\phi}} = \arg\max_{\boldsymbol{\phi}} P(\boldsymbol{\phi}, \mathbf{X}|\boldsymbol{\theta})$. These quantities can be computed efficiently using two closely related dynamic-programming algorithms known as the "forward" and Viterbi algorithms, respectively. The posterior probability that observation $\mathbf{X}_i$ was produced by state s_j, denoted $P(\phi_i = j|\mathbf{X}, \boldsymbol{\theta})$, can be computed for all i and j by combining the forward algorithm with a complementary "backward" algorithm, in a "forward-backward" procedure. Details can be found in [7].

---

[2]For simplicity, transitions to an "end" state are omitted here.

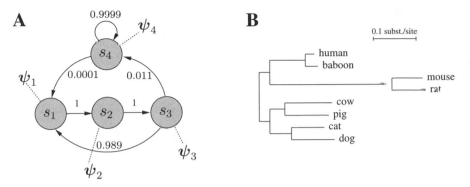

Fig. 12.2. (a) A 4-state phylo-HMM for gene finding. States s_1, s_2, and s_3 represent the three codon positions, and state s_4 represents noncoding sites. The associated phylogenetic models $\psi_1, \ldots, \psi_4$ capture characteristic properties of the different types of sites, such as the higher average rate of substitution and the greater transition/transversion ratio, in noncoding and third-codon-position sites than in first- and second-codon-position sites. (b) The eight mammals and phylogeny assumed for the simulation, with branch lengths drawn in the proportions of the noncoding model (ψ_4). Subsets of species were selected to maximize the sum of the branch lengths of the induced subtree—such as rat and dog for $n = 2$ and rat, dog, and cow for $n = 3$.

Example 12.1 A toy gene finder

This example is meant to demonstrate, in principle, how a phylo-HMM can be used for gene finding. Consider a simple 4-state phylo-HMM, with states for the three codon positions and noncoding sites (Figure 12.2(a)). The problem is to identify the genes in a synthetic data set based on this model using nothing but the aligned sequence data and the model. (This is a multiple-sequence version of the ab initio gene prediction problem.) For simplicity, we assume the model parameters θ are given, along with the data set $\mathbf{X}$. In practice, the parameters have been set to reasonable values for a phylogeny of $n = 8$ mammals (Figure 12.2(b))[3], and the data set has been generated according to these values. The state path was recorded during the generation of the data, so that it could be used to evaluate the accuracy of predictions. The synthetic data set consists of $L = 100000$ sites and 74 genes.

The Viterbi algorithm can be used for prediction of genes in this data set in a straightforward way. For every site i ($1 \leq i \leq L$) and state j ($1 \leq j \leq M$), the emission probability $P(\mathbf{X}_i|\psi_j)$ is computed using Felsenstein's algorithm. These $L \times M$ values, together with the state-transition probabilities $\mathbf{A}$ and initial-state probabilities $\mathbf{b}$, are sufficient to define the joint probability

---

[3]Parameter estimates from [44] were used for the phylogenetic models, and the state-transition probabilities were approximately based on estimates from [43]. (The probability from s_4 to s_1 was inflated so that genes would not be too sparse.) A uniform distribution was assumed for initial-state probabilities.

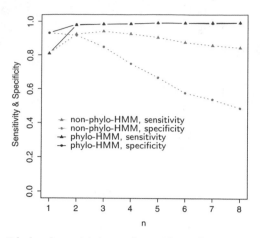

Fig. 12.3. Nucleotide-level sensitivity and specificity for the phylo- and nonphylo-HMMs on the simulated data set of Example 12.1. Results are shown for $n = 1, \ldots, 8$ species.

$P(\phi, \mathbf{X}|\psi)$ for any path ϕ and can be simply plugged into the standard Viterbi algorithm to obtain a maximum-likelihood path, $\hat{\phi}$. This predicted path, in turn, defines a set of predicted genes.

To evaluate the effect on prediction accuracy of the number of species in the data set, subsets of $n = 1, \ldots, 8$ sequences were selected from the full alignment (Figure 12.2(b)), and a separate set of predictions was produced for each subset. Predictions were also produced with an alternative model, in which emission probabilities were based on the assumption that all characters in a column were independently drawn from the background (equilibrium) distribution of each state—in other words, the correlation structure implied by the phylogeny was ignored. This model, which will be called the "nonphylo-HMM," allows the importance of the phylogeny in the phylo-HMM to be assessed.

The nucleotide-level sensitivity (portion correctly predicted of sites actually in genes) and specificity (portion correct of sites predicted to be in genes) for both models are shown in Figure 12.3 as the number of species increases from $n = 1$ to $n = 8$. The two models are identical for $n = 1$ (where there is no phylogeny to consider), but as the number of species increases from $n = 2, \ldots, 8$, the performance of the phylo-HMM rapidly improves, with about 98% sensitivity and specificity achieved by $n = 2$, and 99% sensitivity and specificity achieved by $n = 5$. The nonphylo-HMM, on the other hand, appears to improve slightly then decline in both sensitivity and specificity.[4]

---

[4]It might be expected that the prediction accuracy of the nonphylo-HMM would simply fail to improve as rapidly as that of the phylo-HMM rather than declining. The reason for the decline seems to be that the erroneous assumption of independence causes random fluctuations in base composition to appear more significant

The phylo-HMM is able to capitalize on differences in branch lengths and substitution patterns, while the nonphylo-HMM has to rely completely on more subtle differences in base composition.

This example is obviously a gross simplification of the real gene prediction problem. Here, the model used for prediction exactly matches the model used to generate the data, while in the real problem, the model for prediction tends to fit the data in a much more approximate way. Even if slightly contrived, however, this example should help to illustrate how the information encoded in substitution rates and patterns can be exploited in problems of segmentation, such as gene prediction.   □

Example 12.2 Identification of highly conserved regions

Our second example is concerned with a phylo-HMM in which states correspond to "rate categories"— classes of sites assumed to differ only in overall rate of substitution—rather than "functional categories," as in the previous example. The problem is to identify highly conserved genomic regions in a set of multiply aligned sequences. Such regions are likely to be functionally important, and hence their identification has become a subject of considerable interest in comparative genomics; see Margulies et al. [32] for a recent review and a comprehensive discussion. In this example, we will use a phylo-HMM to identify conserved regions in a subset of the data set analyzed by Margulies et al. It will be shown that a phylo-HMM can be used to obtain results comparable to theirs and has certain potential advantages over their methods.

A phylo-HMM like the one proposed by Felsenstein and Churchill [9] is assumed, with k states corresponding to k rate categories and state transitions defined by a single "autocorrelation" parameter λ (Figure 12.4; a similar model, but with a more complex parameterization of transition probabilities, was proposed by Yang [52]). Regions of the alignment that are likely to have been generated by the "slowest" rate categories will be considered putative "Multi-species Conserved Sequences" (MCSs) [32]. Specifically, we will look at sites i for which the posterior probability $P(\phi_i = 1|\mathbf{X}, \boldsymbol{\theta})$ is high, assuming state s_1 has the smallest rate constant. Posterior probabilities will be computed using the forward-backward algorithm. As in Example 12.1, the $L \times k$ table of emission probabilities—$P(\mathbf{X}_i|\psi_j)$ for every site i ($1 \leq i \leq L$) and state j ($1 \leq j \leq k$)—together with the state-transition and initial-state probabilities (parameters $\mathbf{A}$ and $\mathbf{b}$ of the phylo-HMM), can be plugged into the standard forward-backward algorithm for HMMs. In other words, once the emission probabilities are computed, the phylogenetic models can be ignored, and the phylo-HMM can be treated like an ordinary HMM. Note that inferences about the evolutionary rate at each site could alternatively be based on the Viterbi path. We have opted to use posterior probabilities instead, partly for illustration and partly because they can be conveniently interpreted as a

---

than they really are. These fluctuations are "explained" by changes in state, resulting in errors in the inferred path and a decline in accuracy.

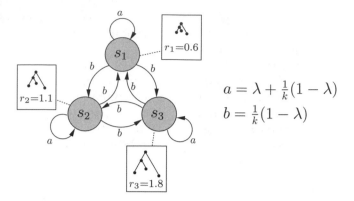

$$a = \lambda + \tfrac{1}{k}(1 - \lambda)$$
$$b = \tfrac{1}{k}(1 - \lambda)$$

Fig. 12.4. State-transition diagram for the autocorrelated rate-variation model of Felsenstein and Churchill [9] with $k = 3$ rate categories and a uniform stationary distribution. The autocorrelation parameter λ defines all transition probabilities, as shown. It takes values between 0 and 1 and describes the degree to which the evolutionary rates at adjacent sites tend to be similar. The values r_1, r_2, and r_3 are applied as scaling constants to the branch lengths of a phylogenetic model; all parameters other than branch lengths are left unchanged. In our case, these "rate constants," as well as λ, are estimated (approximately) from the data (see [42]).

continuous-valued "conservation score" that can be plotted along the genome (see below). With this model, the posterior probabilities also tend to be more robust than the Viterbi path, which is highly sensitive to λ.

The data set consists of about 1.8 Mb of human sequence from chromosome 7 and a homologous sequence from eight other eutherian mammals [46] (we consider only the nine mammals of the 12 species analyzed in [32].) The species and phylogeny are as shown in Figure 12.2(b), except that in this case the chimp is also included and appears in the phylogeny as a sister taxon to the human. Assuming the HKY substitution model and $k = 10$ states, we fitted a phylo-HMM to this alignment, obtaining an estimate of $\hat{\lambda} = 0.94$. Using these parameter estimates, we then computed the posterior probability of each state at each site. The posterior probabilities for s_1 in a selected region of the alignment are shown in Figure 12.5 along with the conservation scores developed by Margulies et al. The known exons in this region all coincide with regions of high posterior probability, as do several conserved intronic features identified by Margulies et al. [32].

A detailed comparison of results is not possible here, but we note that the posterior probabilities based on the phylo-HMM are qualitatively very similar to the binomial- and parsimony-based conservation scores of Margulies et al. [32]. In addition, the phylo-HMM may have certain advantages as a framework for addressing this problem. For example, it requires no sliding window of fixed size and, as a result, is capable of identifying both very short highly conserved sequences and long, not so conserved sequences. In addition, it can

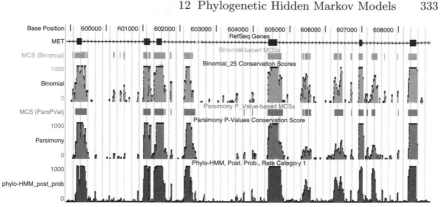

Fig. 12.5. A screen shot from the UCSC Genome Browser [25] showing a selected region of the data set of Example 12.2, including several exons of the *MET* gene (black boxes at top). The binomial-based (light gray) and parsimony-based (medium gray) conservation scores of Margulies et al. [32] are shown as tracks in the browser, as are the posterior probabilities ($\times 1000$) of state s_1 in the phylo-HMM (dark gray). Plots similar to this one, showing phylo-HMM-based conservation scores across the whole human genome, can be viewed online at http://genome.ucsc.edu.

be used with any phylogenetic model, including, for example, ones that allow for nonhomogeneities in the substitution process or context-dependent substitution (see Example 12.3); it extends naturally to the case in which different functional categories of sites, as well as rate categories, are considered [42]; and it could be adapted to model properties such as the length distributions of MCSs (e.g., using techniques from gene finding). □

12.3 Higher-Order Markov Models for Emissions

It is common with (single-sequence) gene-finding HMMs to condition the emission probability of each observation x_i on the observations that immediately precede it in the sequence (e.g., x_{i-2} and x_{i-1}). By taking into consideration the "context" for each observation, emission probabilities become more informative, and the HMM can discriminate more effectively between different classes of observations. For example, in a third-codon-position state, the emission of a base $x_i =$ "A" might have a fairly high probability if the previous two bases are $x_{i-2} =$ "G" and $x_{i-1} =$ "A" (GAA = Glu) but should have zero probability if the previous two bases are $x_{i-2} =$ "T" and $x_{i-1} =$ "A" (TAA = Stop).

Considering the N observations preceding each x_i corresponds to using an Nth-order Markov model for emissions. (Note that such a model does not imply an Nth-order Markov chain for state transitions; indeed, things are kept simpler, and the model remains mathematically valid, if state transitions continue to be described by a first-order Markov chain.) An Nth-order model for emissions is typically parameterized in terms of $(N+1)$-tuples of observations, and conditional probabilities are computed as

$$P(x_i|x_{i-N},\ldots,x_{i-1}) = \frac{P(x_{i-N},\ldots,x_{i-1},x_i)}{\sum_y P(x_{i-N},\ldots,x_{i-1},y)}, \qquad (12.3)$$

with the numerator being the probability of the $(N+1)$-tuple $(x_{i-N},\ldots,x_i)$ and the sum in the denominator being over all possible observations y that could appear in place of x_i.

An Nth-order Markov model for emissions can be incorporated into a phylo-HMM in essentially the same way. In this case, a whole alignment column $\mathbf{X}_i$ is considered in place of each single base x_i. Because we will primarily be concerned below with tuple size, let us also redefine N and speak of $(N-1)$st-order Markov models and N-tuples of observations instead of Nth-order Markov models and $(N+1)$-tuples of observations. With these changes, equation (12.3) can be rewritten as

$$P(\mathbf{X}_i|\mathbf{X}_{i-N+1},\ldots,\mathbf{X}_{i-1}) = \frac{P(\mathbf{X}_{i-N+1},\ldots,\mathbf{X}_{i-1},\mathbf{X}_i)}{\sum_{\mathbf{Y}} P(\mathbf{X}_{i-N+1},\ldots,\mathbf{X}_{i-1},\mathbf{Y})}. \qquad (12.4)$$

Notice that the sum in the denominator is now over all possible alignment columns $\mathbf{Y}$ and has d^n terms, where d is the size of the alphabet ($d = |\mathbf{\Sigma}|$) and n is the number of rows (species) in the alignment. To compute the quantity in the numerator of equation (12.4), we replace an ordinary phylogenetic model, defined with respect to an alphabet $\mathbf{\Sigma}$, with what we will call an "Nth-order" phylogenetic model, defined with respect to $\mathbf{\Sigma}^N$, the alphabet of N-tuples of characters from $\mathbf{\Sigma}$.[5] (The new rate matrix and vector of equilibrium frequencies will have dimensions $d^N \times d^N$ and d^N, respectively.) The N-tuple of columns in the numerator is reinterpreted as a column of N-tuples, and its probability is computed with Felsenstein's pruning algorithm using the Nth-order phylogenetic model. The sum in the denominator can no longer be evaluated directly, but it can be computed efficiently by dynamic programming using a slight adaptation of Felsenstein's algorithm [44, 42]. This new algorithm differs from the original only in its initialization strategy. Thus, the conditional probability $P(\mathbf{X}_i|\mathbf{X}_{i-N+1},\ldots,\mathbf{X}_{i-1})$ can be computed with an Nth-order phylogenetic model and two passes through Felsenstein's algorithm, one for the numerator and one for the denominator of equation (12.4). This procedure is feasible only for small N; so far, for $N \le 3$.

---

[5]Note that the "order" of a phylogenetic model is given by the size of the tuples considered and is not equal to the order of the Markov model for emissions. Here, Nth-order phylogenetic models are used to define an $(N-1)$st-order Markov model.

Once the conditional emission probabilities of equation (12.4) are available, they can be substituted directly into equation (12.2). For example, in the case of $N = 3$, equation (12.2) can be rewritten as

$$P(\phi, \mathbf{X}|\boldsymbol{\theta}) = b_{\phi_1} P(\mathbf{X}_1|\boldsymbol{\psi}_{\phi_1}) a_{\phi_1,\phi_2} P(\mathbf{X}_2|\mathbf{X}_1, \boldsymbol{\psi}_{\phi_2})$$
$$\times \prod_{i=3}^{L} a_{\phi_{i-1},\phi_i} P(\mathbf{X}_i|\mathbf{X}_{i-2}, \mathbf{X}_{i-1}, \boldsymbol{\psi}_{\phi_i}). \qquad (12.5)$$

The forward, Viterbi, and forward-backward algorithms are unaffected by the use of a higher-order Markov model for emissions.

It is important to note that this strategy for incorporating higher order Markov models into a phylo-HMM allows "context" to be considered in the nucleotide substitution process as well as in the equilibrium frequencies of bases. Nth-order phylogenetic models describe the joint substitution probabilities of N-tuples of nucleotides. As a result, the conditional probabilities of equation (12.4) may reflect various important context or neighbor dependencies in the substitution process, such as the tendency for synonymous substitutions to occur at a higher rate than nonsynonymous substitutions in coding regions, or the tendency for a high rate of C→T transitions in CpG dinucleotides. Equations (12.4) and (12.5), as will be shown in Example 12.3, essentially provide a way of "stringing together" context-dependent phylogenetic models so that context dependencies can be considered between every adjacent pair of columns in an alignment.

Example 12.3 Modeling context-dependent substitution

In this example, we will look at how goodness of fit is affected by increasing the order N of a phylogenetic model and by allowing for Markov dependence between sites (as in equation (12.5)). We will consider the goodness of fit of various independent-site ($N = 1$) and context-dependent ($N > 1$) phylogenetic models with respect to about 160,000 sites in aligned noncoding DNA from nine mammalian species. The results presented here are taken from [44]. (The full paper should be consulted for complete details.)

For convenience, let us call the class of phylo-HMMs described by equations (12.4) and (12.5) "Markov-dependent" models because they allow for Markov dependence of columns in the alignment. As will be seen below, these models are actually only approximations of models that properly allow for Markov dependence across sites in the substitution process. Regardless, these Markov-dependent models are valid probability models (the probabilities of all alignments of a given size sum to one), so it is fair to evaluate goodness of fit based on model likelihoods. The way in which these models are approximate is discussed in detail in Section 12.7 and the Appendix.

In this example, there are no functional or rate categories to consider. We assume that the HMM has only a single state, so nothing is actually "hidden"—only one path is possible, and the model reduces to a Markov chain.

As a result, equation (12.5) becomes

$$P(\phi, \mathbf{X}|\theta) = P(\mathbf{X}_1|\psi_1)P(\mathbf{X}_2|\mathbf{X}_1, \psi_1) \prod_{i=3}^{L} P(\mathbf{X}_i|\mathbf{X}_{i-2}, \mathbf{X}_{i-1}, \psi_1). \qquad (12.6)$$

This simplification allows us to focus on the impact of higher-order Markov models and to avoid issues related to the HMM structure. Keep in mind, however, that higher-order Markov models can be used with a nontrivial HMM as easily as with this trivial one.

In [44], various models were fitted to the data set of 160,000 noncoding sites, and their likelihoods were compared. The models differed in the type of phylogenetic model used (its order N and the parameterization of its rate matrix) and whether N-tuples of columns were assumed to be independent or whether Markov dependence was allowed. We will focus here on four types of phylogenetic models: the HKY and UNR first-order models, the U2S second-order model, and the U3S third-order model. The HKY model, introduced in Section 12.2, is treated as a baseline. The UNR, or "unrestricted," model has a separate free parameter for every nondiagonal element of the rate matrix and is the most general model possible for single-nucleotide substitution (see, e.g., [51]). The U2S and U3S models are fully general second and third-order models, respectively, except that they assume strand symmetry (so that, e.g., the rate at which AG changes to AC is the same as the rate at which CT changes to GT), and like most codon models [12], they prohibit instantaneous substitutions of more than one nucleotide. They have 48 and 288 rate-matrix parameters, respectively. We will consider two cases for each phylogenetic model: an "independent tuples" case, in which the data set was partitioned into N-tuples of columns, which were considered independent; and a Markov-dependent case, in which N-tuples were allowed to overlap, and likelihoods were computed with equations (12.4) and (12.6). Note that, with first-order models, the independent tuples and Markov-dependent cases are identical.

Figure 12.6(a) shows the log-likelihoods of the UNR, U2S, and U3S phylogenetic models, with and without Markov dependence, relative to the log-likelihood of the HKY model. Even when N-tuples are considered independent, context-dependent models (here U2S and U3S) produce a striking improvement in likelihood—a far larger increase than is obtained by replacing even a fairly parsimonious first-order model (HKY) with a fully general one (UNR). When Markov dependence between sites is introduced, another large improvement occurs. This improvement appears to be largely a consequence of the fact that, with Markov dependence, every boundary between adjacent sites is considered, while with independent tuples, only every other (U2S) or every third (U3S) such boundary is considered. Notice that, even with Markov dependence, goodness of fit improves significantly when a second-order model (U2S) is replaced with a third-order model (U3S). This is probably partly because of direct context effects that extend beyond the nearest neighbors of each base and partly because the third-order model does a better job than

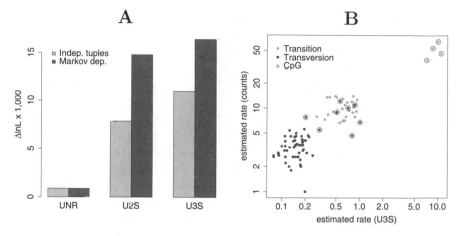

Fig. 12.6. (a) Log-likelihoods of the UNR, U2S, and U3S phylogenetic models, with and without Markov dependence between sites, relative to the log-likelihood of the HKY model. Results are for an alignment of nine species and approximately 160,000 sites of noncoding data, as described in [44]. (b) Parameter estimates of substitution rates for the U3S model vs. estimates based on counts from aligned human genes and pseudogenes [16]. The rates cluster into three groups: transversions, transitions, and CpG transitions. (CpG transversions cluster with non-CpG transitions.) In general, the two sets of estimates agree fairly well, considering the differences in methods and data sets. (See [44] for a detailed discussion.)

the second-order model of accounting for indirect context effects—that is, it provides a better approximation of a proper process-based model of context-dependent substitution (see below).

The observed improvements remain essentially unchanged when a measurement is used that considers the different numbers of parameters in the models and the size of the data set (the Bayesian information criterion) and in cross-validation experiments [44]. Thus, the apparent improvement in goodness of fit is not an artifact of the number of parameters in the models.

The U2S and U3S models allow context-dependent substitution rates to be estimated with full consideration of the phylogeny and allowance for multiple substitutions per site, unlike simpler "counting" methods for estimating context-dependent substitution rates [16]. Parameter estimates indicate a wide variation in rates, spanning a 200-fold range, and, in particular, pronounced CpG effects (Figure 12.6(b)).

Coding regions can be modeled using a simple 3-state phylo-HMM, with a separate third-order phylogenetic model for each codon position. Thus, the state corresponding to the third codon position considers columns of aligned codons, like an ordinary codon model, but the two other states consider columns of nucleotide triples that are out-of-frame, and, consequently, these states can capture context effects that cross codon boundaries. Such a model improves substantially on ordinary codon models, indicating that context effects that cross codon boundaries are important [44] (see also [40]). □

12.4 Phylogenetic Models, HMMs, and Phylo-HMMs as Graphical Models

In recent years, probabilistic models originally developed in various research communities have been unified under the heading of "graphical models." Graphical models provide an intuitively appealing framework for constructing and understanding probabilistic models and at the same time allow for rigorous analysis, in very general statistical and graph-theoretic terms, of algorithms for inference and learning. Many familiar classes of models fit naturally into the graphical models framework, including HMMs and phylogenetic models, as well as mixture models and hierarchical Bayesian models. A phylo-HMM can be seen as a graphical model whose structure is a hybrid of the graphical models for HMMs and phylogenetic models (Figure 12.7). Viewing phylo-HMMs as graphical models helps to provide insight about why they permit efficient inference and why this property may be sacrificed when assumptions such as site independence are relaxed. Our discussion of graphical models will necessarily be brief; other tutorials should be consulted for a more complete introduction to the field (e.g., [31, 14, 23]).

In graphical models, random variables are represented by nodes in a graph, and dependencies between variables are represented by edges (Figure 12.7).[6] Let X be the set of random variables represented by a graph with nodes (vertices) V and edges E such that X_v is the variable associated with $v \in V$. In addition, let X_C be the subset of variables associated with $C \subseteq V$, and let lowercase letters indicate (sets of) instances of variables (e.g., x_v, x_C, and x). Graphical models can be defined in terms of directed or undirected graphs and accordingly are called directed or undirected models; here we will focus on the directed case, which for our purposes is simpler to describe. In a directed model, the edges of the graph correspond to local conditional probability distributions, and the joint probability of a set of instances x is a product of the conditional probabilities of nodes given their parents,

$$P(x) = \prod_{v \in V} P(x_v | x_{\mathcal{P}_v}), \tag{12.7}$$

where $\mathcal{P}_v$ denotes the set of parents of node v and $P(x_v | x_{\mathcal{P}_v})$ is the local conditional probability associated with x_v. It should not be too hard to see, looking at Figure 12.7, that equation (12.7) generalizes the joint probability of a sequence and a particular path in the case of an HMM and the joint

---

[6]The brief introduction to graphical models provided here roughly follows the more detailed tutorial of Jordan and Weiss [23].

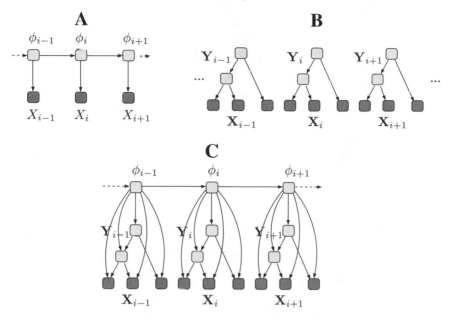

Fig. 12.7. Graphical model representations of (a) an HMM, (b) a phylogenetic model, and (c) a phylo-HMM. In each case, nodes correspond one-to-one with random variables; shaded nodes represent observed variables, and unshaded nodes represent unobserved (latent) variables. These are directed graphical models based on directed acyclic graphs (sometimes called Bayesian networks). The edges between nodes correspond to local conditional probability distributions and can be thought of as implying dependencies between variables. (More precisely, the set of all edges defines a set of conditional independence assertions about the variables.) In (a), each X_i represents an observation in the sequence and each ϕ_i represents a state in the path. The conditional probability distribution for observation X_i given state ϕ_i is incorporated in the directed edge from ϕ_i to X_i, and the conditional probability distribution for state ϕ_i given state ϕ_{i-1} (i.e., of a transition from ϕ_{i-1} to ϕ_i) is incorporated in the directed edge from ϕ_{i-1} to ϕ_i. In (b), each set of nodes collectively labeled $\mathbf{X}_i$ represents an alignment column, and each set collectively labeled $\mathbf{Y}_i$ represents a set of ancestral bases. The conditional probabilities of nucleotide substitutions (based on the continuous-time Markov model) are incorporated in the directed edges from each parent node to its two children. In (c), conventions from (a) and (b) are combined.

probability of an alignment and a particular set of ancestral bases in the case of a phylogenetic model.

The general problem of *probabilistic inference* is to compute marginal probabilities from this joint distribution –probabilities of the form $P(x_U) = \sum_{x_W} P(x_U, x_W)$, where (U, W) is a partitioning of V. The likelihood is an example of such a marginal probability, with x_U being the observed data and X_W being the set of latent variables. When the likelihood of an HMM is

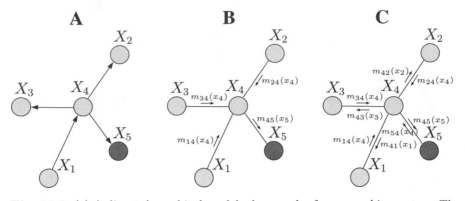

Fig. 12.8. (a) A directed graphical model whose nodes form an arbitrary tree. The marginal probability of an observed value of X_5 is desired. (b) The intermediate values of the elimination algorithm can be seen as "messages" that are passed from one node to another in the direction of X_5. (c) In the belief-propagation algorithm, all possible messages are generated simultaneously; the marginal probability of each node is a product of the incoming messages. (Based on Figure 1 of Jordan and Weiss [23].)

computed, x_U is the (observed) sequence and X_W is the (latent) path. With a phylogenetic model, the procedure is applied independently at each site, and x_U is an (observed) alignment column and X_W is a set of (latent) ancestral bases. Conditional probabilities of interest, such as the posterior probabilities of Example 12.2, can be computed as quotients of marginal probabilities. For instance, suppose x_U is the observed data and X_w ($w \in W$) is a latent variable; then $P(x_w|x_U) = \frac{P(x_{U \cup \{w\}})}{P(x_U)}$.

Marginal probabilities can always be computed from the complete joint distribution by brute-force summation[7]. The problem is to keep these computations tractable as the number of random variables becomes large. It turns out that if a directed graphical model is a tree (or set of trees), as in Figure 12.7(a, b) and Figure 12.8, meaning that every node has at most one parent, then exact inference can be accomplished efficiently by dynamic programming. (As we will see, efficient exact inference is also possible in certain cases in which the directed graph is not a tree.)

The basic algorithm for computing marginal probabilities is known as "elimination", and is most easily described by example. Consider the graph of Figure 12.8(a), with $X = (X_1, X_2, X_3, X_4, X_5)$ and edges as depicted. The elimination algorithm takes advantage of the commutativity of sums and products, and reuse of intermediate computations, to reduce the computational complexity of a marginal summation.

---

[7]This discussion is restricted to discrete random variables, although it extends directly to the continuous case.

Algebraically, the algorithm proceeds as follows,

$$
\begin{aligned}
P(x_5) &= \sum_{x_1,x_2,x_3,x_4} P(x_1,x_2,x_3,x_4,x_5) \\
&= \sum_{x_1}\sum_{x_2}\sum_{x_3}\sum_{x_4} P(x_1)P(x_2|x_4)P(x_3|x_4)P(x_4|x_1)P(x_5|x_4) \\
&= \sum_{x_4} P(x_5|x_4)\sum_{x_3} P(x_3|x_4)\sum_{x_2} P(x_2|x_4)\sum_{x_1} P(x_1)P(x_4|x_1) \\
&= \sum_{x_4} P(x_5|x_4)\sum_{x_3} P(x_3|x_4)\sum_{x_2} P(x_2|x_4)m_{14}(x_4) \\
&= \sum_{x_4} P(x_5|x_4)\sum_{x_3} P(x_3|x_4)m_{24}(x_4)m_{14}(x_4) \\
&= \sum_{x_4} P(x_5|x_4)m_{34}(x_4)m_{24}(x_4)m_{14}(x_4) \\
&= m_{45}(x_5),
\end{aligned}
\tag{12.8}
$$

where the terms of the form $m_{ij}(x_j)$ denote the results of intermediate (nested) summations. (Each $m_{ij}(x_j)$ is the result of a sum over x_i and is a function of x_j.) The algorithm can be described in graph-theoretic terms as a procedure that eliminates one node at a time from the graph until only the node corresponding to the desired marginal probability remains. From the algebraic description, many readers will recognize the similarity to Felsenstein's pruning algorithm [8]. Felsenstein's algorithm, it turns out, is an instance of the elimination algorithm—one of the earliest instances to be discovered. The forward algorithm is another instance of the elimination algorithm, as is the combined forward/Felsenstein algorithm that we used above to compute the likelihood of a phylo-HMM. The Viterbi algorithm is closely related to the elimination algorithm; it can be derived by noting that the "max" operator commutes with products, just as the summation operator does. Note that the elimination algorithm depends on a good "elimination ordering". An optimal ordering is difficult to find for arbitrary graphs but can be determined easily for specific classes of models (such as with HMMs, phylogenetic models, and phylo-HMMs).

Often, not just one but many marginal probabilities are desired. The elimination algorithm can be extended to compute the marginal probabilities for all nodes in two passes across the graph, with conditional probabilities computed in a forward pass and marginals in a backward pass [30]. Typically, this procedure is described as "belief propagation" [38], with node elimination replaced by a "message-passing" metaphor (Figure 12.8(b, c)). The belief-propagation (also called "sum-product") algorithm generalizes the forward-backward algorithm for HMMs and algorithms for phylogenetic models that compute marginal probabilities of ancestral bases [27].

We have focused on directed models, but undirected models are similar. Moreover, the undirected case turns out to be, in a sense, the more general one with respect to inference. In undirected models, the graph is viewed in terms of cliques (maximal fully connected subgraphs), and a *potential function* (essentially an unnormalized probability distribution) is associated with each clique. The joint probability of all variables (equation (12.7)) is now a product over cliques, with a normalizing constant to ensure that $\sum_x P(x) = 1$. Directed graphs can be converted to undirected graphs by a process known as "moralization," wherein the arrowheads of the edges are removed and new edges are added between all parents of each node. (The resulting graph is called the "moral" graph, because it requires that all parents be "married".) By explicitly creating a clique that includes each node and all of its parents, moralization ensures that all dependencies implied by the local conditional distributions of the directed graph are captured in the undirected graph.

The moral graph for a directed tree is simply an undirected tree (i.e., no new edges are added), and the belief-propagation algorithm for this undirected tree is the same as that illustrated in Figure 12.8. For undirected graphs that contain cycles, a generalization of the belief propagation algorithm, called the "junction-tree" algorithm, can be used. The junction-tree algorithm operates on a tree of cliques rather than of nodes and computes (unnormalized) marginal probabilities for cliques. (Marginal probabilities of nodes can be obtained afterwards.) It requires an additional step, called "triangulation," in which new edges are added to the graph to represent certain implicit dependencies between nodes. A complete introduction to the junction-tree algorithm is not possible here. (More details can be found in [31] and [23].) The key point for our purposes is that the computational complexity of the algorithm is exponential in the size of the largest clique. Thus, graphs with cycles can still be handled efficiently if their clique size is constrained to be small. It is for this reason that phylo-HMMs permit efficient inference; their (triangulated) moral graphs have cycles, but the maximum clique size turns out to be three[8]. When the clique size is large, exact inference is intractable, and approximate methods are required. Some of the approximate methods in use include Monte Carlo algorithms and variational methods, which include mean field methods and "loopy" belief propagation. (Approximate methods are partially surveyed in [23]; see also [37, 53, 48, 49].)

With phylo-HMMs, the junction-tree algorithm allows computation not only of the posterior probability that each site was emitted by each state (as in Example 12.2), but also of marginal posterior probabilities of ancestral bases considering all paths. In addition, the algorithm can be used to compute posterior expected values of interest, such as the expected number of substitutions per site, or the expected numbers of each type of substitution (A→C, A→G, etc.) along each branch of the tree (the sufficient statistics for parameter estimation by expectation maximization [10, 44]). Using the junction tree algorithm in the expectation step of an expectation-maximization algorithm,

---

[8]In the case of a phylo-HMM, the parents of each node are already connected (Figure 12.7(c)), so moralization simply amounts to removing the arrowheads from all edges in the graph. Moreover, it turns out that this graph is already triangulated.

A

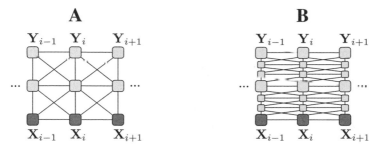

B

Fig. 12.9. (a) The lattice that results when context-dependent substitution is incorporated into a phylogenetic model, shown as an undirected graphical model. For clarity, only a single leaf node is shown for each site, with a chain of ancestral nodes leading to the root. (The phylogeny can be imagined as going into and out of the page.) Each node depends not only on its parent node in the phylogeny but also on its parent's left and right neighbors in the alignment. (b) A version of the graph in (a) with intermediate nodes added to the branches of the tree. As more and more nodes are added, the branch lengths between them approach zero, and the model approaches a true "process-based" model of context-dependent substitution. In both (a) and (b), the untriangulated graph is shown; additional edges appear during triangulation, leading to prohibitively large clique sizes.

it is possible to train a phylo-HMM (including its phylogenetic models) completely from unlabeled data. This technique could be used, for example, for de novo detection of binding-site motifs in aligned sequences.

Once the effect of cycles in graphical models is understood, it becomes clear that efficient exact inference will not be possible with models that accurately describe the *process* of context-dependent substitution, by allowing for dependencies between adjacent bases on all branches of the phylogenetic tree. Figure 12.9(a) illustrates what happens to the graphical structure of a phylogenetic model when this kind of proper contextdependence is introduced. The additional edges in the graph lead to the formation of a kind of lattice of dependency, reminiscent of the classic Ising model from statistical mechanics. (This case is like a two-dimensional Ising model, except that the branching structure of the phylogeny creates a branching structure of two-dimensional sheets, not shown in Figure 12.9(a).) Unless the size of the lattice is constrained to be small, models of this kind are well-known to require approximate methods for inference.

Moreover, for context-dependent substitution to be modeled properly, it should be integrated into the continuous-time Markov model of substitution, so that context effects can propagate indefinitely across sites as substitutions accumulate along each branch of the phylogeny. This behavior can be approximated by introducing intermediate nodes in the phylogeny while keeping total branch lengths constant, as shown in Figure 12.9(b). As more and more nodes are introduced, the branch lengths between them will approach zero, and the model will approach the desired "process-based" model. Exact inference is intractable for such models, even in the case of two sequences and an unrooted tree, but Markov chain Monte Carlo (MCMC) methods have been applied in this special case [21, 39]. The stationary distribution of a related process has also been studied [2]. Extending such process-based models to full phylogenies appears difficult, even with MCMC. However, a model without intermediate nodes (as in Figure 12.9(a)) has been studied by Jojic et al. [22] using variational methods for approximate inference. Jojic et al. have shown experimentally that this model can produce significantly higher likelihoods than the U2S version of the more approximate Markov-dependent model described in Section 12.3.

The model of Section 12.3 essentially works by defining a simple $(N-1)$st-order Markov chain of alignment columns (observed variables), while ignoring the dependencies between the ancestral bases (latent variables) that are associated with overlapping N-tuples of columns. As a result, this model has no reasonable process-based interpretation. Nevertheless, it is a valid probability model that appears to fit the data well, and it allows for exact inference at modest computational cost [44]. The Markov-dependent model is compared with the model of Jojic et al. in more detail in the Appendix.

12.5 Discussion

Phylogenetic hidden Markov models are probabilistic models that describe molecular evolution as a combination of two Markov processes—one that operates in the dimension of *space* (along a genome) and one that operates in the dimension of *time* (along the branches of a tree). They combine HMMs and phylogenetic models, two of the most powerful and widely used classes of probabilistic models in biological sequence analysis. Phylo-HMMs often fit aligned DNA sequence data considerably better than models that treat all sites equally or that fail to allow for correlations between sites. In addition, they are useful for identifying regions of interest in aligned sequences, such as genes or highly conserved regions.

Three examples have been presented to illustrate some of the ways in which phylo-HMMs may be used, and each one deserves additional comment. Applying phylo-HMMs to gene prediction (Example 12.1) is a much harder problem than implied here, for several reasons. First, while coding and noncoding sites have quite different properties on average, both types of sites are heterogeneous mixtures, so that correctly classifying particular sequence segments can be difficult. For example, protein-coding sites show higher average levels of evolutionary conservation than noncoding sites, but mammalian genomes do appear to have many islands of conservation in noncoding regions [4, 32], which can lead to false-positive predictions of exons [43]. Similarly, coding sites in mammalian genomes exhibit higher average G+C content than do noncoding sites, but base composition varies considerably in both kinds of sites from one genomic region to another, which can have the effect of confounding gene prediction software. Second, the gene-finding problem ends up being largely about identifying the boundaries of exons as determined by splice sites, and phylo-HMMs are not necessarily the best tools for detecting these so-called "signals." Gene finders are often based on composite models, with specialized submodels for signal detection; a similar approach may be required for phylo-HMMs to be effective in gene prediction. A third problem is that a straightforward phylo-HMM like that of Example 12.1 induces a geometric distribution of exon lengths, which is known to be incorrect. Some of these problems have been addressed with a "generalized" phylo-HMM that allows for arbitrary length distributions of exons, and also uses different sets of parameters for regions of different overall G+C content [33]. In other recent work, it has been shown that the prediction performance of a phylo-HMM-based exon predictor can be improved significantly by using context-dependent phylogenetic models, and by explicitly modeling both conserved noncoding regions and nucleotide insertions/deletions [43]. Additional challenges in multispecies gene prediction are also discussed in [43], stemming from lack of conservation of exon structure across species and errors in the multiple alignment.

There are many possible ways of identifying conserved regions (Example 12.2), and even quite different methods (e.g., ones that do and do not consider the phylogeny) tend to be fairly concordant in the regions they identify [45, 32]. Perhaps more difficult than proposing a method to identify conserved regions is confirming that it produces biologically useful results. Limited kinds of validation can be done computationally, but this is ultimately an experimental problem and must be addressed in the laboratory. Most likely, phylo-HMMs of the kind described in Example 12.2 will not produce results dramatically different results from other methods, but, as mentioned above, they provide a flexible framework in which to address the problem. It should be noted that, while the original papers introducing phylo-HMMs focused on improving the realism and goodness of fit of models allowing for rate variation [9, 52], they also showed that phylo-HMMs could be used to predict the evolutionary rate at each site.

Modeling context-dependent substitution is an active area of current research, and the Markov-dependent model described here (Example 12.3) represents only one of several possible approaches to this problem. The approach of Jojic et al. [22], discussed at the end of Section 12.4, is another, and we are aware of work in progress on at least two other, completely different methods. At this stage, it remains unclear which models and algorithms for inference will allow for the best compromise between computational efficiency and goodness of fit. It is likely that different approaches will turn out to be appropriate for different purposes.

Space has not allowed for a complete survey of the applications of phylo-HMMs. In particular, we have not discussed their use in the prediction of secondary structure [11, 47, 29] or the detection of recombination [20], nor have we touched on their use in a Bayesian setting [34, 19]. We also have not discussed the models similar in spirit to phylo-HMMs that have been applied to the problems of RNA secondary structure prediction [26] and multiple alignment [36, 18, 17, 15]. It has been noted [41] that phylo-HMMs themselves could be used for multiple alignment in a direct extension of the way pair HMMs are used for pairwise alignment [7]. Indeed, phylo-HMMs provide a natural framework for simultaneously addressing the multiple alignment and gene prediction problems, as has been done in the two-sequence case with pair HMMs [1, 35]. Another area in which phylo-HMMs may prove useful is homology searching. In principle, the profile HMMs that are commonly applied to this problem [7] could be adapted to use phylogenetic models instead of assuming independence of aligned sequences or relying on ad hoc weighting schemes.

Acknowledgments

We thank Nick Goldman, David Heckerman, and Michael Jordan for helpful discussions about context-dependent substitution, and Brian Lucena, Mathieu Blanchette, Robert Baertsch, and Michael Jordan for comments on the manuscript. A. S. is supported by an ARCS Foundation scholarship and NHGRI grant IP41HG02371, and D. H. is supported by the Howard Hughes Medical Institute.

Appendix

In this short appendix, we will examine more closely how the Markov-dependent model for context-dependent substitution that was presented in Section 12.3 (Example 12.3) compares with the graphical models of Section 12.4. We will concentrate on the model studied by Jojic et al. [22], which we will refer to as the "simple-lattice" model, in contrast with the full process-based model of Figure 12.9(b). The undirected graph for the simple-lattice model is shown in Figure 12.10(a), assuming a very small alignment of $n = 3$ sequences and $L = 3$ columns. (The complete graph is shown here, whereas in Figure 12.9(a) only a subgraph was shown.) From Figure 12.10(a), it should be clear that the graph contains an $L \times 2$ lattice of nodes for each branch of the phylogeny.

The Markov-dependent model of Section 12.3 is a graphical model insofar as it is based on a Markov chain of random variables, but it is quite different from the simple-lattice model. The Markov-dependent model actually operates at two levels, as illustrated in Figure 12.10(b). At one level (top of figure), a simple Markov chain of alignment columns is assumed, with each column being treated as an observed random variable. At another level (boxes at bottom of figure), the conditional probability of each column given the previous column is computed according to a phylogenetic model for pairs of columns. (Each of these phylogenetic models is a submodel of the model shown in Figure

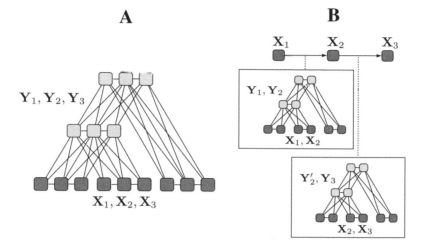

Fig. 12.10. (a) Undirected graph for the "simple-lattice" model of Figure 12.9(a), for an alignment of $L = 3$ sites and $n = 3$ species. Each node in the phylogeny is represented by a sequence of three nodes, corresponding to sites 1, 2, and 3, and each of these nodes is connected not only to its parent but to its parent's neighbors to the left and right. The shaded nodes together represent the three columns of the alignment, $\mathbf{X}_1$, $\mathbf{X}_2$, and $\mathbf{X}_3$, and the unshaded nodes represent the corresponding sets of ancestral bases, $\mathbf{Y}_1$, $\mathbf{Y}_2$, and $\mathbf{Y}_3$. (b) An interpretation of the Markov chain model of Section 12.3 applied to the same alignment. (The case of $N = 2$ is illustrated.) At one level (top), a Markov chain of alignment columns is assumed. At another level (bottom, inside boxes), the conditional probability of each column given the previous column is computed according to a phylogenetic model for pairs of sites.

12.10(a).) When conditional probabilities are computed according to these separate phylogenetic models, multiple versions of the random variables for ancestral bases are effectively introduced (e.g., $\mathbf{Y}_2$ and $\mathbf{Y}_2'$ in Figure 12.10(b)). Moreover, these different versions are not required to be consistent. The effect of this modeling choice is to ignore (indirect) dependencies between latent variables that do not belong to the same "slice" of N columns but at the same time permit exact likelihood computations and to capture what are probably the most important context effects.

By failing to tie together the ancestral nodes of these multiple phylogenetic models, the Markov-dependent model sacrifices any claim of accurately representing the process of context-dependent substitution. Nevertheless, it allows the major *consequences* of this process to be characterized empirically in such a way that valid likelihoods can be extracted, as well as reasonable approximations of the conditional expectations of key quantities.

References

[1] M. Alexandersson, S. Cawley, and L. Pachter. Cross-species gene finding and alignment with a generalized pair hidden Markov model. *Genome Res.*, 13:496–502, 2003.

[2] P. F. Arndt, C. B. Burge, and T. Hwa. DNA sequence evolution with neighbor-dependent mutation. In *Proceedings of the 6th International Conference on Research in Computational Molecular Biology (RE-COMB'02)*, pages 32–38. ACM Press, New York, 2002.

[3] D. Boffelli, J. McAuliffe, D. Ovcharenko, K. D. Lewis, I. Ovcharenko, L. Pachter, and E. M. Rubin. Phylogenetic shadowing of primate sequences to find functional regions of the human genome. *Science*, 299:1391–1394, 2003.

[4] F. Chiaromonte, R. J. Weber, K. M. Roskin, M. Diekhans, W. J. Kent, and D. Haussler. The share of human genomic DNA under selection estimated from human-mouse genomic alignments. *Cold Spring Harbor Symp. Quant. Biol.*, 68:245–254, 2003.

[5] Mouse Genome Sequencing Consortium. Initial sequencing and comparative analysis of the mouse genome. *Nature*, 420:520–562, 2002.

[6] Rat Genome Sequencing Project Consortium. Genome sequence of the Brown Norway Rat yields insights into mammalian evolution. *Nature*, 428:493–521, 2004.

[7] R. Durbin, S. Eddy, A. Krogh, and G. Mitchison. *Biological Sequence Analysis: Probabilistic Models of Proteins and Nucleic Acids*. Cambridge University Press, Cambridge, 1998.

[8] J. Felsenstein. Evolutionary trees from DNA sequences. *J. Mol. Evol.*, 17:368–376, 1981.

[9] J. Felsenstein and G. A. Churchill. A hidden Markov model approach to variation among sites in rate of evolution. *Mol. Biol. Evol.*, 13:93–104, 1996.

[10] N. Friedman, M. Ninio, I. Pe'er, and T. Pupko. A structural EM algorithm for phylogenetic inference. *J. Comp. Biol.*, 9:331–353, 2002.

[11] N. Goldman, J. L. Thorne, and D. T. Jones. Using evolutionary trees in protein secondary structure prediction and other comparative sequence analyses. *J. Mol. Biol.*, 263:196–208, 1996.

[12] N. Goldman and Z. Yang. A codon-based model of nucleotide substitution for protein-coding DNA sequences. *Mol. Biol. Evol.*, 11:725–735,, 1994.

[13] M. Hasegawa, H. Kishino, and T. Yano. Dating the human-ape splitting by a molecular clock of mitochondrial DNA. *J. Mol. Evol.*, 22:160–174, 1985.

[14] D. Heckerman. A tutorial on learning with Bayesian networks. In M. I. Jordan, editor, *Learning in Graphical Models*. MIT Press, Cambridge, MA, 1999.

[15] J. Hein, J. L. Jensen, and C. N. S. Pedersen. Recursions for statistical multiple alignment. *Proc. Natl. Acad. Sci. USA*, 100:14960–14965, 2003.

[16] S. T. Hess, J. D. Blake, and R. D. Blake. Wide variations in neighbor-dependent substitution rates. *J. Mol. Biol.*, 236:1022–1033, 1994.

[17] I. Holmes. Using guide trees to construct multiple-sequence evolutionary HMMs. *Bioinformatics*, 19(Suppl. 1):i147–i157, 2003.

[18] I. Holmes and W. J. Bruno. Evolutionary HMMs: A Bayesian approach to multiple alignment. *Bioinformatics*, 17:803–820, 2001.

[19] D. Husmeier and G. McGuire. Detecting recombination in 4-taxa DNA sequence alignments with Bayesian hidden Markov models and Markov chain Monte Carlo. *Mol. Biol. Evol.*, 20:315–337, 2003.

[20] D. Husmeier and F. Wright. Detection of recombination in DNA multiple alignments with hidden Markov models. *J. Comp. Biol.*, 8:401–427, 2001.

[21] J. L. Jensen and A.-M. K. Pedersen. Probabilistic models of DNA sequence evolution with context dependent rates of substitution. *Adv. Appl. Prob.*, 32:499–517, 2000.

[22] V. Jojic, N. Jojic, C. Meek, D. Geiger, A. Siepel, D. Haussler, and D. Heckerman. Efficient approximations for learning phylogenetic HMM models from data. In *Proceedings of the 12th International Conference on Intelligent Systems for Molecular Biology*. UAI Press, Banff, Canada, 2004.

[23] M. I. Jordan and Y. Weiss. Graphical models: probabilistic inference. In M. Arbib, editor, *The Handbook of Brain Theory and Neural Networks*. MIT Press, Camebridge, MA, second edition, 2002.

[24] M. Kellis, N. Patterson, M. Endrizzi, B. Birren, and E. S. Lander. Sequencing and comparison of yeast species to identify genes and regulatory elements. *Nature*, 423:241–254, 2003.

[25] W. J. Kent, C. W. Sugnet, T. S. Furey, K. M. Roskin, T. H. Pringle, A. M. Zahler, and D. Haussler. The human genome browser at UCSC. *Genome Res.*, 12:996–1006, 2002.

[26] B. Knudsen and J. Hein. RNA secondary structure prediction using stochastic context-free grammars and evolutionary history. *Bioinformatics*, 15:446–454, 1999.

[27] J. M. Koshi and R. M. Goldstein. Probabilistic reconstruction of ancestral protein sequences. *J. Mol. Evol.*, 42:313–320, 1996.

[28] P. Liò and N. Goldman. Models of molecular evolution and phylogeny. *Genome Res.*, 8:1233–1244, 1998.

[29] P. Liò, N. Goldman, J. L. Thorne, and D. T. Jones. PASSML: Combining evolutionary inference and protein secondary structure prediction. *Bioinformatics*, 14:726–733, 1998.

[30] B. Lucena. *Dynamic programming, tree-width, and computation on graphical models*. PhD thesis, Brown University, 2002.

[31] W. P. Maddison and D. R. Maddison. Introduction to inference for Bayesian networks. In M. I. Jordan, editor, *Learning in Graphical Models*. MIT Press, Cambridge, MA, 1999.

[32] E. H. Margulies, M. Blanchette, NISC Comparative Sequencing Program, D. Haussler, and E. D. Green. Identification and characterization of multi-species conserved sequences. *Genome Res.*, 13:2507–2518, 2003.

[33] J. D. McAuliffe, L. Pachter, and M. I. Jordan. Multiple-sequence functional annotation and the generalized hidden Markov phylogeny. *Bioinformatics*, 20:1850–1860, 2004.

[34] G. McGuire, F. Wright, and M. J. Prentice. A Bayesian model for detecting past recombination events in DNA multiple alignments. *J. Comp. Biol.*, 7:159–170, 2000.

[35] I. M. Meyer and R. Durbin. Comparative ab initio prediction of gene structures using pair HMMs. *Bioinformatics*, 18:1309–1318, 2002.

[36] G. J. Mitchison. A probabilistic treatment of phylogeny and sequence alignment. *J. Mol. Evol.*, 49:11–22, 1999.

[37] K. Murphy, Y. Weiss, and M. I. Jordan. Loopy belief-propagation for approximate inference: An empirical study. In K. B. Laskey and H. Prade, editors, *Proceedings of the Fifteenth Conference on Uncertainty in Artificial Intelligence (UAI)*, pages 467–476. Morgan Kaufmann, San Mateo, CA, 1999.

[38] J. Pearl. *Probabilistic Reasoning in Intelligent Systems: Networks of Plausible Inference*. Morgan Kaufmann, San Mateo, CA, 1988.

[39] A.-M. K. Pedersen and J. L. Jensen. A dependent rates model and MCMC based methodology for the maximum likelihood analysis of sequences with overlapping reading frames. *Mol. Biol. Evol.*, 18:763–776, 2001.

[40] A.-M. K. Pedersen, C. Wiuf, and F. B. Christiansen. A codon-based model designed to describe lentiviral evolution. *Mol. Biol. Evol.*, 15:1069–1081, 1998.

[41] J. S. Pedersen and J. Hein. Gene finding with a hidden Markov model of genome structure and evolution. *Bioinformatics*, 19:219–227, 2003.

[42] A. Siepel and D. Haussler. Combining phylogenetic and hidden Markov models in biosequence analysis. *J. Comp. Biol.*, 11(2-3):413–428, 2004.

[43] A. Siepel and D. Haussler. Computational identification of evolutionarily conserved exons. In *Proceedings of the 8th International Conference on Research in Computational Molecular Biology (RECOMB'04)*, pages 177–186. ACM Press, New York, 2004.

[44] A. Siepel and D. Haussler. Phylogenetic estimation of context-dependent substitution rates by maximum likelihood. *Mol. Biol. Evol.*, 21:468–488, 2004.

[45] N. Stojanovic, L. Florea, C. Riemer, D. Gumucio, J. Slightom, M. Goodman, W. Miller, and R. Hardison. Comparison of five methods for finding conserved sequences in multiple alignments of gene regulatory regions. *Nucleic Acids Res.*, 27:3899–3910, 1999.

[46] J. W. Thomas, J. W. Touchman, and R. W. Blakesley et al. Comparative analyses of multi-species sequences from targeted genomic regions. *Nature*, 424:788–793, 2003.

[47] J. L. Thorne, N. Goldman, and D. T. Jones. Combining protein evolution and secondary structure. *Mol. Biol. Evol.*, 13:666–673, 1996.

[48] M. Wainwright, T. Jaakkola, and A. Willsky. Tree-based reparameterization framework for analysis of sum-product and related algorithms. *IEEE Trans. Inf. Theory*, 49:1120–1146, 2001.

[49] M. J. Wainwright and M. I. Jordan. Graphical models, exponential families, and variational inference. Technical Report 649, Department of Statistics, University of California, Berkeley, 2003.

[50] S. Whelan, P. Liò, and N. Goldman. Molecular phylogenetics: State-of-the-art methods for looking into the past. *Trends Genet.*, 17:262–272, 2001.

[51] Z. Yang. Estimating the pattern of nucleotide substitution. *J. Mol. Evol.*, 39:105–111, 1994.

[52] Z. Yang. A space-time process model for the evolution of DNA sequences. *Genetics*, 139:993–1005, 1995.

[53] J. Yedidia, W. Freeman, and Y. Weiss. Bethe free energy, Kikuchi approximations, and belief propagation algorithms. Technical Report TR2001-16, Mitsubishi Electronic Research Laboratories, Camebridge, MA, 2001.

Part IV

Inferences on Molecular Evolution

The Evolutionary Causes and Consequences of Base Composition Variation

Gilean A. T. McVean

Department of Statistics, 1 South Parks Road, Oxford OX1 3TG, UK,
mcvean@stats.ox.ac.uk

Every nucleotide position in a genome experiences a unique set of mutational and selective forces. Local sequence context, location in coding or noncoding regions, structure of the DNA, and timing of replication are just some of the diverse factors important in determining the evolutionary forces likely to act on a given base pair. One consequence of such complexity is that base composition (the relative usage of different DNA nucleotides and motifs) varies considerably both within and among genomes. Drawing accurate evolutionary inferences in the face of base composition variation is hugely challenging. First, we must decide which factors cannot be ignored when trying to model or interpret patterns of molecular evolution. Second, we must develop methods and models that address such factors explicitly. This chapter aims to discuss some of the key processes, both mutational and selective, influencing base composition variation and various models that have been proposed to describe their effects on molecular evolution. I will also explore how to measure base composition variation and discuss some of the pitfalls that can arise if such effects are ignored.

The chapter is split into three sections. The first presents a brief overview of empirical patterns of base composition variation, both within and among genomes. In the second section, I consider the impact of ignoring base composition variation on estimating evolutionary divergence between DNA sequences and discuss some of the models of sequence evolution proposed to correct for such complications. Finally, I look at models of sequence evolution that attempt to model explicitly the different evolutionary factors that influence base composition evolution.

13.1 Empirical Patterns of Base Composition Variation

Summarizing empirical patterns of base composition variation presents a challenge because different biological factors influencing base composition (e.g., selective constraints, mutational influences, replication timing, DNA strand, etc.) will act at different scales and in different contexts. Here I consider three ways of summarizing base composition variation: in terms of raw nucleotide frequencies; in terms of context-dependent variation (or variation in the frequency of sequence motifs); and in terms of variation in synonymous codon usage. For each, I also discuss some of the biological factors that are likely to be involved.

13.1.1 Biased Nucleotide Composition

The simplest way of measuring base composition variation within or among genomes is through the relative usage of the bases GC and AT (note that these are base-pairing, and hence any skew to G or A will apply equally to C or T, though biases may still exist with respect to coding and noncoding strands). Table 13.1 illustrates the huge variation among genomes, with bacteria in the range 25–75% GC, unicellular eukaryotes showing similar variation, and vertebrates relatively little variation. At the genome level, base composition can differ considerably even between related species. For example, *Plasmodium falciparum*, the primary cause of malaria in sub-Saharan Africa, has a largely uniform GC content of 23%, while *P. vivax*, the primary agent of malaria in Asia, has a highly variable GC content (among genomic regions), ranging from 15% to 50% [10].

Table 13.1. Genome-wide base composition variation.

Group	Species	GC content (%)
Bacteria	*Escherichia coli*	51
	Clostridium tetani	29
	Streptomyces coelicolor	72
Eukaryotes	*Saccharomyces cerevisiae*	35
	Plasmodium falciparum	23
	Drosophila melanogaster	42
	Caenorhabditis elegans	35
	Arabidopsis thaliana	35
Vertebrates	*Homo sapiens*	41
	Mus musculus	40
	Fugu rubripes	44
Organelles	*H. sapiens* mitochondrion	44
	A. thaliana chloroplast	49

Nucleotide composition can also vary considerably within genomes. In some prokaryotes, the complimentary DNA strands have differing base compositions, with a shift in the GC content often marking the origin and terminus of replication [43]. Eukaryotes present a diverse, and little-understood, spectrum of base composition variation. In *Drosophila melanogaster*, noncoding regions vary in GC content, but the most marked base composition variation is seen between coding regions in terms of the GC content at silent (or synonymous) sites [54]. Even within genes there is variation in GC content, with a general trend for higher GC at the 5' end of genes [38]. In birds and mammals, base composition variation has long been described as a mosaic structure of long (many kilobases) regions (known as isochores) with differing GC content [5, 4]. However, genome-sequencing projects have revealed a much more complex pattern of continuous variation in base composition acting simultaneously at different scales [50, 42]. Figure 13.1 shows GC-content variation in a 1 Mb stretch of human chromosome 20, demonstrating both small-scale and large-scale fluctuations.

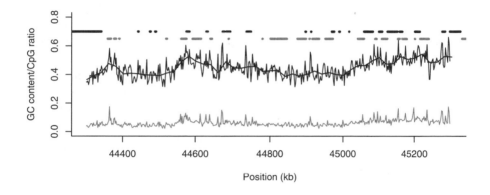

Fig. 13.1. Base composition variation in a 1 Mb stretch of human chromosome 20. Significant variation in GC content is seen both in 2 kb and 20 kb windows (central, superimposed lines). Also shown is the odds ratio for CpG dinucleotides (*lower line*) and the location of genes on the forward (black) and reverse (grey) strands (data from ENSEMBL Build 33). Stretches where CpGs are less under-represented at the 5' end of genes are known as CpG islands [8].

Base composition bias can arise from three processes: biased mutation, biased DNA repair, and natural selection. Biases in mutation arise from differences in mutational environment (e.g., methylation or replication timing) or the relative abundance of nucleotides that may be misincorporated. For example, in prokaryotes, compositional skews in the compliementary strands are understood to be due to the differing mutational environments of the leading and lagging strands during replication [43].

Biased DNA repair mechanisms have a little-understood but probably central role in generating certain biases in base composition. For example, GC/AT base-pairing mismatches (as can occur through mutation or when heterozygotes undergo gene conversion) may preferentially be "repaired" to the GC base in vertebrates [16], a factor that may be important in shaping GC-content variation [21].

The role of natural selection in shaping nucleotide frequencies is much debated. The tendency for thermophilic prokaryotes to have a high GC content in ribosomal and transfer RNA stems may be explained by the increased thermal stability of the G:C pair (with three Hydrogen bonds) compared with the A:T pair (two Hydrogen bonds), though the pattern is not generally observed across the whole genome [20]. Likewise, the tendency for warm-blooded vertebrates to have a higher GC content than cold-blooded ones has also been suggested to reflect selection for increased thermal stability [5]. However, there are both theoretical and empirical grounds for doubting the efficacy of selection to generate the effect in vertebrates [16].

Irrespective of cause, biased nucleotide composition presents many challenges to studies of molecular evolution. When viewed over considerable evolutionary time, base composition cannot be viewed as a stationary process (i.e., genomes are not at base composition equilibrium). However, because any detectable change in genome-wide base composition requires many thousands of nucleotide substitutions, there is considerable phylogenetic inertia in the trait. On the one hand, such inertia means that there is information about evolutionary relationships in base composition [49]. The converse of this argument is that organisms with similar GC content may appear artificially closely related if stationarity is assumed [56]. Statistical methods aimed at estimating evolutionary distances (and relationships) in the face of nonstationarity have been developed [18, 26]. These methods have revealed complex patterns of base composition evolution in the early history of prokaryotes [22].

13.1.2 Context-Dependent Base Composition Bias

Like nucleotide composition, the frequency of nucleotide motifs varies considerably among (and sometimes within) genomes. For example, the under-representation of CpG dinucleotides in species with methylation is largely a result of the C→T hypermutability [12] of methylated C residues in such a context [8]. Figure 13.1 depicts variation in the under-representation of CpG motifs in a region of human chromosome 20.

The bias of nucleotide motifs is best measured relative to the expectation of the constituent submotifs; for example, a CG deficit is measured as the ratio of the frequency of the CG dinucleotide to the product of the frequencies of the C and G nucleotides. More generally, for any motif X, which can be viewed as the union of two submotifs X_1 and X_2, a measure of the relative bias to the abundance of the motif is the odds ratio

$$\rho(X) = \frac{f(X)f(X_1 \cap X_2)}{f(X_1)f(X_2)}, \tag{13.1}$$

where $f(Y)$ is the estimated frequency of the motif Y. By way of example, consider the motif CTAG in the bacterium *Escherichia coli*. In the K12 genome, the frequencies of the trinucleotides CTA and TAG are 0.0058 and 0.0059, respectively, while the frequency of the TA dinucleotide is 0.0457 and the frequency of CTAG is 0.00019. Therefore, the odds-ratio measure of bias is 0.25, indicating a strong under-representation of the motif in the genome, in this case probably due to avoidance of short palindromes and bias in the very short patch (*vsr*) DNA mismatch repair system [33].

Large-scale comparative analyses of the relative motif biases of different genomes have revealed both strong motif over- and under-representations; for example, the dinucleotide TA is under-represented (odds ratio of 0.5 to 0.8) in most prokaryotic sequences. However, other biases show much less evolutionary conservation; for example, CG is under-represented in *Mycoplasma genitalium* but not in *M. pneumoniae* [33].

A consequence of the variation among genomes in the nature and magnitude of composition biases is that, as for nucleotide frequencies, there is phylogenetic information in the extent to which genomes share similar biases. Karlin and colleagues [34] have proposed a simple measure of distance between genomes f and g in terms of dinucleotide odds ratios, referred to as a *genomic signature*,

$$\delta^*(f,g) = \frac{1}{16} \sum_{XY} |\rho_f^*(XY) - \rho_g^*(XY)|, \qquad (13.2)$$

where the $\rho^*(XY)$s are calculated from (13.1) from a concatenation of the forward and reverse complement DNA strands. (Note that other distance metrics, such as Euclidean distance [13], could also be used.) Importantly, such measures tend to show much greater variation among genomes rather than within genomes. Consequently, genomic regions showing anomalous signatures may point to recent horizontal gene transfer events [32], although the extent to which base composition is a reliable indicator of such events has recently been questioned [41].

Motifs longer than a few nucleotides can also show marked variation in frequency among genomes, though clearly the expected number of motifs also diminishes with increased length, leading to greater sampling variance in estimates. Particularly abundant motifs tend to have specific functions; for example, the sequence TTCAGACGGC and its reverse complement are abundant in *Neisseria gonorrhoeae* and are related to the uptake of DNA from the environment [33].

13.1.3 Codon Bias

So far, we have considered measures of base composition bias that take no account of the underlying genome structure. However, the coding nature of genes imposes constraints on base composition through the structure of the genetic code. The physicochemical properties of an amino acid are most strongly correlated with the central nucleotide of the codon, while the third position is often free to vary due to the degeneracy of the genetic code. Hence proteins with identical functions can potentially be encoded for by genes of considerably different nucleotide base composition. The nonrandom usage of different codons (within genes or genomes) is referred to as codon bias.

The nonrandom use of synonymous codons is most easily measured in terms of deviation from equality. For example, a widely used measure of codon bias is the effective number of codons (ENC) [57],

$$\text{ENC} = \sum_a \frac{1}{\sum_c f_{ac}^2},\tag{13.3}$$

where f_{ac} is the frequency of codon c for amino acid a. ENC has a range from 20 (highly biased such that each amino acid is encoded for by a single codon) to 61 (note that the three stop codons are ignored). Another widely used measure is an odds-ratio formulation known as the relative synonymous codon usage (RSCU) [53],

$$\text{RSCU}_{ac} = \frac{f_{ac}}{1/n_a},\tag{13.4}$$

where n_a is the number of codons for amino acid a. Table 13.2 illustrates the bias towards codon usage in the *E. coli* genome. Using these genome-wide codon frequencies, the effective number of codons is 52; in short, while certain amino acids, such as lysine (K), show strong codon bias, the genome-wide picture is for relatively little bias.

Whole-genome analysis obscures any variation in codon bias that might occur between genes within a genome. Within *E. coli* there is significant variation in codon bias among genes, a phenomenon first discovered in the early 1980s [30, 25].

What evolutionary forces might be responsible for the variation in codon bias among genes in the *E. coli* genome? An important clue is the finding that the more commonly used codons typically correspond to the more abundant transfer RNA species [30]. Bacteria (and actually all organisms) typically do not have tRNAs corresponding to every codon. Furthermore, different tRNAs are present in different copy numbers in the genome and are expressed at different levels. A complicating factor is that some tRNAs can recognize more than one codon through modification of the first anticodon nucleotide (corresponding to the third position of the codon); however, such tRNAs typically show differential affinity to the recognized codons.

Variation in the cellular concentration of different tRNA species has potential consequences for the translation process. As mRNA moves through a ribosome, the waiting time for the incorporation of the correct amino acid is, under a variety of models [9], inversely proportional to the concentration of the corresponding tRNA. Genes with a high proportion of codons corresponding to abundant tRNAs will therefore be translated faster. Consequently, if

Table 13.2. Genome-wide synonymous codon usage in *Escherichia coli*. Data from the Codon Usage Database: www.kazusa.or.jp/codon. Standard single-letter abbreviations of amino acids are used. The asterisk indicates the stop codon.

Codon	AA	Freq.	Codon	AA	Freq.	Codon	AA	Freq.	Codon	AA	Freq.
UUU	F	0.57	UCU	S	0.15	UAU	Y	0.57	UGU	C	0.45
UUC	F	0.43	UCC	S	0.15	UAC	Y	0.43	UGC	C	0.55
UUA	L	0.13	UCA	S	0.12	UAA	*	0.64	UGA	*	0.29
UUG	L	0.13	UCG	S	0.15	UAG	*	0.07	UGG	W	1.00
CUU	L	0.10	CCU	P	0.16	CAU	H	0.57	CGU	R	0.38
CUC	L	0.10	CCC	P	0.13	CAC	H	0.43	CGC	R	0.40
CUA	L	0.04	CCA	P	0.19	CAA	Q	0.35	CGA	R	0.06
CUG	L	0.50	CCG	P	0.52	CAG	Q	0.65	CGG	R	0.10
AUU	I	0.51	ACU	T	0.17	AAU	N	0.45	AGU	S	0.15
AUC	I	0.42	ACC	T	0.43	AAC	N	0.55	AGC	S	0.28
AUA	I	0.07	ACA	T	0.13	AAA	K	0.77	AGA	R	0.04
AUG	M	1.00	ACG	T	0.27	AAG	K	0.23	AGG	R	0.02
GUU	V	0.26	GCU	A	0.16	GAU	D	0.63	GGU	G	0.34
GUC	V	0.22	GCC	A	0.27	GAC	D	0.37	GGC	G	0.40
GUA	V	0.15	GCA	A	0.21	GAA	E	0.69	GGA	G	0.11
GUG	V	0.37	GCG	A	0.35	GAG	E	0.31	GGG	G	0.15

translation is a rate-limiting step for an organism (i.e., individuals that have a higher rate of protein production have higher fitness), genes whose products are required in high abundance will be under selection to use codons corresponding to more abundant tRNAs. Similar arguments can be made if selection is mediated by translational accuracy rather than translational efficiency [9].

In many unicellular organisms, a strong correlation between the degree of codon bias and level of gene expression has been found [30, 24, 3]. In multicellular eukaryotes, such patterns are harder to detect because tRNA concentrations may vary across developmental stages and/or tissues, and genes that are normally expressed at low levels may be needed at very high levels for short periods in certain tissues [48]. However, by using counts of ESTs (expressed sequence tags) as a proxy for gene expression level, Duret and colleagues have shown strong correlations between codon usage bias and expression level in several multicellular eukaryotes, including *Caenorhabditis elegans* and *Arabidopsis thaliana* [15].

In addition to correlations between gene expression level and codon bias, indirect evidence for the action of natural selection on codon bias can be sought from patterns of molecular evolution and genetic variation. A striking observation made by Sharp and Li [53], in a comparison of *E. coli* and *Salmonella typhii*, is that genes with strong codon bias tend to have relatively low levels of divergence. This can be understood in terms of constraint: highly expressed genes have strong selection on codon bias, and hence most codons will be in the "optimal" state and mutations will tend to be towards nonoptimal codons and hence deleterious. In contrast, synonymous mutations in genes with low codon bias will tend to be neutral, leading to higher rates of molecular evolution. Similar observations have been made for *Drosophila* [54, 14].

Patterns of polymorphism can also reveal the signature of selection. Deleterious mutations tend to be removed by selection from a population, whereas favorable mutations may reach high frequency and fixation. By comparing the frequency distribution (and also patterns of substitution) of mutations to and from favorable codons, Akashi [1, 2] has demonstrated the action of selection on codon usage in several *Drosophila* species.

13.2 Biased Base Composition and Models of Sequence Evolution

Models of the substitution process employed in phylogenetic tree estimation and other applications of molecular evolution typically make simplifying assumptions to aid computational efficiency. Some of these simplifications are implicitly making (often unrealistic) assumptions about the effects of base composition bias on rates of substitution such as stationarity, independence of substitution processes between sites (or codons), absence of selection, and time-reversibility. In practice, some of these assumptions, though incorrect, do not lead to misleading inferences except in exceptional circumstances. However, other assumptions may have a critical influence on the biological conclusions drawn from an analysis of molecular evolution.

In this section, I consider problems arising from biased base composition in three areas of molecular evolution: estimating evolutionary distances; reporting evolutionary distances; and interpreting evolutionary distances. First, however, I give a brief review of how evolutionary distances are typically estimated.

13.2.1 The General Markov Model of Sequence Evolution

Most molecular evolution analyses make use of explicit models of sequence evolution. At the heart of these models is a continuous-time transition matrix describing how nucleotides (or codons) are likely to change over evolutionary time. For nucleotide substitution processes, the fully parameterized model has 12 parameters, which can be written as the following matrix (bases are in the order TCAG; dashes indicate minus the sum of the parameters in the same row of the matrix):

$$\mathbf{Q} = \begin{pmatrix} - & a\pi_C & b\pi_A & c\pi_G \\ a'\pi_T & - & d\pi_A & e\pi_G \\ b'\pi_T & d'\pi_C & - & f\pi_G \\ c'\pi_T & e'\pi_C & f'\pi_A & - \end{pmatrix}. \tag{13.5}$$

For the transition matrix to be time-reversible (a useful property in the estimation of phylogenetic trees), the requirement is that $a = a'$, $b = b'$, $c = c'$, $d = d'$, $e = e'$, and $f = f'$ [55]. Under these conditions, the equilibrium frequencies of the bases are π_T, π_C, π_A, and π_G, respectively. Matrix exponentiation can be used to derive the conditional probabilities of a nucleotide in state i at $t = 0$ being in state j at time t,

$$\mathbf{P}(t) = e^{\mathbf{Q}t}. \tag{13.6}$$

Note that time cannot be estimated independently of the substitution rate parameters in the transition matrix, and hence $\mathbf{Q}$ is usually normalized such that the sum over bases of the equilibrium frequency of each base times the total rate of mutation for that base is equal to one. (Time is therefore measured in expected substitutions per site.) The probability of observing the states S_1 and S_2 at a homologous nucleotide position in species 1 and 2 separated by time t can be extracted from the matrix given by

$$\mathbf{H}(t) = \mathbf{P}_1(t_1)\mathbf{v}(t)\mathbf{P}_2^T(t_2), \tag{13.7}$$

where $\mathbf{v}$ is a vector of the base frequencies in the most recent common ancestor. (Note that the species might have different mutation rates or biases; hence the use of the subscripts.) Evolutionary distances are typically estimated from empirical data by maximum likelihood or Bayesian approaches.

Many of the most widely used models of sequence evolution are special cases of this general model. For example, the Jukes-Cantor model [31] assumes equal rates of substitution between all bases, the Kimura two-parameter model [37] assumes equal base frequencies but allows for a nonunity transition-transversion ratio, and the HKY model [28] allows for unequal base frequencies. For the simplest evolutionary models, explicit analytical expressions exist for the maximum likelihood estimate of divergence time; for more complex models, numerical methods are used.

13.2.2 Estimating Evolutionary Divergence

The effect of base composition bias on estimating evolutionary divergence (the product of the time separating sequences and the mutation rate) is quite simple: if the model is wrong, the estimate of divergence is also wrong. For example, in comparisons of GC-rich genomes, a model that ignores biased nucleotide frequencies will underestimate the true degree of divergence (because in such genomes multiple substitutions are more likely to return the base to the original state). Such considerations might explain the strongly negative relationship observed between codon usage bias and rate of substitution in studies of bacteria [53] and *Drosophila* [54]. These analyses made use of methods for estimating evolutionary divergence that do not account for biased codon usage. Consequently, in genes with stronger bias, the correction for multiple substitutions becomes more problematic and divergence is typically underestimated.

There are many ways in which models of sequence evolution can be wrong. Nonstationarity (base composition changing over time), nonindependence (substitution patterns influenced by neighboring base composition), and natural selection are perhaps the most important factors missing from most models of sequence evolution. The extent to which each factor may be important will depend on the biological question being asked and the degree of divergence between the genomes being compared. (Many of the problems are only really important when correcting for multiple substitutions.) Methods for estimating divergence under models that incorporate nonstationarity [18, 19, 26], nonindependence [27, 40], and selection [47] have been developed, but computational demands mean that such factors are not yet regularly considered in molecular evolution.

A related and important question is whether simple models of sequence evolution that have been developed for genomes with biased nucleotide frequencies, such as [28], have any biological credibility. These models, and similar ones used in codon-based methods [23], use a substitution matrix of the form

$$\mathbf{Q} = \begin{pmatrix} - & \kappa\pi_C & \pi_A & \pi_G \\ \kappa\pi_T & - & \pi_A & \pi_G \\ \pi_T & \pi_C & - & \kappa\pi_G \\ \pi_T & \pi_C & \kappa\pi_A & - \end{pmatrix}. \tag{13.8}$$

In effect, the model assumes infrequent substitutions to nucleotides (or codons) that are rare. However, a nucleotide (or codon) may be rare either because substitutions to the state are rare or because it mutates (or is substituted) relatively faster. Which model is more correct? In *Drosophila*, G- and C-ending codons are more frequent [54], despite a general AT bias in mutation [52, 47], due to the action of natural selection (in line with the model above). Substitutions to the more common codons are therefore expected to occur at a higher rate than substitutions to the rare codons. In contrast, CpG residues are rare not because they are not generated by substitution but because they mutate rapidly. In short, whether the model is appropriate or not depends on the biology of the system; ideally, different models should be formally compared to find the most appropriate one for the system in question.

13.2.3 Reporting Evolutionary Distances

Estimates of evolutionary divergence might be used for several different purposes: to estimate phylogenetic relationships; to estimate the absolute time separating two species; or to ask whether one gene, genomic region, or species has a higher rate of evolution than another. If we could be certain that the model of sequence evolution employed was accurate, it would be natural to think that the reporting of evolutionary distances between DNA sequences should be straightforward. However, biased base composition raises several issues in the reporting of evolutionary distance.

The central issue is that because time and substitution rate cannot be estimated separately, the transition matrix is typically scaled such that at equilibrium (when base composition is unchanging) one unit of time corresponds to an expectation of one substitution (averaged over nucleotides); see Section 13.2.1. Clearly, if a genome is not at equilibrium, an estimate of one unit of time may not actually correspond to an average of one substitution per site. For example, in a genomic region that has experienced a dramatic shift in forces influencing base composition, an estimate of one unit of equilibrium time is likely to correspond to more than one substitution on average (because systems out of equilibrium tend to evolve faster). It is therefore important to test for nonstationarity in molecular evolution analyses.

A related issue arises in the reporting of estimates of synonymous site divergence *per synonymous site* in codon-based models [23, 7]. In coding regions, different amino acids have different potentials for synonymous mutation due to the degeneracy of the genetic code. This potential is further influenced by the relatively higher rate of transition mutations, compared with transversions, and also by skews in base composition, particularly at the third position. In codon-based models of sequence evolution, one unit of time represents an expectation of one substitution, be it synonymous or nonsynonymous. To report synonymous divergence per synonymous site, the expected number of synonymous substitutions to which the estimated time corresponds (averaged over all codons) is divided by the average mutation potential of nucleotides (again averaged over all codons). However, most synonymous substitutions occur at the third position, which in highly biased genes has the least mutation potential under the assumed model. That estimates of synonymous divergence are influenced by the mutation potential of nonsynonymous positions is problematic for comparisons of divergence among genes of differing base composition [7].

13.2.4 Interpreting Evolutionary Distances

The substitution matrix reflects the joint effects of biased mutation, biased DNA repair, and natural selection. However, by comparing patterns of molecular evolution in parts of the genome with different degrees of base composition bias we might be able to ask which factor is most important. For example, if genes with higher codon bias evolve more slowly, we might attribute this to the action of purifying selection in highly biased genes [53, 54]. However, what would we expect of a model in which variation in mutation bias is the primary cause of base composition variation? It seems perfectly plausible that we might expect similar patterns under a purely neutral model.

The key point is that in order to interpret estimated evolutionary distances, we need an understanding of what different biological explanations for base composition biases would predict. Most widely used models of sequence evolution correct for base composition bias not by modelling the factors generating base composition biases explicitly but in an ad hoc manner. In the final section, I outline a simple model for the joint effects of biased mutation and natural selection on patterns of sequence evolution. (Different types of biased DNA repair can be modelled as either biased mutation or selection.) This model produces some surprising results about the relationship among base composition biases, selection, and rates of molecular evolution.

13.3 Explicit Models of Base Composition Evolution

The demonstration that natural selection can influence synonymous codon usage and other aspects of base composition has important implications for the analysis of patterns of molecular evolution and genetic variation: models that assume neutrality may potentially give misleading inferences. However, the level of selective constraint acting on base composition and codon usage is considerably less than that acting on protein-coding positions [51]. So just how strong are the selection coefficients acting on base composition, and how does selection acting on them influence patterns of molecular evolution? More generally, what happens to our view of molecular evolution if we try to model the factors influencing DNA composition explicitly?

13.3.1 A Two-State Model

The simplest model we might consider is a genome where there are two types of states (e.g., GC/AT or preferred/unpreferred codon), which we call A and B [9, 45]. At some starting point, we have a population of genomes, each composed of As and Bs. Over time there is mutation from one state to another: let A mutate to B with probability u per replication and B mutate to A with probability v. The probability that a single mutation reaches fixation in the population depends on the strength of selection acting on the mutation, the rate of mutation, population demography, and many other factors. However, we can approximate the probability through Kimura's formula [35]

$$u(1/2N) \approx \frac{2N_e/Ns}{1 - e^{-4N_e s}}, \tag{13.9}$$

where N is the diploid population size (assumed to be constant), N_e is the effective population size, and s is the selective differential between the novel mutation and the ancestral state. The key point about this formula is that the fixation probability depends on the product of the selection coefficient and the effective population size, a compound parameter often referred to as $\sigma = 4N_e s$. When $\sigma = 0$, the formula reduces to $1/2N$.

What questions might we ask of such a model? Important properties might include the equilibrium frequency of the two alleles (if I pick a genome at random when the population is at equilibrium, what is the expected frequency of the A allele?), the rate of substitution (at equilibrium), and the time taken to achieve equilibrium. The equilibrium position is defined as the point where the overall frequencies of the two alleles are not changing (though individual substitutions are still occurring). Of course, in any population, the stochastic properties of mutation and genetic drift mean that the population is constantly changing, but if we assume that the genome is very long, such subtle effects can effectively be ignored if we are just interested in the average properties of the genome. In order to make the solution more tractable, we must also assume an infinite-sites model [36] (polymorphic sites cannot mutate again until fixation of one allele) and assume that the timescale of the fixation process, governed by (13.9), is effectively instantaneous. In effect, we are considering the fixation process as a continuous-time Markov process, as do all models of sequence evolution.

If we let the frequency of the A state or allele in the genome at time t be $f_A(t)$ (note $f_A(t) + f_B(t) = 1$), the rate of change is

$$\frac{df_A(t)}{dt} = -f_A(t)u\frac{-\sigma}{1 - e^\sigma} + [1 - f_A(t)]v\frac{\sigma}{1 - e^{-\sigma}}, \tag{13.10}$$

where the A state or allele has a selective advantage of s (which can be negative) over the B allele. At equilibrium, the rate of change in frequency is zero, giving the solution

$$f_A(\infty) = \frac{v}{v + ue^{-\sigma}}. \tag{13.11}$$

When there is no selection, the equilibrium frequency becomes $v/(u+v)$. The rate of substitution at equilibrium is therefore

$$\tilde{k} = f_A(\infty)u\frac{-\sigma}{1 - e^\sigma} + [1 - f_A(\infty)]v\frac{\sigma}{1 - e^{-\sigma}} \tag{13.12}$$

$$= \frac{2uv\sigma}{(u + ve^\sigma)(1 - e^{-\sigma})}. \tag{13.13}$$

Finally, the rate of approach to equilibrium is of the order of the mutation rate; for a given starting frequency of the A allele, $f_A(0)$, the frequency of the mutation at time t is

$$f_A(t) = f_A(0)e^{-\Delta t} + f_A(\infty)(1 - e^{-\Delta t}) \text{ where } \Delta = \sigma\left(\frac{u}{e^\sigma - 1} + \frac{v}{1 - e^{-\sigma}}\right). \tag{13.14}$$

Some examples of the equilibrium base composition and rate of evolution for different values of the mutation bias (u/v) are shown in Figure 13.2.

What can we learn from such a model? In the absence of selection base composition is entirely determined by mutation bias. Because mutation biases are expected to be similar in both coding and noncoding DNA, this suggests that comparisons of codon usage with base composition bias in nearby noncoding regions may provide important clues as to whether selection is

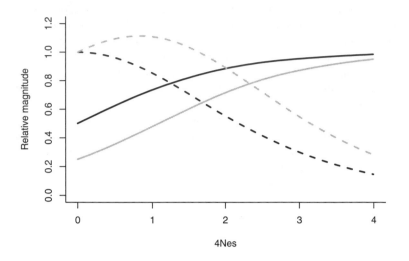

Fig. 13.2. The relationship between the strength of selection acting on codon bias (as measured by the scaled selection differential $4N_es$ between the preferred and unpreferred alleles), the frequency of the preferred allele (solid lines), and the equilibrium substitution rate (dashed lines). When there is no mutation bias ($u/v = 1$; black lines) there is a monotonic relationship between the selection parameter and both properties. However, when mutation bias acts in the direction opposite selection ($u/v = 3$; grey lines), the rate of substitution reaches a maximum at intermediate selection coefficients.

acting on codon usage. For example, there is a strong correlation between nearby noncoding GC content and synonymous codon GC content in mammalian genomes [4] but a very much weaker correlation in the genome of *D. melanogaster* [39]. If selection does act on base composition, only if the scaled selection parameter is within a fairly narrow range, approximately $0.5 < \sigma < 2$, do we expect to see a balance among selection, mutation, and genetic drift [9, 45]. Weaker selection coefficients are indistinguishable from neutrality, and stronger selection coefficients will lead to near-complete fixation of the advantageous allele. That codon bias seems to be at such a balance in a huge variety of organisms with very different census population sizes is therefore something of a paradox. Possible explanations are that effective population sizes may vary much less than census population size or that there may be opposing selective forces acting on codon usage [46].

The rate of molecular evolution is also affected by selection and mutation bias. Contrary to the argument outlined in Section 13.1.3, selection on base composition can actually increase the rate of substitution if selection and mutation bias act in opposite directions [45]. A further important complication is that the rate at which a population approaches equilibrium is very slow relative to the rate of change in population size, and other demographic processes, that species may experience. Patterns of synonymous substitution between closely related species may therefore reflect more the demographic history of the populations than equilibrium expectations. For example, detailed analysis of patterns of synonymous substitution in *Drosophila* sibling species have shown that most substitutions in the *D. melanogaster* lineage have been from optimal to nonoptimal codons, suggesting a complete absence of selection on codon usage over the most recent 3–5 million years [2, 47]. Complete loss of selective constraint after a period of long-term equilibrium leads to an instantaneous increase in the rate of substitution by a factor of

$$\lambda_{\text{loss}} = \frac{e^{\sigma} - e^{-\sigma}}{2\sigma}, \tag{13.15}$$

where σ is the scaled selective differential prior to loss of selective constraint. For $\sigma = 2$, the relative increase in rate is 81%. In contrast, a gain of selective constraint after a period of long-term equilibrium leads to an instantaneous increase in the rate of substitution by a factor of

$$\lambda_{\text{gain}} = \frac{\sigma}{2} \frac{e^{\sigma/2} + e^{-\sigma/2}}{e^{\sigma/2} - e^{-\sigma/2}}. \tag{13.16}$$

For $\sigma = 2$, this leads to a 31% increase in rate; see Figure 13.3. In short, substantial changes in substitution rate can be induced by changes in the selection pressure acting on base composition or the effective population size.

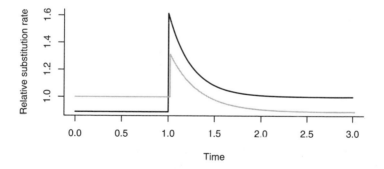

Fig. 13.3. Changes in substitution rate associated with changes in the selective constraint on base composition. Following a period of equilibrium, both the loss of selective constraint ($\sigma_0 - 2, \sigma_\infty = 0$; black line) and the gain of selective constraint ($\sigma_0 = 0, \sigma_\infty = 2$; grey line) at $t = 1.0$ (expressed in arbitrary units) lead to an instantaneous increase in substitution rate followed by a gradual decay to the new equilibrium.

Finally, variation in other factors, for example mutation bias or mutation rate, can also influence rates of synonymous substitution. For example, the variation in synonymous substitution rate seen in bacterial genomes is considerably greater than the simple model might predict given the range of codon bias observed and probably results from a poorly understood relationship between expression level and mutation rate [17].

13.3.2 More Realistic Models

The two-state model is a simple caricature of base composition evolution. More realistic models have to incorporate mutational biases among all four bases, which may also be context-dependent, and the selective differences among multiple codons. A further complication when dealing with codon usage evolution is that nonsynonymous substitutions will cause a change in the selective context: a G-ending codon may be optimal for one amino acid, but for another it may be nonoptimal. Yet a further complication is that selection does not act on each site independently. In particular, selective interference or the Hill-Robertson effect [29] (either through hitch-hiking [44] or background selection [11]) reduces the efficacy of selection at any site, a factor shown to be important in *Drosophila* codon usage evolution [6]. Incorporating simultaneous selection at multiple sites into models of sequence evolution is both computationally unrealistic (at least currently) and probably unnecessary for addressing many biological questions arising in molecular evolution. However, unravelling the causes of biased base composition is also essential for a full understanding of the factors influencing genome evolution.

Acknowledgments

Many thanks to Graham Coop, Chris Spencer, and an anonymous reviewer for their helpful comments on the manuscript.

References

[1] H. Akashi. Inferring weak selection from patterns of polymorphism and divergence at "silent" sites in *Drosophila* DNA. *Genetics*, 139:1067–1076, 1995.

[2] H. Akashi. Molecular evolution between *Drosophila melanogaster* and *D. simulans*: Reduced codon bias, faster rates of amino acid substitution and larger proteins in *D. melanogaster*. *Genetics*, 151:221–238, 1996.

[3] J. L. Bennetzen and B. D. Hall. Codon selection in yeast. *J. Biol. Chem.*, 257:3026–3031, 1982.

[4] G. Bernardi. The isochore structure of the human genome. *Annu. Rev. Genet.*, 23:637–661, 1989.

[5] G. Bernardi, B. Olofsson, J. Filipski, M. Zerial, J. Salinas, G. Cuny, M. Meunier-Rotival, and F. Rodier. The moasic genome of warm-blooded vertebrates. *Science*, 228:953–958, 1985.

[6] A. J. Betancourt and D. C. Presgraves. Linkage limits the power of natural selection in *Drosophila*. *Proc. Natl. Acad. Sci USA*, 99:13616–13620, 2002.

[7] N. Bierne and A. Eyre-Walker. The problem of counting sites in the estimation of the synonymous and nonsynonymous substitution rates. Implications for the correlation between the synonymous substitution rates and codon usage bias. *Genetics*, 165:1587–1597, 2003.

[8] A. P. Bird. DNA methylation and the frequency of CpG in animal DNA. *Nucleic Acids Res.*, 8:1499–1504, 1980.

[9] M. G. Bulmer. The selection-mutation-drift theory of synonymous codon usage. *Genetics*, 129:897–907, 1991.

[10] J. Carlton. The *Plasmodium vivax* genome sequencing project. *Trends Parasitol.*, 19:227–231, 2003.

[11] B. Charlesworth, M. T. Morgan, and D. Charlesworth. The effect of deleterious mutations on neutral molecular variation. *Genetics*, 134:1289–1303, 1993.

[12] C. Coulondre, J. H. Miller, P. J. Farabaugh, and W. Gilbert. Molecular basis of base substitution hotspots in *Escherichia coli*. *Nature*, 278:775–780, 1978.

[13] P. J. Deschavanne, A. Giron, J. Vilain, G. Fagot, and B. Fertil. Genomic signature: Characterization and classification of species assessed by chaos game representation of sequences. *Mol. Biol. Evol.*, 16:1391–1399, 1999.

[14] K. A. Dunn, J. P. Bielawski, and Z. Yang. Substitution rates in *drosophila* nuclear genes: Implications for translational selection. *Genetics*, 157:295–305, 2001.

[15] L. Duret and D. Mouchiroud. Expression pattern and, surprisingly, gene length shape codon usage in *Caenorhabditis*, *Drosophila*, and *Arabidopsis*. *Proc. Natl. Acad. Sci. USA*, 96:4482–4487, 1999.

[16] A. Eyre-Walker. Recombination and mammalian genome evolution. *Proc. R. Soc. London B*, 252:237–243, 1993.

[17] A. Eyre-Walker and M. Bulmer. Synonymous substitution rates in enterobacteria. *Genetics*, 140:1407–1412, 1995.

[18] N. Galtier and M. Gouy. Inferring phylogenies from DNA sequences of unequal base compositions. *Proc. Natl. Acad. Sci. USA*, 92:11317–11321, 1995.

[19] N. Galtier and M. Gouy. Inferring pattern and process: Maximum-likelihood implementation of a nonhomogeneous model of DNA sequence evolution for phylogenetic analysis. *Mol. Biol. Evol.*, 15:871–879, 1998.

[20] N. Galtier and J. R. Lobry. Relationships between genomic G+C content, RNA secondary structures, and optimal growth temperature in prokaryotes. *J. Mol. Evol.*, 44:632–636, 1997.

[21] N. Galtier, G. Piganeau, D. Mouchiroud, and L. Duret. GC-content evolution in mammalian genomes: The biased gene conversion hypothesis. *Genetics*, 159:907–911, 2001.

[22] N. Galtier, N. Tourasse, and M. Gouy. A nonhyperthermophilic common ancestor to extant life forms. *Science*, 283:220–221, 1999.

[23] N. Goldman and Z. Yang. A codon-based model of nucleotide substitution for protein-coding DNA sequences. *Mol. Biol. Evol.*, 11:725–736, 1994.

[24] M. Gouy and C. Gautier. Codon usage in bacteria: Correlation with gene expressivity. *Nucleic Acids Res.*, 10:7055–7074, 1982.

[25] R. Grantham, C. Gautier, M. Gouy, M. Jacobzone, and R. Mercier. Codon catalog usage is a genome strategy modulated for gene expressivity. *Nucleic Acids Res.*, 9:r43–r79, 1981.

[26] X. G. Gu and W. H. Li. Estimation of evolutionary distances under stationary and nonstationary models of nucleotide substitution. *Proc. Natl. Acad. Sci. USA*, 95:5899–5905, 1998.

[27] B. Gulko and D. Haussler. Using multiple alignments and phylogenetic trees to detect RNA secondary structure. In *Pacific Symposium on Biocomputing*, pages 350–367. World Scientific, Singapore, 1996.

[28] M. Hasegawa, H. Kishino, and T. Yano. Dating of the human-ape splitting by a molecular clock of mitochondrial DNA. *J. Mol. Evol.*, 22:160–174, 1985.

[29] W. G. Hill and A. R. Robertson. The effect of linkage on limits to artificial selection. *Genet. Res.*, 8:269–294, 1966.

[30] T. Ikemura. Correlation between the abundance of *Escherichia coli* transfer RNAs and the occurrence of the respective codons in its protein genes: A proposal for a synonymous codon choice that is optimal for the *E. coli* translation system. *J. Mol. Biol.*, 151:389–409, 1981.

[31] T. H. Jukes and C. R. Cantor. *Mammalian Protein Metabolism*, pages 21–123. Academic Press, New York, 1969.

[32] S. Karlin. Detecting anomalous gene clusters and pathogenicity islands in diverse bacterial genomes. *Trends Microbiol.*, 9:335–343, 2001.

[33] S. Karlin, A. M. Campbell, and J. Mrázek. Comparative DNA analyses across diverse genomes. *Annu. Rev. Genet.*, 32:185–225, 1998.

[34] S. Karlin, I. Ladunga, and B. E. Blaisdell. Heterogeneity of genomes: Measures and values. *Proc. Natl. Acad. Sci. USA*, 91:12837–12841, 1994.

[35] M. Kimura. On the probability of fixation of mutant genes in a population. *Genetics*, 47:713–719, 1962.

[36] M. Kimura. Theoretical foundation of population genetics at the molecular level. *Theor. Pop. Biol.*, 2:174–208, 1971.

[37] M. Kimura. A simple method for estimating evolutionary rates of base substitution through comparative studies of nucleotide sequences. *J. Mol. Evol.*, 16:111–120, 1980.

[38] R. M. Kliman and A. Eyre-Walker. Patterns of base composition within the genes of *Drosophila melanogaster*. *Mol. Biol. Evol.*, 46:534–541, 1998.

[39] R. M. Kliman and J. Hey. The effects of mutation and natural selection on codon bias in the genes of *Drosophila*. *Mol. Biol. Evol.*, 137:1049–1056, 1994.

[40] B. Knudsen and J. Hein. RNA secondary structure prediction using stochastic context-free grammars and evolutionary history. *Bioinformatics*, 15:446–454, 1999.

[41] L. B. Koski, R. A. Morton, and G. B. Golding. Codon bias and base composition are poor indicators of horizontally transferred genes. *Mol. Biol. Evol.*, 18:404–412, 2001.

[42] W. H. Li. Are isochore sequences homogeneous? *Gene*, 300:129–139, 2002.

[43] J. R. Lobry and J. M. Louarn. Polarisation of prokaryotic chromosomes. *Curr. Opin. Microbiol.*, 6:101–108, 2003.

[44] J. Maynard Smith and J. Haigh. The hitch-hiking effect of a favourable gene. *Genet. Res.*, 23:23–35, 1974.

[45] G. A. T. McVean and B. Charlesworth. A population genetic model for the evolution of synonymous codon usage: Patterns and predictions. *Genet. Res.*, 74:145–158, 2000.

[46] G. A. T. McVean and B. Charlesworth. The effects of Hill-Robertson interference between weakly selected sites on patterns of molecular evolution and variation. *Genetics*, 155:929–944, 2001.

[47] G. A. T. McVean and J. Vieira. Inferring parameters of mutation, selection and demography from patterns of synonymous site evolution in *drosophila*. *Genetics*, 157:245–257, 2001.

[48] E. N. Moriyama and J. R. Powell. Codon usage bias and tRNA abundance in *Drosophila*. *J. Mol. Evol.*, 45:514–523, 1997.

[49] A. Muto and S. Osawa. The guanine and cytosine content of genomic DNA and bacterial evolution. *Proc. Natl. Acad. Sci. USA*, 84:166–169, 1987.

[50] A. Nekrutenko and W. H. Li. Assessment of compositional heterogeneity within and between eukaryotic genomes. *Genome Res.*, 10:1986–1995, 2000.

[51] T. Ohta. Synonymous and nonsynonymous substitutions in mammalian genes and the nearly neutral theory. *J. Mol. Evol.*, 40:56–63, 1995.

[52] D. Petrov and D. Hartl. Patterns of nucleotide substitution in *Drosophila* and mammalian genomes. *Proc. Natl. Acad. Sci. USA*, 96:1475–1479, 1999.

[53] P. M. Sharp and W.-H. Li. The codon adaptation index—a measure of directional synonymous codon bias, and its potential application. *Nucleic Acids Res.*, 15:1281–1295, 1987.

[54] D. Shields, P. M. Sharp, D. G. Higgins, and F. Wright. "Silent" sites in *Drosophila* are not neutral: Evidence of selection among synonymous codons. *Mol. Biol. Evol.*, 5:704–716, 1988.

[55] S. Tavaré. Some probabilistic and statistical problems in the analysis of DNA sequences. *Lect. Math. Life Sci.*, 17:57–86, 1986.

[56] W. G. Weisburg, S. J. Giovannoni, and C. R. Woese. The Deinococcus-Thermus phylum and the effect of rRNA composition on phylogenetic tree reconstruction. *Syst. Appl. Microbiol.*, 11:128–134, 1989.

[57] F. Wright. The 'effective number of codons' used in a gene. *Gene*, 87:23–29, 1990.

14

Statistical Alignment: Recent Progress, New Applications, and Challenges

Gerton Lunter,[1] Alexei J. Drummond,[1,2] István Miklós,[1,3] and Jotun Hein[1]

[1] Bioinformatics group, Department of Statistics, Oxford University, Oxford OX1 3TG, UK, {lunter,hein}@stats.ox.ac.uk

[2] Current affiliation: Department of Zoology, Oxford University, Oxford OX1 3PS, UK, alexei.drummond@zoology.oxford.ac.uk

[3] Current affiliation: Theoretical Biology and Ecology Group, Hungarian Acadademy of Science and Eötvös Loránd University, Budapest H-1117, Hungary, miklosi@ramet.elte.hu

Summary

Two papers by Thorne, Kishino, and Felsenstein in the early 1990s provided a basis for performing alignment within a statistical framework. Here we review progress and associated challenges in the investigation of models of insertions and deletions in biological sequences stemming from this early work. In the last few years, this approach to sequence analysis has experienced a renaissance, and recent progress has given this methodology the potential for becoming a practical research tool. The advantages of a statistical approach to alignment include the possibility of parameter inference, hypothesis testing, and assessment of uncertainty, none of which are possible using the score-based methods that currently predominate.

Recent progress in statistical alignment includes better models, the extension of pairwise alignment algorithms to many sequences, faster algorithms, and the increased use of MCMC methods to handle practical problems. In this chapter, we illustrate the statistical approach to multiple sequence alignment on a series of increasingly large data sets.

14.1 Introduction

Although bioinformatics is perceived as a new discipline, certain aspects have a long history and could be viewed as classical bioinformatics. For example, the application of string comparison algorithms to sequence alignment has a history spanning the last three decades, beginning with the pioneering paper by Needleman and Wunsch [36]. They used dynamic programming to maximize a similarity score based on a matching score for amino acids and a cost function for insertions and deletions. Independently, Sankoff and Sellers in 1972 introduced an approach comparing sequence pairs by minimizing a distance function. Their algorithm is very similar to the algorithm maximizing similarity. Sankoff and Cedergren generalized the distance-minimizing approach to multiple sequences related by a phylogenetic tree. In the last three decades, these algorithms have received much attention from computer scientists and have been generalized and accelerated. Despite knowledge of exact algorithms, essentially all current multiple alignment programs rely on heuristic approximations to handle practical-sized problems. An example is the very popular Clustal family of programs. A completely different approach to alignment was introduced in 1994 by Krogh et al., who used hidden Markov models (HMMs) to describe a family of homologous proteins. This statistical approach has proved very successful; however, it was not based on an underlying model of evolution or phylogeny.

In 1981, Smith and Waterman introduced a local similarity algorithm for finding homologous DNA subsequences that has so far remained the gold standard for the local alignment problem. The main use of local alignment algorithms is to search databases, and in this context the Smith-Waterman algorithm has proved too slow. A series of computational accelerations have been proposed, with the BLAST family of programs being the de facto standard in this context [1].

At the same time that score-based methods were being developed for sequence alignment, parsimony methods were being used to solve the problem of phylogenetic reconstruction. The method of parsimony, which finds the minimum number of evolutionary events that explain the data, can be viewed as a special case of score-based methods. Over the last two decades, the parsimony method of phylogenetic reconstruction has been criticized, and it has essentially been replaced by methods based on stochastic modelling of nucleotide, codon, or amino acid evolution. This probabilistic treatment of evolutionary processes is based on explicit models of evolution and thus gives rise to meaningful parameters. In addition, these parameters can be estimated by maximum likelihood or Bayesian techniques, and the uncertainty in these estimates can be readily assessed. This is in contrast with score-based methods, where the weight or cost parameters cannot be easily estimated or necessarily even interpreted. Because this probabilistic treatment of phylogenetic evolution is based on explicit models, it also allows for hypothesis testing and model comparison.

Despite the increased statistical awareness of the biological community in the case of phylogenetic inference, which is now fundamentally viewed as a statistical inference problem [9], the corresponding problem of alignment has not undergone the same transformation, and score-based methods still predominate in this field. However, recent theoretical advances have opened up the possibility of a similar statistical treatment of the alignment inference problem. A pioneering paper by Thorne, Kishino, and Felsenstein from 1991 proposed a time-reversible Markov model for insertions and deletions (termed the TKF91 model) that allowed a proper statistical analysis for two sequences. This model provides methods for obtaining pairwise maximum likelihood sequence alignments and estimates of the evolutionary distance between two sequences. The model can also be used to define a test of homology that is not predicated on a particular alignment of the sequences. At present, this is a test of global similarity, and although analogues of local alignment methods are possible, they have not yet been developed in the statistical alignment framework.

The recent extension of the TKF91 model to multiple sequences, and algorithmic improvements to the analysis of this model, have considerably increased the practical applicability of the model. Along with the evolutionary processes of insertion, deletion, and mutation, analyzing multiple sequences additionally requires the consideration of their phylogeny. Most current alignment programs treat alignment and phylogeny separately, whereas in fact they are interdependent. A more principled approach is to estimate both simultaneously (see, e.g., [11, 45]). In this chapter, we show some preliminary results on the co-estimation of phylogeny and alignment under the TKF models of evolution. For up to about four sequences, a full probabilistic treatment is feasible (see Section 14.4). For larger data sets, it is necessary to use approximative methods such as MCMC (see Section 14.5).

In conclusion, the statistical alignment framework enables a coherent probabilistic treatment of both the sequence alignment and phylogenetic inference problems. However, challenges still remain, especially with respect to the computational problems inherent in using larger data sets and the biological realism of the evolutionary models. In this chapter, we shall review the basic model in some detail and sketch out some recent developments and current directions of research.

14.2 The Basic Model

The pioneering paper by Thorne, Kishino, and Felsenstein [42] proposed a continuous-time evolutionary model (TKF91) for sequence insertions and deletions, as well as substitutions, that allowed a proper statistical analysis of the alignment of two sequences. This model treats insertions and deletions (indels) as single-nucleotide events and is arguably the simplest possible continuous-time model for sequence evolution in the presence of nucleotide insertions and deletions. A major advantage of the model is that it can be treated analytically, and in fact it can be reformulated as a hidden Markov model (HMM). This leads to alignment procedures that, using the standard HMM algorithms, are as fast as score-based approaches.

In this section, we describe the TKF91 model and sketch the derivation of the transition probabilities. We introduce the extension of TKF91, termed TKF92 [43], which is able to deal with arbitrary-length nonoverlapping indels

(a)
$\sim$T$\sim$A$\sim$T$\sim$A$\sim$A$\sim$A$\sim$A$\sim$A$\sim$G$\sim$G$\sim$G$\sim$

(b)
```
 - A T - - A A C
 G A T C C - - G
```

Fig. 14.1. (a) In the TKF91 model, a sequence is viewed as nucleotides separated by *links* ($\sim$). Deletions originate from nucleotides, while insertions originate from links. The leftmost link is never deleted and is called the *immortal link*. (b) Example of a five-nucleotide sequence that evolved into a six-nucleotide sequence through a series of indel and substitution events. The evolutionary outcome is summarized by an alignment showing that three of the ancestral nucleotides (top line) share homology with descendant nucleotides, while other nucleotides have been either deleted or inserted.

and can be viewed as the statistical analogue of "affine gap penalties" in the score-based setting. Finally, we introduce the "long indel" model, a stochastic indel process that allows for overlapping indels of arbitrary length, and discuss some approaches that approximate this process.

14.2.1 The TKF91 Model

In the TKF91 model, a nucleotide sequence is modelled as a finite string of nucleotides, or *letters*, separated by *links*. The string both starts and ends with a link, so that there is always one more link than there are nucleotides; see Figure 14.1(a). The insertion and deletion events are modelled as continuous-time Markov processes. Insertions of single letters originate from *links* and occur at a rate of λ per unit of time and per link. Deletions, also of a single letter at a time, originate from the *letters* and occur at a rate μ per unit of time per letter. Models like these are known as *birth-death processes*. We may view the sequence as consisting of a single link followed by letter-link pairs that get inserted and deleted as little modules. In this view, the leftmost link is never deleted and is called the *immortal link*. This immortal link ensures that the empty sequence is not a sink for the process.

Parallel to this birth-death process, the individual nucleotides are subject to a continuous-time substitution process. The original paper used Felsenstein's one-parameter model [8], but this can be generalized to other models without difficulty. Similarly, in case alignments of proteins are desired, a substitution model on the amino acid alphabet is used.

Birth-death processes in which only singlet births and deaths occur, of which the TKF91 model is an example, are automatically time-reversible by virtue of the state graph's linear topology. This fact considerably simplifies calculations. Saying that a model is time-reversible is equivalent to saying that the *detailed balance condition* holds, and this can be used to work out the equilibrium length distribution. Suppose that, at equilibrium, the probability of observing a sequence of length k is q_k. The transition rate from a length-k sequence to one of length $k-1$ is μk since each individual nucleotide contributes a deletion rate μ. Since a sequence of length $k - 1$ has k links, the transition rate in the other direction is similarly λk. Detailed balance now requires that

$$\mu k q_k = \lambda k q_{k-1} \qquad \Leftrightarrow \qquad \frac{q_k}{q_{k-1}} = \frac{\lambda}{\mu}. \qquad (14.1)$$

Since the q_k are probabilities, $\sum_{k=0}^{\infty} q_k = 1$, and we have

$$q_k = \left(1 - \frac{\lambda}{\mu}\right)\left(\frac{\lambda}{\mu}\right)^k. \qquad (14.2)$$

This means that $\lambda < \mu$ is a requirement to have an equilibrium length distribution. This is not surprising since otherwise the birth rate of a length-k sequence, $\lambda(k + 1)$ (there are $k + 1$ links), always exceeds the death rate μk, so that sequences would tend to grow indefinitely.

Now suppose we let the TKF91 process act on a given initial sequence. After time t, the process will have resulted in a descendant sequence through a series of insertion, deletion, and substitution events (see Figure 14.2). Some nucleotides will have survived (though they may have undergone substitutions), and others will have been deleted or inserted. The latter will not be homologous to any nucleotide in the other sequence. This outcome can be summarized by an alignment of the ancestral and descendant sequences, where the homologous nucleotides are aligned in columns (see Figure 14.1(b)).

Because all nucleotides evolve independently, the probability of a particular outcome at time t, conditioned on the ancestral sequence, can be calculated by simply multiplying the probabilities of the outcomes of the individual nucleotides. For a given nucleotide, there are two sets of possible outcomes we want to distinguish, namely those where the ancestral nucleotide survives and those where it is deleted. To complete the description, we also need the probabilities for births emanating from the immortal link:

Outcome:		Probability:
# − ⋯ − # # ⋯ # (Homologous nucleotide survives, with $n - 1$ new ones)		$p_n^H(t) \quad (n = 1, 2, \ldots)$
# − ⋯ − − # ⋯ # (Ancestor was deleted, leaving n new nucleotides)		$p_n^N(t) \quad (n = 0, 1, \ldots)$
⋆ − ⋯ − ⋆ # ⋯ # (Immortal link gives rise to n new nucleotides)		$p_n^I(t) \quad (n = 0, 1, \ldots)$

Here # denotes a nucleotide, and we adopt the usual convention that nucleotides appearing in a column are homologous, with the ancestor appearing above the descendant. We do not explicitly write the links, except the immortal link, which is denoted by a ⋆. It is now possible to set up differential equations, known as *Kolmogorov's forward equations*, for the time-dependent outcome probabilities by considering the rate at which a state is populated from other states and the rate at which it populates other states. For instance, the equations for $p_n^I(t)$ are

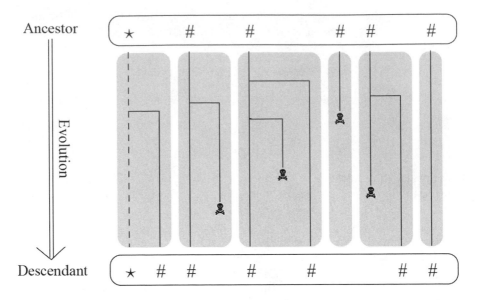

Fig. 14.2. One possible evolution of a sequence under the TKF91 model, resulting in the outcome represented in Figure 14.3(a), and summarized by the alignment of Figure 14.3(b). In this example, the immortal link ($\star$) gave birth to a new nucleotide that survived, its neighboring ancestral nucleotide gave rise to a new nucleotide that did not survive, and so on. Note that this detailed evolution contains far more information than the outcome as depicted in Figure 14.3(a) (and far more than we can observe). The associated outcome probability includes contributions of all possible evolutions compatible with the outcome.

$$\frac{d}{dt}p_n^I(t) = (n+1)\mu p_{n+1}^I + n\lambda p_{n-1}^I - [n\mu + (n+1)\lambda]\, p_n^I(t), \qquad (14.3)$$

$$p_n^I(0) = 1 \text{ for } n = 0, \quad 0 \text{ otherwise}, \qquad (14.4)$$

where p_{-1}^I is defined to be 0. These equations for a classic birth-death process are solved by

$$p_n^I(t) = (1 - \lambda\beta(t))\,[\lambda\beta(t)]^n, \qquad \text{where} \quad \beta(t) = \frac{1 - e^{(\lambda - \mu)t}}{\mu - \lambda e^{(\lambda - \mu)t}}. \qquad (14.5)$$

The differential equations for the other probabilities are more involved but can also be solved analytically [42]. In terms of the following abbreviations,

$$B_\tau = \lambda\beta(\tau), \qquad\qquad\qquad E_\tau = \mu\beta(\tau),$$

$$N_\tau = (1 - e^{-\mu\tau} - \mu\beta(\tau))(1 - \lambda\beta(\tau)), \qquad H_\tau = e^{-\mu\tau}(1 - \lambda\beta(\tau)),$$

$$I_\tau = 1 - \lambda\beta(\tau), \qquad\qquad\qquad\qquad\qquad\qquad\qquad (14.6)$$

the solutions are

(a)
$$
\begin{array}{ll}
t = 0: & \left\| \begin{array}{c|c|c|c|c|c|c|c|c} \star & - & \# & \# & - & \# & \# & - & \# \end{array} \right. \\
t = \tau: & \left. \begin{array}{c|c|c|c|c|c|c|c|c} \star & \# & \# & \# & \# & - & - & \# & \# \end{array} \right\| \\
\hline
\text{Probability:} & \left\| I_\tau \ B_\tau \ H_\tau \ H_\tau \ B_\tau \ E_\tau \ N_\tau \ H_\tau \right.
\end{array}
$$

(b)
$$
\left\| \begin{array}{c c c c c c c c} - & \# & \# & - & - & \# & \# & \# \\ \# & \# & \# & \# & \# & - & - & \# \end{array} \right\|
$$

(c)
$$
\begin{array}{ll}
t = -\infty: & \left\| \begin{array}{c c c c c c c c c} \star & & & & & & \# & \# & \# \end{array} \right. \cdots \\
t = 0: & \left\| \begin{array}{c c c c c c c c c} \star & \# & \# & \# & \# & \# & - & - & - \end{array} \right. \cdots \\
\hline
\text{Probability:} & \left\| I_\infty \ B_\infty \ B_\infty \ B_\infty \ B_\infty \ B_\infty \ E_\infty \ E_\infty \ E_\infty \right. \cdots
\end{array}
$$

Fig. 14.3. Example of an evolutionary history for two sequences, and the associated probability according to the TKF91 model. (a) Example history for five nucleotides evolving into a length-6 sequence. Note that the event N_τ, where a nucleotide dies but not before giving birth to a new, nonhomologous nucleotide, is represented by two columns in an alignment. Conditional on the ancestral sequence, the probability for this history is $I_\tau B_\tau^2 H_\tau^2 N_\tau E_\tau$. (b) The alignment resulting from this history. Since alignments summarize only the homology relationships between sequences, certain columns can be swapped without altering the meaning of the alignment (and different evolutionary histories may give rise to the same alignment). (c) Summary of the probabilities for a length-5 sequence at equilibrium (that is, after an infinitely long time). The last columns are added for illustration; the ancestral sequence at $t = -\infty$ is unknown, but this makes no difference since $E_\infty = 1$ (the probability of a nucleotide being deleted tends to 1 if we wait long enough). Therefore, the probability of observing a length-5 sequence at equilibrium is $I_\infty B_\infty^5 = (1 - \lambda/\mu)(\lambda/\mu)^5 = q_5$.

$$
p_0^N(t) = E_\tau, \tag{14.7}
$$
$$
p_n^N(t) = N_\tau B_\tau^{n-1}, \qquad (n > 0), \tag{14.8}
$$
$$
p_n^H(t) = H_\tau B_\tau^{n-1}, \qquad (n > 0), \tag{14.9}
$$
$$
p_n^I(t) = I_\tau B_\tau^n. \tag{14.10}
$$

See Figure 14.3 for an example of how to calculate the probability of a particular evolutionary history.

14.2.2 The TKF92 Model

The most obvious drawback of the TKF91 model, as already noted in the original paper, is that insertions and deletions occur one letter at a time. In reality, many indel events involve more than a single nucleotide. In 1992, Thorne, Kishino, and Felsenstein introduced an improved version of their model, designed to model indel events of more than a single letter [43]. This model, referred to as TKF92, differs from the TKF91 model by acting on sequence *fragments* instead of single nucleotides. The fragments themselves are not observed, and their length is randomly distributed according to a geometric distribution with parameter ρ. This approach leads to a model that can still be treated analytically and is a reasonably good approximation of the actual observed indel length distribution.

The approximation that is made in the model, and that makes it possible to analytically compute the probabilities, is that the fragments (and their sizes) are supposed to stay fixed over the entire evolutionary history of the sequence. This assumption, made for technical reasons, is clearly not realistic. However, things are not as bad as they might seem. In the same way that the TKF91 model sums over all possible alignments, the TKF92 model also sums over all possible *fragment* assignments. Effectively, this means that indels of any length may occur at any position in the sequence, but such indels may not, in the course of evolutionary history, overlap. See Figure 14.8 for an alignment under the TKF91 and TKF92 models.

14.2.3 Parameters of the TKF Models

Although the TKF91 model has two parameters, λ and μ, their ratio is in practice fixed by the sequence length. Indeed, if we maximize the likelihood

$$q_L = \left(1 - \frac{\lambda}{\mu}\right) \left(\frac{\lambda}{\mu}\right)^L \tag{14.11}$$

in terms of λ/μ, for a fixed sequence length L, we find that the maximum is obtained for

$$\frac{\lambda}{\mu} = \frac{L}{L+1}. \tag{14.12}$$

For maximum likelihood parameter estimates, it is therefore not meaningful to estimate λ and μ independently but rather to fix their ratio based on the average of the sequence lengths that are to be aligned and estimate just one free parameter.

The TKF92 model has one extra parameter, ρ, parameterizing the geometric fragment length distribution. Fragments drawn from this distribution have an expected length of $\frac{1}{1-\rho}$. In the TKF92 model, the parameters λ and μ refer to the indel rate *per fragment*. To allow a meaningful comparison, it is useful to introduce new parameters λ' and μ' that specify the average indel rates *per site* and are related to the parameters λ and μ by

$$\lambda = (1 - \rho)\lambda', \qquad \mu = (1 - \rho)\mu'. \tag{14.13}$$

Note that TKF91 is a special case of TKF92, obtained by setting $\rho = 0$. This corresponds to a degenerate fragment length distribution where all fragments have length 1 (see, e.g., Figure 14.4). For an example of how to calculate the likelihood given a fragmentation, see Figure 14.5.

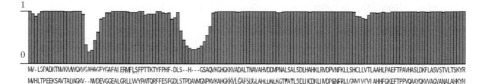

Fig. 14.4. The most likely pairwise alignment of human α and β hemoglobins, according to the TKF91 model. The vertical bars indicate posterior column probabilities (i.e., the proportion of alignments that include that particular column, weighted according to the posterior probability under the model). See Section 14.3.3 for the algorithms used to calculate the alignment and posterior probabilities. The log-likelihood of observing this alignment under the TKF91 model, using maximum likelihood parameters for these sequences ($\lambda = 0.03718$, $\mu = 0.03744$, $t = 0.91618$, see [14]), is -735.859. This low likelihood reflects the relatively high sequence divergence and the fact that it is very unlikely for the ancestor of these sequences to have evolved by chance; however, the log-likelihood of observing both sequences by chance independently is far smaller still, $-401.372 - 418.764 = -820.136$, giving strong support to the hypothesis that these sequences are homologous.

(a)

$$
\begin{array}{c|c|c c c|c c c c|c c c c|c c}
t = 0: & \star & \# & \# & \# & - & - & - & - & \# & \# & \# & \# & \# & \# \\
t = \tau: & \star & \# & \# & \# & \# & \# & \# & \# & \# & \# & \# & \# & \# & \# \\
\hline
\text{Probability:} & I_\tau & H_\tau\rho^2(1-\rho) & & & B_\tau\rho^3(1-\rho) & & & & H_\tau\rho^3(1-\rho) & & & & H_\tau\rho(1-\rho) &
\end{array}
$$

(b)

$$
\begin{array}{|c c c c c c c c c c c c|}
\# & \# & \# & - & - & - & - & \# & \# & \# & \# & \# & \# \\
\# & \# & \# & \# & \# & \# & \# & \# & \# & \# & \# & \# & \# \\
\end{array}
$$

Fig. 14.5. An evolutionary history according to the TKF92 model. (a) One possible fragmentation into fragments of sizes 3, 4, 4, and 2, respectively, and the associated probability for this evolutionary history. (b) The alignment resulting from this history. Many different fragmentations contribute to this alignment.

14.2.4 The "Long Indel" Model

The TKF92 model is a substantial improvement over the TKF91 model, as it allows indel events involving more than one nucleotide. The main assumptions that go into the model are (1) that indel events do not overlap and (2) that the indel lengths are geometrically distributed. A natural, more general evolutionary model would relax these two assumptions, specifically by allowing indel events to overlap and by allowing an arbitrary indel length distribution. Here we focus on relaxing the former assumption, although the proper modelling of the actual indel length distribution (see, e.g., [38]) is probably at least as important for alignment accuracy. We refer to the more general model as the "long indel" model. In its general form, no closed-form solution of the outcome probabilities is known, even for a geometric indel length distribution.

The main difficulty is that by allowing overlapping indel events, the fates of neighboring nucleotides become entangled over time, so that the probability of the total outcome does not factorize into individual nucleotide outcome probabilities, as is the case for the TKF models.

To arrive at a tractable implementation of this model, some kind of approximation is necessary. Knudsen and Miyamoto [21] develop an approximation that is analytically no more complex than the TKF models: their pairwise alignment algorithm takes $O(L^2)$ time, where L is the sequence length. In fact, their model is formulated as an HMM with the same topology as that in which TKF models are commonly formulated, and differs only in the transition probabilities. It is satisfying that, in contrast with TKF92, this indel model is derived from first principles, but given its similar structure, it is unclear how much it improves upon TKF92.

If one is willing to use computationally more demanding algorithms, then an even more realistic approximation to the long indel model is possible. In [32] an approximation is used that allows each indel event to overlap with up to two others, and allows an arbitrary indel length distribution to be used. The corresponding pairwise alignment algorithm has time complexity $O(L^4)$, making the algorithm unsuitable for large database searches, for example. However, single pairwise alignments can still be computed relatively quickly, and on a set of trusted alignments based on known 3D protein structure, this model outperformed TKF92. See [32] for more details.

14.3 Pairwise Alignment

In this section, we describe how the TKF models are used in practical pairwise sequence alignment algorithms. First, we describe an intuitive dynamic programming recursion, which, however, has a high computational complexity. More efficient recursions exist, and we describe in detail one that is based on the formulation of the TKF models in terms of hidden Markov models. The additional structure makes it easier to describe the various algorithms that are based on it and paves the way for the multiple alignment algorithms later on.

14.3.1 Recursions for the Likelihood of Two Homologous Sequences

Let us now turn to the task of calculating the likelihood of homology; that is, the likelihood that two sequences have evolved from a common ancestor. Because of the time-reversibility of the TKF91 model, this is equivalent to the likelihood that one sequence evolved into the other in twice the time that separates the ancestor from the two descendants (referred to as τ below). This likelihood is, by definition, the total probability corresponding to all evolutionary histories that are consistent with the observed sequences. Obviously, there are extremely many of these evolutionary histories, so a direct evaluation of this sum is impractical. However, a dynamic programming approach is possible that computes this sum in reasonable time.

In the following, $P(i,j)$ is the likelihood of the length-i prefix of the ancestral sequence evolving into the length-j prefix of the descendant sequence. For instance, $P(0,0) = I_\infty I_\tau$ since the probability of observing the empty ancestral sequence is I_∞, while the probability of the empty sequence evolving, in time τ, into the empty sequence again is I_τ. The dynamic programming solution now consists of computing $P(i,j)$ in terms of previously computed $P(i',j')$. By filling a table, all values can then be computed in reasonable time.

As we saw in Section 14.2.1, each ancestral nucleotide evolves independently of the others, and a single ancestral nucleotide can evolve into $0, 1, 2, \ldots$ descendant nucleotides (which may or may not be homologous to the ancestor). For the recursion, this means that we can express $P(i,j)$ in terms of $P(i-1, j-k)$ with $k = 0, 1, \ldots, j$ (corresponding to outcomes with k descendant nucleotides) multiplied by the probability of a particular evolution of the last ancestral nucleotide. The contribution of the indel process to these probabilities is given in Section 14.2.1. This must be multiplied by (1) the probability B_∞ of observing one additional ancestral nucleotide; (2) the equilibrium probability of the particular nucleotide observed; (3) the probability of that nucleotide evolving into the descendant nucleotide (in case of a homologous descendant nucleotide); and (4) the nucleotide equilibrium probabilities of any nonhomologous descendant nucleotides. The resulting algorithm is illustrated in Figure 14.6.

As the algorithm is formulated here, its running time is cubic in the sequence length. However, due to the geometric tails of the outcome probabilities p_n^N, p_n^H, and p_n^I as functions of n, the recursion may be reformulated so that it only uses a bounded lookback, resulting in an algorithm that has quadratic time complexity [42]. In this context, this is just an algebraic trick and is reminiscent of the method used by Gotoh in 1982 for reducing the time complexity of a score-based sequence alignment algorithm with affine gap penalties (see [10]). However, there is a more meaningful and conceptual way to look at this. It turns out that the TFK91 model can be viewed as an instance of what is known as a hidden Markov model (HMM). From that point of view, the algorithm derived using the algebraic trick becomes the well-known forward algorithm for HMMs, and more algorithms are immediately applicable, such as the Viterbi algorithm for determining the most likely path through the chain, corresponding to the most likely alignment supported by the model. In the next section, we will develop this point of view in more detail.

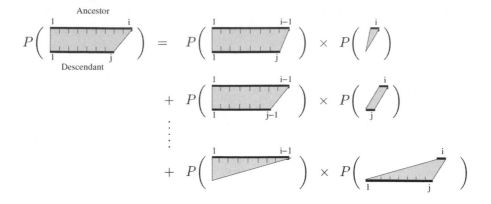

Fig. 14.6. A graphical illustration of a dynamic programming recursion for the TKF91 likelihoods. The left-hand side is the likelihood $P(i, j)$ defined in the text; horizontal black bars represent sequences, and the grey area represents "evolution." Since individual nucleotides evolve independently, the likelihood for the length-i ancestral prefix to evolve into the length-j descendant prefix factorizes into several other prefix likelihoods and probabilities of the ith residue to evolve into descendant subsequences.

14.3.2 A Hidden Markov Model Formulation of TKF models

The outcome probabilities $p_n^H(t)$ and $p_n^N(t)$ of the TKF91 model are geometric functions of n. As a result, we can construct a graph, with probabilities on each of its edges, such that each path through the graph corresponds uniquely to a particular outcome, and the product of all the probabilities encountered is precisely the probability of that outcome. The graph shown in Figure 14.7(a) has all these properties and generates alignments according to the TKF91 model. Such graphs are called *Markov models* if the outgoing probabilities sum to 1 for each of the states. This is accomplished in Figure 14.7 by multiplying and dividing by a factor $1 - B_\tau$ at certain positions in such a way that the total probability of any closed path from the start state to the end state does not change. We use this Markov model as a *hidden* Markov model (HMM) because we treat only the sequences as known, while the alignment structure is regarded as unknown. This unknown information is encoded by the path taken through the Markov chain, while the emitted sequences of nucleotides are given. We refer to [6] for more information about HMMs.

By manipulating the graph of Figure 14.7(a), the number of states can be reduced to just three (apart from the Start and End states) see Figure 14.7(b). This reduces the time and memory complexity of the HMM algorithms (especially when more sequences are considered, see below). Because of the algebraic manipulations, the transition probabilities take a more com-

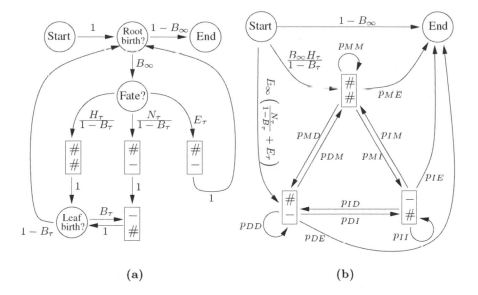

(a)   (b)

Fig. 14.7. Two HMM formulations of the TKF91 model. (a) Direct translation of TKF91 probabilities into an HMM. A factor $1 - B_\tau$ is needed for three transitions to make all outgoing transition probabilities add up to 1. (b) Another HMM that is emission-equivalent to (a). The two states $^{\#}$ were merged into one and all non-emitting states removed, leaving a fully connected three-state HMM. Note that the evolutionary indel model by Knudsen and Myamoto [21] is formulated using an HMM with exactly the same topology.

plicated form and are listed in Table 14.1. Henceforth, when we refer to the TKF91 HMM, we are referring to the reduced HMM of Figure 14.7(b).

Starting from this HMM formulation of TKF91, it is straightforward to transform it into the HMM for TKF92. This is done by adding a self-transition (with probability ρ) to each state, which accounts for the geometric fragment length distribution. To compensate, all other transition probabilities (including the existing self-transitions) are multiplied by $1 - \rho$, making outgoing probabilities add up to 1 again. See Table 14.1 for the explicit transition probabilities.

14.3.3 Algorithms

The formulation of the TKF models above in terms of HMMs allows us to use standard HMM algorithms, such as the forward-backward algorithm, and the Viterbi algorithm. For a detailed explanation of these algorithms, we refer to [6]; here we focus on their application.

Applied to the TKF HMMs, the forward (or backward) algorithm calculates the total likelihood of one sequence to have evolved from another. In

Edge:	Transition probabilities	
	TKF91	TKF92
p_{MM}	$B_\infty H_\tau$	$\rho + (1 - \rho)B_\infty H_\tau$
p_{MI}	B_τ	$(1 - \rho)B_\tau$
p_{MD}	$B_\infty(N_\tau + E_\tau(1 - B_\tau))$	$(1 - \rho)B_\infty(N_\tau + E_\tau(1 - B_\tau))$
p_{ME}	$(1 - B_\tau)(1 - B_\infty)$	$(1 - \rho)(1 - B_\tau)(1 - B_\infty)$
p_{II}	B_τ	$\rho + (1 - \rho)B_\tau$
p_{IM}	$B_\infty H_\tau$	$(1 - \rho)B_\infty H_\tau$
p_{ID}	$B_\infty(N_\tau + E_\tau(1 - B_\tau))$	$(1 - \rho)B_\infty(N_\tau + E_\tau(1 - B_\tau))$
p_{IE}	$(1 - B_\tau)(1 - B_\infty)$	$(1 - \rho)(1 - B_\tau)(1 - B_\infty)$
p_{DD}	$E_\tau B_\infty$	$\rho + (1 - \rho)E_\tau B_\infty$
p_{DM}	$E_\tau H_\tau B_\infty[N_\tau + E_\tau(1 - B_\tau)]^{-1}$	$(1 - \rho)E_\tau H_\tau B_\infty[N_\tau + E_\tau(1 - B_\tau)]^{-1}$
p_{DI}	$N_\tau[N_\tau + E_\tau(1 - B_\tau)]^{-1}$	$(1 - \rho)N_\tau[N_\tau + E_\tau(1 - B_\tau)]^{-1}$
p_{DE}	$E_\tau(1 - B_\tau)(1 - B_\infty)[N_\tau + E_\tau(1 - B_\tau)]^{-1}$	$(1 - \rho)E_\tau(1 - B_\tau)(1 - B_\infty)[N_\tau + E_\tau(1 - B_\tau)]^{-1}$

Table 14.1. Transition probabilities in the HMM of Figure 14.7(b) for the TKF91 and TKF92 models. The probabilities for TKF91 are obtained from those of TKF91 by multiplying all transition probabilities by $1 - \rho$ and adding a self-transition with probability ρ to every state.

fact, in most cases, we want to calculate the likelihood of an unknown root sequence having evolved, independently, into two observed modern sequences. Because of the time-reversibility of the model, the position of the root on the branch connecting these two sequences does not influence the likelihood, and therefore these two likelihoods are equal. This symmetry property is known as Felsenstein's pulley principle [8].

The forward and backward algorithms compute the total probability of all paths through the Markov chain that emit the observed sequences. This can be used for homology testing [12] and to estimate evolutionary parameters [42], such as the divergence time and the indel rate, by maximum likelihood. The Viterbi algorithm is the HMM analogue of the Needleman-Wunsch [36] score-based alignment algorithm and is traditionally the main workhorse for doing inference in hidden Markov models. The algorithm finds the most probable (that is, the maximum likelihood) path to emit the given sequences, and this path codifies the alignment of the sequences.

From the intermediate results from both the forward and backward algorithms, it is possible to compute the posterior probability of passing through any given state, conditional on emitting the observed sequences. Figures 14.4 and 14.8 show examples of Viterbi alignments and corresponding posterior state probabilities on the Viterbi paths. For the alignment models, these are interpreted as the posterior probability of observing an individual column in the alignment. These posteriors are therefore indicators of the local "reliability" of an alignment. They add important information to the simple "best" answer obtained for example by the Viterbi algorithm and can be seen as the alignment equivalent of confidence intervals for simple numerical parameter

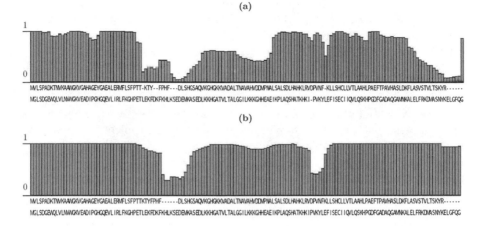

Fig. 14.8. Viterbi alignment of human α hemoglobin with human myoglobin under (a) the TKF91 model and (b) the TKF92 model (with parameter $\rho = 0.44$). Clearly, the TKF92 model fits the data much better, generally assigning higher posterior probabilities to the alignment. (Maximum log-likelihoods for TKF91 and TKF92 are -825.25 and -817.25, respectively.) Note the ≈ 6 aligned columns that show a sudden decrease in posterior probability in the TKF92 alignment, where the corresponding TKF91 alignment has two small indels. The TKF92 model is reluctant to include many individual indels, preferring a single large one. Although the maximum likelihood path is the one without any indels in that region, alignments with indels contribute significantly to the total likelihood, indicating that the homology implied by the alignment there should be treated with caution. This is a good example of what additional information can be obtained from the posterior column probabilities.

estimates. In practical examples, there are very many alignments that contribute to the total likelihood, and the most likely alignment may contribute only a very small fraction. This makes a single best answer not very informative, and the local reliability measure indicates which parts of the alignment can be trusted and which parts are essentially random, giving a quantitative underpinning of the notion of "unalignable region" [23].

Although the Viterbi algorithm, computing the maximum likelihood path, is ubiquitously used for alignment inference, it should here be mentioned that there is no one-to-one relationship between paths through the Markov chain and alignments. More than one evolutionary history can give rise to a single alignment, see Figure 14.3 for an example. Note that for the output of the Viterbi algorithm, the exact topology of the HMM is important, and, in general, two HMMs may be *emission*-equivalent without being *path*-equivalent. An example is provided by the TKF92 versions of the HMMs of Figure 14.7, which are derived by adding self-transitions with probability ρ to each (emitting) state. Paths through Figure 14.7(a) codify the sequence fragmentation, while in Figure 14.7(b) the sequence fragmentation is analytically summed out and cannot be deduced from the path. Their hidden information differs, but the observables (the emitted nucleotides) follow exactly the same distribution. The result is that the forward or backward algorithms give the same answers, but the Viterbi algorithm is biased toward alignments with more indels if the HMM of Figure 14.7(a) is used.

Although not much of a problem for pairwise alignment, the nonequivalence of paths and alignments turns up again, and more seriously, in the case of alignments on trees. One way of dealing with this problem is to explicitly look for the most probable alignment and keep track of all paths that contribute to it [22]. Unfortunately, the resulting algorithm is very slow. Another method that recovers a "best" alignment from an HMM, without relying on path reductions, is *posterior decoding* [6]. The idea is first to compute posterior probabilities for each possible column that may appear in the alignment and then find the alignment that maximizes the combined posterior column probability. This can be done efficiently using dynamic programming, which is the same strategy that underlies the forward, backward, and Viterbi algorithms. Although there is no guarantee that the alignment obtained in this way is the most probable one, in practice this method gives very good results. Another advantage of the method is that it is also applicable in Markov chain Monte Carlo settings (see Section 14.5), where the Viterbi algorithm cannot be used but estimates of posterior column probabilities are available.

14.4 Multiple Statistical Alignment

The simultaneous alignment of several sequences can reveal conserved motifs much more sensitively than a pairwise alignment can. This assists in the alignment of more distantly related sequences and the detection of functional sites. Unfortunately, multiple alignment is a computationally hard problem, and certain particular cases are known to be NP-hard [46]. Furthermore, the problems of multiple alignment and phylogenetic inference are closely interlinked: to properly align a set of homologous sequences, it is necessary to know their phylogeny, and vice versa [11, 45]. Keeping this interrelatedness in mind, we will nonetheless focus mostly on alignments. We do not discuss the various interesting approaches developed for phylogenetic reconstruction and will return to this topic only at the end of this section, where we discuss co-estimation of alignment and phylogeny.

In the 1970s, Sankoff introduced the first multiple-alignment algorithm [39], and since then many other algorithms have been proposed. Most of these are "score-based" and use a score function that assigns a "goodness" to particular multiple alignments (and sometimes phylogenies). The algorithms then find the best alignment by optimizing this score function. Because of the large number of possible alignments, full optimizations are practically impossible, and several clever heuristics have been introduced to find reasonable solutions in reasonable time. Successful programs include ClustalW [41], PSI-Blast [2], DiAlign [34], and T-Coffee [37].

A drawback of score-based approaches is that it is hard to justify the parameter settings of the score function—or indeed the score function itself. This is one reason why probabilistic approaches are becoming more popular. Instead of assigning a score, these methods assign a probability to alignments, making it easier to train a model on data and find parameters by techniques such as maximum likelihood. Two popular probabilistic approaches, both based on HMMs, are HMMER [7] and SAM [20]. Another important advantage of probabilistic models is that they provide estimates of the uncertainty in the final answer, such as posterior column probabilities for alignments and confidence intervals for parameter estimates. An example of a probabilistic progressive multiple-alignment method is [25], which has since been extended to include structure-dependent evolution (Löytynoja and Goldman, pers. comm.). Another example is by Mitchison [33], who estimates phylogeny and alignment simultaneously using an MCMC sampler in a probabilistic framework.

However, probabilistic models also have some problems. Such models are mostly phenomenological, describing the data but not explicitly making statements about the process that generated them. In particular, the evolutionary relationships between the sequences are often treated heuristically. Parameters of phenomenological models are linked to observables, not to the evolutionary process, making it difficult to interpret parameter values. For correct modelling, one should ideally reestimate parameters for every data set with different evolutionary parameters.

An evolutionary approach is based on a model of sequence evolution from which a probabilistic model for the observed sequences is derived. In this way, the parameters of the model (such as indel and substitution rates and divergence time) are meaningful and can be estimated using the same methods as for probabilistic models. The TKF91 and TKF92 models fit in this framework. Algorithmically, the approach is not very different from probabilistic or even score-based methods, and it encounters the same problems. Full-likelihood methods are possible only for a very limited number of sequences, after which approximations and heuristics are necessary. One particularly useful approximation method is Markov chain Monte Carlo (MCMC). This method generates samples from the posterior distribution of alignments, thereby disregarding alignments that are very unlikely.

14.4.1 Multiple Alignment and Multiple HMMs

The first step in extending statistical alignment to multiple sequences was taken by Steel and Hein [40], who provided an algorithm to align sequences related by a star tree (a tree with a single internal node). This was soon extended to arbitrary phylogenetic trees [12] with an algorithm with time complexity $O(L^{2n})$, where L is the mean sequence length and n the number of sequences. For star trees, this running time was subsequently reduced to

$O(4^n L^n)$ [31]. These results used rather complicated algebraic manipulations to derive the algorithms, and when it was realized that for two sequences the TKF91 model can be described as a pair HMM [29, 12, 18], the extension to multiple sequences became much easier. Holmes and Bruno [18] showed how to construct a multiple HMM describing the evolution of an ancestral sequence and its two descendants. Subsequently, Hein, Jensen, and Pedersen showed how to generate a multiple HMM for TKF91 on an arbitrary phylogenetic tree [13]. The details concerning the construction of these multiple HMMs are beyond the scope of this book, but to give a flavor of the techniques involved we give a single example for three sequences in Figure 14.9.

We can loosely argue that this multiple HMM correctly generates multiple alignments according to TKF91. First, note that each path from the start state to the end state corresponds to a multiple alignment. From the start state, the chain first jumps to a silent state next to the state emitting a character to all the sequences, which models "births" emanating from the immortal link. Eventually the process reaches the rightmost silent state, where a decision is made whether there is a new root birth. If there is, a decision tree with transition probabilities α_i and $1 - \alpha_i$ decides on which branches this nucleotide survives, after which subsequent births associated with the surviving nucleotides are introduced. It can be verified that the path probabilities equal the probabilities that the TKF91 model assigns to the corresponding alignments, a task we gladly leave to the reader.

In the same vein, TKF92 can be extended to multiple alignments on trees. The simplest way to do this is by adding self-transitions to the HMM of Figure 14.9. This fixes fragmentations over the entire phylogenetic tree, so that indels cannot overlap even if they occur on separate branches, clearly creating undesirable correlations between independent subtrees. A better behavior is obtained if the three-state TKF92 HMM is used as a building block on each of the branches and communicates sequences (not fragmentations) at internal nodes. Holmes introduced the concept of *transducers*, or conditionally normalized pair HMMs describing the evolution along a branch, which provides an algorithmic way to construct multiple HMMs on a tree [17]. This leads to an HMM with the same number of states as before, but one that allows overlapping indels as long as they occur on separate branches.

As an aside, note that, in contrast with the fixed-fragmentation TKF92 model, likelihoods now depend on the number and position of internal nodes along a branch. In fact, even introducing a node of degree 2 (i.e., a node with one incoming and one outgoing branch) changes the model. By increasing the density of such degree-2 nodes, the model eventually converges to the long indel model, allowing arbitrary overlapping indels. Unfortunately, the number of HMM states increases exponentially with the number of nodes, so that adding such degree-2 nodes is an impractical way of approximating the long indel model.

A technical problem with the multiple HMMs generated above is that they may contain *silent states* that do not emit any characters or emit only into

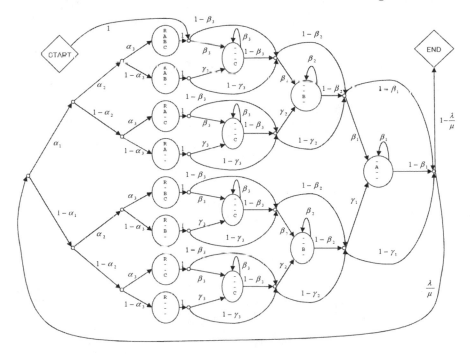

Fig. 14.9. Multiple HMM describing the evolution of three sequences related to a star tree, under the TKF91 model. The following abbreviations are used: $\alpha_i = e^{-\mu t_i}$, $\beta_i = (\lambda - \lambda e^{(\lambda-\mu)t_i})/(\mu - \lambda e^{(\lambda-\mu)t_i})$, and $\gamma_i = (1 - \alpha_i)^{-1}(1 - e^{-\mu t_i} - \beta_i)$, where t_i is the length of the branch descending to tip i in the phylogenetic tree. Big circles are states that emit the column shown according to the underlying substitution model; R, A, B, and C represent characters in the root sequence and the three observed sequences, respectively. Small ellipses represent silent states, (see [18]).

(unobserved) internal nodes. (An example of a silent state is the R/-/-/- state in Figure 14.9.) These states create self-references (or loops) in the state graph and need to be eliminated before the Markov chains can be used in algorithms. The technique of silent-state elimination is well-known in the HMM literature [7], and it involves solving a set of linear equations. See [27] for more details.

14.4.2 Algorithms for Multiple Sequence Alignment

After eliminating silent states, we can calculate the joint probability of a set of sequences related by a phylogeny by the standard forward and backward algorithms, calculate the posterior probability of particular alignment columns, and can find the most likely alignment with the Viterbi algorithm [6]. An example is presented in Figures 14.10 and 14.11.

A practical problem that besets algorithms for multiple HMMs is that the time and memory complexity increase rapidly with the number of sequences.

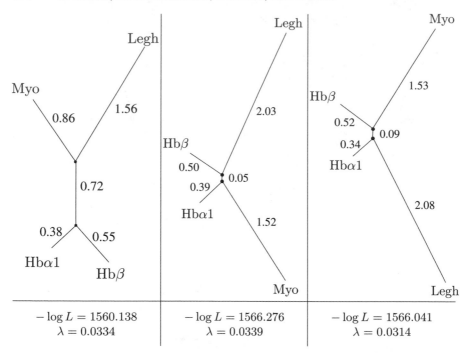

Fig. 14.10. Maximum likelihood trees relating human $\alpha 1$ and β hemoglobins, myoglobin, and bean leghemoglobin for all three topologically distinct trees, total likelihood values (L), and insertion rates (λ) under the TKF91 model. The numbers next to the branches refer to branch lengths in units of expected number of substitutions per site. Dayhoff's PAM matrix was used as the substitution rate matrix. As expected, the most likely tree is the one that groups human alpha and beta hemoglobins together. The other trees are close to degenerate, with only a very short segment connecting the internal nodes, again suggesting that these phylogenies are incorrect. The tree likelihoods combine all possible alignments of the four sequences, in contrast with most other methods, which rely on a single alignment, preventing inaccuracies in a single alignment from biasing the phylogeny inference (see [27]).

Two factors contribute to this rapid increase: (1) The dimension of the dynamic programming (DP) table is equal to the number of sequences n, and (2) the number of states S of the multiple HMM itself grows exponentially with the number of sequences. Generally, the basic algorithms have time and memory complexities of $O(S^2 L^n)$ and $O(SL^n)$, respectively. For TKF91 and TKF92, the number of states S is of the order $\sqrt{5}^n$ [27]. One implementation of the TKF91 model for four sequences uses 47 states and 1293 transitions between them, so that for sequences of length 150, naive implementations would require about $2 \cdot 10^{10}$ memory positions and 10^{12} floating point operations.

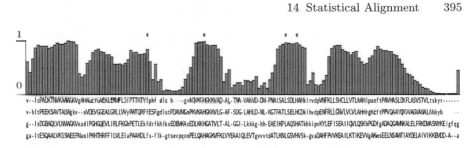

Fig. 14.11. The Viterbi alignment of α and β human hemoglobin, human myoglobin, and leghemoglobin (Lupinus luteum) for the first phylogenetic tree in Figure 14.10. The log-likelihood of this alignment (one of those included in the tree likelihood of Figure 14.10) is -1593.223. The column posterior probabilities vary considerably and clearly point to several highly conserved domains, punctuated by much less conserved regions. Amino acids that participate in α helices are shown in uppercase; asterisks denote the four conserved residues that coordinate the heme group.

The TKF91 model has some surprising symmetries that allow the forward algorithm based on the three-state pair HMM to be reduced to a one-state recursion [14]. This algebraic reduction results in a recursion that contains negative coefficients, so that it cannot be interpreted as a Markov chain anymore. Nevertheless, similar reductions are possible on trees, also resulting in a one-state recursion, resulting in an algorithm to compute the total likelihood using L^n memory positions with a running time of order $O(2^n L^n)$. See [27] for details.

The tricks involved in the reduction seem unique to TKF91, and for TKF92 and similar models, we have to resort to general algorithms. In the following section, we discuss a number of modifications to the original forward-backward and Viterbi algorithms and some corner-cutting methods that make full likelihood methods possible in practice.

Multiple forward-backward algorithm

In practice, memory resources are often the limiting factor, and strategies to reduce memory usage are therefore of great practical importance. For the forward and backward algorithms, if only the total likelihood is required, one can relinquish DP table entries dynamically, resulting in memory requirements of the order SL^{n-1}, not SL^n. To compute posterior column probabilities, the straightforward algorithm computes the full DP table using both the forward and backward algorithms and therefore requires order SL^n memory. However, if the posterior probabilities for a particular alignment are required, a more careful implementation can still compute this using order SL^{n-1} memory by relinquishing during the iterations all DP table entries that are not referenced by either the alignment of interest or new DP table entries [15, 30]. Even with these leaner implementations, the computational complexity is still very high. Further reduction in space can be achieved by heuristic *corner-cutting* methods. Such methods are well-known in score-based alignment approaches [44, 35, 3, 24, 47], and here we describe their counterparts for HMMs.

In practice, only a small region of the DP table is responsible for the dominant contribution to the total likelihood. This region often consists of a well-defined "spine" corresponding to the maximum likelihood alignment and close neighbors. If this "contributing region" were known, the recursion could be confined to it, resulting in a considerable speedup [14] and a negligible loss of total likelihood. The problem is clearly circular, however, as the contributing region can only be determined after the full DP table has been computed. Having said this, heuristic methods for selecting the contributing region exist that work very well in practice, for example based on full pairwise alignments.

The likelihood that is computed in this way is, by construction, a lower bound for the actual likelihood since each time the DP recursion refers outside the contributing region, probability 0 is used instead of the true (small but nonzero) probability. It is possible also to compute an upper bound using the same contributing region. This sandwiches the actual likelihood between two bounds and, if these bounds are tight enough, provides an effective a posteriori proof that the maximum likelihood alignment lies within the contributing region. The method is based on calculating the alignment likelihood of m known sequences and $n - m$ sequences of unknown composition and length on an n-leaved phylogenetic tree. A recursion of memory complexity SL^m exists that computes the sum of alignment probabilities over all possible alignments of these sequences, where for a particular alignment this probability is maximized over the sequence composition of the unknown sequences. This obviously gives an upper bound for the total alignment likelihood of the n sequences, and one that is considerably better than the likelihood of simply aligning the m known sequences on an m-leaved tree. Moreover, it gives upper bounds for each of the DP table entries in the n-dimensional table by projecting to the smaller m-dimensional table. By taking the minimum over all combinations of m sequences out of the n given ones, good upper bounds are obtained for the entire DP table. The final upper bound for the alignment probability is obtained by performing the DP recursion on the contributing region and using the m-sequence-based upper bound whenever the recursion refers to an entry outside that region. This approach was used to compute the alignment likelihoods and maximum likelihood trees depicted in Figure 14.10.

Multiple Viterbi algorithm

The method of "shaving off" a dimension of the DP table in the forward-backward algorithm cannot be used for the Viterbi algorithm, as it contains a backtracking loop to find the most likely path, which may visit any part of the DP table. A clever idea due to Hirschberg [15] reduces the memory requirements to order SL^{n-1} in a different way at the cost of an increase in time complexity by only a constant factor. The algorithm consists of a standard Viterbi algorithm that, however, does not retain its DP table and stops halfway. A "backward" Viterbi algorithm then starts at the other end and again stops halfway. Using their outputs, the central state of the Viterbi path is determined, but no backtracking is possible. However, with the central state known, the Viterbi recursion can be performed again but is now constrained to two DP tables of size roughly $(L/2)^n$. The same strategy is used again in the smaller tables until after several recursive divisions the full Viterbi path is found. The algorithm runs in time proportional to

$$
S^2 L^n \times \left[1 + \left(\frac{1}{2} \right)^{n-1} + \left(\frac{1}{2} \right)^{2(n-1)} + \cdots \right] = S^2 L^n \frac{2^{n-1}}{2^{n-1} - 1}, \quad (14.14)
$$

an increase of at most a factor 2. Unfortunately, Hirschberg's algorithm does not perform so well if it is combined with constraints to a contributing region. Such regions usually lie close to the diagonal of the DP table, and the Hirschberg halving strategy takes off almost nothing from such an essentially one-dimensional contributing region. The use of table constraints is highly desirable, however, as the algorithm otherwise becomes impractical already for as little as four sequences.

Another strategy, termed "bushy Viterbi," has the same memory usage as Hirschberg's algorithm and the same constant time penalty but can be combined with the contributing region strategy as well. The idea is to combine the two stages of the Viterbi algorithm into one and do backtracking on-the-fly. For this to work, each state requires an additional pointer to the state it refers to and a reference count. The algorithm keeps optimal paths for each state in the current $n - 1$-dimensional DP table slice. Whenever a slice is completed, all reference counts in the previous slice are decreased by one, and those that are not referenced by states in the current slice are removed. The table entries to which these states refer have their reference counts decreased as well, and when they reach zero, the entries are removed in turn, and so on. Since the optimal paths for the various states quickly coalesce, the set of all paths is in practice very tree-like, as most coalescence events occur close to the tips, and requires not much more memory beyond the L^{n-1} DP table entries. By doing the garbage collection only occasionally, the time complexity is also not much more than for the ordinary Viterbi algorithm. This algorithm was used to calculate the Viterbi alignment of Figure 14.11.

14.5 Monte Carlo Approaches

The major difficulty with statistical alignment has been in extending it to practical problem sizes. Alignments of tens or hundreds of sequences are routinely required in standard bioinformatics and phylogenetics settings. Exact techniques for statistical alignment are restricted to four or five sequences [27].

In this section, we review Monte Carlo approaches that promise to considerably extend the domain of application of statistical alignment.

14.5.1 Statistical Alignment Using MCMC and TKF91

A number of researchers have been motivated to develop MCMC sampling algorithms to extend the use of the TKF91 model into the realms of practical multiple-sequence alignment. The first such effort was by Holmes and Bruno [18], who produced an MCMC approach to statistical alignment under the TKF91 model conditional on a fixed tree topology and branch lengths. They used data-augmentation techniques to include paired-sequence alignments (henceforth referred to as branch alignment) on each branch of the tree as well as inferred sequences at internal nodes. The proposal distribution they used consisted of two Gibbs sampling moves that resampled (1) a branch alignment conditional on the two adjacent sequences (one of which might be an inferred sequence) and (2) a sequence at an internal node conditional on the three adjacent branch alignments (while allowing insertion of characters unaligned with any of the three neighbors). Both of these moves involve sampling a subspace of the augmented problem from the exact conditional probability. This method was followed by another Gibbs sampler [13] that reduced the state-space by not requiring the branch alignments to be retained between successive states. This was achieved by using a more computationally intensive Gibbs move that resampled an internal sequence conditional only on the three neighboring sequences. The algorithm of Hein et al. is $O(L^3)$ in the length of the sequence, as opposed to the $O(L^2)$ move of Holmes and Bruno. However, Hein et al. demonstrated that their algorithm's superior mixing more than made up for the extra computational time. In terms of effectively independent samples per CPU second, the Hein et al. method appeared to be an improvement. Both of these methods relied on EM optimization for values of the rates of substitution, insertion, and deletion. Theoretically, these parameters could easily be Metropolis sampled as part of the algorithm.

A third group has used MCMC to sample pairs of sequences [29, 28]. This work focuses on including alignment uncertainty into estimates of branch lengths. While these authors do not address the full problem of multiple alignments, they were the first to demonstrate the feasibility of a full Bayesian approach to co-sampling alignments and evolutionary parameters.

14.5.2 Removing the Requirement for Data Augmentation

One of the reasons that data augmentation was required for the MCMC methods above was that the likelihood of the whole tree could not be efficiently calculated without internal sequences. A better solution would be to have an analogue of the Felsenstein peeling algorithm [8], which would analytically sum out the sequences and gaps at internal nodes. With such an algorithm, the state could simply consist of the tree topology together with the homology structure (multiple sequence alignment) at the tips. No branch alignments or internal sequences are then required. Not only would this considerably simplify the extension of the statistical alignment problem to co-estimation (the tree topology can be sampled without worrying about disturbing augmented data), but it should also reap computational benefits in the same way that the Hein et al. method did over the Holmes and Bruno one.

Surprisingly, a peeling method for the TKF91 model on a binary tree is not only possible, but is also computationally very cheap. We used this method to include indels as informative events in phylogenetic inference [26], and it is the basis of the co-estimation method described below.

14.5.3 Example of Co-estimation

Previous methods applying MCMC to statistical alignment problems did not sample evolutionary trees. The recent development of the TKF91 peeling method mentioned above removes the requirement for data augmentation, making tree-change proposals very simple. However, this ease of manipulating the tree comes with a drawback: without data augmentation, it does not appear to be possible to perform Gibbs sampling on the alignment. Instead, other sampling methods are required, and careful design is needed for good performance. We have developed a partial importance sampler, which has good mixing properties in terms of estimated sample size (ESS) per CPU cycle. This method uses a stochastic score-based approach to propose new alignments. The proposal distribution is reshaped into the posterior distribution by standard Metropolized importance sampling techniques. We used the program BEAST written in Java as the MCMC inference engine [4, 5].

In more detail, the method works as follows. Given a multiple alignment, a random window is selected for modification, and a new subalignment in this window is proposed by a stochastic version of a score-based progressive alignment method. In this stochastic alignment method, sequences and profiles are progressively aligned using a pairwise algorithm, guided by the tree of the current MCMC state. In each iteration of the stochastic alignment, the dynamic programming table is filled as in the deterministic case by using linear gap penalties and standard similarity matrices. The stochastic element appears during the traceback phase. At each step during traceback, a random decision is made, biased toward the highest-scoring alternative. If the three alternatives have scores a, b, and c, respectively, the algorithm chooses among the alternatives with probabilities proportional to x^a, x^b, and x^c, respectively, where $x > 1$. The stochastic path chosen determines the proposed alignment. It can be shown that all possible alignments of the subsequences can be proposed in this manner, and the proposal and back-proposal probabilities can be calculated relatively easily.

To get a reversible Markov chain, all window sizes must be proposed with a nonzero probability. We used a truncated geometric window-size distribution, but other distributions can also be used. The parameters that appear in this

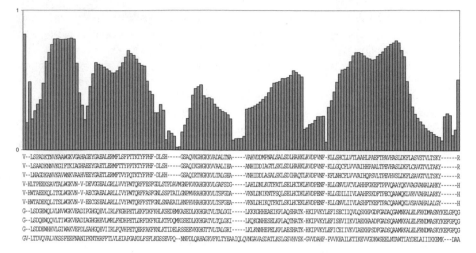

Fig. 14.12. The maximum posterior decoding of an alignment of ten globins: alpha hemoglobin (human, chicken, turtle), beta hemoglobin (human, etc.), myoglobin (human, etc.), and bean leghemoglobin). Estimates of posterior column probabilities were obtained by co-sampling phylogenetic trees and alignments through MCMC using an alignment proposal distribution in windows of varying sizes and a linear-time likelihood calculator for the TKF91 model in trees. For the MCMC run on which these results are based, the estimated sample size was about 80. The column posterior probabilities qualitatively agree with the analytic posterior probabilities for the maximum likelihood alignment, based on just four of the ten globins (see Figure 14.11).

stochastic alignment algorithm, such as the average window size, determine the proposal distribution but do not influence the resulting posterior distribution. However, they do influence the efficiency of the MCMC sampler. For example, if the basis of exponentiation, x, is small, the proposal distribution will be flat, leading to a small acceptance ratio. When x is too large, the proposal distribution will be too narrow, resulting in bad mixing behavior if the distribution is far from the target distribution. The gap penalty value has a similar effect: if it is small, many alignments have a similar probability of being proposed, while a big penalty results in a proposal distribution containing very few alignments.

Figures 14.12 and 14.13 illustrate the results of this co-estimation method on a set of ten globin sequences. These pictures were produced from two MCMC runs with a total chain length of 10,000,000 and a burn-in of 500,000. The basis of exponentiation x was chosen to be 1.5, and the mean window size was 40 amino acids. We used the BLOSUM62 matrix and gap penalty -10.

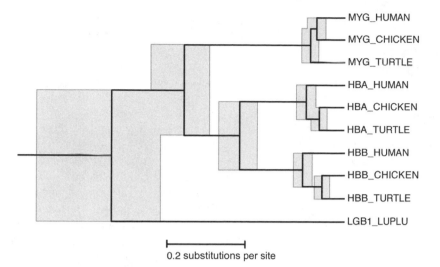

Fig. 14.13. The maximum posterior tree (black) relating the ten globins of Figure 14.12 and 95% confidence intervals of the node heights (grey boxes). Most of the tree's topology is well-determined, with the exception of the myoglobin subtree. Note that this relatively unresolved topology differs from the more well-defined topologies down the alpha and beta hemoglobin branches, both of which conform to the accepted phylogenies of human, chicken, and turtle.

14.6 Discussion

Recent progress in the development of statistical alignment methods, and especially the emergence of practical algorithms, has made it possible to treat the problem of sequence alignments as a statistical inference problem, estimate evolutionary indel parameters, and quantify alignment uncertainties. This development shows parallels with the success of statistical methods for phylogenetic inference since the 1980s.

Several aspects of statistical alignment methods have seen important progress: methods for pairwise alignment have been generalized to multiple sequences; more realistic insertion/deletion models have been proposed; hidden Markov model theory has conceptually simplified many algorithms; and MCMC methods have considerably extended the domain of application. These successes are due to, and resulted in, a growing interest in statistical alignment problems [14, 18, 12, 40, 29, 31, 19, 26, 27, 28, 13, 21, 17, 32]. In 2001, pairwise alignment was just about a feasible task for statistical alignment. At present, the limit has been pushed up to about ten sequences. Much larger data sets are routinely of interest, and there is clearly a need for cleverly designed MCMC algorithms to tackle such problems.

The possibility of assessing the goodness-of-fit of a given statistical alignment model is a strength of probabilistic approaches and allows for data-driven model improvements. Many such challenges remain, such as the inclusion of more biological realism in the models, incorporating, for example, indel rate heterogeneity and variable substitution rates. Although heterogeneity of substitution processes has been extensively explored in the context of phylogenetic inference, it is largely unexplored in the context of sequence alignment. Perhaps even more importantly, the development of user-friendly software will be essential to make the methods appeal to a wider audience.

Sequence alignment is often just the first step in any analysis. Most current methods, such as comparative gene finding and RNA secondary structure prediction, but also phylogenetic inference, assume a prior and fixed alignment. These methods can be combined with statistical alignment either by a full co-estimation procedure, simply by using a sample of alignments, or by incorporating the column reliabilities as weights. Such a hybrid approach would reduce the bias introduced by assuming exact knowledge of sequence homology and at the same time increase the sensitivity by focussing on reliable data, and work in this direction is already in progress (see ,e.g., [16]).

The understanding of molecular evolution today owes much to the development of adequate evolutionary models. We hope that statistical alignment will contribute to this fundamental understanding in the coming years.

Acknowledgements

The authors wish to thank Ian Holmes, Yun Song, Arnt von Haeseler, Bjarne Knudsen, Dirk Metzler, Korbinian Strimmer, and Anton Wakolbinger for helpful remarks and stimulating discussions. The authors acknowledge support from EPSRC and the MRC grant HAMKA. I.M. was further supported by a Békésy György postdoctoral fellowship.

References

[1] S. F. Altschul, W. Gisha, W. Miller, E. W. Meyers, and D. J. Lipman. Basic local alignment search tool. *J. Mol. Biol.*, 215(3):403–410, 1990.

[2] S. F. Altschul, T. L. Madden, A. A. Schaffer, J. Zhang, Z. Zhang, W. Miller, and D. J. Lipman. Gapped-BLAST and PSI-BLAST: A new generation of protein database search programs. *Nucleic Acids Res.*, 25:3389–3402, 1997.

[3] H. Carillo and D. Lipman. The multiple alignment problem in biology. *SIAM J. Appl. Math.*, 48:1073–1082, 1988.

[4] A. J. Drummond, G. K. Nicholls, A. G. Rodrigo, and W. Solomon. Estimating mutation parameters, population history and genealogy simultaneously from temporally spaced sequence data. *Genetics*, 161(3):1307–1320, 2002.

[5] A. J. Drummond and A. Rambaut. BEAST v1.0.3. http://evolve.zoo.ox.ac.uk/beast/, 2003.

[6] R. Durbin, S. Eddy, A. Krogh, and G. Mitchison. *Biological sequence analysis.* Cambridge University Press, Camebridge, 1998.

[7] S. Eddy. HMMER. Profile hidden Markov models for biological sequence analysis (http://hmmer.wustl.edu/), 2001.

[8] J. Felsenstein. Evolutionary trees from DNA sequences: A maximum likelihood approach. *J. Mol. Evol.*, 17:368–376, 1981.

[9] J. Felsenstein. The troubled growth of statistical phylogenetics. *Syst. Biol.*, 50:465–467, 2001.

[10] O. Gotoh. An improved algorithm for matching biological sequences. *J. Mol. Biol.*, 162:705–708, 1982.

[11] J. Hein. A unified approach to phylogenies and alignments. *Meth. Enzym.*, 183:625–644, 1990.

[12] J. Hein. An algorithm for statistical alignment of sequences related by a binary tree. In *Pacific Symposium on Biocomputing*, pages 179–190. World Scientific, Singapore, 2001.

[13] J. Hein, J. L. Jensen, and C. N. S. Pedersen. Recursions for statistical multiple alignment. *Proc. Natl. Acad. Sci. USA*, 100(25):14960–14965, 2003.

[14] J. Hein, C. Wiuf, B. Knudsen, M. B. Møller, and G. Wibling. Statistical alignment: Computational properties, homology testing and goodness-of-fit. *J. Mol. Biol.*, 302:265–279, 2000.

[15] D. S. Hirschberg. A linear space algorithm for computing maximal common subsequences. *Communi. ACM*, 18:341–343, 1975.

[16] A. Hobolth and J. L. Jensen. Applications of hidden Markov models for comparative gene structure prediction. Technical Report MPS-RR 2003-35, MaPhySto, Aarhus, Denmark, 2003.

[17] I. Holmes. Using guide trees to construct multiple-sequence evolutionary HMMs. *Bioinformatics*, 19:i147–i157, 2003.

[18] I. Holmes and W. J. Bruno. Evolutionary HMMs: A Bayesian approach to multiple alignment. *Bioinformatics*, 17(9):803–820, 2001.

[19] J. L. Jensen and J. Hein. Gibbs sampler for statistical multiple alignment. *Stat. Sinica*, 2004+. (In press).

[20] K. Karplus, C. Barrett, and R. Hughey. Hidden Markov Models for detecting remote protein homologies. *Bioinformatics*, 14:846–856, 1998.

[21] B. Knudsen and M. M. Miyamoto. Sequence alignments and pair Hidden Markov Models using evolutionary history. *J. Mol. Biol.*, 333:453–460, 2003.

[22] A. Krogh. Two methods for improving performance of a HMM and their application for gene finding. In T. Gaasterland, P. Karp, K. Karplus, C. Ouzounis, C. Sander, and A. Valencia, editors, *Proceedings of the Fifth International Conference on Intelligent Systems for Molecular Biology*, pages 179–186, Menlo Park, CA, 1997. AAAI Press.

[23] M. S. Y. Lee. Unalignable sequences and molecular evolution. *Trends Ecol. Evol.*, 16:681–685, 2001.

[24] D. J. Lipman, S. F. Altschul, and J. D. Kececioglu. A tool for multiple sequence alignment. *Proc. Natl. Acad. Sci. USA*, 86:4412–4415, 1989.

[25] A. Löytynoja and M. C. Milinkovitch. A hidden Markov model for progressive multiple alignment. *Bioinformatics*, 19:1505–1513, 2003.

[26] G. A. Lunter, I. Miklós, A. Drummond, J. L. Jensen, and J. Hein. Bayesian phylogenetic inference under a statistical indel model. In *Proceedings of WABI'03*, volume 2812 of *Lecture Notes in Bioinformatics*, pages 228–244, Heidelberg, 2003. Springer-Verlag.

[27] G. A. Lunter, I. Miklós, Y. S. Song, and J. Hein. An efficient algorithm for statistical multiple alignment on arbitrary phylogenetic trees. *J. Comp. Biol.*, 10(6):869–889, 2004.

[28] D. Metzler. Statistical alignment based on fragment insertion and deletion models. *Bioinformatics*, 19(4):490–499, 2003.

[29] D. Metzler, R. Fleissner, A. Wakolbinger, and A. von Haeseler. Assessing variability by joint sampling of alignments and mutation rates. *J. Mol. Evol.*, 53:660–669, 2001.

[30] I. M. Meyer and R. Durbin. Comparative ab initio prediction of gene structures using pair HMMs. *Bioinformatics*, 18(10):1309–1318, 2002.

[31] I. Miklós. An improved algorithm for statistical alignment of sequences related by a star tree. *Bull. Math. Biol.*, 64:771–779, 2002.

[32] I. Miklós, G. A. Lunter, and I. Holmes. A "long indel" model for evolutionary sequence alignment. *Mol. Biol. Evol.*, 21(3):529–540, 2004.

[33] G. Mitchison. A probabilistic treatment of phylogeny and sequence alignment. *J. Mol. Evol.*, 49:11–22, 1999.

[34] B. Morgenstern. DIALIGN 2: Improvement of the segment-to-segment approach to multiple sequence alignment. *Bioinformatics*, 15:211–218, 1999.

[35] E. W. Myers. An O(ND) difference algorithm and its variations. *Algorithmica*, 1:251–266, 1986.

[36] S. B. Needleman and C. D. Wunsch. A general method applicable to the search for similarities in the amino acid sequences in two proteins. *J. Mol. Biol.*, 48:443–453, 1970.

[37] C. Notredame, D. Higgins, and J. Heringa. T-Coffee: A novel method for multiple sequence alignments. *J. Mol. Biol.*, 302:205–217, 2000.

[38] B. Qian and R. A. Goldstein. Distribution of indel lengths. *Proteins Struct. Func. Gen.*, 45:102–104, 2001.

[39] D. Sankoff. Minimal mutation trees of sequences. *SIAM J. Appl. Math.*, 28:35–42, 1975.

[40] M. Steel and J. Hein. Applying the Thorne-Kishino-Felsenstein model to sequence evolution on a star-shaped tree. *Appl. Math. Lett.*, 14:679–684, 2001.

[41] J. D. Thompson, D. G. Higgins, and T. J. Gibson. CLUSTAL W: Improving the sensitivity of progressive multiple sequence alignment through

sequence weighting, position-specific gap penalties and weight matrix choice. *Nucleic Acids Res.*, 22:4673–4680, 1994.

[42] J. L. Thorne, H. Kishino, and J. Felsenstein. An evolutionary model for maximum likelihood alignment of DNA sequences. *J. Mol. Evol.*, 33:114–124, 1991.

[43] J. L. Thorne, H. Kishino, and J. Felsenstein. Inching toward reality: An improved likelihood model of sequence evolution. *J. Mol. Evol.*, 34:3–16, 1992.

[44] E. Ukkonen. Algorithms for approximate string matching. *Inf. Control*, 64:100–118, 1985.

[45] A. von Haeseler and M. Vingron. Towards integration of multiple alignment and phylogenetic tree construction. *J. Comp. Biol.*, 4(1):23–34, 1997.

[46] L. Wang and T. Jiang. On the complexity of multiple sequence alignment. *J. Comp. Biol.*, 1:337–348, 1994.

[47] S. Wu, U. Manber, G. Myers, and W. Miller. An O(NP) sequence comparison algorithm. *Inf. Process. Lett.*, 35:317–323, 1990.

15

Estimating Substitution Matrices

Von Bing Yap[1] and Terry Speed[2]

[1] Department of Mathematics, University of California, Berkley, CA 94720-3840, USA, vonbing@math.berkeley.edu.
[2] Department of Statistics, University of California,Berkeley, CA 94720-3840, USA, terry@stat.berkeley.edu

15.1 Introduction

An amino acid substitution matrix is used in sequence alignment to assign a measure of evolutionary similarity between every pair of amino acids. The substitution score $S(a, b)$ for amino acids a and b can be interpreted probabilistically as being proportional to the logarithm of a probability ratio

$$S(a, b) = s \log \frac{f(a, b)}{\pi(a)\pi(b)}, \tag{15.1}$$

where f is a joint distribution that reflects some evolutionary relation, π is the background distribution, and s is a constant. Thus the product in the denominator is the probability that two unrelated aligned residues happen to be a and b, and the fraction is the likelihood ratio of homology (relatedness) against nonhomology. Usually, the joint probability is chosen to be symmetric (i.e., for any a and b, $f(a, b) = f(b, a)$), and the background frequencies are the same as the row or column sum or the marginal distribution of the joint distribution: $\pi(a) = \sum_b f(a, b) = \sum_b f(b, a)$. As a consequence, S is symmetric, and this has the advantage that it is not necessary to distinguish between the sequences. Another consequence is that construction of S is reduced to that of a symmetric joint distribution. In this chapter, we only deal with symmetric substitution matrices. Asymmetric matrices are useful for certain database searches [32]. Since amino acid matrices have been used for a longer time than DNA matrices, and almost all of the theory is common to both, we will only discuss the former, except in Section 15.7.

The result of sequence alignment depends substantially on the substitution matrix, and much work has gone into choosing optimal matrices [16, 36, 5, 23]. Moreover, sequence similarity varies considerably, so that a series of substitution matrices tuned to different evolutionary distances are necessary. The most widely used matrices are the PAM [9] and the BLOSUM [24] series. They are empirical in the sense that they were estimated from aligned amino acid sequences without any extra information. The PAM matrices and similar matrices [31, 22, 27, 26] were derived on the assumption that the substitution process was like a reversible homogeneous Markov process (i.e., the substitution rates were constant in time, and the substitution process looked the same going forward and backward in time). The data used consisted of closely related sequences, so the fact that the sequences were separated by different evolutionary distances could be ignored. More recently, methods capable of dealing rigorously with more distantly related sequences were used, including the resolvent method by Müller and Vingron [34] and the maximum likelihood methods of Müller et al. and Holmes and Rubin [33, 25]. There are several other methods for estimating a reversible substitution process [2, 11], but these will not be discussed because they were not used to construct substitution matrices. A very different approach was taken by Henikoff and Henikoff [24], who employed a heuristic weighting technique on multiple alignments to produce the BLOSUM series. Unlike the former methods, the BLOSUM method is not explicitly based on any substitution model, though it quite clearly takes into account nonconstant substitution rates (inhomogeneity). In principle, this seems to us a good idea since nonconstant substitution rates had been observed [3]; however, see [35] for a different point of view. Indeed, BLO-SUM matrices often outperform PAM matrices in database searches. There are many substitution matrices that were derived with some additional information such as physical or chemical properties of amino acids and secondary structures of proteins [24, 29]. Generally, these matrices seem no better than the empirical matrices.

The aim of this chapter is to describe some estimation methods for substitution matrices and to discuss their relative strengths and weaknesses. We conclude that the BLOSUM method is preferable for constructing substitution matrices at large evolutionary distances, while methods based on homogeneous substitution processes are better at small evolutionary distances. The outline of the chapter is as follows. In Section 15.2, the theory of Markov processes is briefly reviewed. The PAM and BLOSUM methods are explained in Sections 15.3 and 15.5. Section 15.4 discusses two consistent estimation methods for homogeneous Markov processes: maximum likelihood and the resolvent method. Section 15.6 is a comparison of methods on simulated data from homogeneous and inhomogeneous processes. Finally, Section 15.7 is a brief description of DNA substitution matrices.

15.2 Markov Substitution Processes

The substitution process on a site is almost always modeled as a Markov process $\{X_t\}_{t\geq0}$ (i.e., if $t_1 < t_2 < \cdots < t_n$, then the conditional distribution of X_{t_n} given $X_{t_1}, \ldots, X_{t_{n-1}}$ depends only on $X_{t_{n-1}}$). Equivalently, given the present state, the past and future substitutions are conditionally independent. The *Markov property* is probably a good approximation to reality and greatly simplifies the calculation of probabilities. For $s < t$, let $P(s, t)$ denote the *transition probability* from s to t (i.e., its (a, b) entry $P(s, t, a, b) = \Pr(X_t = b | X_s = a)$). In a Markov process, the family of transition probabilities is sufficient to compute the probabilities of all events of interest. Another assumption is that sites evolve independently, which is known to be false in general but is not a great concern here since most sites seem to behave independently. A Markov process can be realized on a rooted tree by starting at the root and splitting into independent copies whenever a branching point is encountered until the leaf nodes are reached.

A *homogeneous* Markov process has the property that $P(s, t)$ depends only on $t - s$, so that it makes sense to define $P(t - s) = P(s, t)$. Without loss of applicability, we may assume that the transition probabilities have a right derivative at $t = 0$ (defining $P(0) = I$, the identity matrix). This derivative Q is the infinitesimal generator, or *rate matrix*, of the process. Thus, the off-diagonal entries of Q are nonnegative, and every row sums to 0. Furthermore, we have $P(t) = \exp(Qt)$. Consider only those rate matrices with strictly positive off-diagonal entries. Such a matrix Q has a unique probability vector π satisfying $\pi Q = 0$, or equivalently, $\pi P(t) = \pi$ for any t. π is called the *equilibrium distribution* of Q. Let the process start with π (i.e., $X_0 \sim \pi$). Then the marginal distribution of each X_t is also π (i.e., the process is *stationary*). A canonical construction of the process proceeds as follows. Given $X_s = a$, the waiting time to the next substitution is distributed exponentially with rate $-Q(a, a)$ (i.e., the probability that the next substitution happens after time $t > s$ is $\exp(Q(a, a)(t - s)))$. If the next substitution happens at time $u > s$, then the process jumps from a to $b \neq a$ with probability $-Q(a, b)/Q(a, a)$. A homogeneous Markov process is *reversible* if for any $t_1 < \cdots < t_n$, $(X_{t_1}, \ldots, X_{t_n})$ has the same distribution as $(X_{t_n}, \ldots, X_{t_1})$. A reversible process is stationary, and furthermore Q satisfies the detailed balance equations

$$\pi(a)Q(a, b) = \pi(b)Q(b, a) \quad \forall a, b \tag{15.2}$$

or equivalently

$$\pi(a)P(t, a, b) = \pi(b)P(t, b, a) \quad \forall t, a, b.$$

Conversely, if the detailed balance equations hold and the process is stationary, then it is reversible [28]. Consider the *joint distribution* F for two states X_α and X_β at leaf nodes α and β at distance t apart: $F(a, b) = \Pr(X_\alpha = a, X_\beta = b)$. A nice property of the reversible process is that F is independent of where the common ancestor is located on the path connecting the leaf nodes (i.e., $F(t) := F$ is well-defined). In particular, treating α and β as the ancestor, respectively, gives

$$F(t, a, b) = \pi(a)P(t, a, b) = \pi(b)P(t, b, a) = F(t, b, a), \tag{15.3}$$

or $F(t) = \Pi P(t)$ is symmetric, where Π is the diagonal form of π. Plugging $F(t)$ into the numerator of (15.1) gives a substitution matrix at distance t.

All substitution matrices, except for the BLOSUM series, were constructed based on the reversible process, and hence they are symmetric. The reversible process is also widely used in the study of molecular evolution [30].

If a set of extant sequences are homologous (i.e., they share a common ancestor), then their evolutionary history can be represented as a tree with the ancestor at the root and the sequences at the leaf nodes. Moreover, the distance (time) d between the root and any leaf node is the same. Consider the problem of inferring the tree from a multiple alignment of the sequences. Clearly, without knowing d, this is impossible, but if the molecules behave like clocks, then we will get branch lengths proportional to the real distances. However, real molecules are not clock-like (i.e., the substitution rates vary across lineages). One is almost forced to estimate *evolutionary distance*, which is proportional to the number of substitutions per site, instead of chronological time. Clearly, if the process is sped up by a factor of 2, then the estimated evolutionary distance is halved. In order to compare distances estimated from different processes, it is necessary to calibrate the processes so that they all have the same expected number of substitutions per unit time. For the homogeneous process determined by Q, the expected number of substitutions per site in an interval of length t is $\kappa(Q)t$, where

$$\kappa(Q) = \sum_a \pi(a)Q(a,a).$$

Dayhoff et al. [7] essentially chose the calibration $\kappa(Q) = 0.01$ (explained fully in the next section) and called the associated unit of evolutionary time PAM.

15.3 PAM Matrices

The PAM (accepted point mutations) matrices by Dayhoff et al. [7, 8, 9] were the first substitution matrices. The dataset in [9] consisted of families of closely related amino acid sequences, such that every pair of homologous sequences were more than 85% identical. The aim was to estimate a transition probability P_1 with equilibrium distribution π such that the probability that a site is unchanged, $\sum_a \pi(a)P_1(a,a)$, is 0.99. This calibration is essentially the same as that in the previous section (i.e., if $P_1 = \exp(Qt)$ for some calibrated Q and some $t > 0$, then $t \approx 1$). Indeed, since Qt is small, $P_1 \approx I + Qt$, so we have

$$0.99 = \sum_a \pi(a)(1 + Q(a,a)t),$$

which reduces to $\kappa(Q)t = 0.01$, or $t = 1$. In view of this, we will use $P(1)$ instead of P_1. The PAM1 substitution matrix, by (15.1) and (15.3), is

$$\text{PAM1}(a,b) = 10\log_{10}\frac{F(1,a,b)}{\pi(a)\pi(b)} = 10\log_{10}\frac{P(1,a,b)}{\pi(b)}.$$

Matrices at larger distances are derived by raising $P(1)$ to the appropriate power. In practice, the entries of substitution matrices are rounded.

A tree was constructed for each family, and the ancestral sequences were inferred by parsimony [15]. The number of occurrences of all 400 types of substitutions between neighboring sequences were collected into a 20×20 frequency table C, which was symmetrized by adding its transpose to itself. After C is normalized (i.e., divided by the sum of all entries), we have a symmetric joint distribution. Associated with C is a transition probability P obtained by dividing each row of C by its sum. One way of going from P to $P(1)$ is to compute $Q = \log P$ followed by calibration and exponentiation. Using octave, these operations produce Q with negative off-diagonal entries. It turns out that the phenomenon is general: if Q has many small off-diagonal entries, then going from P to Q will likely yield negative off-diagonal entries; a simple fix is to set every negative off-diagonal entry to 0. Dayhoff et al. avoided taking logarithms, perhaps for this reason. Instead they relied on the fact, without explicit calculation involving a rate matrix, that the family of transition probabilities $\{P_\lambda\}_{\lambda>0}$ defined by

$$P_\lambda(a, b) = \begin{cases} 1 - \lambda m(a) & \text{if } a = b \\ \lambda P(a, b) & \text{otherwise} \end{cases}$$

approximately correspond to a reversible process, where $m(a)$ is the mutability of amino acid a (i.e., $m(a) = \sum_{d \neq a} P(a, d)$). Now λ could be chosen so that the probability that a site does not change is 0.99:

$$\lambda = \frac{0.01}{\sum_a \pi(a)m(a)}.$$

Then set $P(1) = P_\lambda$.

Wilbur [37] pointed out some deficiencies of the PAM matrices based on codon substitution considerations. Here, we are more concerned with the statistical aspect of going from the data to the estimate $P(1)$. Two steps of the method could introduce bias:

1. Substitution events over branches of different lengths are aggregated. This is correct if all branch lengths are the same.
2. The approximation $P(t) = I + Qt$ was used for calibration: going from P to $P(1)$. This is good if Qt is small enough.

As explained before, step 2 can be accomplished reasonably via the logarithm of P and setting negative off-diagonal entries to 0. Let D2 be a new procedure, which uses step 1 and exact calibration. Thus, the only source of bias in D2 is step 1. Comparing the PAM method and D2 gives some idea about the bias introduced by step 2.

We performed simulations to evaluate the bias introduced by the Dayhoff and D2 procedures. In fact, we only investigate the asymptotic bias, by assuming that the amount of data is infinite, so that all transition probabilities are known. A slightly modified version of the rate matrix estimated by Dayhoff et al. was taken as the true rate matrix Q. Assuming that on average each of the 71 trees used by Dayhoff et al. has five leaf sequences, the total number of branches is about $71 \times 8 = 568$. We generated 600 branches of lengths between 0 and 15 PAMs according to six schemes:

1. All branches are of length 1.
2. All branches are of length 5.
3. All branches are of length 10.
4. All branches are of length 15.
5. Branch lengths are drawn from the uniform distribution on $(0, 15)$, $U(0, 15)$.
6. 300, 150, and 50 branch lengths are drawn from $U(0, 5)$, $U(5, 10)$ and $U(10, 15)$, respectively. This case is perhaps close to the original dataset.

Since the amount of data is infinite, the frequency table for each branch is proportional to the appropriate joint distribution. The two methods were applied to get substitution matrices at 1, 50, 100, 160, and 250 PAMs, which were then compared with the true substitution matrices computed from Q. It turns out that in almost all the cases, the entrywise absolute difference is 0 or 1, so the number of different entries is a good measure of distance. The results are summarized in Table 15.1. For the last two schemes, 100 simulations were performed and the rounded averages were reported; the standard deviations are at most 7% of the averages and hence are not shown.

Table 15.1. Number of different entries between estimated and true substitution matrices. a: PAM method; b: D2 method.

branch lengths	50		100		160		250	
	a	b	a	b	a	b	a	b
all equal to 1	0	0	0	0	0	0	0	0
all equal to 5	58	0	26	0	34	0	10	0
all equal to 10	128	0	76	0	69	0	32	0
all equal to 15	192	0	131	0	103	0	58	0
scheme 5	129	36	75	18	69	20	31	6
scheme 6	102	38	52	19	46	22	22	6

In the first four rows of Table 15.1, as expected, D2 makes no error, and the PAM method deteriorates with increasing branch length. Since branch lengths are the same in each case, the biases are entirely due to step 2. The last two rows suggest that the bias caused by both steps 1 and 2 is about 3 to 4 times larger than that due to step 1 alone. The results are the same when the number of branches is increased up to 5000. Thus, it appears that step 2 is culpable for most of the bias in the Dayhoff method. Interestingly, in all the simulations, the bias at larger PAM distances tends to become smaller.

This is due to the fact that both methods estimate the equilibrium distribution well, so as the distance gets larger, the estimated transition probabilities become closer to the true transition probability. The largest bias at 160 PAMs is 103 for the worst case, where each branch length is 15 PAMs. Henikoff and Henikoff suggested that BLOSUM62 is comparable to PAM160. The maximum entrywise difference is 8, while the sum of the absolute differences is 808. The corresponding numbers for comparing PAM160 to BLOSUM45, which is more appropriate in our opinion (see Section 15.6), are 7 and 803. These are quite large compared with the largest deviation at 160 PAMs (103), suggesting that correcting the bias in the PAM method does not seem to matter much in practice. The corrected PAM matrices should be very similar to the original ones, relative to the BLOSUM matrices.

15.4 Consistent Estimation of a Reversible Rate Matrix

The PAM method is inconsistent (i.e., it is biased even with an infinite amount of data). Besides the two issues discussed in the previous section, inferring tree topologies by parsimony, which is inconsistent [13], also contributes to its inconsistency, though this is not serious for the dataset of Dayhoff et al. Maximum likelihood (ML) is consistent but is computationally expensive. A promising approach is via Markov chain Monte Carlo in a Bayesian framework [12]. From now on, we assume that the tree topology is known and consider the estimation of branch lengths and a reversible substitution process. A simple approach is to toggle between two simpler procedures: (a) estimating the calibrated rate matrix given the current branch-length estimates and (b) estimating the branch lengths given the current rate matrix estimate. It is not hard to see that if (a) and (b) are consistent, then the whole procedure is consistent. (b) can be done quite easily by ML. We describe three consistent methods of doing (a). The first, called ML I, is ML for a reversible process. The second, ML II, is also ML, but for a more general substitution process, which can be implemented via an EM algorithm. Finally, there is a deterministic method based on resolvents (RES).

15.4.1 ML I

To find the ML estimate of the calibrated rate matrix, it is necessary to compute the probabilities of the observed sequences at the leaf nodes. Suppose for the moment that the ancestral sequences at the internal nodes as well as the root are observed. Let t_i be the length of the ith branch, and let $\nu(i)$ be the frequency table for the pair of sequences connected by this branch:

$$\nu(i, a, b) = |\{\text{ancestor} = a, \text{descendant} = b\}|.$$

Denote the row sum of the frequency table corresponding to the root by r. Then, by the Markov property, the probability of all sequences is

$$\prod_a \pi(a)^{r(a)} \prod_i \prod_{a,b} P(t_i, a, b)^{\nu(i,a,b)}.$$

By reversibility, this probability is independent of the location of the root. Summing probabilities like that above over all possible ancestral sequences gives the probability of the observed data at the leaf nodes. This can be efficiently done by the up algorithm (Subsection 15.4.5) due to Felsenstein [14], which recursively moves up the tree from the leaf nodes to the root. The probabilities of the data under different calibrated rate matrices constitute the likelihood function, and the rate matrix with the highest likelihood is the ML estimate.

To maximize the likelihood, one can use any of various parameterizations of a calibrated rate matrix. A natural choice is obtained by noting that the detailed balance equation (15.2) implies that $Q = R\Pi$, with R symmetric. Then the parameters are the equilibrium frequencies π and the top right off-diagonal elements of R, with the constraints

$$\pi(a) \geq 0, \quad \forall a,$$
$$R(a, b) \geq 0, \quad \forall a < b,$$
$$\sum_a \pi(a) = 1,$$
$$2 \sum_{a<b} \pi(a)\pi(b)R(a, b) = 0.01.$$

The last constraint calibrates Q. Another parameterization was used by Müller et al. [33]. Even for the simplest case, where each tree consists of a pair of sequences, it is difficult to get closed-form expressions for the ML estimate; numerical maximization needs to be used.

The relation $Q = R\Pi$ can be used to diagonalize Q, and this is useful for quick computation of the transition probabilities from Q. Let A be the positive square root of Π. Then $Q = R\Pi = A^{-1}(ARA)A$. Since ARA is symmetric, there exist orthogonal V and diagonal Λ such that $ARA = V\Lambda V'$, so

$$Q = A^{-1}V\Lambda V'A \qquad (15.4)$$

(i.e., Q is diagonalizable and its eigenvalues are the diagonal entries of Λ).

15.4.2 ML II

Holmes and Rubin [25] showed that an EM algorithm can be used to estimate a reversible rate matrix from a rooted tree. We will show that in general this algorithm finds the most likely substitution process defined by an initial distribution π and rate matrix Q, where π need not be the equilibrium distribution of Q and Q need not be reversible. If the data are such that the frequency table for any pair of leaf nodes is symmetric, and the initial rate matrix estimate is reversible, then the final estimate is guaranteed to be reversible. Thus, to force the algorithm to output a reversible rate matrix, some symmetrization of data is required. Moreover, a root position has to be specified, though this seems unimportant if the data are symmetrized; the root can be put at any leaf node.

If the root is put at a leaf node, then in some sense the estimates $\hat{\pi}$ and $\hat{Q}$ are separately inferred: $\hat{\pi}$ is simply the observed frequency at the root, and $\hat{Q}$ maximizes the conditional probability of the data given the root; this quantity is known as the partial likelihood of Q. Maximizing the partial likelihood is consistent [4], though less efficient than ML on the reversible process.

It is interesting to note that ML I implicitly symmetrized the data. This is most clearly seen by considering a pair of sequences separated by t PAMs with frequency table ν. Since the joint distribution $F(t)$ is symmetric, the probability of the data is

$$\prod_{a,b} F(t,a,b)^{\nu(a,b)} = \prod_{a,b} F(t,b,a)^{\nu(a,b)},$$

and this is in turn the same as

$$\prod_{a,b} F(t,a,b)^{\nu^*(a,b)},$$

where $\nu^*(a,b) = (\nu(a,b) + \nu(b,a))/2$. In other words, we can assume that the symmetric frequency table ν^* is observed; this ν^* is to be used in the EM algorithm to guarantee a reversible estimate. Now we can see that, given multiple independent pairs of sequences of similar compositions, both ML I and ML II give very similar estimates of a reversible rate matrix. Indeed, ML I gives an estimate $\hat{Q}_I$ whose equilibrium distribution $\hat{\pi}_I$ is very close to the overall composition. Let the ML II estimates be $(\hat{\pi}_{II}, \hat{Q}_{II})$. Since $\hat{\pi}_{II}$ is exactly the overall composition, $\hat{\pi}_I \approx \hat{\pi}_{II}$, and it follows that $\hat{Q}_I$ maximizes the partial likelihood (i.e., $\hat{Q}_I \approx \hat{Q}_{II}$). The claim that ML I and ML II are essentially equivalent for estimating a reversible rate matrix for the more general cases where the pairs have different compositions, and for more than two sequences, is plausible, but we will not need it.

If sequences at all internal nodes are observed, all branches have the same length t, and the branch-specific frequency tables ν_i are symmetrized, then the partial likelihood is

$$\prod_i \prod_{a,b} P(t_i,a,b)^{\nu(i,a,b)}$$

$$= \prod_{a,b} P(t,a,b)^{C(a,b)},$$

where $C = \sum_i \nu(i)$ is exactly the frequency table in the PAM method. Thus, the PAM method may be viewed as an approximation to maximizing partial likelihood.

We present the EM algorithm for two sequences generated by a discrete-time Markov chain, then the continuous-time version, and finally the tree version.

15.4.3 Pair EM: Discrete-Time Version

Let T be a positive integer. Consider independent realizations of the Markov chain $\{X_0, X_1, \ldots, X_T\}$ with initial distribution π and transition probability P. The full data $\mathbf{F}$ refer to the complete record of states at each time. Suppose that the observed data $\mathbf{O}$ consist of states only at times 0 and T (i.e., only the ancestral and descendant sequences are observed, but not the intermediate sequences). The probability of the realization $X_0 = x_0, \ldots, X_T = x_T$ is clearly

$$\pi(x_0) \prod_{t=1}^{T} P(x_{t-1}, x_t).$$

Let $\mathbf{X_i} = (X_{i,0}, X_{i,1}, \ldots, X_{i,T})$ denote the realization at site i. The log-likelihood for all sites can be expressed as

$$L(\pi, P; \mathbf{F}) = \sum_{a=1}^{s} \pi(a) Y(a) + \sum_{a,b=1}^{s} \log P(a,b) Z(a,b), \qquad (15.5)$$

where Y and Z are the frequencies of initial states and transitions:

$$Y(a) = \sum_i |\{X_{i,0} = a\}|,$$

$$Z(a,b) = \sum_i \sum_{t=1}^{T} |\{X_{i,t-1} = a, X_{i,t} = b\}|.$$

It follows that the ML estimate of (π, P) based on $\mathbf{F}$ is

$$\hat{\pi}(a) = \frac{Y(a)}{\sum_{c=1}^{s} Y(c)}, \qquad (15.6)$$

$$\hat{P}(a,b) = \frac{Z(a,b)}{\sum_{d=1}^{s} Z(a,d)}. \qquad (15.7)$$

The ML estimates based on $\mathbf{O}$ are not expressible in closed forms. However, using the EM algorithm, we can start at any estimate, say, (π_0, P_0), and get new estimate (π_1, P_1) such that the log-likelihood based on $\mathbf{O}$ increases:

$$L(\pi_0, P_0; \mathbf{O}) \leq L(\pi_1, P_1; \mathbf{O}).$$

The new estimates are given by formulas similar to (15.6) and (15.7), with Y and Z substituted by the respective conditional expectations $E_{\pi_0, P_0}[Y|\mathbf{O}]$ and $E_{\pi_0, P_0}[Z|\mathbf{O}]$. In fact, the inequality is strict unless $(\pi_0, P_0) = (\pi_1, P_1)$, and iterating the algorithm gives a sequence of estimates that converges to a local maximum [10].

The EM algorithm consists of an iteration between two steps.

1. **E step.** Evaluate the conditional expectation of the log-likelihood for the full data at (π_1, P_1), given the observed data, under the distribution specified by (π_0, P_0):

$$G := E_{\pi_0, P_0}\left[\log L(\pi_1, P_1; \mathbf{F})|\mathbf{O}\right].$$

2. **M step.** Find the (π_1, P_1) that maximizes G.

Clearly, G is a sum of similar terms over the independent observations. Thus, it suffices to consider a single site i with $(X_{i,0}, X_{i,T}) = (c, d)$. Let Y_i and Z_i be frequencies of initial states and transitions for this site.

$$Y_i(a) = |\{X_{i,0} = a\}|,$$

$$Z_i(a, b) = \sum_{t=1}^{T} |\{X_{i,t-1} = a, X_{i,t} = b\}|.$$

Then,

$$E_{\pi_0, P_0}[Y_i(a)|X_{i,0} = c, X_{i,T} = d] = \mathbf{1}_{\{a=c\}},$$

$$E_{\pi_0, P_0}[Z_i(a, b)|X_{i,0} = c, X_{i,T} = d] = P_0(a, b) \cdot \frac{u(a, b; c, d)}{P_0^T(c, d)},$$

where u is given by

$$u(a, b; c, d) = \sum_{t=1}^{T} P_0^{t-1}(c, a) P_0^{T-t}(b, d) \tag{15.8}$$

and the s^4 components sum to exactly Ts^2. Summing over all sites gives

$$G = \sum_{a=1}^{s} \log \pi_1(a) y(a) + \sum_{a,b=1}^{s} \log P_1(a, b) z(a, b), \tag{15.9}$$

where

$$y(a) = \sum_{i} |\{X_{i,0} = a\}|,$$

$$z(a, b) = P_0(a, b) \sum_{c,d=1}^{s} \nu(c, d) \frac{u(a, b; c, d)}{P_0^T(c, d)},$$

$$\nu(c, d) = \sum_{i} |\{X_{i,0} = c, X_{i,T} = d\}|.$$

This completes the E step. Notice that (15.9) has exactly the same form as (15.5), with Y and Z substituted by their respective conditional expectations y and z. Incidentally, since Y is observed, $y = Y$. The M step is easy:

$$\pi_1(a) = \frac{y(a)}{\sum_{c=1}^{s} y(c)},$$

$$P_1(a,b) = \frac{z(a,b)}{\sum_{d=1}^{s} z(a,d)}.$$

We make two remarks about the pair EM algorithm. First, only half of the s^4 u quantities need to be computed. Indeed, let the $s^2 \times s^2$ matrix M be defined by

$$M((c-1)s+a, (b-1)s+d) = u(a,b;c,d).$$

Then it is easy to see that M is symmetric. Second, if P_0 is reversible,

$$x(a)P_0(a,b) = x(b)P_0(b,a), \quad \forall a,b,$$

where x is the equilibrium distribution of P_0, and the frequency matrix ν is symmetric, then P_1 is also reversible. To show this fact, notice that

$$x(c)u(a,b;c,d)P_0(a,b)$$

$$= \sum_{t=0}^{T} x(c)P_0^{t-1}(c,a)P_0(a,b)P_0^{T-t}(b,d) \quad \text{by (15.8)}$$

$$= \sum_{t=0}^{T} P_0^{t-1}(a,c)P_0(b,a)P_0^{T-t}(d,b)x(d) \quad \text{(reversibility)}$$

$$= x(d)u(b,a;d,c)P_0(b,a).$$

It follows from this calculation, the reversibility of P_0^T, and the symmetry of ν that z is a symmetric matrix:

$$z(a,b) = \sum_{c,d=1}^{s} \nu(c,d) \frac{x(c)u(a,b;c,d)P_0(a,b)}{x(c)P_0^T(c,d)}$$

$$= \sum_{c,d=1}^{s} \nu(c,d) \frac{x(d)u(b,a;d,c)P_0(b,a)}{x(d)P_0^T(d,c)}$$

$$= \sum_{d,c=1}^{s} \nu(d,c) \frac{u(b,a;d,c)P_0(b,a)}{P_0^T(d,c)}$$

$$= z(b,a).$$

It is then easy to verify that the equilibrium distribution of P_1 is proportional to the row or column sum of z and that P_1 is reversible.

15.4.4 Pair EM: Continuous-Time Version

Let T be a fixed positive number, let N be a fixed large positive integer, and set $h = T/N$. Approximate the Markov process $\{X_t\}_{t\geq 0}$ defined by (π, Q) by the

Markov chain $\{X_0, X_h, \ldots, X_{Nh}\}$. The E step evaluates G_h, which is exactly as in (15.9), except that P_0 and P_1 are substituted by $P_0(h) = \exp(Q_0 h)$ and $P_1(h) = \exp(Q_1 h)$, respectively. Clearly, the first term of G_h is the same as G so that the estimate of π is given by (15.6). It remains to consider the estimation of Q. Since h is small, the following are good approximations:

$$P(h, a, b) \approx \begin{cases} 1 + Q(a,a)h & \text{if } a = b \\ Q(a,b)h & \text{otherwise,} \end{cases}$$

$$\log P(h, a, b) \approx \begin{cases} Q(a,a)h & \text{if } a = b \\ \log Q(a,b) + \log h & \text{otherwise.} \end{cases}$$

Hence, maximizing the second term of G_h is equivalent to maximizing

$$\sum_{a=1}^{s} Q_1(a,a) z_h(a,a) + \sum_{a,b=1, a \neq b}^{s} \log Q_1(a,b) Q_0(a,b) z_h(a,b), \qquad (15.10)$$

where

$$z_h(a,b) = \sum_{c,d} \nu(c,d) \frac{u_h(a,b;c,d)h}{P(T,c,d)},$$

$$u_h(a,b;c,d) = \sum_{t=1}^{N} P_0((t-1)h, c, a) P_0((N-t+1)h, b, d).$$

The first and second derivatives of (15.10) with respect to $Q_1(a,b)$, where $a \neq b$, are

$$-z_h(a,a) + Q_0(a,b) \frac{z_h(a,b)}{Q_1(a,b)},$$

$$-Q_0(a,b) \frac{z_h(a,b)}{Q_1(a,b)^2} < 0.$$

It follows from elementary calculus that setting

$$Q_1(a,b) = Q_0(a,b) \frac{z_h(a,b)}{z_h(a,a)}$$

maximizes the second term of G_h. Define

$$u_0(a,b;c,d) = \int_0^T P_0(x,c,a) P_0(T-x,b,d) dx = \lim_{h \downarrow 0} u_h(a,b;c,d),$$

$$z_0(a,b) = \sum_{c,d=1}^{s} \nu(c,d) \frac{u_0(a,b;c,d)}{P(T,c,d)}.$$

The continuous-time version of the EM algorithm is given by

$$Q_1(a, b) = \begin{cases} Q_0(a, b)\frac{z_0(a,b)}{z_0(a,a)} & \text{if } a \neq b \\ -\sum_{d \neq a} Q_1(a, d) & \text{if } a = b. \end{cases}$$

Since the approximations can be made as accurate as possible and we know that the discrete-time EM increases the log-likelihood, we conclude that the continuous-time version does also:

$$L(\pi_0, Q_0; \mathbf{O}) \leq L(\pi_1, Q_1; \mathbf{O}).$$

As in the discrete-time case, only half of the u_0 need to be evaluated. If Q_0 is diagonalizable, then u_0 can be computed exactly. Write

$$Q_0 = E\Lambda E^{-1}$$

for invertible E and diagonal $\Lambda = \text{diag}(\lambda_1, \ldots, \lambda_s)$. It follows that

$$P_0(x, c, a) = \sum_k E(c, k)E^{-1}(k, a)\exp(\lambda_k x),$$

$$P_0(T - x, b, d) = \sum_l E(b, l)E^{-1}(l, d)\exp(\lambda_k(T - x)),$$

and hence

$$u_0(a, b; c, d) = \sum_{k,l} E(c, k)E^{-1}(k, a)E(b, l)E^{-1}(l, d)e(k, l),$$

where

$$e(k, l) = \begin{cases} T\exp(\lambda_k T) & \text{if } k = l, \\ \frac{\exp(\lambda_k T) - \exp(\lambda_l T)}{\lambda_k - \lambda_l} & \text{otherwise.} \end{cases}$$

In fact, in this case, u_0 is much faster to compute than the u in the discrete-time version, especially when T is large, although this is true only when Q_0 is diagonalizable.

15.4.5 Tree EM

Let r denote the root of a phylogenetic tree. Suppose that only the sequences at the leaf nodes are observed (the observed data $\mathbf{O}$). Proceeding as in the pair EM, we consider evaluating G_i at a site i for which the full and observed data are denoted by F_i and O_i. Let Y_i be the frequencies of states at the root r. For a branch j of length t_j, let ϕ_j and ψ_j be, respectively, the nodes immediately above and below it, and let F_i^j denote the full data corresponding to this branch, including X_{i,ϕ_j} and X_{i,ψ_j}. Since F_i^j is conditionally independent of O_i given X_{i,ϕ_j} and X_{i,ψ_j}, for site i, we have

$$E_{\pi_0,Q_0}[L_i(\pi_1, Q_1; F_i)|O_i] = \sum_{a=1}^{s} \log \pi_1(a)\Pr(Y_i = a|O_i)$$

$$+ \sum_j \sum_{a,b=1}^{s} G_j(a,b)\Pr(X_{i,\phi_j} = a, X_{i,\psi_j} = b|O_i),$$

where

$$G_j(a,b) = E_{\pi_0,Q_0}[L_i(\pi_1, Q_1; F_j^i)|X_{i,\phi_j} = a, X_{i,\psi_j} = b].$$

Let

$$y(a) = \sum_i \Pr(Y_i = a|O_i),$$

$$\mu(j,a,b) = \sum_i \Pr(X_{i,\phi_j} = a, X_{i,\psi_j} = b|O_i).$$

It follows from summing over the sites that

$$G = \sum_{a=1}^{s} \log \pi_1(a)y(a) + \sum_j \sum_{a,b=1}^{s} G_j(a,b)\mu(j,a,b).$$

Thus, the first term corresponds to π, and the others, corresponding to Q, are branch-specific G_j terms such as G in the continuous-time pair EM, weighted by μ_j, which are analogs of ν. Each G_j can be computed as before, and the M step is given by

$$\pi_1(a) = \frac{y(a)}{\sum_{c=1}^{s} y(c)},$$

$$Q_1(a,b) = \begin{cases} Q_0(a,b)\frac{\sum_j z_0(j,a,b)}{\sum_j z_0(j,a,a)} & \text{if } a \neq b \\ -\sum_{d\neq a} Q_1(a,d) & \text{if } a = b, \end{cases}$$

where

$$z_0(j,a,b) = \sum_{c,d=1}^{s} \mu(j,c,d)\frac{u_0(j;a,b;c,d)}{P(t_j,c,d)},$$

$$u_0(j;a,b;c,d) = \int_0^{t_j} P_0(x,c,a)P_0(t_j - x,b,d)dx.$$

The probability $\Pr(X_{\phi_j} = a, X_{\psi_j} = b|O)$ can be efficiently computed by the up-down algorithm. Let the nodes of the tree be labeled $1, 2, \ldots$ so that every parent has a larger label than all its children. For a node α, let χ_α denote the set of its children. Let ϕ and γ denote its parent and grandparent, respectively, and let the distance between α and β be $t_{\alpha\beta}$. Let $U(\alpha, a)$ be the probability that the states at the leaf nodes that are descendants of α are as in the observed data, given that $X_\alpha = a$. As shown by Felsenstein [14], U can be evaluated recursively starting from the leaf nodes. If α is a leaf node and $X_\alpha = b$, then $U(\alpha, a) = \delta_{a,b}$. If α is not a leaf node, then

$$U(\alpha, a) = \prod_{\kappa \in \chi_\alpha} \sum_{c=1}^{s} P(t_{\alpha\kappa}, a, c) U(\kappa, c).$$

The joint probability of the observed data is given by

$$\Pr(O) = \sum_{a=1}^{s} \pi(a) U(r, a).$$

Next, let $D(\alpha, a)$ be the probability that the states at the nodes that are not α or its descendants are as in the observed data and that the parent node ϕ is in state a. D is evaluated from the children of the root down the tree. If α is a child of r, then

$$D(\alpha, a) = \prod_{\kappa \in \chi_\phi, \kappa \neq \alpha} \sum_{c=1}^{s} P(t_{\phi\kappa}, a, c) U(\kappa, c) \pi(a).$$

If α is not a child of r, then

$$D(\alpha, a) = \prod_{\kappa \in \chi_\phi, \kappa \neq \alpha} \sum_{c=1}^{s} P(t_{\phi\kappa}, a, c) U(\kappa, c) \sum_{d=1}^{s} D(\phi, d) P(t_{\gamma\phi}, d, a).$$

It is then easy to verify that

$$\Pr(X_{\phi_j} = a, X_{\psi_j} = b | O) = \frac{D(\psi, a) P(t_{\phi_j \psi_j}, a, b) U(\psi_j, b)}{\Pr(O)}.$$

Finally, we remark that combining sequences from different trees to estimate the parameters is very easy: simply add the G quantities corresponding to each tree, and then do the M step.

15.4.6 The Resolvent Method

Müller and Vingron [34] proposed a fast estimation method on sequence pairs based on resolvents. It will be apparent that it can in fact be applied to multiple alignments generated by a reversible Markov process without knowledge of the tree topology. For $\alpha > 0$, the *resolvent* R_α of a rate matrix Q is defined as

$$R_\alpha = (\alpha I - Q)^{-1}.$$

Solving for Q gives the following formula, valid for any α:

$$Q = \alpha I - R_\alpha^{-1}. \tag{15.11}$$

It turns out (see [20] or [34]) that the resolvent is the Laplace transform of the transition probabilities:

$$R_\alpha = \int_0^\infty e^{-\alpha t} P(t) dt. \tag{15.12}$$

This is the key formula for the resolvent method. Clearly, given many pairs of sequences (not necessarily disjoint) that are separated by t PAMs, an unbiased estimate of $P(t)$ is obtained by normalizing the symmetrized sum of frequency tables. If $P(t)$ can be estimated for a wide range of t's, then we get an estimate of Q via (15.12) and (15.11).

In practice, there are two issues: (1) the distances are unknown and are estimated by maximum likelihood, and (2) the estimated distances are discrete, so interpolation is used to estimate the rate matrix. Let the estimated distances be $0 < t_1 < \cdots < t_N$. The integral (15.12) is approximately equal to the sum of N pieces:

$$R_\alpha \approx \left(\int_0^{t_1} + \cdots \int_{t_{N-1}}^{t_N} \right) e^{-\alpha t} P(t) dt.$$

The kth integral is approximated by linear interpolation,

$$\int_{t_{k-1}}^{t_k} e^{-\alpha t} \left(P(t_{k-1}) + \frac{t - t_{k-1}}{t_k - t_{k-1}} \left[P(t_k) - P(t_{k-1}) \right] \right) dt,$$

which can be evaluated exactly after replacing the P's by their estimates. Summing these integrals gives an estimate of R_α, and a new estimate of Q, denoted by Q_1, is obtained by using (15.11). In principle, R_α, and hence Q_1, is independent of α, but this is not so in practice. Müller and Vingron recommended choosing the α that maximizes the likelihood of all data. Another issue is that Q_1 may not be reversible. In our implementation, we force it to be reversible by first deriving its equilibrium distribution π_1, and let $R_1 = Q_1 \Pi_1^{-1}$. If R_1 is symmetric, then Q_1 is reversible. If not, let $R_2 = R_1 + R_1'$ be the sum of R_1 and its transpose. Then $Q_2 = \Pi_1 R_2$ is reversible and upon calibration is our RES estimate.

15.5 BLOSUM Matrices

Henikoff and Henikoff [24] used an ad hoc method that takes inhomogeneity into account to construct the BLOSUM (BLOck SUbstitution Matrix) matrices. The input is a set of *blocks*, which are gap-free multiple alignments of segments of homologous amino acid sequences. A frequency table is derived from the blocks by summing over the frequency tables from all within-block pairwise comparisons. Since a mismatch, such as an A aligned with an S, can be written in two ways, AS and SA, we get rid of the ambiguity by using only AS. In general, a mismatch is represented by XY, where X precedes Y alphabetically. For example, suppose that in a block with six sequences, two columns are as follows:

```
..AD..
..AD..
..AE..
..AE..
..AD..
..SD..
```

There are a total of 15 pairwise comparisons. The left column contributes 10 AA and 5 AS pairs to the frequency table. Similarly, the right column contributes 6 DD, 1 EE, and 8 DE pairs. Adding these column contributions within the block, and then across all blocks, gives a triangular frequency table. The matrix is symmetrized by adding itself to its transpose. Dividing the matrix by its sum yields a symmetric joint distribution and a substitution matrix via (15.1).

To capture the substitution patterns from the more distantly related sequences, the more closely related ones are downweighted by clustering. Let θ be a fixed number between 0 and 100. Sequences that are more than $\theta\%$ similar are "greedily" clustered. In other words, any two sequences that are more than $\theta\%$ similar are put in the same cluster, and if each sequence already belongs to some cluster, then the two clusters are combined to form a larger cluster. In the end, the sequences within a block are partitioned into disjoint clusters so that any two sequences from distinct clusters are less than $\theta\%$ similar. It is clear that the clusters are independent of the initial choice of sequences. Sequences in the same cluster are downweighted by the cluster size in cross-cluster pairwise comparisons, and pairwise comparisons of sequences in the same cluster do not contribute to the frequency table. In the example, suppose that the first four sequences are clustered while the last two sequences are not. Then the contribution of the left column is the same as an A-A-S column: 1 AA, 2 AS pairs. The right column is effectively (D/E)-D-D, where D/E represents half a D and half an E. Its contribution is 2 DD ($1 + 1/2 + 1/2$) and 1 DE ($1/2 + 1/2$) pairs. Equivalently, sequences in the same cluster are replaced by an "average" sequence with a fractional number of residues at each position. Then the frequency table is derived as if the average sequences are real sequences; blocks that have only one cluster are left out.

Let the symmetric joint distribution at threshold θ be denoted by f_θ. Let π be the row of column sum of f_θ. Then the substitution matrix BLOSUMθ is given by

$$S(\theta, a, b) = 10 \log_{10} \frac{f_\theta(a, b)}{\pi(a)\pi(b)}.$$

If θ is 100, then every cluster is of size 1, so f_θ is an average of the substitution patterns over all distances. If θ is a small value, such as 20, then there are a small number of large clusters, and the similarity between sequences from distinct clusters ranges from 0 to 20%, so f_θ only depends on the substitution patterns of the distantly related sequences. Thus, reducing the threshold θ gives substitution matrices that are more suitable for aligning distantly related sequences. For example, when aligning distantly related protein sequences, BLOSUM40 is preferred to BLOSUM80.

The BLOSUM method does not explicitly involve the phylogenetic tree relating the sequences in a block. This has the advantage that it is not necessary to know the tree in order to estimate the substitution matrices. Nevertheless, it has a ready interpretation when the tree is made explicit, provided the tree has the molecular clock property: the distance between the root and any leaf node is the same; this distance T is the depth of the tree. An example is illustrated in Figure 15.1.

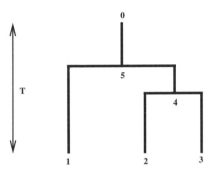

Fig. 15.1. A phylogenetic tree with the molecular clock property. The ancestral sequence at 0 and the intermediate sequences at the internal nodes 4 and 5 are not observed, but the sequences at the leaf nodes 1, 2, and 3 are observed. The depth of the tree is T.

Suppose that the sequences were generated by a Markov substitution process with the property that the average percentage identity of a pair of observed sequences is a strictly decreasing function ϕ of the distance between them. Both the well-known reversible and the inhomogeneous processes described in Subsection 15.6.1 have this property. This strict monotonicity allows us to restate the clustering step in the BLOSUM method in terms of tree distances.

Let T be the depth of the tree. Fix a distance t with $0 < t < T$. Consider cutting the tree at distance $T - t/2$ below the root. Any pair of observed sequences that descended from the same cut point are separated by at most t, so they are more than $\theta\%$ similar, where

$$\theta = \phi(t).$$

Conversely, any pair of observed sequences more than $\theta\%$ similar share a common cut point at distance $t/2$ from both of them. Hence, the clustering step is exactly the same as replacing the observed sequences below each cut point by an average sequence.

15.6 Comparison of Methods

Five methods of estimating symmetric substitution matrices have been described: PAM, ML I, ML II, RES, and BLOSUM. The first four methods assume that sequences are generated by a reversible substitution process. These methods are consistent, except for PAM, which will be dropped from the ensuing exposition. The BLOSUM method is not explicitly based on any substitution model but seems able to take inhomogeneity in substitution rates into account. Since all our models produce sequence pairs of the same composition, we can appeal to the discussion in paragraph 3 of Subsection 15.4.2 to conclude that ML I and ML II behave equivalently. Hence, only ML I will be used, and this will be referred to as ML. We will compare the performance of ML, RES, and BLOSUM theoretically and also by simulations on independent sequence pairs generated by a reversible (homogeneous) process and two inhomogeneous processes.

Instead of comparing the estimated substitution matrices directly, we compare the joint distributions used to compute the substitution matrices. Consider a giant "tree of proteins" that relates all existing proteins. For a distance t, collecting all the protein pairs separated by t units yields a frequency table, symmetrized if necessary. We assume that, for any t, the substitution matrix constructed from the frequency table via (15.1) is the best one for aligning proteins separated by t. Then we just need to compare, for various evolutionary distances, the joint distributions obtained by the different methods to the "real joint distribution" in order to assess their accuracy. Unlike ML and RES, the BLOSUM joint distribution f_θ at threshold θ is not constructed explicitly from an evolutionary distance. Thus, in order to compare the methods, it is necessary to associate an *effective divergence time* with f_θ. This issue was dealt with in [35], but we chose a simpler approach. If a symmetric joint distribution f can be written as

$$f = \Pi^* \exp(Q^* t^*)$$

for some calibrated rate matrix Q^* with equilibrium distribution π^* and some $t^* > 0$, then we say f is *embeddable* in a reversible process, and t^* is the effective divergence time of f. Although not all symmetric joint distributions are embeddable, in practice, almost all realizations of f_θ are. If $f = F(t)$, a joint distribution at t PAMs from a calibrated reversible process as in (15.3), then its effective divergence is clearly t. The effective divergence time provides another way of matching PAM and BLOSUM matrices. Henikoff and Henikoff used Altschul's idea of relative entropy [1, 24] to conclude that BLOSUM45, BLOSUM62, and BLOSUM80 are, respectively, comparable to PAM250, PAM160, and PAM120. We found that the effective divergence times of these BLOSUM matrices are 168, 138, and 101 PAMs, respectively. Thus, our matching is clearly different.

We can now imagine assessing the performance of the methods on sequence data generated by a substitution process in the following way: for a fixed threshold θ, compare the BLOSUM f_θ, $f_{ml}(t_\theta^*)$, and $f_{res}(t_\theta^*)$, the joint distributions based on the ML and RES estimates of Q, where t_θ^* is the effective divergence time of f_θ, with the "true" joint distribution. Before we embark on the comparison study, we first provide some concrete examples of the kind of inhomogeneous processes that will be considered.

15.6.1 Inhomogeneous Processes

An inhomogeneous process can be easily postulated starting with the homogeneous process: replace the constant rate matrix Q with a family of rate matrices $\{Q(t)\}_{t \geq 0}$. Then the construction of the process is the same except that now, given $X_s = a$, the probability that the next substitution happens after $t > s$ is $\exp(\int_s^t Q(x, a, a)dx)$, and if the next substitution happens at time $u > s$, then the process jumps from a to b with probability $-Q(u, a, b)/Q(u, a, a)$. Thus, one technical condition for this to work is that the rate matrix family should be integrable. The analogous formula for the transition probability, namely $P(s, t) = \exp(\sum_s^t Q(x)dx)$, is correct if every pair of rate matrices in the family commutes [21]. In addition to commutativity, we shall also assume that the $Q(t)$'s have a common distribution π, so that the process is stationary. Under the inhomogeneous process, the joint distribution of two leaf states depends on the location of the common ancestor. However, if the ancestor is equidistant to the leaf nodes, then the joint distribution is symmetric. If each $Q(t)$ is calibrated, then the expected number of substitutions in any time interval of length 1 is 0.01, and we say that the process is calibrated. Hence we may compare estimated distances from homogeneous and inhomogeneous processes.

Fix a tree with the molecular clock property with depth T. We only consider stationary inhomogeneous processes such that the percentage identity between two observed sequences is a strictly decreasing function of the distance separating them. This will be effected by imposing some constraints on the family of rate matrices, which will be described after several examples are presented.

Example 15.1

Let Q_1 be a calibrated reversible rate matrix, with all off-diagonal entries strictly positive, and let π be its equilibrium distribution. Define a reversible rate matrix Q_0 as the calibrated version of the following:

$$Q_0(a, b) = \begin{cases} \pi(b) & \text{if } a \neq b, \\ -\sum_{d \neq a} \pi(d) & \text{if } a = b. \end{cases}$$

Thus, Q_0 is the simplest reversible rate matrix with equilibrium distribution π. Since Q_1 is irreducible, π is strictly positive, so all off-diagonal entries of Q_0 are positive. It can be readily checked that Q_0 commutes with Q_1: $Q_0 Q_1 = Q_1 Q_0$. Consider the process defined by the family

$$Q(x) = \begin{cases} Q_0 & 0 \le x < 100, \\ Q_1 & 100 \le x \le 200. \end{cases}$$

Since both Q_0 and Q_1 have the same equilibrium distribution π, the inhomogeneous process is also stationary with the same equilibrium distribution. Moreover, since both Q_0 and Q_1 are calibrated, the new process is also calibrated. The process is sketched in Figure 15.2.

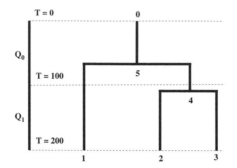

Fig. 15.2. An illustration of the process in Example 15.1 generating sequences on a tree with the molecular clock property. The distance between nodes 2 and 3 is less than 200 PAMs, while that between nodes 1 and 2 is more than 200 PAMs.

Example 15.2

Let Q_0 and Q_1 be as in Example 1, and let $g : [0, \infty) \to [0, 1]$ be an integrable function. Then the process defined by the family

$$Q(x) = Q_0 + g(x)(Q_1 - Q_0), \qquad x \ge 0, \tag{15.13}$$

is calibrated and has equilibrium distribution π. Clearly, this generalizes Example 15.1.

Example 15.3

Let Q_1 be as before. We will construct Q_0 by diagonalizing Q_1. Let

$$Q_1 = B\Lambda_1 B^{-1},$$

where $B = A^{-1}V$, as in (15.4). Since the off-diagonal entries of Q_1 are strictly positive, the Perron-Frobenius theorem [17] implies that exactly one diagonal entry of Λ_1 is zero. Let Λ_0 be the diagonal matrix with zero at the same position as in Λ_1 and with the other diagonal entries equal to $-a$ with $a > 0$. Then it is easy to verify that $B\Lambda_0 B^{-1}$, after calibration, is exactly Q_0. Furthermore, the diagonalized forms of Q_0 and Q_1 immediately imply that they are commutative. This device can be extended to generate a set of commutative reversible rate matrices all sharing the same equilibrium distribution as Q_1. Denote this set by $r(Q_1)$. Let Λ be a diagonal matrix with zero at the same position as in Λ_1 and with the other diagonal entries restricted to be negative, and let $Q = B\Lambda B^{-1}$. Such a Q might not be a rate matrix; some of its off-diagonal entry may be negative. But if the diagonal entries of Λ are not too different from Λ_1, then Q is a rate matrix by continuity. In any case, Q_0 is a rate matrix, so $r(Q_1)$ has at least two elements, and by interpolating between Q_0 and Q_1, we have infinitely many calibrated rate matrices in $r(Q_1)$. All these rate matrices share the same equilibrium distribution and are reversible. Any integrable function from $[0, \infty)$ into $r(Q_1)$ gives a family of desired rate matrices. Hence this example generalizes Examples 15.2 and 15.1.

Example 15.3 represents the class of inhomogeneous processes that we intend to study. Thus, we are interested in families of rate matrices whose members all belong to a set $r(Q)$ of all calibrated reversible rate matrices that have the same eigenvectors as some reversible rate matrix Q. As a consequence, the rate matrices are commutative: $Q(s)Q(t) = Q(t)Q(s)$ for any s and t, so

$$P(s,t) = \exp\left(\int_s^t Q(x)dx\right). \tag{15.14}$$

Reversibility implies that each transition probability is reversible with respect to the equilibrium distribution. Finally, we have a monotonicity property, proved in the Appendix: the percentage identity between two observed sequences is a strictly decreasing function of the distance separating them.

Let α and β be leaf nodes separated by t PAMs, where $0 < t \le 2T$ and T is the depth of the tree. Then, the distance between either node and their common ancestor γ is $t/2$. Let $f(t)$ denote the joint distribution of X_α and X_β. $f(t)$ is a well-defined function of t: it is the joint distribution of states at any pair of leaf nodes separated by t PAMs since the tree has the molecular clock property. By the Markov property, $f(t)$ given X_γ is independent of the history before time $T - t/2$. It follows that

$$f(t,a,b) = \sum_z \pi(z)P(T - t/2, T, z, a)P(T - t/2, T, z, b)$$

$$= \pi(a)\sum_z P(T - t/2, T, a, z)P(T - t/2, T, z, b). \text{ (reversibility)}$$

In matrix notation, using (15.14) and commutativity, we have

$$f(t) = \Pi P(T - t/2, T)P(T - t/2, T)$$

$$= \Pi \exp\left(\int_{T-t/2}^T Q(x)dx\right)\exp\left(\int_{T-t/2}^T Q(x)dx\right)$$

$$= \Pi \exp\left(2\int_{T-t/2}^T Q(x)dx\right). \tag{15.15}$$

Clearly, $f(t)$ is a symmetric matrix with effective divergence time t. Since the process is stationary, the sum of its ath row is $\pi(a)$. Similarly, the sum of the bth column is also $\pi(b)$. Dividing the ath row of $f(t)$ by $\pi(a)$, for all rows, gives a transition probability $p(t)$:

$$p(t) = \Pi^{-1} f(t) = \exp\left(2 \int_{T-t/2}^{T} Q(x) dx \right). \tag{15.16}$$

$p(t)$ is also well-defined and is not equal to any of the transition probabilities $P(s, t)$ of the inhomogeneous process in general. Clearly, if the process is reversible, then $f(t) = F(t)$ and $p(t) = P(t)$. Given a pair of observed sequences known to be separated by distance t, $f(t)$ and $p(t)$ can be estimated easily from the frequency table without knowledge of the tree topology.

Equations (15.15) and (15.16) can be simplified for simple inhomogeneous processes. In Example 15.1, let t_{23} be the distance between nodes 2 and 3 in Figure 15.2, with common ancestor 4. Since $200 - t_{23}/2 > 100$, it follows that

$$\int_{200-t_{23}/2}^{200} Q(x) dx = Q_1 t_{23}/2,$$
$$f(t_{23}) = \Pi \exp(t_{23} Q_1),$$
$$p(t_{23}) = \exp(t_{23} Q_1).$$

15.6.2 Theoretical Comparison

Consider a phylogenetic tree with the molecular clock property and so many leaf nodes that the distances between all sequence pairs $0 < t_1 < \cdots < t_n$ cover a wide range of values, n being the number of distinct distances. Imagine running (1) a reversible (homogeneous) process and (2) an inhomogeneous process down this tree to give leaf sequences. By the monotonicity property, the percentage identity of sequence pairs also covers the range $(0, 100)$. Assume that the sequences are long enough that the frequency tables yield almost exact estimates of the joint distributions $\{f(t_k) : 1 \leq k \leq n\}$ and hence the transition probabilities $\{p(t_k) : 1 \leq k \leq n\}$, as well as the distances $t_1 < \cdots < t_n$.

With the distances considered known, the Laplace transform (15.12) is approximately

$$R_\alpha \approx \sum_{k=1}^{n} \int_{t_{k-1}}^{t_k} e^{-\alpha t} \left(p(t_{k-1}) + \frac{t - t_{k-1}}{t_k - t_{k-1}} [p(t_k) - p(t_{k-1})] \right) dt, \tag{15.17}$$

where $t_0 = 0$.

Let $\phi(t_n) < \theta < 100$ be a BLOSUM threshold. For each k, let $w(t_k)$ be the weight given to $f(t_k)$ after clustering, so that f_θ is given by

$$f_\theta = \sum_{k=1}^{n} f(t_k) w(t_k)$$

$$= \sum_{t_k \geq \phi^{-1}(\theta)} f(t_k) w(t_k) \tag{15.18}$$

since $w(t_k)$ is zero whenever $t_k < \phi^{-1}(\theta)$.

(1) If the sequences were generated by a reversible model, then the estimated rate matrix $Q_{res} = \alpha I - R_\alpha^{-1}$ by RES is close to Q since there is a large number of distances spread out on the positive real line. It follows that the joint distribution $f_{res}(t)$ is close to $f(t)$ for any t.

Suppose that f_θ is embeddable, so $f_\theta = \Pi^* \exp(Q^* t^*)$. Then, in general, $Q^* \neq Q$, and the effective divergence time t^* is quite a bit larger than $\phi^{-1}(\theta)$. The average of $d(f_\theta, f(t^*))$ is bound to be larger than the average of $d(f_{res}(t^*), f(t^*))$. We conclude that BLOSUM is worse than RES at any threshold θ.

(2). Let the sequences be generated by an inhomogeneous model. In (15.17), if α is large, then since $\exp(-\alpha t)$ decays rapidly, R_α is independent of $p(t)$ at large distances. If α is very small, then $\exp(-\alpha t)$ is almost flat, and R_α is almost like a simple average over the smaller distances. In both cases, R_α only depends on the $p(t)$ at small distances.

On the other hand, for BLOSUM, the weights in (15.18) seem to be able to capture the $f(t)$ at large distances. In conclusion, BLOSUM should produce an f_θ that is better than RES, especially when θ is small.

Although it is impossible to analyze ML explicitly, some of its qualitative behavior may be roughly described. When the process is reversible, ML seems to perform better than RES [33], so it is better than BLOSUM. On the other hand, if the process is inhomogeneous and the data are such that there are more closely related sequences than distantly related sequences, then computational experiments suggest that the estimated rate matrix tends to reflect the substitution patterns of closely related sequences. In this case, it is plausible that ML is worse than BLOSUM for estimating joint distributions at large distances.

15.6.3 Simulations

We will compare BLOSUM, RES, and ML in the following way. Let $\theta \in \{30, 35, \ldots, 85\}$ be fixed. Given the sequence blocks generated by a substitution process, run BLOSUM at threshold θ to get f_θ, with effective divergence time t_θ^*. Apply RES and ML to get the estimated rate matrices f_{res} and f_{ml}. Define the deviations

$$d_{blo} = d(f_\theta, f(t_\theta^*)),$$
$$d_{res} = d(f_{res}(t_\theta^*), f(t_\theta^*)),$$
$$d_{ml} = d(f_{ml}(t_\theta^*), f(t_\theta^*)),$$

where d, the total variation distance between two probability distributions μ and ν on the same finite set X, is defined as

$$d(\mu, \nu) := \sum_{x \in X} |\mu(x) - \nu(x)|.$$

Suppose that the substitution model is fixed (i.e., the substitution process, the trees, and the sequence lengths are all fixed). Then the average of the deviations over all realizations of the sequences is a measure of the relative performance of the methods. In the simulations, we approximate the averages by averaging the outcomes of 100 random samples. Using 200 samples gives almost the same averages and standard deviations (SD), so we are quite confident that the sample size 100 is large enough.

Let Q_1 be a calibrated rate matrix estimated by ML from 13,255 pairwise amino acid alignments of length at least 100 from the SYSTERS database, courtesy of Tobias Müller. Let Q_0 be the simplest calibrated reversible rate matrix with the same equilibrium distribution as Q_1, as constructed in Subsection 15.6.1. In every simulation, 30 sequence pairs of length 5000 separated by $10, 20, \ldots, 300$ PAMs are generated according to a substitution model. In other words, there are 30 blocks, and each block has two sequences. Also, all trees have depth 150 PAMs, so that a sequence pair separated by 100 PAMs have a most recent common ancestor 50 PAMs ago, and the most recent common ancestor of a sequence pair separated by 300 PAMs sits at the top of the tree. The simulations differ in the substitution model, as described below.

1. A reversible process generated by Q_1.
2. An inhomogeneous process with

$$Q(t) = Q_0 + (Q_1 - Q_0)t/150, \qquad 0 \le t \le 150.$$

3. An inhomogeneous process with

$$Q(t) = \begin{cases} Q_0 & 0 \le t < 100 \\ Q_1 & 100 \le t \le 150. \end{cases}$$

The average deviations are displayed in the following graphs. Since the SDs are typically not larger than one-tenth of the averages, they are not represented in the plots.

Figure 15.3 shows that under the reversible process, ML is better than RES, which in turn is better than BLOSUM, for all the thresholds $30, 35, \ldots,$ 85. While the deviations of ML and RES are quite constant, BLOSUM's decreases from 0.08 at 85 to 0.04 at 30, and at the lower thresholds, it is almost as good as the resolvent. This may be explained by observing that, at a high threshold, sequence pairs at a wider range of similarity are summed, causing more error in the joint distribution compared with a low threshold.

Under the inhomogeneous process with smooth rates (Figure 15.4), ML is the best, followed by BLOSUM, while RES is rather more worse off than both ML and BLOSUM.

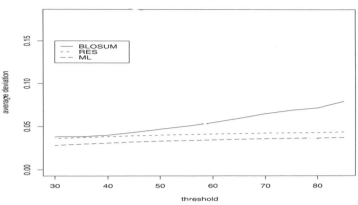

Fig. 15.3. Simulation 1: reversible process.

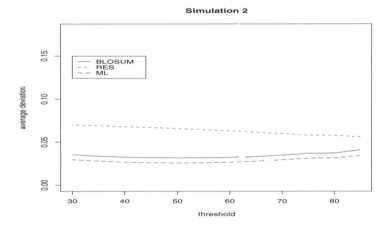

Fig. 15.4. Simulation 2: inhomogeneous process with smooth rates.

Finally, under the inhomogeneous process with discontinuous rates (Figure 15.5), at high thresholds, BLOSUM is worse than both ML and RES, but at low thresholds, BLOSUM is significantly better than ML and RES. The BLOSUM and RES curves cross at 80, while those of BLOSUM and ML cross at about 65.

In summary, for an inhomogeneous process, BLOSUM is rather more stable and better than both ML and RES at low thresholds, but not so good at high thresholds. On the other hand, for a reversible process, BLOSUM is only slightly worse than ML and RES at high thresholds. Thus, the simulations suggest that for pair data (1) BLOSUM is a robust method for deriving substitution matrices at large evolutionary distances, regardless of the homogeneity

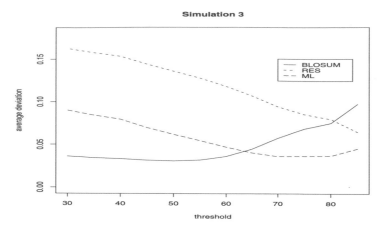

Fig. 15.5. Simulation 3: inhomogeneous process with discontinuous rates.

of the substitution process, while (2) ML and RES are better than BLOSUM (and ML appreciably better than RES) at small distances. It seems plausible that our conclusion for pair data should still hold for the general case where the trees have an arbitrary number of observed sequences.

15.7 DNA Substitution Matrices

As more genomes are getting completely sequenced, DNA substitution matrices are expected to play an important role in the great challenge to understand the evolutionary history and biological functions of genomic DNA sequences. While the previous theory applies readily to DNA substitution matrices, in order for the result of alignment to be independent of the choice of strands, it makes sense to construct strand-symmetric matrices, where the score is invariant under complementation. For example, $S(A, C) = S(T, G)$, and so on. The reversible rate matrices Q that give strand-symmetric substitution matrices at all evolutionary distances are precisely those that have a strand-symmetric R, where $Q = R\Pi$ is the factorization described at the end of Subsection 15.5.1. Both strand symmetry in primates and the lack of it in bacterial genes have been described [18, 19]. A priori, strand symmetry is not expected to hold due to the asymmetry of the replication process and the exposure of the complementary strand of genes to mutagens during protein synthesis; however, intergenic substitution patterns between human and mouse seem approximately strand-symmetric [38]. The $HOXD70$ matrix, constructed by Chiaromonte et al. [6] from reliable human-mouse alignments with BLOSUM and ML methods, outperformed several default matrices, thus confirming the utility of the empirical approach.

15.8 Appendix

Let $\{Q(t)\}_{t\geq 0}$ be integrable $n \times n$ rate matrices of the form

$$Q(t) = A^{-1}V\Lambda(t)V'A, \tag{15.19}$$

where A is the positive square root of the diagonal matrix Π with a strictly positive probability distribution π on its diagonal, V is orthogonal, and for each t, $\Lambda(t)$ is diagonal with $\Lambda(t, 1, 1) = 0$ and other diagonal entries strictly negative.

It is readily checked that the rate matrices are commutative. Hence, the percentage identity between two observed sequences separated by t is

$$\sum_{a=1}^{n} f(t, a, a) = \sum_{a=1}^{n} \pi(a)p(t, a, a),$$

where $f(t)$ and $p(t)$ are given by (15.15) and (15.16), respectively. To show that the percentage identity is strictly decreasing in t, it suffices to examine each $p(t, a, a)$.

Substituting (15.19) into (15.16) gives

$$p(t) = \exp\left(2\int_{T-t/2}^{T} Q(x)dx\right)$$

$$= A^{-1}V \exp\left(2\int_{T-t/2}^{T} \Lambda(x)d(x)\right) V'A,$$

implying that in terms of the entries of V and $\Lambda(x)$,

$$p(t, a, a) = V(a, 1)^2 + \sum_{b=2}^{n} V(a, b)^2 \exp\left(2\int_{T-t/2}^{T} \Lambda(x, b, b)d(x)\right).$$

Since the integral is strictly decreasing, each $p(t, a, a)$ is either constant or strictly decreasing in t. If all are constant, then the matrix V consists of 1's on its first column and 0's elsewhere, contradicting the assumption that V is orthogonal. Hence, all $p(t, a, a)$ are monotone decreasing, some strictly, and the percentage identity is indeed strictly decreasing in t.

Acknowledgments

The authors thank Ian Holmes, Tobias Müller, and an anonymous referee for many comments and suggestions.

References

[1] S. F. Altschul. Amino acid substitution matrices from an information theoretic perspective. *J. Mol. Biol.*, 219:555–565, 1991.

[2] L. Arvestad and W. J. Bruno. Estimation of reversible substitution matrices from multiple pairs of sequences. *J. Mol. Evol.*, 45:696–703, 1997.

[3] S. A. Benner, M. A. Cohen, and G. H. Gonnet. Amino acid substitution during functionally constrained divergent evolution of protein sequences. *Protein Eng.*, 7:1323–1332, 1994.

[4] P. Billingsley. *Statistical Inference for Markov Processes*. University of Chicago Press, Chicago, 1961.

[5] S. E. Brenner, C. Chothia, and T. J. P. Hubbard. Assessing sequence comparison methods with reliable structurally identified distant evolutionary relationships. *Proc. Natl. Acad. Sci. USA*, 95:6073–6078, 1998.

[6] F. Chiaromonte, Yap V. B., and W. Miller. Scoring pairwise genomic sequence alignments. In R. B. Altman, A. K. Dunker, L. Hunter, K. Lauderdale, and T. E. Klein, editors, *Proceedings of the Pacific Symposium on Biocomputing*, pages 115–126. World Scientific, Singapore, 2002.

[7] M. O. Dayhoff and R. V. Eck. A model of evolutionary change in proteins. In M. O. Dayhoff, editor, *Atlas of Protein Sequence and Structure*. National Biomedical Research Foundation, Silver Spring, MD, 1968.

[8] M. O. Dayhoff, R. V. Eck, and C. M. Park. A model of evolutionary change in proteins. In M. O. Dayhoff, editor, *Atlas of Protein Sequence and Structure*, volume 5. National Biomedical Research Foundation, Washington, DC, 1972.

[9] M. O. Dayhoff, R. M. Schwartz, and B. C. Orcutt. A model of evolutionary change in proteins. In M. O. Dayhoff, editor, *Atlas of Protein Sequence and Structure*, volume 5. National Biomedical Research Foundation, Washington, DC, 1979.

[10] A. P. Dempster, N. M. Laird, and D. B. Rubin. Maximum likelihood from incomplete data via the EM algorithm. *J. Roy. Stat. Soc. B*, 39:1–38, 1977.

[11] C. Devauchelle, A. Grossmann, A. Hénaut, M. Holschneider, M. Monnerot, J. L. Risler, and B. Torrésani. Rate matrices for analyzing large families of protein sequences. *J. Comput. Biol.*, 8:381–399, 2001.

[12] R. Durbin, S. Eddy, A. Krogh, and G. Mitchison. *Biological Sequence Analysis*. Cambridge University Press, Cambridge, 1998.

[13] J. Felsenstein. Cases in which parsimony or compatibility methods will be positively misleading. *Syst. Zool.*, 27:401–410, 1978.

[14] J. Felsenstein. Evolutionary trees from DNA sequences. *J. Mol. Evol.*, 18:368–376, 1981.

[15] J. Felsenstein. *Inferring Phylogenies*. Sinauer Associates, Inc., Sunderland, MA, 2004.

[16] D. F. Feng, M. S. Johnson, and R. F. Doolittle. Aligning amino acid sequences: Comparison of commonly used methods. *J. Mol. Evol.*, 21:112–125, 1985.

[17] J. B. Fraleigh and R. A. Beauregard. *Linear Algebra*. Addison-Wesley, Reading, MA, 3rd edition, 1994.

[18] M. P. Francino and H. Ochman. Strand asymmetries in DNA evolution. *Trends Genet.*, 13:240–245, 1997.

[19] M. P. Francino and H. Ochman. Strand symmetry around the β-globin origin of replication in primates. *Mol. Biol. Evol.*, 17:416–422, 2000.

[20] M. Fukushima. *Dirichlet Forms and Markov Processes*. North Holland, Amsterdam, 1980.

[21] R. D. Gill and S. Johansen. A survey of product-integration with a view towards application in survival analysis. *Ann. Stat.*, 18:1501–1555, 1990.

[22] G. H. Gonnet, M. A. Cohen, and S. A. Benner. Exhaustive matching of the entire protein sequence database. *Science*, 256:1433–1445, 1992.

[23] R. E. Green and S. E. Brenner. Bootstrapping and normalization for enhanced evaluations of pairwise sequence comparison. *Proc. IEEE*, 9:1837–1847, 2002.

[24] S. Henikoff and J. G. Henikoff. Amino acid substitution matrices from protein blocks. *Proc. Natl. Acad. Sci. USA*, 89:10915–10919, 1992.

[25] I. Holmes and G. M. Rubin. An expectation maximization algorithm for training hidden substitution models. *J. Mol. Biol.*, 317:753–764, 2002.

[26] M. S. Johnson and J. P. Overington. A structural basis for sequence comparisons. *J. Mol. Biol.*, 233:716–738, 1993.

[27] D. T. Jones, W. R. Taylor, and J. M. Thornton. The rapid generation of mutation data matrices from protein sequences. *Comput. Appl. Biosci.*, 8:275–282, 1992.

[28] F. P. Kelly. *Reversibility and Stochastic Networks*. John Wiley & Sons, New York, 1979.

[29] J. M. Koshi and R. A. Goldstein. Context-dependent optimal substitution matrices. *Protein Eng.*, 8:641–645, 1994.

[30] P. Lió and N. Goldman. Models of molecular evolution and phylogeny. *Genome Res.*, 8:1233–1244, 1998.

[31] A. D. McLachlan. Tests for comparing related amino acid sequences. *J. Mol. Biol.*, 61:409–424, 2002.

[32] T. Müller, S. Rahmann, and M. Rehmsmcier. Non-symmetric score matrices and the detection of homologous transmembrane proteins. *J. Mol. Evol.*, 17:182–189, 2001.

[33] T. Müller, R. Spang, and M. Vingron. Estimating amino acid substitution models: A comparison of dayhoff's estimator, the resolvent approach and a maximum likelihood method. *Mol. Biol. Evol.*, 19:8–13, 2002.

[34] T Müller and M. Vingron. Modeling amino acid replacement. *J. Comput. Biol.*, 7:761–776, 2000.

[35] S. Veerassamy, A. Smith, and E. R. M. Tillier. A transition probability model for amino acid substitutions from blocks. *J. Comput. Biol.*, 10:997–1010, 2003.

[36] G. Vogt, T. Etzold, and P. Argos. An assessment of amino acid exchange matrices in aligning protein sequences: The twilight zone revisited. *J. Mol. Biol.*, 249:816–831, 1995.

[37] W. J. Wilbur. On the PAM matrix model of protein evolution. *Mol. Biol. Evol.*, 2:434–447, 1985.

[38] V. B. Yap and T. P. Speed. Modeling DNA base substitution in large genomic regions from two organisms. *J. Mol. Evol.*, 58:12–18, 2004.

16

Posterior Mapping and Posterior Predictive Distributions

Jonathan P. Bollback

Section of Ecology, Behavior, and Evolution, Division of Biological Sciences, University of California, San Diego, La Jolla, CA 92093-0116, USA, bollback@biomail.ucsd.edu

> If we view statistics as a discipline in the service of science, and science as being an attempt to understand (i.e., model) the world around us, then the ability to reveal sensitivity of conclusions from fixed data to various model specifications, all of which are scientifically acceptable, is equivalent to the ability to reveal boundaries of scientific uncertainty. When sharp conclusions are not possible without obtaining more information, whether it be more data, new theory, or deeper understanding of existing data and theory, then it must be scientifically valuable and appropriate to expose this sensitivity and thereby direct efforts to seek the particular information needed to sharpen conclusions. (Rubin [38])

16.1 Introduction

Bayesian statistical approaches are becoming increasingly common in the field of molecular evolution and phylogenetics. Rubin [38] makes an eloquent argument for the value of Bayesian approaches through the identification of sensitivity to our assumptions and the potential uncertainty in our conclusions given our data at hand. While many may see Bayesian approaches as flawed by their dependence on prior distributions and sensitivity to model specifications, others, as with Rubin, will view this as a beneficial property of the method—not accounting for uncertainty can lead to overconfidence in the conclusions. This chapter will review two Bayesian approaches that in the last few years have seen important developments: posterior mapping of characters and posterior predictive distributions. These methods clearly identify and accommodate uncertainty while providing valuable solutions to our questions. It is this author's opinion that these methods will provide invaluable contributions to our understanding of molecular evolution and phylogenetics in the future.

16.1.1 Character Mapping

The mapping of characters on genealogies has been invaluable in answering questions in evolutionary biology since the 1970s; studies such as testing for a molecular clock [21], detecting the signature of positive selection [28], and looking for associations between characters (see [10] for a review) have all employed character mapping. Traditionally, parsimony has been the mainstay—although approaches that combine the methods of maximum likelihood and parsimony and Bayesian inference and parsimony have been developed [21, 16]. Parsimony as a method for mapping characters, while straightforward in its application, has a number of serious drawbacks. First, it underestimates the number of character transformations, often severely. This underestimation arises because parsimony does not account for evolutionary time along branches of a phylogeny: as evolutionary time increases, the number of inferred changes at a site is either zero or one. Second, parsimony underestimates the variance in ancestral states, placing all of the support on one reconstruction when they are not known with certainty. Lastly, parsimony provides no framework for accommodating uncertainty in genealogical relationships.

The drawbacks inherent in parsimony have long been recognized both by molecular evolutionists [8] and phylogeneticists [10]. For example, Langley and Fitch [21], in a study testing the molecular clock hypothesis, employed a mixed method of parsimony to assign ancestral states and maximum likelihood to estimate the rates along the branches. While this early approach acknowledged the underestimation of character changes by parsimony and accommodated it using maximum likelihood, it still left the problem of uncertainty in the phylogeny and ancestral states unresolved.

Recently, methods for accommodating uncertainty in the ancestral states and topology have been devised. For example, one approach to accommodating uncertainty in ancestral states is to use maximum likelihood to estimate the probabilities of each possible state and parsimony to reconstruct the character changes weighted by their probabilities [39, 40, 29, 34]. Uncertainty in topology has also been addressed in a number of ways. Some authors have used a set of reasonable trees and evaluated mappings on each of them (e.g., [42]). Others have evaluated mappings on trees generated under a stochastic process, such as birth-death [24, 26], or evaluated mappings on trees weighted by the probability of the tree being true [25, 33, 16].

While these approaches have made significant advances in accommodating different sources of uncertainty none of them accommodate all sources of uncertainty. In addition, due to their reliance on parsimony, none of these approaches is able to provide detailed information on the timing, order, and types of multiple changes—if any—occurring along a branch. Nielsen [30, 31] has developed a stochastic method for mapping characters using a Bayesian statistical framework. This approach of sampling from the posterior distribution of character histories (also referred to in this chapter as mappings or maps) successfully addresses the drawbacks inherent in parsimony and provides a statistically valid framework for accommodating uncertainty in the phylogeny and model parameters. This approach is the topic of the next section and will be discussed in detail.

16.1.2 Posterior Predictive Distributions

Posterior predictive distributions evolved from concerns regarding the dependence on the prior distribution in prior predictive distributions. Instead of integrating out nuisance parameters using the specified prior distribution of the parameters, the posterior approach integrates with respect to the posterior distribution of the parameters. The justification for, the particular implementation of, and other issues surrounding the use of posterior predictive distributions are rather contentious among statisticians, resulting in an active and healthy research program. Because of this there exists a diversity of different approaches—prior, posterior, and their use in approximating Bayes' factors, to name a few—and opinions regarding these predictive distributions. Much of the discussion revolves around the appropriate formulation of a p-value. The discussion here will deal mostly with posterior predictive distributions and their related p-values. Differences between the approaches and the shortcomings of the posterior method will be highlighted in the relevant places, and a brief account of the controversy will be discussed at the end of the chapter.

Within evolutionary biology, posterior predictive distributions appeared simultaneously with those of posterior mapping. While they can be used to test a variety of hypotheses, their first application was to character histories [32]. A similarity between posterior mapping and posterior predictive distributions is their ability to naturally accommodate uncertainty in the phylogeny and model parameters by treating them as nuisance parameters. (This aspect of both methods is not unique to them and has been a motivating factor behind many of the Bayesian developments in biology; see [18] for a review.) Posterior predictive distributions have in the last few years seen application to hypotheses such as detecting positive selection [32], evaluating substitution model adequacy [3], testing for nucleotide frequency heterogeneity [18], correlated character change [14], concordance between genes [46], and patterns in protein evolution.

The use of predictive distributions in Bayesian hypothesis testing in general and evolutionary biology in particular is appealing for a number of reasons. First, the generality of the approach makes it applicable to a wide variety of questions in molecular evolution and phylogenetics. Second, the method provides a rigorous statistical framework for accommodating uncertainty in model parameters and genealogical (or phylogenetic) relationships. This alone may be the strongest argument for the use of predictive distributions over methods such as the parametric bootstrap. Third, predictive probabilities (called posterior predictive p-values) are constructed using tail areas of the predictive distribution and are straightforward in their implementation and interpretation. Unlike classical frequency probabilities, posterior predictive probabilities do not evaluate observed values relative to a fixed set of values under the null model but averages over probable sets. Lastly, predictive p-values produce a Type I frequentist error at a given α similar to the expected α (often lower but never greater than 2α) [27]. While these reasons make predictive distributions appealing, a number of concerns and potential drawbacks exist and will be discussed at the end of Section 16.3. Briefly, this approach requires the description of a probabilistic model (null hypothesis), specification of a prior distribution for the model, an estimation of the model's posterior distribution, and a little ingenuity on the part of the researcher in determining appropriate test statistics (see [38] for a general review). Each of these will be dealt with in detail in Section 16.3. Of these requirements, the last is clearly the most difficult to accomplish: a good test statistic needs to be a relevant summary of the hypothesis being tested, and each question will require a different sort of test statistic. The logic behind the posterior predictive approach is similar to that underlying the parametric bootstrap. In fact, the parametric bootstrap sampling distribution may be indistinguishable from the posterior predictive sampling distribution when maximum likelihood estimates are used and the posterior is concentrated. The fit of a hypothesis is tested by comparison of the observed test statistic—often referred to as the realized value—with the distribution of that statistic under the null model. If our realized value falls within the 95% confidence region of the null distribution, we are unable to reject the null hypothesis—otherwise, we reject it.

The remainder of this chapter will explore the underlying methodology of these two approaches, review a number of their recent applications, demonstrate how posterior predictive distributions can be used to test hypotheses about character histories, and discuss how predictive distributions can be used to address a wealth of different questions in molecular evolution and phylogenetics.

16.2 Posterior Mapping

In this section, I will try to answer four questions: (1) What are character histories?; (2) How do we go about sampling character histories?; (3) How do we accommodate uncertainty in model parameters and topologies?; and (4) What types of questions can we address with posterior mapping? The second and third questions will be answered by introducing the method of posterior mapping first proposed by Nielsen [30, 31] and then later extended by Huelsenbeck et al. [14] to morphological characters. The last question will be answered by briefly reviewing examples from the literature.

First, let us tackle the question of what a character history is by providing a definition. A character history is a description of the historical pattern of state occurrences and transformations along a phylogeny. The history is more than just a simple description of the ancestral reconstructions at the internal nodes of the tree. It includes information about the placement (timing), order of states, types of character state transformations (e.g., A ⇔ G), and direction (or bias; e.g., A → G versus G → A) of transformations when the root of the phylogeny is known (see Figure 16.1d for an example of a character history). What we would like is to sample possible character histories (individual character histories will also be referred to as a map) in which they are sampled in proportion to their posterior probabilities. More often we will be interested in a function of these sampled histories and not individual histories. For example, we may wish to determine the number of radical amino acid changes relative to conservative changes [32]. In addition, we may be interested not only in the relative types of changes but also the order and timing of changes. For example, contingency tests of neutrality rely on being able to determine types of changes (silent/replacement) and their placement on the tree [30].

But, before we get into the details of the method (questions 2 and 3), we might wonder why we should not rely on parsimony and what the differences are. To illustrate these differences, we will explore four different mappings of a single site for four species shown in Figure 16.1. We will ask: (1) How does the placement of character transformations along a branch differ?; (2) How does the number of character transformations along a branch differ?; and (3) How probable are nonparsimonious mappings? Two of the trees in Figure 16.1 are parsimony mappings (trees a and b) and two are posterior mappings, one of which is consistent with parsimony (trees c and d).

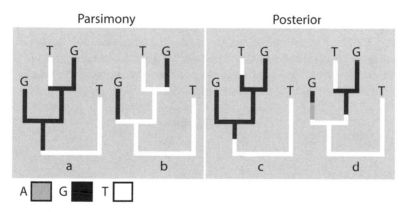

Fig. 16.1. A comparison between parsimony and two representative realizations from the posterior distribution of mappings. Trees a and b are parsimony reconstructions, while c and d are from the posterior distribution of mappings. The inferred number of changes in tree *c* is consistent with parsimony. The posterior mappings were generated with SIMMAP, a program that implements the posterior mapping method and can be downloaded at http://www.simmap.com.

The position along a branch at which an inferred change occurs under parsimony is shown directly following a bifurcation. This was done for convenience—we could have placed the changes equidistant along the branches. This illustrates the first difference between parsimony and posterior mappings—their placement of transformations along branches. Parsimony provides no information about the timing of changes along a branch; parsimony simply concludes that a single change has occurred. Posterior mapping, however, does provide information about placement and order of multiple changes along a branch. (In addition, the timing of changes between different sites can be compared. See the discussion on correlated character evolution, in Subsection 16.3.5 , for an example.) For example, in trees c and d, in Figure 16.1, we can clearly see when the events occurred and the order in the case of tree d. In many cases, the order of changes is of interest. For example, we might wish to know whether a burst of amino acid replacements immediately follows speciation or whether it is evenly distributed after the split.

To illustrate the difference in the number of transformations considered by each method, let us compare the posterior mapping on tree d in Figure 16.1 with the parsimony mappings (trees a and b). First, we should note that the map on tree d is not consistent with parsimony; four changes have been inferred, compared with two changes required by parsimony. Sampling from the posterior distribution of mappings has produced a map in which two additional changes have occurred. While, admittedly, I have not shown you that mappings with two additional changes have a large or small probability, it does have a probability greater than zero. Under parsimony this is not even considered plausible, let alone probable, while the posterior method is not constrained to minimizing the number of changes.

Let's consider the final difference between parsimony and posterior mappings—how probable are nonparsimonious mappings? In this example, we will compare the probability of parsimonious and nonparsimonious mappings. In effect, we will be evaluating two assumptions of parsimony: the minimization of changes and the reduction in variance associated with ancestral state reconstruction at the root. This example should also provide an introduction to the underlying logic of posterior mapping. To address this difference, we will first calculate the overall probability of the data and then conditional probabilities given the branch lengths and the number of character changes along the trees shown in Figure 16.2.

In this particular example of two species, there is only a single phylogeny relating the two sites. This is the equivalent of assuming that the tree is known in cases of more than three species. (Later, it will be shown that the method allows us to accommodate uncertainty in the phylogeny and model parameters.) To compare the mappings, we are interested in calculating

$$\Pr(M_i|D) = \frac{\Pr(M_i, D)}{\Pr(D)}, \tag{16.1}$$

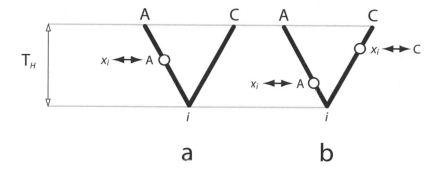

Fig. 16.2. Comparison of the probabilities associated with parsimonious and non-parsimonious character histories. T_H is the tree height, from the root, in the expected number of substitutions per site and will be used to evaluate an increase in branch lengths (0.5, 1.0, and 2.0, respectively; see the text). x_i is the state we are changing from at the root and is dependent on whether we are observing one or two changes; under one change $x_i \in \{C\}$, while under two changes $x_i \in \{G, T\}$.

where M_i is a character map and D is the observed data. This is the probability of the map given the data. Calculation of the probability of the data, $\Pr(D)$, requires a model that describes substitution probabilities from one state to the next. We will assume the Jukes and Cantor [19] model, which is a time-reversible Markov model. Under the JC69 model, the stationary nucleotide frequencies are $\pi_i = 1/4$ for all i, and the probability of a change from nucleotide i to j along a branch of length t is

$$P_{ij}(t) = \begin{cases} 1/4 + (3/4)e^{-(4/3)t} & \text{if } i = j, \\ 1/4 - (1/4)e^{-(4/3)t} & \text{if } i \neq j. \end{cases} \tag{16.2}$$

We can now calculate $\Pr(D)$ by considering all possible state assignments at the root i as

$$\Pr(D) = \sum_{i \in \{A,C,G,T\}} \pi_i P_{iA}(t) P_{iC}(t). \tag{16.3}$$

When $T_H = 0.5$, then $\Pr(D) = 0.04602$ for the data and phylogeny shown.

Next we want to calculate the probability of histories a and b conditional on the data at the tips of the trees. For the mapping shown on tree a (M_a), we want to calculate $\Pr(M_a, D)$. This can easily be done using the fact that for the JC69 model and other continuous-time Markov chain models, the number of changes along a branch is Poisson-distributed. For example, along the left lineage of tree a, the conditional probability of observing a single change is $0.5e^{-0.5} \times (1/3)$. The last term represents the probability of a change between nucleotides, which is 1/3 under the JC69 model. Therefore, we calculate $\Pr(M_a, D)$ as

$$\Pr(M_a, D) = \frac{e^{-0.5} \times (0.5e^{-0.5}/3)}{4} = 0.0153, \qquad (16.4)$$

where the probability of not observing a change along a branch of length $t = 0.5$ is $e^{-0.5}$ and again the probability of observing a single change along a branch of this length is $0.5e^{-0.5} \times 1/3$ under the JC69 model.

The root state for tree a must be a C, given the states at the tips and a single change occurring along the branch leading to the state A. However, in tree b the state of the root is uncertain. An observation of a T or a G at the root of tree b would be consistent with the mapping shown and the states at the tips of the tree. Given these possible root states, we can calculate the probability as

$$\Pr(M_b, D) = \frac{(0.5e^{-0.5}/3)^2}{4} \times \frac{(0.5e^{-0.5}/3)^2}{4} = 0.0051. \qquad (16.5)$$

Using these probabilities and $\Pr(D)$, we can calculate the conditional probabilities for the character histories on trees a and b as 0.333 and 0.111, respectively. The parsimony-consistent history is three times as probable. However, what happens as the time from the root to the tips increases? Table 16.1 shows the probabilities for the trees and mappings in Figure 16.2 given three different sets of branch lengths.

Table 16.1. A comparison of the probabilities associated with the parsimony consistent mapping in tree a with that of the nonparsimonious mapping of tree b (see Figure 16.2) and the cumulative probability of mappings greater than two substitutions ($\Pr(M_{i>b}|D)$).

			Tree a		Tree b	Changes > 2
T_H	$\Pr(D)$	$\Pr(M_a, D)$	$\Pr(M_a|D)$	$\Pr(M_b, D)$	$\Pr(M_b|D)$	$\Pr(M_{i>b}|D)$
0.5	0.046025	0.015328	0.333	0.005109	0.111	0.556
1.0	0.058157	0.011277	0.194	0.007519	0.129	0.677
2.0	0.066239	0.003052	0.034	0.004070	0.045	0.921

A couple of things should be noticed in Table 16.1. First, as the branch lengths increase, the probability of the mapping consistent with parsimony (tree a) decreases. Second, the parsimony mapping decreases from a threefold higher probability to a probability lower than the mapping with two changes (tree b) as branch lengths increase. As expected, as time increases, the probability of multiple changes increases, making mappings with one, and even two, changes much less probable (although they probably have the largest probabilities). The cumulative probability of more than two changes increases with increasing time, reaching 0.921 at divergences of 2.0 expected substitutions per site. Hopefully, I have been able o show that even for the simplest phylogeny, nonparsimonious mappings should be considered.

16.2.1 Sampling Character Histories

How do we go about sampling character histories using the method of posterior mapping? The following is a description of simulating a map for a site. Complete gene sequences can be sampled by repeating this approach for each site. Four steps are involved in sampling a character map: (1) define a substitution model in which probabilities of state changes can be calculated; (2) calculate the conditional likelihood for each state at each node of the tree, including the root; (3) simulate ancestral states; and (4) simulate a substitution (mutational) history, conditional on the ancestral states and states at the tips of the tree. (Often the states at the tips of the tree are unknown or uncertain (e.g., N, R, etc.). This type of uncertainty can easily be accommodated by revisiting these nodes after simulating ancestral states for the internal nodes and repeating step 3 for the tips.)

First, we need to define a model of nucleotide (or morphological) change (step 1). Any number of continuous-time Markov models are available, that accommodate a variety of different plausible aspects of sequence evolution. Available models and their uses have been extensively described elsewhere, and a detailed treatment is beyond the scope of this chapter [48, 11]. Briefly, many commonly used models are special cases of the general time-reversible (GTR) model of sequence evolution [20, 37]. With this model, we can describe the instantaneous rates of changing from state i to state j using the rate matrix

$$\mathbf{Q} = \{q_{ij}\} = \begin{pmatrix} - & a\pi_C & b\pi_G & c\pi_T \\ a\pi_A & - & d\pi_G & e\pi_T \\ b\pi_A & d\pi_C & - & f\pi_T \\ c\pi_A & e\pi_C & f\pi_G & - \end{pmatrix}, \qquad (16.6)$$

where a–f represent the rates of changing from one nucleotide to the next, and π_i represent the stationary nucleotide frequencies. Using this matrix, we can easily calculate substitution probabilities for a change from nucleotide i to j over a branch of length t as $\mathbf{P} = \{p_{ij}(t)\} = e^{\mathbf{Q}t}$. In many cases, such as the JC69 model described above, analytical solutions are available. In those cases in which solutions are not available, standard linear algebra approaches are available for exponentiating the matrix $\mathbf{Q}$.

With these probabilities, step 2 can be easily accomplished using the *pruning algorithm* of Felsenstein [4]. Given a tree with branch lengths τ, a set of observations D at the tips of the tree, and a vector θ containing a set of model parameters describing sequence evolution, we can calculate the conditional likelihood for each internal node and the root using a post-order traversal of the tree.

Next, we simulate a state at the root of the tree (step 3). Let us denote the root as σ and the simulated observation as d. The new state at the root will then be denoted d_σ. (All s descendant nodes and branches are indexed as $\sigma - 1, \ldots, \sigma - (2s - 3)$.) A site can be simulated by sampling from the posterior distribution

$$\Pr(d_\sigma = i | D, \tau, \theta) = \frac{l_{\sigma,i} \pi_i}{\sum_{j \in \{A,C,G,T\}} l_{\sigma,j} \pi_j}, \tag{16.7}$$

where $l_{\sigma,i}$ is the conditional likelihood of being in state i—we are conditioning on the observations at the tips of the tree, model parameters, and topology. Now, in a preorder traversal of the tree from the root, we visit a node directly above, $\sigma - 1$, and simulate an ancestral state by sampling from

$$\Pr(d_{\sigma-1} = j | d_\sigma = i, D, \tau, \theta) = \frac{l_{\sigma-1,i} P_{ij}(t_{\sigma-1})}{\sum_{k \in \{A,C,G,T\}} l_{\sigma-1,k} P_{ik}(t_{\sigma-1})}, \tag{16.8}$$

where j represents the recently simulated state at the ancestral node (in this case the root) and $P_{ij}(t_{\sigma-1})$ is the transition probability from state i to state j over a length of $t_{\sigma-1}$. We proceed with the traversal and simulate ancestral states for the remaining nodes. As noted above, often we find that a site may be unknown or uncertain for some sequences. Using this approach, we can also simulate a tip state. In this way, we treat the uncertainty at the tips in a fashion identical to that for internal nodes. Now we have sampled and assigned ancestral states from the posterior distribution for each internal node of the phylogeny.

The final step is to generate a character history for each branch of the tree given the previously simulated ancestral states and observed states at the tips of the tree (step 4). This, perhaps, is the most challenging step, and Nielsen [31] provides an elegant and computationally efficient solution. We simulate a realization of a continuous-time Markov chain conditional on the starting state and ending states along a branch. The waiting times between substitution events along a branch are drawn from an exponential distribution

$$\lambda e^{-\lambda t} \tag{16.9}$$

with the rate $\lambda = -q_{ii}$. This rate is taken from the diagonal elements of our $\mathbf{Q}$ matrix, which are interpreted as the rate of moving away from a state i. Waiting times can be obtained from this distribution using the inverse transformation method. If the exponential waiting time is longer than the branch length t and the states at each end of the branch are the same, then the process is terminated; no changes have occurred along this branch. If the waiting time is smaller than the branch length t, then a character transformation is determined by $\Pr_{ij} = \frac{q_{ij}}{-q_{ii}}$, and the process is continued with the new length, $t - t_1$, by drawing another exponential waiting time. If the next waiting time is longer than the remaining time along the branch and the states are the same, the process ends for that branch. On the other hand, if the states are different, the process is repeated from the ancestral node, not the previous simulated transformation. If we were to proceed from the previous transformation, the waiting times would no longer be exponentially distributed.

Nielsen [30] has pointed out that this approach is not computationally efficient when the reconstructed ancestral states are not the same and the length l is small. Nielsen [30] proposed conditioning the first waiting time on being less than the length of the branch as

$$f(t_1|t_1 < t) = \frac{\lambda e^{-\lambda t_1}}{1 - e^{-\lambda t}}, \quad 0 \le t_1 < t, \tag{16.10}$$

where $\lambda = -q_{ii}$. Waiting times can also be drawn from this distribution using the inverse transformation method. This approach enhances the computational efficiency of the algorithm by reducing the number of realizations that are rejected. Using this approach, the first draw always produces a waiting time less than t and thus is consistent with at least one change occurring along the branch. The next draw uses the unconditional distribution as above. Once all internal nodes of the tree have been visited, we have successfully simulated a single realization of a map from $\Pr(M|D, \theta, \tau)$.

16.2.2 Integrating over Topologies and Model Parameters

In general, parameter values of the substitution model θ and the topology τ are not known with certainty. We would like to evaluate $\Pr(M|D)$ and not $\Pr(M|D, \theta, \tau)$. The Bayesian approach permits a natural way of accommodating uncertainty in these values. We wish to sample from

$$\Pr(M|D) = \sum_{k=1}^{\psi} \int_{v_k} \int_{\theta} \Pr(M|D, \theta, \tau) p(\tau_k, v_k, \theta|D) dv_k d\theta, \tag{16.11}$$

where ψ is the set of possible trees and v_k are the branch lengths associated with tree k. While this cannot be solved analytically due to its complexity, numerical approximations can be obtained using MCMC methods [35, 22, 17] (see Chapters 3 and 7).

In practice, how do we go about sampling character histories not dependent on fixed values for these parameters? The answer is quite simple. As described above, we have a method for sampling a map along a phylogeny. Using a program such as MrBayes or BAMBE, we can easily obtain an approximation of $p(\tau_k, v_k, \theta|D)$. With this distribution in hand, we can simulate a map for each posterior sample producing a valid approximation of $\Pr(M|D)$.

As mentioned previously, what we are most often interested in is some function of the histories, $h(M, D)$. These functions might evaluate the number of nonsynonymous substitutions, radical amino acid changes, relative timing of changes, correlation in the timing of transformations between two sites, or covariation of states between sites. We now have all the pieces necessary to evaluate any desired function and its expectation. For example, if we wish to evaluate the expected number of nonsynonymous changes, $n_{NSYN}(M, D)$, we can evaluate the expectation numerically from the distribution of character histories as

$$E[n_{NSYN}(M, D)|D] \approx \frac{1}{N} \sum_{i=1}^{N} n_{NSYN}(M_i, D), \qquad (16.12)$$

where N is the number of simulated character histories and $n_{NSYN}(M_i, D)$ is the observed number of nonsynonymous changes along map i.

16.2.3 Examples from the Literature

This section is intended to direct the reader to the most recent applications of posterior mapping in the literature. A brief overview of the specific questions addressed in the literature should provide a better understanding of the power of this approach.

The first application of this method in the literature [30] used it to address a number of questions pertinent to molecular evolution and population genetics. First, the author made inferences regarding the population parameter θ, which is the product of the population size and mutation rate, to a data set of 63 human mtDNA sequences from the Nuu-chah-Nulth tribe (see [50] in [30]) demonstrating the method's utility in population genetics. In addition, the method was applied to estimating the ages of mutations and then specifically the ages of synonymous and nonsynonymous mutations in a test of neutrality proposed by Templeton [49].

The method was further used to address how the parsimony method compared with the posterior method in estimating the number of mutations across two genes: β-globin and influenza hemagglutinin-A [31]. An analysis of the complete gene sequences found that the parsimony method greatly underestimated the total number of substitutions compared with the posterior method. Nielsen argued that the large discrepancy was likely due to differences in lineages; for example, rate heterogeneity among lineages, mutational biases among lineages, such as a transition/transversion bias, or biases among lineages in synonymous and nonsynonymous evolutionary rates. To address these questions, he tested for rate homogeneity among lineages, finding that there appeared to be considerable variance among lineages, particularly in the β-globin data set.

Finally, this method was extended to mapping morphological characters [14] using the Mk series of stochastic models [23]. While possibly of little interest to molecular evolutionists, this represents a major advancement in the phylogeneticist's ability to address questions about morphological character evolution using a statistical approach not relying on parsimony. Not only does this paper extend the method of stochastic mapping to morphological characters, using the Nielsen [31] method, but it provides a novel approach to looking for correlated character evolution using predictive distributions (see Section 16.3).

16.3 Predictive Distributions

Often we are confronted with situations in which the data, or some aspect of an analysis, do not meet the assumptions of a standard statistical test (e.g., the use of improper prior distributions in calculating Bayes factors). In cases like these in molecular evolution and phylogenetics, we rely on alternative methods, such as permutation tests (e.g., randomization tests), resampling approaches (e.g., the nonparametric bootstrap), the parametric bootstrap, and, in the Bayesian framework, predictive distributions. The latter approach is operationally analogous to the parametric bootstrap but has a number of differences and potential advantages over the traditional parametric bootstrap. This potential will hopefully become clear in the remainder of the chapter.

16.3.1 Posterior Predictive Simulations

Bayesian approaches to hypothesis testing come in two general forms: Bayes factors and predictive distributions. While hypothesis tests using Bayes factors have received a fair amount of attention in the phylogenetics literature [44, 13, 43, 45], the alternative, predictive distributions, only recently have been applied to methods in molecular evolution and phylogenetics [32, 31, 3, 46]. In this section, I will provide background on what predictive distributions are and how to use them, explore some recent applications from the literature, and discuss the pros and cons of their use. Predictive distributions provide a very general and flexible framework for Bayesian hypothesis testing, making them likely to be applied to a broad array of questions. In addition, they provide a natural way of accommodating uncertainty in the substitution model parameters and topology. This being said, the method isn't free of problems. The specifics of these issues will be reviewed at the end of this section. In evaluating a hypothesis, we would like to know how well it fits the underlying process that generated the data at hand. If a hypothesis is adequate, then it should perform well in predicting the distribution of data observations or some summary value relevant to the hypothesis being scrutinized. These distributions of future observations are called *predictive distributions* (also called reference distributions or densities). Most often we are not directly interested in the predictive distribution of the data but a summary statistic, referred to as a test statistic in this chapter, that captures relevant features of predictive data and our observed data given the hypothesis. Test statistics are dealt with in Section 16.3.2 but, for the moment let us assume we have some function, $T(\cdot)$, that summarizes an aspect of our data.

An analogy: parametric bootstrap

Before we get into the details of how to sample from posterior predictive distributions, I want to develop an operational analogy with the parametric bootstrap. Since many readers are already familiar with the use of the parametric bootstrap, it will hopefully serve as a useful heuristic to understanding predictive methods. The thought experiment will be a test of the molecular clock. While I don't advocate the test described below, as it is untested, it does provide a useful heuristic for understanding the differences between the two methods. (Note: There are numerous other well-established ways of testing the molecular clock.)

Let θ_c be a vector containing our model parameters (which include the substitution model parameters, topology, and associated node depths) under the clock hypothesis and θ_{nc} be the similar vector of parameters under the unconstrained hypothesis. Under the parametric bootstrap, these values are chosen to be the maximum likelihood estimates (MLE) for these quantities. Since we wish to test the molecular clock, we can generate our reference distribution using these $\hat{\theta}_c$ values and simulate n data sets (see Figure 16.3). These are the predictive outcomes we might expect to observe in future data collection expeditions, given that the values of $\hat{\theta}_c$ are true. Next, we need to summarize the data (observed and predictive) in some relevant way. We can use the difference in maximum likelihood estimates between the constrained (clock) and unconstrained branch length hypotheses [4], but for this example we will take an alternate approach. Let's assume that we have an outgroup that establishes the placement of the root and use the standard deviation of distance of the tips to the root under each hypothesis. The reference distribution, simulated under the clock, allows us to check the degree to which the clock would appear violated (magnitude of the standard deviation), given that the underlying process is truly clock-like. If the observed, or realized, value falls outside of this distribution, we might be inclined to reject the clock hypothesis or, more precisely, we reject that the observed deviation could have arisen under our null hypothesis—a molecular clock and the particulars of the substitution model.

In comparison, how might this be accomplished using posterior predictive simulations, and what are the possible differences in outcome with the parametric bootstrap? The first difference is immediately apparent: values of θ are not point estimates but averaged over samples from the posterior distribution of θ (see Figure 16.3) under the clock and unconstrained hypotheses. Samples from the posterior distribution under the clock model (θ_c) and unconstrained model (θ_{nc}) can be obtained using a program such as MrBayes [17]. Using these models, we can evaluate the expectation of our standard deviation test statistic, under the unconstrained hypothesis. This reveals a–second difference with the parametric bootstrap: we have accommodated uncertainty in the θ_{nc}, and therefore uncertainty in the value of the realized test statistic, by averaging over values sampled from the posterior distribution. To obtain the null distribution of the test statistic under the clock hypothesis, we will simulate data by sampling the posterior distribution of θ_c under the clock hypothesis (see Figure 16.3). For each of the predictive data sets sampled, we will need to perform another round of MCMC to sample from the posterior distributions

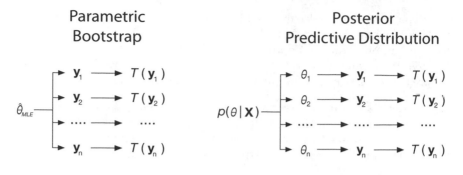

Fig. 16.3. Comparison of the parametric bootstrap and posterior predictive simulation. Values of θ are used to simulate n new data sets ($\{y_1, y_2, \ldots, y_n\}$). These are then evaluated using our chosen test statistic, $T(\cdot)$, giving us the reference distribution under the hypothesis, which is compared with the realized test statistic, $T(\mathbf{X})$.

under the unconstrained hypothesis. The null distribution is summarized from these samples. In this case, the standard deviation for each of these replicates is the predictive distribution of standard deviations expected under the clock hypothesis (conditional on the data and chosen model). As with the parametric bootstrap, we can compare the expectation of the realized deviation to the predictive values under the molecular clock. If the realized value falls outside of the predictive distribution under the clock, then we are tempted to consider the observed deviations as unexplained by a strict molecular clock.

Now, hopefully, you have a feel for the mechanics of predictive tests and some of the differences with the parametric bootstrap, and we are ready to move on and look more closely at the method of posterior predictive simulations.

Sampling from posterior predictive distributions

First, we need a method for generating the predictive distribution of the data before evaluating some function of it. Let $\mathbf{Y} = \{y_1, y_2, \cdots, y_n\}$ be a vector containing n future observations and $\mathbf{X} = \{x_1, x_2, \cdots, x_n\}$ be a vector containing our current observations. What we would like to sample is the predictive distribution of $\mathbf{Y}$ conditional on the hypothesis H,

$$p(\mathbf{Y}|H, \mathbf{X}) = \int_{\theta} p(\mathbf{Y}|\theta)p(\theta|\mathbf{X})d\theta, \qquad (16.13)$$

where θ is a vector containing model parameters under the hypothesis under scrutiny, and $p(\theta|\mathbf{X})$ is the posterior distribution of these parameters. Unfortunately, we can't analytically determine $p(\mathbf{Y}|H, \mathbf{X})$ because the posterior distribution, $p(\theta|\mathbf{X})$, the source of a reasonable set of values for θ under the hypothesis being scrutinized, is impossible to determine analytically for all but the simplest cases in molecular evolution. Furthermore, we can use sampling methods, such as Markov chain Monte Carlo (MCMC), to sample from this distribution, providing an approximation of $p(\theta|\mathbf{X})$ [35, 22, 17] (see Chapters 3 and 7). With values of θ from the posterior distribution, we can approximate the predictive distribution by sampling using the following algorithm:

1) Draw a set of parameter values, θ_i, from the joint posterior distribution of parameters under the null model being tested. (In practice, this can be accomplished by sampling the posterior output of a program that approximates posterior distributions using MCMC, such as MrBayes [17].)

2) Using the values of θ_i (which may include values for the parameters of the substitution process, topology, branch lengths, etc.), simulate data, $\mathbf{y}_i$.

3) Repeat steps one and two N times to create a collection of data sets, $\mathbf{y}_1, \mathbf{y}_2, \cdots, \mathbf{y}_n$, corresponding to samples from the posterior distribution of $\theta_1, \theta_2, \cdots, \theta_n$.

4) These simulated data sets are samples from the posterior predictive distribution shown in (16.13) and can be used to evaluate our hypothesis of interest.

The precision of the sampling approximation is a function of the number of draws from the posterior distribution, the precision of our posterior estimate, and the appropriateness of the underlying prior distributions. Fortunately, we are guaranteed by the law of large numbers that we will converge on the target distribution. What exactly is "large" is not clear and is likely to be dependent on the particular parameters of the distribution.

16.3.2 Test Statistics

By sampling we now have an approximation of the posterior predictive distribution of the data simulated under the null model being scrutinized. But we are still left with the following problem: How can we use the posterior predictive distribution to assess our hypothesis H? As already mentioned, we are generally not interested in the predictive distribution of the data directly but some function of it (in our case, a function of the sampling distribution), or more concisely the predictive distribution of the function of interest. Our functions will most often be a descriptive test statistic (often referred to as a summary or discrepancy variable [5]) that quantifies some aspect of the data. The test statistic is referred to as a realized value when summarizing the observed data. In principle, an appropriate test statistic can be defined to measure any aspect of the predictive distribution of the data, but in practice the issue of defining an appropriate statistic for a given hypothesis may not be straightforward [6] and is considered contentious [2].

I follow the general notation $T(\cdot)$, where this is some function of the data. To emphasize that our interests are in sampling from the predictive distribution of $T(\cdot)$, equation (16.13) could be rewritten as

$$p[T(\mathbf{Y})|\mathbf{X}] = \int_\theta p[T(\mathbf{Y})|\theta]p(\theta|\mathbf{X})d\theta, \qquad (16.14)$$

where $\mathbf{Y}$ is a set of future or predictive observations of the data, $\mathbf{X}$. Using the algorithm outlined above, we can sample this distribution with one additional step; for each simulated data set, we evaluate the function $T(\mathbf{Y})$. (Examples of different test statistics will be described later.) In this way, we now have a sampling approximation of the predictive distribution of the test quantity in which we are directly interested. Importantly, it should be noted that this distribution is averaged over samples from the posterior distribution, allowing us to accommodate uncertainty in our parameter estimates. This frees the test from dependence on any particular set of parameter values by evaluating them in accordance with their probabilities. Whether this is a benefit of the method is yet unclear. (The effects of accommodating uncertainty in parameters in Bayesian molecular evolution studies has not been looked at closely.) This distribution can then be compared with the realized test statistic, $T(\mathbf{X})$, which is calculated from the original data, and the predictive probability of the null hypothesis can then be evaluated.

16.3.3 Predictive p-Values

Recently, much research has been directed at the use, properties, and interpretation of p-values as measures for predictive distributions and we direct the reader to [2, 36]. Predictive p-values are often denoted p_T to indicate their dependence on the test statistic and have an operational interpretation similar to classical p-values, as they are both derived from tail area probabilities; values that lie in the extremes of the null distribution of the test quantity are considered significant to reject the null hypothesis. Under classical statistics, the distributions are conditioned on point estimates for model parameters. Predictive densities, on the other hand, are not because parameter values are sampled from the posterior distribution in proportion to their probabilities. This sampling scheme allows them to be treated as nuisance parameters—values not of direct interest—and to be integrated out. Samples from the predictive distribution of the test statistic allow us to evaluate the posterior predictive probability as

$$p_T = \Pr[T(\mathbf{y_{rep}}) \geq T(\mathbf{X})|\mathbf{X}, \theta]. \qquad (16.15)$$

The posterior predictive p-value for the test statistic is calculated as

$$p_T = \frac{1}{N}\sum_{i=1}^{N} I(T(\mathbf{y}_i) \geq T(\mathbf{X})), \qquad (16.16)$$

where I is an indicator function that takes on the value of 1 when the equality is satisfied and 0 otherwise, $T(\mathbf{y}_i)$ is the test statistic for the ith simulated data set, and $T(\mathbf{X})$ is the realized test statistic. Probabilities less than the critical threshold, say $\alpha = 0.05$, suggest that the hypothesis under examination is inadequate. Predictive p-values are interpreted as the probability that the hypothesis would produce as extreme a test value as that observed for the data [6]. This approach evaluates the practical fit of the hypothesis to our observations and is dependent on the test statistic employed. These p-values should not be interpreted as frequentist error probabilities or as the probability of our hypothesis. Sellke, Bayarri, and Berger [41] have suggested that p-values can be calibrated to allow for a Bayes factor interpretation (i.e., the odds of H_0 to an unspecified alternative H_1),

$$B(p) = -ep\log(p), p < e^{-1},$$ (16.17)

or a frequentist error probability,

$$\alpha(p) = (1 + [-ep\log(p)]^{-1})^{-1}.$$ (16.18)

While an extremely powerful and appealing aspect of predictive distributions is the ease and flexibility in test statistics that can be employed, not all test statistics are appropriate. Careful consideration of the hypothesis and its underlying assumptions, and the test statistic, should be made prior to decisions about the hypothesis under scrutiny.

16.3.4 Issues Concerning the Use of Predictive Distributions

Practitioners should be aware of a number of issues surrounding the application of posterior predictive distributions. First, there is an apparent double use of the data. The data are used in the estimation of the posterior distribution during simulation of the predictive distribution and are used again during calculation of the tail area probabilities. A number of general solutions have been suggested by various authors (see [27, 6, 7]). Second, the results are dependent on the choice of test statistic. While the ability of the method to accommodate many different statistics is a benefit, poorly chosen statistics may lead to incorrect conclusions and unpredictable behavior. Third, there are concerns over the properties and interpretation of the different predictive p-values that are available (see [2, 36]), particularly in situations for which composite null models are being entertained. Finally, posterior predictive methods may be highly conservative, resulting in a failure to detect problems with, or deviations from, the null model.

16.3.5 Examples from the Literature

Predictive distributions are a new introduction to studies in molecular evolution and phylogenetics although they have been extensively discussed in the statistical literature (see [38]). Yet they have seen a rapid application to a diverse array of questions in the last few years. In this section, I will briefly review a few different applications from the literature. This should give us some insight into what types of questions have been addressed and can be addressed in the future.

Substitution model adequacy

While substitution model testing in phylogenetics and molecular evolution has been an area of extensive research, until recently little had been done within the Bayesian framework, and many researchers relied on classical approaches, such as the likelihood-ratio test (for a review, see [15]), parametric bootstrap [9], or Akaike information criterion [1], to select models for Bayesian analysis. One drawback to these approaches is that they do not easily accommodate uncertainty in parameter estimates and the topology used in the test. As we have seen, predictive distributions provide a natural approach to accommodating uncertainty. (This is not the only Bayesian approach to model testing that accommodates uncertainty; see the use of Bayes factors in model selection [44].) This approach has been applied to determining model adequacy and choice [3], testing for homogeneity of base frequencies among lineages [18], and testing for lineage rate heterogeneity [31].

Bollback [3] proposed that we could evaluate a substitution model's adequacy using predictive distributions and that this would naturally lead to selection through refinement or enhancement of the model to be used in further analysis. This approach differed most importantly from likelihood-based approaches by taking into account uncertainty in topology, branch lengths, and model parameters. Therefore, model choice has been freed from conditioning on these parameters and has resulted in a more accurate estimate of model variance. The multinomial test statistic was used to evaluate how well a model was able to generate data similar to existing data. Further, the study found that a number of factors affected an increase in the power of the test statistic: (1) increasing the number of sites; (2) increasing sequence divergence (expected number of substitutions per site); and (3) the degree of violation of a model's assumptions.

In a review of Bayesian inference, Huelsenbeck et al. [18] tested for homogeneity of nucleotide frequencies among lineages of the *Drosophila* alcohol dehydrogenase (*Adh*) locus. They used the following test statistic to evaluate the deviation from homogeneity among 58 lineages over time:

$$\chi^2 = \sum_{i=1}^{58} \sum_{j \in \{A, C, G.T\}} \frac{(f_{ij} - \bar{f}_j)^2}{f_j}. \tag{16.19}$$

The authors were able to strongly reject the null hypothesis of nucleotide frequency homogeneity among lineages.

In the final example of evaluating substitution models, Nielsen [31] evaluated lineage rate variation for two data sets: β-globin and influenza hemagglutinin-A. He used the variance in expected number of substitutions, (V_k), as the test statistic and tested the null hypothesis of homogeneity of variances among lineages. By examining the posterior and predictive distributions, he concluded that, because of their small overlap, the null hypothesis of homogeneity could be rejected. This study is important because it used the method of posterior mapping to obtain estimates of V_k for each lineage and used predictive distributions to evaluate significance.

Positive selection

A diverse array of methods for detecting positive selection at sites within a gene is available to molecular evolutionists and phylogeneticists alike, ranging from parsimony-based methods [47] to likelihood-based methods (see Chapter 5) and Bayesian methods (e.g., [32, 12]). The use of posterior mapping and predictive distributions to detect positive selection was introduced by Nielsen [32]. I will focus on this paper because it demonstrates both posterior mapping and predictive distributions to test the null hypothesis of no selection. The authors evaluated the number of nonsynonymous substitutions as their test statistic for an influenza hemagglutinin-A data set. They observed that 11 sites had significant p-values ($p_T \leq 0.01$), suggesting these sites had an excess of nonsynonymous substitutions. They concluded that these sites were under positive selection. To further strengthen their argument, they compared their results with the results of Yang et al. [51], showing a strong concordance between the posterior predictive p-values and posterior probabilities according to the M3 model. None of the 11 sites determined to be under positive selection showed posterior probabilities lower than 0.975.

Correlated character evolution

In this last section, I will review a recent study in which the authors used posterior mapping and predictive distributions to determine correlation among evolving characters [14]. Because the paper deals with morphological characters, it may seem on the surface to have little importance to studies in molecular evolution. But, quite the contrary, it demonstrates how these methods can be extended to studies of correlated molecular evolution. For example, the methods could be applied to looking for correlated change among nucleotides, such as RNA stem partners, or interactions among amino acid sites. Huelsenbeck et al. [14] analyzed the coincidence of states for two morphological characters: self-incompatibility and flower reproductive structure morphology in the family Pontederiaceae.

The phylogeny was estimated using molecular data, and then characters were mapped using the Mk class of models of Lewis [23]. In addition, because the branch lengths of the topology do not reflect the evolutionary rates of the morphological traits and the bias parameter of the morphology model is unknown, a variety of prior distributions were explored for these parameters to reduce dependence on a particular set of values. They used two different test statistics to evaluate coincidence or correlation among the states of the two traits. The first evaluated each character individually, while the second looked for coincidence summed over all state comparisons between the two characters. The basic form of the statistics is

$$d_{ij} = a_{ij}^{(o)} - a_{ij}^{(e)}, \qquad (16.20)$$

where $a_{ij}^{(o)}$ is the observed coincidence and $a_{ij}^{(e)}$ is the expected coincidence. The authors found that when evaluating overall coincidence among states they were unable to detect a significant coincidence between the states of the traits. However, by looking at states individually, there was support for a strong coincidence between tristylous flowers and self-incompatibility. This demonstrates an important point about test statistics: a test statistic is only as good as it is a relevant summary of the data with respect to the hypothesis being tested. In the case of the overall coincidence measure, it masked the effect.

16.4 Conclusions

Two recent developments, posterior mapping and predictive distributions, have been developed and applied to questions on molecular evolution and phylogenetics. These methods provide a natural way to address and accommodate uncertainty in various model parameters by sampling with respect to the model's posterior distribution. Posterior mapping provides a powerful method for addressing questions in which detailed data (e.g., type, timing, and order) about the history of a character(s) is required. The dependence on the method of parsimony and its assumptions is no longer necessary. Predictive distributions offer a new approach to hypothesis testing that is general and flexible. Application of these new methods has just begun and will undoubtedly play an ever-increasing role in future studies in molecular evolution and phylogenetics.

References

[1] H. Akaike. A new look at statistical model identification. *IEEE Transactions on Automatic Control*, 19:716–723, 1974.

[2] M. J. Bayarri and J. O. Berger. *P* values for composite null models. *Journal of the American Statistical Association*, 95:1127–1142, 2000.

[3] J. P. Bollback. Bayesian model adequacy and choice in phylogenetics. *Molecular Biology and Evolution*, 19:1171–1180, 2002.

[4] J. Felsenstein. Evolutionary trees from DNA sequences: A maximum likelihood approach. *Journal of Molecular Evolution*, 17:368–376, 1981.

[5] D. Gelfand and X. L. Meng. Model checking and model improvement. In *Markov Chain Monte Carlo in Practice*, pages 189–198. Chapman and Hall, London, 1996.

[6] A. Gelman, J. B. Carlin, H. S. Stern, and D. B. Rubin. *Bayesian Data Analysis*. Chapman and Hall, London, 1995.

[7] A. Gelman, X. L. Meng, and H. Stern. Posterior predictive assessment of model fitness via realized discrepancies. *Statistica Sinica*, 6:733–807, 1996.

[8] J. Gillespie. *The Causes of Molecular Evolution*. Oxford University Press, Oxford, 1991.

[9] N. Goldman. Statistical tests of models of DNA substitution. *J Mol Evol*, 36:182–198, 1993.

[10] P. H. Harvey and M. D. Pagel. *The Comparative Method in Evolutionary Biology*. Oxford University Press, 1991.

[11] J. P. Huelsenbeck and J. P. Bollback. Application of the likelihood function in phylogenetic analysis. In *Handbook of Statistical Genetics*, pages 415–439. John Wiley and Sons, Inc., New York, 2001.

[12] J. P. Huelsenbeck and K. A. Dyer. Detecting adaptive molecular evolution when selection changes over time. *Genetics*, In Press.

[13] J. P. Huelsenbeck and N. S. Imennov. Geographic origin of human mitochondrial DNA: Accommodating phylogenetic uncertainty and model comparison. *Systematic Biology*, 51:155–165, 2002.

[14] J. P. Huelsenbeck, R. Nielsen, and J. P. Bollback. Stochastic mapping of morphological characters. *Systematic Biology*, 52:131–158, 2003.

[15] J. P. Huelsenbeck and B. Rannala. Phylogenetic methods come of age: Testing hypotheses in a phylogenetic context. *Science*, 276:174–180, 1997.

[16] J. P. Huelsenbeck, B. Rannala, and J. P. Masly. Accommodating phylogenetic uncertainty in evolutionary studies. *Science*, 288:2349–2350, 2000.

[17] J. P. Huelsenbeck and F. Ronquist. MRBAYES: Bayesian inference of phylogenetic trees. *Bioinformatics Applications Note*, 17:754–755, 2001.

[18] J. P. Huelsenbeck, F. Ronquist, R. Nielsen, and J. P. Bollback. Bayesian inference of phylogeny and its impact on evolutionary biology. *Science*, 294:2310–2314, 2001.

[19] T. Jukes and C. Cantor. Evolution of protein molecules. In *Mammalian Protein Metabolism*, pages 21–132. Academic Press, New York, 1969.

[20] C. Lanavé, G. Preparata, C. Saccone, and G. Serio. A new method for calculating evolutionary substitution rates. *Journal of Molecular Evolution*, 20:86–93, 1984.

[21] C. H. Langley and W. M. Fitch. An estimation of the constancy of the rate of molecular evolution. *Journal of Molecular Evolution*, 3:161–177, 1974.

[22] B. Larget and D. Simon. Markov chain Monte Carlo algorithms for the Bayesian analysis of phylogenetic trees. *Molecular Biology and Evolution*, 16:750–759, 1999.

[23] P. O. Lewis. A likelihood approach to estimating phylogeny from discrete morphological character data. *Systematic Biology*, 50:913–925, 2002.

[24] J. B. Losos. An approach to the analysis of comparative data when a phylogeny is unavailable or incomplete. *Systematic Biology*, 43:117–123, 1994.

[25] J. B. Losos and D. B. Miles. Ecological Morphology: Integrative Organismal Biology. In P. C. Wainwright and S. M. Reilly, editors, *Adaptation, constraint, and the comparative method: Phylogenetic issues and methods*, pages 60–98. University of Chicago Press, Chicago, 1994.

[26] E. P. Martins. Conducting phylogenetic comparative studies when the phylogeny is not known. *Evolution*, 50:12–22, 1996.

[27] X-L. Meng. Posterior predictive p-values. *Annals of Statistics*, 22:1142–1160, 1994.

[28] W. Messier and C-B. Stewart. Episodic adaptive evolution of primate lysomzymes. *Nature*, 385:151–154, 1997.

[29] A. Ø. Mooers and D. Schluter. Support for one and two rate models of discrete trait evolution. *Systematic Biology*, 48:623–633, 1999.

[30] R. Nielsen. Mutations as missing data: Inferences on the ages and distributions of nonsynonymous and synonymous mutations. *Genetics*, 159:401–411, 2001.

[31] R. Nielsen. Mapping mutations on phylogenies. *Systematic Biology*, 51:729–732, 2002.

[32] R. Nielsen and J. P. Huelsenbeck. Detecting positively selected amino acid sites using posterior predictive p-values. In *Pacific Symposium on Biocomputing, Proceedings*, pages 576–588. World Scientific, Singapore, 2001.

[33] M. D. Pagel. Detecting correlated evoluton on phylogenies: A general method for the comparative analysis of discrete characters. *Proceedings of the Royal Society B*, 255:37–45, 1994.

[34] M. D. Pagel. The maximum likelihood approach to reconstructing ancestral character states of discrete characters on phylogenies. *Systematic Biology*, 48:612–622, 1999.

[35] B. Rannala and Z. Yang. Probability distribution of molecular evolutionary trees: A new method of phylogenetic inference. *Journal of Molecular Evolution*, 43:304–311, 1996.

[36] J. R. Robins, A. van der Vaart, and V. Ventura. Asymptotic distribution of p-values in composite null models. *Journal of the American Statistical Association*, 95:1143–1156, 2000.

[37] F. Rodríguez, J. Oliver, A. Marín, and J. Medina. The general stochastic model of nucleotide substitution. *Journal of Theoretical Biology*, 142:485–501, 1990.

[38] D. B. Rubin. Bayesianly justifiable and relevant frequency calculations for the applied statistician. *Annals of Statistics*, 12:1151–1172, 1984.

[39] D. Schluter. Uncertainty in ancient phylogenies. *Nature*, 377:108–109, 1995.

[40] D. Schluter, T. Price, A. Ø. Mooers, and D. Ludwig. Likelihood of ancestor states in adaptive radiation. *Evolution*, 51:1699–1711, 1997.

[41] T. Sellke, M. J. Bayarri, and J. O. Berger. Calibration of *p*-values for precise null hypotheses. In *ISDS Discussion Paper 99-13*, Durham, NC, 1999. Duke University.

[42] B. Sillen-Tullberg. Evolution of gregariousness in aposematic butterfly larvae: A phylogenetic analysis. *Evolution*, 42:293–305, 1988.

[43] M. A. Suchard, R. E. Weiss, K. S. Dorman, and J. S. Sinsheimer. Oh brother, where art thou? A Bayes factor test for recombination with uncertain heritage. *Systematic Biology*, 51:715–728, 2002.

[44] M. A. Suchard, R. E. Weiss, and J. S. Sinsheimer. Bayesian selection of continuous-time Markov chain evolutionary models. *Molecular Biology and Evolution*, 18:1001–1013, 2001.

[45] M. A. Suchard, R. E. Weiss, and J. S. Sinsheimer. Testing a molecular clock without an outgroup: Derivations of induced priors on branch length restrictions in a Bayesian framework. *Systematic Biology*, 52:48–54, 2003.

[46] M. A. Suchard, R. E. Weiss, J. S. Sinsheimer, K. S. Dorman, P. Patel, and E. R. B. McCabe. Evolutionary similarity among genes. *Journal of the American Statistical Association*, 98:653–662, 2003.

[47] Y. Suzuki and T. Gojobori. A method for detecting positive selection at single amino acid sites. *Molecular Biology and Evolution*, 16:1315–1328, 1999.

[48] D. L. Swofford, G. J. Olsen, P. J. Waddell, and D. M. Hillis. Phylogenetic inference. In *Molecular Systematics*, pages 407–514. Sinauer Associates, Sunderland, MA, 2nd edition, 1996.

[49] A. R. Templeton. Contingency tests of neutrality using intra/interspecific gene trees: the rejection of neutrality for the evolution of the cytochrome oxidase ii gene in the hominoid primates. *Genetics*, 144:1263–1270, 1996.

[50] R. H. Ward, B. L. Frazier, K. Dew-Jager, and S. Pääbo. Extensive mitochondrial diversity within a single Amerindian tribe. *Proceedings of the National Academy of Sciences USA*, 88:8720–8724, 1991.

[51] Z. Yang, R. Nielsen, N. Goldman, and A.-M. K. Pedereon. Codon-substitution models for variable selection pressure at amino acid sites. *Genetics*, 155:431, 2000.

Assessing the Uncertainty in Phylogenetic Inference

Hidetoshi Shimodaira[1] and Masami Hasegawa[2]

[1] Department of Mathematical and Computing Sciences, Tokyo Institute of Technology, Ookayama, Meguro-ku, Tokyo 152-8552, Japan, shimo@is.titech.ac.jp

[2] Institute of Statistical Mathematics, Minami-Azabu, Minato-ku, Tokyo 106-8569, Japan, hasegawa@ism.ac.jp

17.1 Introduction

Since the seminal works of Cavalli-Sforza and Edwards [6] and Felsenstein [14], the maximum likelihood (ML) method has been widely used for inferring molecular phylogenies. None of the common methods, for estimating phylogenies, such as distance matrix methods, parsimony methods, and ML, can estimate the true phylogenetic tree with one hundred percent confidence. This is because the amount of information in molecular sequences of finite length is limited. Erroneous estimates of the tree topology can occur because of the statistical sampling error. It is therefore quite important to assess the uncertainty in the tree topology estimation. We discuss this problem in terms of likelihood, bootstrap, and testing.

Our arguments are based on the probability function of molecular sequences, which is obtained by specifying the substitution process and the tree topology. Estimating these specifications is formulated as a problem of statistical model selection. We mostly focus on tree topology selection and assume that the substitution process has been estimated in a preliminary analysis.

Model selection procedures using information criteria have been developed to overcome limitations of conventional hypothesis testing such as the likelihood ratio test. However, they do not have a built-in measure to assess the uncertainty of model selection. Therefore, model selection tests, such as the Shimodaira-Hasegawa test, have been developed by incorporating the idea of testing into the model selection procedure. The model selection uncertainty is then measured by the probability value (p-value).

There are several other measures of statistical confidence used in phylogenetics beyond the p-values calculated in the model selection tests, including Bayesian posterior probabilities, bootstrap probabilities, and the approximately unbiased p-value. These measures of statistical uncertainty can lead

to very different conclusions in practice. With respect to bias and robustness
we compare these measures with model misspecifications.

17.2 Likelihood Methods

17.2.1 Maximum Likelihood Estimate

Likelihood function of a tree

The data sets in our analysis consist of aligned molecular sequences from
different species. The number of species is denoted by s, and the sequence
length is denoted by n. Let x_{ih} be the nucleotide or amino acid of the ith
species at the hth site for $i = 1, \ldots, s$ and $h = 1, \ldots, n$. Then the data set
is the $s \times n$ matrix $\mathbf{X} = (\mathbf{x}_1, \ldots, \mathbf{x}_n)$, where each s-dimensional vector $\mathbf{x}_h$
consists of elements $x_{1h}, \ldots, x_{sh}$.

We assume that the pattern at site h, $\mathbf{x}_h$, is an observed value of random
variable $\mathbf{x}$. The probability function of $\mathbf{x}$ may be denoted as $p(\mathbf{x}; S, T, \boldsymbol{\theta}_S, \boldsymbol{\theta}_T)$,
where S specifies the model of evolution (i.e., the substitution process) and T
specifies the branching order of the labeled tree (i.e., the tree topology). $\boldsymbol{\theta}_S$ and
$\boldsymbol{\theta}_T$ denote parameter vectors for S and T, respectively. We assume that there
is an underlying true probability function $q(\mathbf{x})$ of the random variable $\mathbf{x}$. The
parametric model $p(\mathbf{x}; S, T, \boldsymbol{\theta}_S, \boldsymbol{\theta}_T)$ is considered an attempt to approximate
$q(\mathbf{x})$ by specifying S and T as well as $\boldsymbol{\theta}_S$ and $\boldsymbol{\theta}_T$. For example, S may be the
HKY model [23] for nucleotides with gamma rate heterogeneity [55]; then $\boldsymbol{\theta}_S$
may consist of the transition rate, the transversion rate, base frequencies, and
gamma shape parameter. On the other hand, T is one of $1 \times 3 \times 5 \times \cdots \times (2s-5)$
possible unrooted tree topologies, and $\boldsymbol{\theta}_T$ consists of $2s - 3$ branch lengths.
The pair (S, T) specifies the parametric model of $\mathbf{x}$.

The probability function of $\mathbf{X}$ is simply the product of all the probability
functions of $\mathbf{x}_1, \ldots, \mathbf{x}_n$, because we assume these random variables are independent. The likelihood function of $(S, T, \boldsymbol{\theta}_S, \boldsymbol{\theta}_T)$ is nothing but the probability function of $\mathbf{X}$, but the roles of $\mathbf{X}$ and $(S, T, \boldsymbol{\theta}_S, \boldsymbol{\theta}_T)$ are exchanged:

$$L(S, T, \boldsymbol{\theta}_S, \boldsymbol{\theta}_T; \mathbf{X}) = \prod_{h=1}^{n} p(\mathbf{x}_h; S, T, \boldsymbol{\theta}_S, \boldsymbol{\theta}_T).$$

The likelihood function indicates the plausibility of $(S, T, \boldsymbol{\theta}_S, \boldsymbol{\theta}_T)$ for a given
data set $\mathbf{X}$.

Once we give the model specification (S, T), the ML estimate of the para-
meter vector $(\boldsymbol{\theta}_S, \boldsymbol{\theta}_T)$ is obtained by finding the value $(\hat{\boldsymbol{\theta}}_S, \hat{\boldsymbol{\theta}}_T)$ that maximizes
the likelihood function. Because the likelihood function is very complicated
in phylogenetic inference, $(\hat{\boldsymbol{\theta}}_S, \hat{\boldsymbol{\theta}}_T)$ is obtained by numerical optimization algorithms using computer software such as PAML [56]. We often work on the
log-likelihood $\ell = \log L$ for computational convenience. Although the computation can be demanding, the ML estimate of the parameter vector can be
shown to have certain optimal statistical properties.

Akaike information criterion

It is often the case that we have to estimate the model specification (S, T) from a list of candidate models. This is an example of statistical model selection. The likelihood function of (S, T) may be defined as the maximum value of the likelihood function over the parameter space

$$L(S, T; \mathbf{X}) = L(S, T, \hat{\boldsymbol{\theta}}_S, \hat{\boldsymbol{\theta}}_T; \mathbf{X}),$$

and the ML estimate of (S, T) can be obtained by maximizing $L(S, T; \mathbf{X})$ over the candidate models. However, this naive ML approach does not work in general because we can always increase the likelihood value by adding extra parameters in the model. In other words, complicated models are always favored over simpler models.

The information criterion of Akaike [2] is defined as

$$\text{AIC}(S, T; \mathbf{X}) = -2 \times \left\{ \log L(S, T; \mathbf{X}) - (\dim \boldsymbol{\theta}_S + \dim \boldsymbol{\theta}_T) \right\},$$

which estimates the expected performance of the model-based prediction in terms of the Kullback-Leibler (KL) divergence. The penalty term $\dim \boldsymbol{\theta}_S + \dim \boldsymbol{\theta}_T$ is the number of parameters in the model. AIC balances the increase in the log-likelihood value with the loss of estimating many parameters. A wrong model with few parameters can be better on average in approximating $q(\mathbf{x})$ than a correct model with many parameters. AIC does not intend to find a correct model but to find a good model for approximating reality. Instead of maximizing $L(S, T; \mathbf{X})$, we may find an optimal model by minimizing $\text{AIC}(S, T; \mathbf{X})$.

Substitution process selection

Let us consider that we have J candidates for S (i.e., $S_1, S_2, \ldots, S_J$) and that we have K candidates for T (i.e., $T_1, T_2, \ldots, T_K$). We select S and T from these candidates. A straightforward application of the model selection procedure to this problem is to consider all the $J \times K$ combinations (S_j, T_k), $j = 1, \ldots, J$, $k = 1, \ldots, K$, and find the combination that minimizes $\text{AIC}(S_j, T_k; \mathbf{X})$. This joint estimation of S and T is computationally demanding, and interpretation of the confidence limits becomes rather complicated.

What we do in practice is to specify S first and fix S when selecting T from the candidates. There are several ways of specifying S. One may specify S from experience somewhat arbitrarily. Otherwise, we compare $S_1, \ldots, S_J$ by AIC or the likelihood ratio (LR) tests as implemented in MODELTEST [40, 39].

For comparing the substitution processes, we usually specify a guiding tree topology obtained in advance by a fast distance matrix method. The likelihood function of S is $L(S, \hat{T}; \mathbf{X})$, where $\hat{T}$ denotes the guiding T. Then we select

S_j that minimizes $\mathrm{AIC}(S_j, \hat{T}; \mathbf{X})$. The estimated S is denoted by $\hat{S}$. Once $\hat{S}$ is obtained, we discard the temporary $\hat{T}$ and then estimate T using $\hat{S}$.

Instead of AIC, we can also use the LR tests for estimating $\hat{S}$. As described in Section 17.4.1, the LR test procedure judges whether S is appropriate against an alternative S'. The model S is accepted if $L(S', \hat{T}; \mathbf{X})/L(S, \hat{T}; \mathbf{X})$ is smaller than a threshold value. In the hierarchical LR tests, we repeatedly apply the LR test to the candidate models by taking advantage of the relationship among the models. However, this may cause a problem of multiplicity of testings. For constructing a confidence set of models, the closed testing procedure of Marcus et al. [34] suggests that general models should automatically be accepted if any one of their restricted models is accepted by the LR test.

In the following argument, $\hat{S}$ is assumed to specify the substitution process properly, and thus we focus on tree topology selection.

Tree topology selection

The likelihood function of T is defined by dropping S from $L(S, T; \mathbf{X})$ as

$$L(T; \mathbf{X}) = L(\hat{S}, T; \mathbf{X}),$$

and the log-likelihood function of T is $\ell(T; \mathbf{X}) = \log L(T; \mathbf{X})$. They will be written $L(T)$ and $\ell(T)$ for the sake of brevity. $\ell(T)$ is expressed as

$$\ell(T) = \sum_{h=1}^{n} \ell_h(T),$$

where $\ell_h(T)$ is the sitewise log-likelihood

$$\ell_h(T) = \log p(\mathbf{x}_h; S, T, \hat{\boldsymbol{\theta}}_S, \hat{\boldsymbol{\theta}}_T).$$

The ML estimate of T is obtained by searching the tree topology $\hat{T}_{\mathrm{ML}}$ that maximizes $L(T)$ among $T_1, \ldots, T_K$. This is equivalent to maximizing $\ell(T)$, and thus to minimizing $\mathrm{AIC}(\hat{S}, T; \mathbf{X})$, since $\dim \boldsymbol{\theta}_T = 2s - 3$ is the same for all bifurcating trees. See Figure 17.1 for an example of the ML inference.

The ML estimate $\hat{T}_{\mathrm{ML}}$ converges to the true topology as n goes to infinity. This property of an estimator is called asymptotic consistency. The consistency is rather easily verified under the assumption that S is correctly specified in advance and the true topology, denoted $\bar{T}$, is included in the K candidates. Let $D(q, p)$ be the KL divergence between probability functions $q(\mathbf{x})$ and $p(\mathbf{x})$,

$$D(q, p) = \sum_{\mathbf{x}} q(\mathbf{x}) \log q(\mathbf{x}) - \sum_{\mathbf{x}} q(\mathbf{x}) \log p(\mathbf{x}),$$

which indicates the separation between the two functions and becomes zero if they are the same. The KL divergence between the true distribution and the parametric model specified by T is then denoted as

$$D(q, \bar{p}(T)) = \min_{\boldsymbol{\theta}_S, \boldsymbol{\theta}_T} D(q, p(S, T, \boldsymbol{\theta}_S, \boldsymbol{\theta}_T)),$$

where the optimal parameter value that attains the minimum is denoted by $(\bar{\boldsymbol{\theta}}_S, \bar{\boldsymbol{\theta}}_T)$. $\bar{p}(\mathbf{x}; T) = p(\mathbf{x}; S, T, \bar{\boldsymbol{\theta}}_S, \bar{\boldsymbol{\theta}}_T)$ is the best approximating probability function defined for each T. $D(q, \bar{p}(T))$ has the minimum value zero at $T = \bar{T}$, and $D(q, \bar{p}(T)) > 0$ for $T \neq \bar{T}$ unless $\bar{p}(T) = \bar{p}(\bar{T})$. It is shown in the derivation of the AIC [42] that the expected value of $(1/n) \log L(T; \mathbf{X})$ is approximately $-D(q, \bar{p}(T)) + \sum_{\mathbf{x}} q(\mathbf{x}) \log q(\mathbf{x}) + (\dim \boldsymbol{\theta}_S + \dim \boldsymbol{\theta}_T)/2n$, which is maximized at $\bar{T}$. It follows from the law of large numbers that $(1/n) \log L(T; \mathbf{X})$ converges to the expected value, and thus $P(\hat{T}_{\mathrm{ML}} = \bar{T}) \to 1$ as $n \to \infty$.

It is known for the parameter estimation that the ML estimate $(\hat{\boldsymbol{\theta}}_S, \hat{\boldsymbol{\theta}}_T)$ has the minimum variance for sufficiently large n among possible asymptotically unbiased estimates. Estimators with this property are called asymptotically efficient. However, it is not clear that a similar property also holds for $\hat{T}_{\mathrm{ML}}$, although there is a related result of Shibata [44] for the optimality of AIC model selection.

17.2.2 Bayesian Inference

Posterior probability

Bayesian inference is another form of likelihood-based inference in which the likelihood function is averaged instead of maximized. In addition to the likelihood function, we also specify $\pi(\boldsymbol{\theta}_S; S)$ and $\pi(\boldsymbol{\theta}_T; T)$ (i.e., the prior density functions of parameters given model specifications). They are used as weights for averaging the likelihood function over the parameter space to calculate the probability function of $\mathbf{X}$:

$$P(\mathbf{X}; S, T) = \int\int L(S, T, \boldsymbol{\theta}_S, \boldsymbol{\theta}_T; \mathbf{X}) \pi(\boldsymbol{\theta}_S; S) \pi(\boldsymbol{\theta}_T; T) \, d\boldsymbol{\theta}_S \, d\boldsymbol{\theta}_T. \quad (17.1)$$

We further specify $\pi(S)$ and $\pi(T)$ (i.e., the prior probability functions of the model specifications). According to Bayes' theorem, the posterior joint-probability function of (S, T) is given by

$$P(S, T; \mathbf{X}) \propto P(\mathbf{X}; S, T) \pi(S) \pi(T),$$

where $\propto$ indicates "proportional to" as a function of (S, T). In practice, we may specify $\hat{S}$ in a preliminary analysis and assume a uniform prior $\pi(T_k) = 1/K$ for tree topologies. Then the posterior probability (PP) of T is

$$P(T; \mathbf{X}) \propto P(\mathbf{X}; \hat{S}, T),$$

which will be denoted $P(T)$ for the sake of brevity. Here the proportional constant is defined from the relation $P(T_1) + \cdots + P(T_K) = 1$. $P(T)$ indicates the plausibility of tree topology T. The computation of $P(T)$ has become

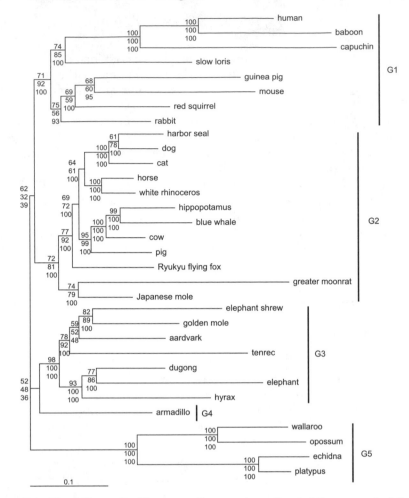

Fig. 17.1. The ML tree for 32 mammalian species estimated from a part of the mitochondrial protein sequences of Nikaido et al. [37]. A biological discussion is given in Section 17.5. This ML topology is represented as ((G1, G2), (G3, G4), G5) using the five groups of taxa defined above. The MAP topology of Bayesian inference (Section 17.2.2) is represented as ((G1,(G2,G3)),G4,G5), where the subtrees are the same as above except for (((elephant shrew, golden mole), aardvark), tenrec) in G3 being changed to ((tenrec, aardvark), (elephant shrew, golden mole)). Our data set consists of the sequences of $n = 3392$ amino acids for $s = 32$ species. The mtREV model [1] was used for amino acid substitutions, and the site heterogeneity was modeled by the discrete gamma distribution [55]. A limited list of 2502 candidate topologies was obtained from heated MCMCMC simulations as explained in Figure 17.2 and Figure 17.3, and their log-likelihood values were calculated by PAML to find the ML topology. Three numbers near the branches are the approximately unbiased p-value (AU) (top), the bootstrap probability (BP) (middle) calculated by CONSEL (Section 17.3), and the posterior probability (PP) (bottom) calculated by MrBayes (Section 17.2.2).

possible by the Markov chain Monte Carlo (MCMC) method implemented in, for example, MrBayes [27]. MCMC is a stochastic simulation procedure for generating a sequence of trees in frequencies proportional to $P(T)$.

The maximum a posteriori (MAP) estimate of T is defined as the tree topology $\hat{T}_{MAP}$ that maximizes $P(T)$ among $T_1, \ldots, T_K$. In Bayesian inference, we are confident that T_{MAP} is the true tree topology if the value of $P(\hat{T}_{MAP})$ is close to one, say, larger than 0.95. Otherwise, the uncertainty in phylogenetic inference is so large that other tree topologies with nonnegligible $P(T)$ values could be true.

A phylogenetic hypothesis is often represented as a collection of tree topologies. For example, we can collect tree topologies from $T_1, \ldots, T_K$ such that a specified group of taxa is monophyletic. Let H denote such a composite hypothesis. The PP of H is then defined as the sum of all the PPs of the tree topologies in the hypothesis

$$P(H) = \sum_{T \in H} P(T).$$

Bayesian information criterion

The computation of $P(\mathbf{X}; S, T)$ is difficult since it involves integration over the parameter space as seen in eq. (17.1). MCMC approximates integration by simulation. There is another approximation method that does not rely on simulation. This method utilizes the fact that, for sufficiently large n, the integration is dominated by a small region centered around $(\hat{\boldsymbol{\theta}}_S, \hat{\boldsymbol{\theta}}_T)$. This idea leads to

$$\log P(\mathbf{X}; S, T) = -\frac{1}{2}\text{BIC}(S, T; \mathbf{X}) + O(1), \qquad (17.2)$$

where BIC is the Bayesian information criterion of Schwarz [43] defined by

$$\text{BIC}(S, T; \mathbf{X}) = -2 \times \left\{ \log L(S, T; \mathbf{X}) - \frac{\log n}{2}(\dim \boldsymbol{\theta}_S + \dim \boldsymbol{\theta}_T) \right\}.$$

The approximation error term $O(1)$ in eq. (17.2) involves the prior density functions. Here we have used the notation $O(n^a)$ to indicate a term proportional to n^a for sufficiently large n. Even if we consider the limit of $n \to \infty$, the approximation error does not vanish, but it will become negligibly smaller than BIC since the magnitude of the log-likelihood is $O(n)$.

For tree topology selection, we again assume the specification of $\hat{S}$ and a uniform prior distribution of trees. Then the BIC approximation states that $\log P(T)$ is approximately $\log L(T)$ plus a constant term independent of T, and thus $\hat{T}_{MAP}$ is approximately $\hat{T}_{ML}$. This also implies consistency of $\hat{T}_{MAP}$ since $P(\hat{T}_{MAP} = \hat{T}_{ML}) \to 1$ as $n \to \infty$. Moreover, we have the approximation formula

$$P(T) \approx \frac{L(T)}{\sum_{k=1}^{K} L(T_k)}, \qquad (17.3)$$

which is indicated in eq. (5) of Hasegawa and Kishino [21]. This formula provides a bridge between ML inference and Bayesian inference. Unfortunately, the BIC approximation is very rough, and the numerical values obtained from eq. (17.3) will not agree very well with the results of MCMC as shown in the left panel of Figure 17.2. This is partially because the $O(1)$ term in eq. (17.2) becomes a multiplier factor to $L(T)$ in eq. (17.3).

It is not a good idea to calculate $P(T)$ from $L(T)$ for a Bayesian inference of topology. However, eq. (17.3) enables us to utilize MCMC for ML inference. $\hat{T}_{\mathrm{ML}}$ should be found in candidates with relatively high $P(T)$ values. Thus we do not have to calculate $L(T)$ for all the K topologies but only for those appearing in the sequence generated using MCMC. We may prepare a limited list of candidate tree topologies consisting of those with $P(T)$ greater than a small threshold value. This greatly speeds up the ML inference by reducing the number of tree topologies to a manageable size.

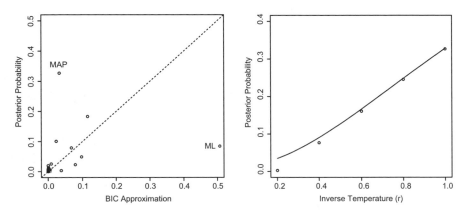

Fig. 17.2. Plots of the PP values. In the left panel, the PP values calculated by the MCMCMC simulation for the 2502 candidate topologies are plotted against those calculated by the BIC approximation of eq. (17.3). The right panel shows the PP values of the MAP topology; the PP values are calculated by heated MCMCMC simulations for five inverse temperature values. The theoretical curve of eq. (17.4) is also shown. For $r = 1$, we ran MCMCMC for 1,080,000 generations after 30,000 burn-in generations. For each $r < 1.0$, we ran MCMCMC for 540,000 generations similarly. Four chains are generated in each MCMCMC simulation. We modified the source code of MrBayes to allow any value of r. The initial tree was obtained by a preliminary MCMCMC run of 300,000 generations with $r = 1$.

Changing the "temperature"

When utilizing MCMC for ML inference, we may need to control the range of the candidate topologies. Sometimes $P(T)$ is concentrated on a very small number of topologies, and it may be necessary to increase the range of the candidates so that a larger collection of topologies are included in the limited list of candidates.

This is possible by reducing the sequence length so that the likelihood surface becomes flatter. Let us consider downsampling the sites of the aligned molecular sequences; a data set $\mathbf{X}'$ of length n' is obtained by resampling n' sites from $\mathbf{X}$. $r = n'/n$ is the proportion of the sampled sites. The expected value of $\log L(T; \mathbf{X}')$ is approximately $r \log L(T; \mathbf{X})$ since it is the sum of the sitewise log-likelihoods. Therefore, we have $L(T; \mathbf{X}') \approx L(T; \mathbf{X})^r$ and thus the PP of T becomes proportional to $P(T)^r$. This makes small $P(T)$ values become larger, while large $P(T)$ values become smaller. A problem of downsampling is that the center of the candidate range (i.e. $\hat{T}_{\mathrm{MAP}}$) may be different from that of the original data set.

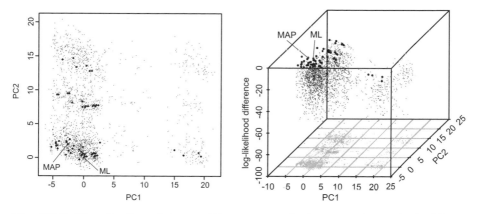

Fig. 17.3. Visualization of the candidate tree topologies using the method of Shimodaira [48]. In the left panel, the first two principal components (PC) of the sitewise log-likelihood vector $(\ell_1(T), \ldots, \ell_n(T))$ are plotted for the 2502 candidate topologies. The axes are adjusted so that the ML topology is centered. Each point represents a tree topology, and the distance between points indicates the square root of the symmetric Kullback-Leibler divergence between the probability functions for the two topologies. In the right panel, the vertical axis indicates the log-likelihood difference from the ML topology. The limited list of candidate topologies was obtained from the MCMCMC runs of the five r values explained in Figure 17.2. The numbers of distinct topologies are 197, 239, 572, 2483, and 15,495, respectively, for $r = 1.0$, 0.8, 0.6, 0.4, and 0.2. Only the topologies with PP values larger than 0.00025 were collected for the candidates from each run; they number 74, 86, 188, 672, and 2086, respectively. Some topologies are overlapped, and the total number is 2502. Large points indicate the topologies obtained from $r = 1.0$, and small points are those from $r < 1.0$.

A sophisticated way to do the same thing yet fix the problem of downsampling is simply to replace $\log L(T; \mathbf{X})$ with $r \log L(T; \mathbf{X})$ in the MCMC computation. This is called "heating" in the   Metropolis-coupled MCMC

(MCMCMC) implemented in MrBayes [27] since r is called the inverse temperature in statistical physics. In MCMCMC, several chains of trees with different r values are generated, and trees are occasionally swapped between the chains. We expect heated chains easily to escape deep valleys of the likelihood surface so that we will not miss the highest peak. Usually the chain of the largest r value (i.e., the cold chain) uses $r = 1$ in MCMCMC, but we will consider other values of r.

Let $P_r(T)$ denote the PP of T obtained by the modified MCMC with the inverse temperature r. According to eq. (17.3), we then have

$$P_r(T) \approx \frac{P(T)^r}{\sum_{k=1}^{K} P(T_k)^r}. \tag{17.4}$$

The PP value changes by altering r as shown in the right panel of Figure 17.2. We can control the range of candidates by choosing r. Smaller r values make the range wider, as shown in Figure 17.3.

17.3 Bootstrap Methods

17.3.1 Bootstrap Resampling

Sampling error

Let us recall that the ML estimate depends on the data set. For tree topology selection, the ML topology $\hat{T}_{\mathrm{ML}}$ is in fact a function of $\mathbf{X}$, and thus we may write $\hat{T}_{\mathrm{ML}}(\mathbf{X})$; it maximizes $L(T; \mathbf{X})$ among $T_1, \ldots, T_K$. If we have used another data set $\mathbf{X}'$, the ML inference will give $\hat{T}_{\mathrm{ML}}(\mathbf{X}')$, which may be different from $\hat{T}_{\mathrm{ML}}(\mathbf{X})$. Since $\mathbf{X}$, if regarded as a random variable, varies due to the sampling error, $\hat{T}_{\mathrm{ML}}(\mathbf{X})$ also varies, and it is not necessarily the true topology $\bar{T}$ even if all the assumptions in the ML inference are satisfied correctly. This is the uncertainty in phylogenetic inference. Thus it is important to measure how much confidence we have in $\hat{T}_{\mathrm{ML}}(\mathbf{X})$.

This issue applies not only to the ML inference but also to any other inference methods. Here we do not intend to discuss which inference method is most accurate, although there are differences in the estimate accuracy. We rather assume that we have chosen the best inference method available, and the tree topology estimation is written as $\hat{T}(\mathbf{X})$ in general. Then, we discuss methods to elucidate how much $\hat{T}(\mathbf{X})$ is influenced by the sampling error.

The mechanism causing the sampling error is explained in the following manner. The data set $\mathbf{X}$ consists of elements $\mathbf{x}_1, \ldots, \mathbf{x}_n$ for n sites. The elements $\mathbf{x}_h$, $h = 1, \ldots, n$ are observed values of a random variable $\mathbf{x}$ distributed according to the underlying true probability function $q(\mathbf{x})$. This is mathematically equivalent to sampling n sites randomly from imaginary molecular sequences of infinite length in which the frequency of pattern $\mathbf{x}$ is proportional to $q(\mathbf{x})$. The observed frequency of pattern $\mathbf{x}$ in $\mathbf{X}$ will converge to $q(\mathbf{x})$ as n goes to infinity, but it varies randomly around $q(\mathbf{x})$ at finite n.

Let us suppose that we can obtain another data set by sampling n sites again randomly from the infinite length sequences, and let $\mathbf{X}'$ be a replicate data set obtained this way. If we repeat it B times, say $B = 10{,}000$, then we get replicates $\mathbf{X}'_1, \ldots, \mathbf{X}'_B$. The sampling error will be observed directly by looking at differences in $\hat{T}(\mathbf{X}'_1), \ldots, \hat{T}(\mathbf{X}'_B)$. Unfortunately, we cannot perform this procedure because only the original data set $\mathbf{X}$ is available in practice.

Bootstrap probability

The bootstrap resampling of Efron [11] is a stochastic simulation technique for observing the sampling error using only the original data set. A replicate data set is obtained by sampling n sites randomly from $\mathbf{X}$ instead of the infinite length sequences. Let $h_1, \ldots, h_n$ be n integers sampled randomly from $1, \ldots, n$. We pick $\mathbf{x}_{h_1}, \ldots, \mathbf{x}_{h_n}$ from $\mathbf{X}$ for constructing a bootstrap replicate data set, denoted $\mathbf{X}^* = (\mathbf{x}_1^*, \ldots, \mathbf{x}_n^*)$, by setting $\mathbf{x}_1^* = \mathbf{x}_{h_1}, \ldots, \mathbf{x}_n^* = \mathbf{x}_{h_n}$. The same integers are allowed to be sampled several times in $h_1, \ldots, h_n$, otherwise $\mathbf{X}^*$ differs from $\mathbf{X}$ only in the order of the elements. In other words, $\mathbf{X}^*$ is obtained by "resampling with replacement" from $\mathbf{X}$. This is mathematically equivalent to sampling from imaginary molecular sequences in which $\mathbf{X}$ is copied infinite times.

The bootstrap resampling is repeated B times to get bootstrap replicates $\mathbf{X}_1^*, \ldots, \mathbf{X}_B^*$. We count how many times each tree topology is found in $\hat{T}(\mathbf{X}_1^*), \ldots, \hat{T}(\mathbf{X}_B^*)$ to measure the sampling error. Let $C(T)$ be the frequency of observing $\hat{T}(\mathbf{X}^*) = T$:

$$C(T) = \#\{\hat{T}(\mathbf{X}_b^*) = T, b = 1, \ldots, B\}.$$

The bootstrap probability (BP) of Felsenstein [15] is then obtained by $C(T)/B$ for T. We are confident that $\hat{T}(\mathbf{X}) = \bar{T}$ if $C(\hat{T}(\mathbf{X}))/B$ is close to one (say, larger than 0.95).

There is a sampling error also in the bootstrap resampling. As we increase B, $C(T)/B$ converges to the limiting value denoted as $\tilde{\alpha}(T; \mathbf{X})$. This is the conditional probability of observing $\hat{T}(\mathbf{X}^*) = T$ given $\mathbf{X}$,

$$\tilde{\alpha}(T; \mathbf{X}) = P\left(\hat{T}(\mathbf{X}^*) = T; \mathbf{X}\right),$$

which will be denoted $\tilde{\alpha}(T)$ for the sake of brevity. The standard error of $C(T)/B$ is $\sqrt{\tilde{\alpha}(T)(1 - \tilde{\alpha}(T))/B}$, and it is only $\sqrt{0.05 \times 0.95/10{,}000} = 0.002$ for $\tilde{\alpha}(T) = 0.05$ and $B = 10{,}000$. This error is ignored in the following argument, and $\tilde{\alpha}(T)$ and $C(T)/B$ are used interchangeably to denote the BP of T.

The BP of the composite hypothesis H is the conditional probability of observing $\hat{T}(\mathbf{X}^*) \in H$ given $\mathbf{X}$,

$$\tilde{\alpha}(H; \mathbf{X}) = P\left(\hat{T}(\mathbf{X}^*) \in H; \mathbf{X}\right),$$

which is denoted $\tilde{\alpha}(H)$. This is the sum of the BPs of tree topologies in H,

$$\tilde{\alpha}(H) = \sum_{T \in H} \tilde{\alpha}(T).$$

Speeding up the bootstrap

When the computation of $\hat{T}(\mathbf{X})$ is demanding, it is almost impractical to compute $\hat{T}(\mathbf{X}^*)$ for thousands of replicates. This is the case for $\hat{T}_{\mathrm{ML}}$, and an approximation is required for making it practical.

Let us consider a vector $\mathbf{w} = (w_1, \ldots, w_n)$ of real-valued elements and a function

$$f(\mathbf{w}, T) = \max_{\boldsymbol{\theta}_S, \boldsymbol{\theta}_T} \sum_{h=1}^{n} w_h \log p(\mathbf{x}_h; S, T, \boldsymbol{\theta}_S, \boldsymbol{\theta}_T).$$

The log-likelihood function of T is expressed as $\log L(T; \mathbf{X}) = f(\mathbf{w}_0, T)$ using $\mathbf{w}_0 = (1, \ldots, 1)$. We can also write $\log L(T; \mathbf{X}^*) = f(\mathbf{w}^*, T)$ using $\mathbf{w}^* = (w_1^*, \ldots, w_n^*)$, where w_h^* is the number of times that $\mathbf{x}_h$ is resampled in $\mathbf{X}^*$. By considering the Taylor expansion of $f(\mathbf{w}, T)$ with respect to $\mathbf{w}$ around $\mathbf{w}_0$, we obtain the linear approximation formula

$$\log L(T; \mathbf{X}^*) \approx \sum_{h=1}^{n} w_h^* \ell_h(T).$$

This is the linear approximation of bootstrap replicates [8] applied to the log-likelihood function and has been implemented as the resampling of estimated log-likelihoods (RELL) of Kishino, Miyata, and Hasegawa [30] in phylogenetics. We can improve the approximation by including the quadratic term of the Taylor expansion [48], but the linear approximation is often accurate enough [22]. By using the RELL method, we can avoid time-consuming recalculation of $(\hat{\boldsymbol{\theta}}_S, \hat{\boldsymbol{\theta}}_T)$ for the replicates.

17.3.2 Approximately Unbiased Tests

Multiscale bootstrap method

The BP is biased, as discussed later, when it is interpreted as the probability value (p-value) of statistical testing. The multiscale bootstrap (MB) method of Shimodaira [49] is an attempt to reduce the test bias of the BP. The MB method calculates an approximately unbiased (AU) p-value. Although we only work on a composite hypothesis H instead of each T below, our argument applies to T by taking $H = \{T\}$ (i.e., the hypothesis consisting of a single T). In this case, we will write T, instead of $\{T\}$, for H to simplify the notation.

The key idea of the MB is to alter the sequence length of bootstrap replicates. Let n' be a revised sequence length and $r = n'/n$ be the ratio. We sample $h_1, \ldots, h_{n'}$ from $1, \ldots, n$ with replacement and construct a bootstrap replicate $\mathbf{X}^* = (\mathbf{x}_1^*, \ldots, \mathbf{x}_{n'}^*)$ by setting $\mathbf{x}_1^* = \mathbf{x}_{h_1}, \ldots, \mathbf{x}_{n'}^* = \mathbf{x}_{h_{n'}}$. We can choose arbitrary positive integers for n' to generate bootstrap replicates. The BP of H with this modification is denoted as $\tilde{\alpha}_r(H)$ to indicate the ratio r. The value of $\tilde{\alpha}_r(H)$ changes by altering r from 1 because the random variation inherent in $\mathbf{X}^*$ is rescaled by the factor $1/\sqrt{r}$.

In the MB method, we calculate the BPs for several r values. Let M be the number of bootstrap simulations, say $M = 10$. We first specify arbitrary values $r_1, \ldots, r_M$ for r. For example, we will use $r_1 = 0.5, r_2 = 0.6, \ldots, r_{10} = 1.4$ for the examples in Figure 17.4. Then, M sets of B bootstrap replicates are generated using these r values to obtain BP values $\tilde{\alpha}_{r_1}(H), \ldots, \tilde{\alpha}_{r_M}(H)$. The total number of replicates is $M \times B$. By fitting a theoretical curve

$$\tilde{\alpha}_r(H) \approx 1 - \Phi(\hat{d}_H \sqrt{r} + \hat{c}_H/\sqrt{r}) \tag{17.5}$$

to the observed BP values, the two coefficients $\hat{d}_H$ and $\hat{c}_H$ in eq. (17.5) are estimated. $\Phi(\cdot)$ is the standard normal distribution function. Finally, we calculate the AU p-value for testing H by

$$\hat{\alpha}_{\mathrm{AU}}(H) = 1 - \Phi(\hat{d}_H - \hat{c}_H). \tag{17.6}$$

This MB method is implemented in the software CONSEL [51].

The hypothesis H is rejected if $\hat{\alpha}_{\mathrm{AU}}(H) < \alpha$, where the threshold value α is a prespecified level of significance for statistical testing; $\alpha = 0.05$ is used conventionally. H is not rejected if $\hat{\alpha}_{\mathrm{AU}}(H) \geq \alpha$, and we may say that H is accepted. However, this acceptance does not mean any strong support for H but could be a consequence of too little information in $\mathbf{X}$ (i.e., the sequence length is too short). The confidence set of trees is obtained by collecting T for which $\hat{\alpha}_{\mathrm{AU}}(T) \geq \alpha$. The number of trees in the confidence set will decrease as n becomes larger.

A large value of $\hat{\alpha}_{\mathrm{AU}}(H)$ provides evidence in support of H. In fact, H is supported significantly if $\hat{\alpha}_{\mathrm{AU}}(H) > 1 - \alpha$. This statement needs explanation since it is not a standard property of statistical testing but holds for the $\hat{\alpha}_{\mathrm{AU}}(H)$ calculated by the MB method. Let H^c be the complement of H (i.e., the composite hypothesis consisting of all $T_1, \ldots, T_K$ but not in H). It is easy to verify from $\tilde{\alpha}_r(H^c) = 1 - \tilde{\alpha}_r(H)$ and eq. (17.5) that $\hat{d}_{H^c} = -\hat{d}_H$ and $\hat{c}_{H^c} = -\hat{c}_H$, and thus $\hat{\alpha}_{\mathrm{AU}}(H^c) = 1 - \hat{\alpha}_{\mathrm{AU}}(H)$. H^c is rejected if $\hat{\alpha}_{\mathrm{AU}}(H^c) < \alpha$, or equivalently $\hat{\alpha}_{\mathrm{AU}}(H) > 1 - \alpha$. This implies support for H.

Simplified working model

There is an extension of the geometric theory of Efron, Halloran, and Holmes [12] behind the MB method. This theory is easily understood by working on

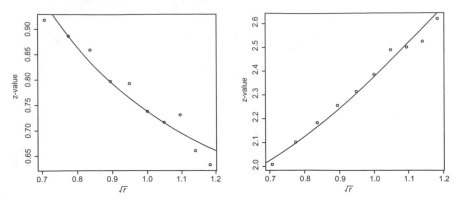

Fig. 17.4. Illustration of the MB method for testing $H = \{T_1\}$ (left panel) and $H = \{T_8\}$ (right panel) of the constrained candidate trees of Table 17.2. The BP values are calculated for the ten r-values. The vertical axis indicates the z-value of the BP; $z = \Phi^{-1}(1 - \tilde{\alpha}_r(H))$, where $q = \Phi^{-1}(p)$ is the inverse function of $p = \Phi(q)$. The horizontal axis indicates $\sqrt{r}$. The theoretical curve of eq. (17.5) becomes $z = \hat{d}_H\sqrt{r} + \hat{c}_H/\sqrt{r}$, where the coefficients are estimated by fitting the curve to the observed z-values by the ML method as described in Shimodaira [49]. The slope of the z curve at $\sqrt{r} = 1$ (i.e., the differentiation $z'(1) = dz/d\sqrt{r}\,|_1$ as a function of $\sqrt{r}$) indicates the AU p-value since eq. (17.6) is also expressed as $\hat{\alpha}_{\mathrm{AU}}(H) = 1 - \Phi(z'(1))$. Thus $\hat{\alpha}_{\mathrm{AU}}(H) > 0.5$ if $z'(1) < 0$ (left panel) and $\hat{\alpha}_{\mathrm{AU}}(H) < 0.5$ if $z'(1) > 0$ (right panel).

a simplified model of the multivariate normal distribution with the identity covariance matrix

$$\mathbf{y} \sim N_p(\boldsymbol{\mu}, \mathbf{I}_m). \tag{17.7}$$

In other words, the elements $y_1, \ldots, y_m$ of an m-dimensional random vector $\mathbf{y}$ are normally distributed with means $\mu_1, \ldots, \mu_m$, and they are independent of each other. We assume that there is a smooth, possibly nonlinear, transformation from $\mathbf{X}$ to $\mathbf{y}$ at least approximately, where m is chosen for convenience of the transformation. The data set $\mathbf{X}$ is now represented by $\mathbf{y}$. The unknown true distribution $q(\mathbf{x})$ is represented by $\boldsymbol{\mu}$.

The hypothesis H is represented by a region $\mathcal{R}_H$ in the m-dimensional space of the mean vector $\boldsymbol{\mu}$. If H is true (i.e., $\bar{T} \in H$) then $\boldsymbol{\mu}$ is included in the region (i.e., $\boldsymbol{\mu} \in \mathcal{R}_H$). Observing $\hat{T}(\mathbf{X}) \in H$ is equivalent to $\mathbf{y} \in \mathcal{R}_H$. Let $\mathcal{R}_T$ denote the region $\mathcal{R}_H$ for $H = \{T\}$. The m-dimensional space of $\boldsymbol{\mu}$ is divided into K regions $\mathcal{R}_{T_1}, \ldots, \mathcal{R}_{T_K}$ when comparing the K alternative tree topologies: $\boldsymbol{\mu} \in \mathcal{R}_T$ for $T = \bar{T}$, and $\mathbf{y} \in \mathcal{R}_T$ for $T = \hat{T}(\mathbf{X})$. The composite hypothesis H is now written as the union $\mathcal{R}_H = \bigcup_{T \in H} \mathcal{R}_T$.

The bootstrap replicate $\mathbf{X}^*$ is represented by a vector $\mathbf{y}^*$, which is distributed as

$$\mathbf{y}^* \sim N_m(\mathbf{y}, \tfrac{1}{r}\mathbf{I}_m). \tag{17.8}$$

In other words, the elements $y_1^*, \ldots, y_m^*$ are distributed around $y_1, \ldots, y_m$, and the variance is $1/r$. The BP is now written as

$$\tilde{\alpha}_r(H) = P\left(\mathbf{y}^* \subset \mathcal{R}_H; \mathbf{y}\right).$$

It has been shown in Shimodaira [49] that the BP satisfies eq. (17.5), where $\hat{d}_H$ and $\hat{c}_H$ are interpreted geometrically. Let $\partial\mathcal{R}_H$ denote the boundary surface of $\mathcal{R}_H$, and let $\hat{\boldsymbol{\mu}}_H$ be the point on $\partial\mathcal{R}_H$ that is closest to $\mathbf{y}$. Then, $\hat{d}_H$ is the signed distance $\hat{d}_H = \pm\|\mathbf{y} - \hat{\boldsymbol{\mu}}_H\|$. The sign is positive for $\mathbf{y} \notin \mathcal{R}_H$ and negative for $\mathbf{y} \subset \mathcal{R}_H$. $\hat{c}_H$ represents the curvature of $\partial\mathcal{R}_H$ at $\hat{\boldsymbol{\mu}}_H$. $\hat{c}_H = 0$ when the boundary is flat, and $\hat{c}_H > 0$ if the boundary is curved toward $\mathcal{R}_H$.

We use $\hat{d}_H$ as a statistic for testing H. If $\hat{d}_H$ is very large, or equivalently if $\mathbf{y}$ is very far from $\mathcal{R}_H$, then H will be rejected. To examine whether $\hat{d}_H$ is significantly large or not, we may generate replicates around $\hat{\boldsymbol{\mu}}_H$ instead of $\mathbf{y}$,

$$\mathbf{y}^{**} \sim N_m(\hat{\boldsymbol{\mu}}_H, \mathbf{I}_m), \tag{17.9}$$

and calculate the signed distance, denoted $\hat{d}_H^{**}$, for each $\mathbf{y}^{**}$. The distribution of $\hat{d}_H^{**}$ simulates that of $\hat{d}_H$ by assuming that H is true. A p-value for testing H is the probability of observing $\hat{d}_H^{**}$ larger than $\hat{d}_H$:

$$\hat{\alpha}_{\mathrm{AU}}(H) = P\left(\hat{d}_H^{**} > \hat{d}_H; \mathbf{y}\right).$$

If this p-value is smaller than α, then we consider that such a large $\hat{d}_H$ value would not be observed by chance under the assumed hypothesis H, and we reach the conclusion that the assumption is wrong. The MB method calculates the same p-value approximately from eq. (17.6), bypassing the theoretical argument given above.

Biases of probability values

There are several definitions of p-values for testing hypotheses. They may differ in the test statistic to detect the discrepancy between the hypothesis and the data set and also differ in the type of mathematical approximation used to calculate the p-value. One of the properties used to compare different p-values is unbiasedness. The p-value calculated by the MB method is approximately unbiased, as explained below.

Let $\hat{\alpha}(H; \mathbf{y})$ denote a p-value, in general, for testing H. The probability of rejecting H is a function of $\boldsymbol{\mu}$ written as

$$\beta(H, \boldsymbol{\mu}) = P\left(\hat{\alpha}(H; \mathbf{y}) < \alpha; \boldsymbol{\mu}\right).$$

If H is true, $\beta(H, \boldsymbol{\mu})$ is the probability of false rejection; the p-value should be designed so that $\beta(H, \boldsymbol{\mu}) \leq \alpha$ for any value $\boldsymbol{\mu} \in \mathcal{R}_H$. If H is not true, $\beta(H, \boldsymbol{\mu})$ is the probability of correct rejection; the p-value may have an additional property that $\beta(H, \boldsymbol{\mu}) \geq \alpha$ for any value $\boldsymbol{\mu} \notin \mathcal{R}_H$. The p-value is said to be unbiased if these two properties hold. Since $\beta(H, \boldsymbol{\mu})$ changes continuously as $\boldsymbol{\mu}$ changes, an unbiased p-value should satisfy

$$\beta(H, \boldsymbol{\mu}) = \alpha, \quad \boldsymbol{\mu} \in \partial\mathcal{R}_H. \tag{17.10}$$

The p-value calculated by the MB method is approximately unbiased. Equation (17.10) holds approximately for the $\hat{\alpha}_{\mathrm{AU}}(H)$ of eq. (17.6), where the bias reduces as n becomes larger. The p-value is said to be third-order accurate, meaning that the bias is of order $O(n^{-3/2})$. The third-order accuracy usually implies that the p-value is very accurate and almost unbiased, while it does not guarantee the accuracy for a finite n.

The geometric theory of the MB method assumes that the boundary surface of the region is smooth and that the curvature is not extremely large. This assumption, however, does not hold in practice, and the bias can be larger than expected from the theory. It has been argued in Shimodaira [49] that the method works well in typical cases of phylogenetic analysis.

The BP is also approximately unbiased, although the bias of the BP can be very large. It follows from eq. (17.5) that $\tilde{\alpha}(H) \approx 1 - \Phi(\hat{d}_H + \hat{c}_H)$, which differs from eq. (17.6) only by the sign of $\hat{c}_H$. The curvature term $\hat{c}_H$ explains the difference between $\tilde{\alpha}(H)$ and $\hat{\alpha}_{\mathrm{AU}}(H)$. The difference is $O(n^{-1/2})$, and it vanishes when $\partial\mathcal{R}_H$ is flat. However, $\partial\mathcal{R}_H$ is often curved in practice, leading to the large bias of the BP. The BP is said to be first-order accurate since the bias is $O(n^{-1/2})$. For sufficiently large n, $n^{-3/2} \ll n^{-1/2}$ so that the MB method is less biased than the BP. The bias of the BP has also been discussed from other viewpoints in the literature by Zharkikh and Li [57], Felsenstein and Kishino [16], and Hillis and Bull [25].

17.4 Testing Methods

17.4.1 Hypothesis Testing

Parametric bootstrap method

In the bootstrap resampling, we have generated the replicates from the data set itself. If we have a reason to believe the parametric model, we can generate the replicates from the model instead of the data set. The elements of a replicate data set $\mathbf{X}^*$ are generated using pseudo-random numbers according to

$$\mathbf{x}_1^*, \ldots, \mathbf{x}_n^* \sim p(\mathbf{x}; S, T, \hat{\boldsymbol{\theta}}_S, \hat{\boldsymbol{\theta}}_T).$$

This is a computer simulation of the evolution along the tree and implemented in the software SeqGen [41]. This procedure is called the parametric bootstrap (PB) method, whereas the previously explained bootstrap procedure is sometimes referred to as nonparametric bootstrapping to indicate the difference.

The distribution of the replicates $\mathbf{X}_1^*, \ldots, \mathbf{X}_B^*$ is specified by (S, T), which represents a null hypothesis to be tested. Let $\delta(\mathbf{X})$ be a test statistic to detect the discrepancy between the hypothesis and the data set. For example, we use the log-likelihood difference

$$\delta(\mathbf{X}) = \ell(S, T'; \mathbf{X}) - \ell(S, T; \mathbf{X}) \tag{17.11}$$

for testing T against an alternative T' given S, and we may use

$$\delta(\mathbf{X}) = \ell(S', T; \mathbf{X}) - \ell(S, T; \mathbf{X}) \tag{17.12}$$

for testing S against an alternative S' given T. The test statistic is designed so that the deviation from the hypothesis toward the alternatives is easily detected. We generate $\delta(\mathbf{X}_1^*), \ldots, \delta(\mathbf{X}_B^*)$ by the PB and count the frequency of observing $\delta(\mathbf{X}^*) \geq \delta(\mathbf{X})$ in the replicates. This frequency

$$\hat{\alpha}_{\mathrm{PB}}(S, T; \mathbf{X}) = P\left(\delta(\mathbf{X}^*) \geq \delta(\mathbf{X}); \mathbf{X}\right) \tag{17.13}$$

is the p-value for testing the null hypothesis. We reject the pair (S, T) if $\hat{\alpha}_{\mathrm{PB}}(S, T; \mathbf{X}) < \alpha$. This PB method is introduced for phylogenetics in Goldman [17].

The model S is said to be nested in S' if S is obtained as a special case of S' by specifying some constraints on $\boldsymbol{\theta}_{S'}$. In this case, $2\delta(\mathbf{X}^*)$ of eq. (17.12) is asymptotically distributed as χ^2 with degrees of freedom $\dim \boldsymbol{\theta}_{S'} - \dim \boldsymbol{\theta}_S$, and the p-value is calculated approximately from the χ^2 distribution function instead of the PB. This procedure is often referred to as the LR test.

The SOWH test

For tree topology selection, the parametric models for two bifurcating tree topologies T and T' are nonnested. In this case, $\delta(\mathbf{X}^*)$ of eq. (17.11) is normally distributed asymptotically under some regularity conditions, and the normal approximation test of Cox [7] can be used instead of the PB.

In practice, however, T' is not specified in advance, but T' is set to $\hat{T}_{\mathrm{ML}}$. The test statistic is then

$$\delta(\mathbf{X}) = \ell(S, \hat{T}_{\mathrm{ML}}(\mathbf{X}); \mathbf{X}) - \ell(S, T; \mathbf{X})$$
$$= \max_{k=1}^{K} \ell(S, T_k; \mathbf{X}) - \ell(S, T; \mathbf{X}) \tag{17.14}$$

for testing T against alternatives $T_1, \ldots, T_K$ given S. $\delta(\mathbf{X}^*)$ of (17.14) is equal to that of (17.11) with $T' = \hat{T}_{\mathrm{ML}}(\mathbf{X}^*)$, which is distributed differently from $\delta(\mathbf{X}^*)$ of (17.11) with $T' = \hat{T}_{\mathrm{ML}}(\mathbf{X})$. In order to calculate $\hat{\alpha}_{\mathrm{PB}}(S, T; \mathbf{X})$ for (17.14), we must use the PB method rather than the Cox test to adjust for the effect of taking the maximum in (17.14), called "selection bias." This test is repeated K times for $T = T_1, \ldots, T_K$, and the confidence set of trees is obtained by collecting nonrejected T's. This procedure is the Swofford-Olsen-Waddell-Hillis (SOWH) test of Swofford et al. [52].

The p-values of tree topologies calculated by the SOWH test are often very small, and the confidence set of trees may include very few trees. The SOWH test is prone to reject too many trees, as discussed in Goldman, Anderson, and Rodrigo [18], Buckley [4], and Aris-Brosou [3]. We know that all the parametric models are merely ideal representations of reality, and none of them are exactly correct. There is nothing wrong with the SOWH test, but it may with high probability detect misspecification of S, resulting in rejection of the pair (S, T) even for the true tree topology.

Difference in the null hypothesis

An intuitive interpretation is given for the SOWH test using the simplified working model of the multivariate normal distribution. First, note that a probability function of $\mathbf{x}$ is represented as a point in the m-dimensional space of $\boldsymbol{\mu}$. For the parametric model of $\mathbf{x}$ specified by (S, T), we can consider a surface or a curve, denoted $\mathcal{M}_T$, consisting of all the points of $\boldsymbol{\mu}$ obtained by changing the parameter value $(\boldsymbol{\theta}_S, \boldsymbol{\theta}_T)$. The K tree topologies are represented by $\mathcal{M}_{T_1}, \ldots, \mathcal{M}_{T_K}$. This picture is exactly true if we consider the multinomial distribution of $\mathbf{x}$ so that any (S, T) becomes a submodel nested in it. Considering the central limit theorem, the simplified model suffices for explanation, and it is accurate enough in practice for large n.

Let $\hat{\boldsymbol{\mu}}(\mathcal{M}_T)$ denote the point on $\mathcal{M}_T$ closest to $\mathbf{y}$, known as the projection of $\mathbf{y}$ onto the surface $\mathcal{M}_T$. This point represents $p(\mathbf{x}; S, T, \hat{\boldsymbol{\theta}}_S, \hat{\boldsymbol{\theta}}_T)$, and thus

$$\mathbf{y}^* \sim N_m(\hat{\boldsymbol{\mu}}(\mathcal{M}_T), \mathbf{I}_m) \qquad (17.15)$$

represents the PB replicate $\mathbf{X}^*$. Since the negative of twice the log-likelihood corresponds to the squared distance in $\boldsymbol{\mu}$, the test statistic $\delta(\mathbf{X})$ of eq. (17.14) is equivalent to

$$\delta(\mathbf{y}) = -\frac{1}{2} \times \left(\min_{k=1}^{K} \|\mathbf{y} - \hat{\boldsymbol{\mu}}(\mathcal{M}_{T_k})\|^2 - \|\mathbf{y} - \hat{\boldsymbol{\mu}}(\mathcal{M}_T)\|^2 \right),$$

and the p-value of eq. (17.13) becomes the frequency of observing $\delta(\mathbf{y}^*) \geq \delta(\mathbf{y})$ in the replicates. We reject the null hypothesis that $\boldsymbol{\mu} \in \mathcal{M}_T$ if this p-value is smaller than α.

The null hypothesis for the AU test is also explained using the notation above. The AU test for $H = \{T\}$ rejects the null hypothesis that $\boldsymbol{\mu} \in \mathcal{R}_T$ if $\hat{\alpha}_{\mathrm{AU}}(T) < \alpha$. When the ML method is used for topology estimation, the region $\mathcal{R}_T$ consists of all the points $\mathbf{y}$ for which $\delta(\mathbf{y}) = 0$. If $\delta(\mathbf{y}) = 0$, then $\hat{T}_{\mathrm{ML}} = T$, and $\mathbf{y} \in \mathcal{R}_T$. If $\delta(\mathbf{y}) > 0$, then $\hat{T}_{\mathrm{ML}} \neq T$, and $\mathbf{y} \notin \mathcal{R}_T$. In fact, $\delta(\mathbf{y})$ becomes analogous to the signed distance $\hat{d}_H$ for $H = \{T\}$, denoted $\hat{d}_T$, if the range of T_k in taking the maximum of eq. (17.14) skips T so that the original $\delta(\mathbf{y})$ is written as $\max(0, \delta(\mathbf{y}))$. This modification does not change the result of the SOWH test except for $T = \hat{T}_{\mathrm{ML}}$. Using this modified $\delta(\mathbf{y})$, the region $\mathcal{R}_T$ is characterized by $\delta(\mathbf{y}) \leq 0$ for $\mathbf{y} \in \mathcal{R}_T$, and the boundary is characterized by $\delta(\mathbf{y}) = 0$ for $\mathbf{y} \in \partial \mathcal{R}_T$. This property of the modified $\delta(\mathbf{y})$ is the same as that of $\hat{d}_T$. The two test statistics $\delta(\mathbf{y})$ and $\hat{d}_T$ are similar in nature, but contours of them are different except for those on $\partial \mathcal{R}_T$.

The point $\hat{\boldsymbol{\mu}}_H$ for $H = \{T\}$ is also written as $\hat{\boldsymbol{\mu}}(\partial \mathcal{R}_T)$, where $\partial \mathcal{R}_T$ consists of points equidistant from $\mathcal{M}_T$ and some other $\mathcal{M}_{T'}$. Therefore, the replicate y^{**} of eq. (17.9) is different from y^* of eq. (17.15) only in the mean vector. The null hypothesis of the AU test is represented by $\mathcal{R}_T$, but that of the PB test is represented by $\mathcal{M}_T$. This difference in the null hypothesis together with that in the test statistic leads to the difference in the p-value. If S is correctly specified, $\boldsymbol{\mu} \in \mathcal{M}_{\bar{T}}$. However, we cannot expect that S is exactly correct in practice. For small deviations from S, $\boldsymbol{\mu} \notin \mathcal{M}_{\bar{T}}$, yet $\boldsymbol{\mu} \in \mathcal{R}_{\bar{T}}$. Thus the PB test is sensitive to misspecifications, while the AU test is more robust.

Combining nonnested models

It is possible to apply the LR test to tree topology selection if the alternative hypothesis is represented by a parametric model in which all the K models specified by $(S, T_1), \ldots, (S, T_K)$ are nested. One such "supermodel" is the multinomial distribution of $\mathbf{x}$ in which any (S, T) is nested. However, it is better to keep the number of free parameters of the supermodel as small as possible to avoid unwanted detection of misspecifications of S. We can combine the log-likelihood functions of these submodels to construct a supermodel with minimum dimensions as described in Shimodaira [48]. The method is analogous to the split decomposition of Dopazo et al. [10] and is considered an extension of the spectral analysis of Hendy et al. [24] for the ML analysis based on a general class of the substitution model.

Let us consider the case where $T_1, \ldots, T_K$ are the unrooted bifurcating tree topologies of g groups of taxa, where $K = 1 \times 3 \times 5 \times \cdots \times (2g - 5)$. For each group of taxa, the subtree topology is estimated in advance so that we obtain the constrained candidate topologies; see Table 17.2 for an example of $g = 5$.

The split or bipartition of the g groups of taxa divides these groups into two sets. Each split is represented by a multifurcating tree topology with only one internal branch among the g groups. There are $m = 2^{g-1} - (g + 1)$ nontrivial splits, and their topologies are denoted by $U_1, \ldots, U_m$. The star-shaped tree topology of the g groups is denoted by U_0.

The branch lengths of the supermodel are $\theta_1, \ldots, \theta_m$. They constitute the "spectrum" in the sense of the spectral analysis. All of them are zero for U_0, and only one of them is nonzero for each $U_1, \ldots, U_m$. There are $g - 3$ nonzero values for each $T_1, \ldots, T_K$. Each tree topology is obtained from the supermodel by specifying constraints on the branch lengths. Let $\hat{\theta}_{U_i}$ be the ML estimate of the nonzero branch length of U_i and $\hat{\theta}_i$ be the ML estimate of the corresponding branch length under the supermodel. Then we can calculate the latter estimate approximately from the former by

$$\hat{\theta}_i \approx v_i \hat{\theta}_{U_i},$$

where the coefficients v_i, $i = 1, \ldots, m$, are

$$\begin{pmatrix} v_1 \\ \vdots \\ v_m \end{pmatrix} = \begin{pmatrix} \mathbf{a}'_1 \mathbf{a}_1 & \cdots & \mathbf{a}'_1 \mathbf{a}_m \\ \vdots & & \vdots \\ \mathbf{a}'_m \mathbf{a}_1 & \cdots & \mathbf{a}'_m \mathbf{a}_m \end{pmatrix}^{-1} \begin{pmatrix} \mathbf{a}'_1 \mathbf{a}_1 \\ \vdots \\ \mathbf{a}'_m \mathbf{a}_m \end{pmatrix},$$

and $\mathbf{a}_i$ is an n-dimensional vector consisting of elements $\ell_h(U_i) - \ell_h(U_0)$, $h = 1, \ldots, n$. Here $\mathbf{a}'_i \mathbf{a}_j$ denotes the inner product of the two vectors. The site-wise log-likelihood of the supermodel is calculated approximately by

$$\ell_h(\text{supermodel}) \approx \sum_{i=1}^{m} v_i(\ell_h(U_i) - \ell_h(U_0)) + \ell_h(U_0).$$

This procedure can be applied to any combination of the splits, not just the m splits. The tree topology selection is then treated as the subset selection problem of multiple regression, and even sophisticated searching techniques, such as the branch-and-bound algorithm, can be used for topology selection, although the accuracy of the approximation may be a problem in practice.

The LR test statistic for testing T against the supermodel is $2\Delta\ell = 2 \times \ell(\text{supermodel}) - 2 \times \ell(T)$, which is asymptotically distributed as χ^2 with degrees of freedom $m - (g - 3)$. This LR test is also sensitive to misspecifications of S. It sometimes rejects all the bifurcating tree topologies, including even $\hat{T}_{\text{ML}}$; see Table 17.1.

The null hypothesis for testing T is that $\theta_i = 0$ for $m - (g - 3)$ splits not included in T, whereas the branch lengths for the $g - 3$ splits in T can change freely as long as $\theta_i \geq 0$; the $\mathcal{M}_T$ of the simplified working model is represented as a $(g-3)$-dimensional surface in the m-dimensional space of branch lengths. We ignore the complication caused by the constraints $\theta_i \geq 0$ [38]. A general point in the m-dimensional space does not necessarily correspond to a tree but may correspond to a network in the sense of split decomposition. It represents a mixture of the splits that is the weighted sum of their probability functions. A tree is a special case of the network. The region $\mathcal{R}_T$ consists of general points for which the surface $\mathcal{M}_T$ is closer than $\mathcal{M}_{T'}$, $T' \neq T$.

17.4.2 Model Selection Tests

The Kishino-Hasegawa test

For comparing two bifurcating trees T and T', the test statistic $\delta(\mathbf{X})$ of eq. (17.11) indicates how much better T' is than T. The data set $\mathbf{X}$ suggests that T' is better than T if $\delta(\mathbf{X}) > 0$, and that T is better than T' if

Table 17.1. The LR test for the 15 tree topologies of Table 17.2.

model	θ_1	θ_2	θ_3	θ_4	θ_5	θ_6	θ_7	θ_8	θ_9	θ_{10}	$2\Delta\ell$	p-value
super	58	138	164	117	53	60	86	70	24	60	0.0	
T_1	62	118	0	0	0	0	0	0	0	0	89.5	10^{-15}
T_2	0	0	167	108	0	0	0	0	0	0	92.5	10^{-16}
T_3	101	0	156	0	0	0	0	0	0	0	92.9	10^{-16}
T_4	0	153	0	0	60	0	0	0	0	0	93.3	10^{-16}
T_5	0	147	0	0	0	43	0	0	0	0	94.8	10^{-16}
T_6	101	0	0	0	0	0	67	0	0	0	102.0	10^{-18}
T_7	0	0	0	85	0	0	0	60	0	0	103.2	10^{-18}
T_8	0	0	0	111	44	0	0	0	0	0	106.2	10^{-19}
T_9	0	0	0	0	0	0	96	97	0	0	106.9	10^{-19}
T_{10}	0	0	0	0	0	53	0	100	0	0	109.3	10^{-19}
T_{11}	0	0	190	0	0	0	0	0	8	0	115.0	10^{-20}
T_{12}	0	0	0	0	0	0	102	0	0	27	121.3	10^{-22}
T_{13}	0	0	0	0	69	0	0	0	0	31	126.7	10^{-23}
T_{14}	0	0	0	0	0	62	0	0	4	0	127.1	10^{-23}
T_{15}	0	0	0	0	0	0	0	0	5	34	132.5	10^{-24}

Note: $\Delta\ell$ denotes the log-likelihood difference from the supermodel. The p-value is calculated by assuming $2\Delta\ell$ is distributed as χ^2 with 8 degrees of freedom. Branch lengths $(\theta_1, \ldots, \theta_{10})$ are multiplied by 10,000. The ten splits $U_1, \ldots, U_{10}$ represents clades (G1, G2), (G3, G4), (G1, G2, G3), (G2, G3), (G2, G3, G4), (G1, G3, G4), (G1, G2, G4), (G1, G4), (G1, G3), and (G2, G4), respectively.

$\delta(\mathbf{X}) < 0$. We may choose either T or T' depending on the sign of $\delta(\mathbf{X})$. However, this model selection procedure is influenced by the sampling error.

The expected value of $\delta(\mathbf{X})$, denoted $E(\delta(\mathbf{X}))$, indicates how much better T' is than T on average, if $\mathbf{X}$ is sampled repeatedly from the underlying true distribution. It follows from the consistency argument of $\hat{T}_{\mathrm{ML}}$ in Section 17.2.1 that $E(\delta(\mathbf{X}))$ is minimized at $T = \bar{T}$ against any fixed T' if S is correctly specified. Therefore, the model selection procedure becomes error-free if $E(\delta(\mathbf{X}))$ is used instead of $\delta(\mathbf{X})$, although we never know the value of $E(\delta(\mathbf{X}))$ in practice.

If a positive value of $\delta(\mathbf{X})$ is observed, we check whether this value is sufficiently large to reject the null hypothesis $E(\delta(\mathbf{X})) \leq 0$. If $\delta(\mathbf{X})$ is larger than a certain threshold, then we conclude that $E(\delta(\mathbf{X})) > 0$ and that T' is better than T. Otherwise, the decision is inconclusive; the two topologies are left for further analysis. This test is directional, and the roles of the two topologies are exchanged if a negative value of $\delta(\mathbf{X})$ is observed.

This procedure is the test of Kishino and Hasegawa [29], which has been used widely in phylogenetics and is referred to as the Kishino-Hasegawa (KH)

Table 17.2. p-values for the 15 constrained candidate tree topologies.

model	$\Delta\ell$	PP1	PP2	BP	AU	KH	SH	WSH	tree topology
T_1	0.0	**0.28**	**0.61**	**0.23**	**0.69**	**0.55**	**0.97**	**0.95**	((G1,G2),(G3,G4),G5)
T_2	1.5	**0.49**	**0.14**	**0.28**	**0.60**	**0.46**	**0.83**	**0.86**	((G1,(G2,G3)),G4,G5)
T_3	1.7	**0.15**	**0.12**	**0.16**	**0.47**	**0.41**	**0.84**	**0.84**	(((G1,G2),G3),G4,G5)
T_4	1.9	**0.06**	**0.09**	**0.13**	**0.45**	**0.33**	**0.84**	**0.81**	(G1,(G2,(G3,G4)),G5)
T_5	2.6	0.01	0.04	**0.09**	**0.37**	**0.27**	**0.80**	**0.73**	((G1,(G3,G4)),G2,G5)
T_6	6.2	0.00	0.00	0.02	**0.16**	**0.15**	**0.64**	**0.54**	(((G1,G2),G4),G3,G5)
T_7	6.8	0.00	0.00	0.03	**0.25**	**0.28**	**0.58**	**0.61**	((G1,G4),(G2,G3),G5)
T_8	8.3	0.00	0.00	0.01	**0.08**	**0.23**	**0.51**	**0.40**	(G1,((G2,G3),G4),G5)
T_9	8.7	0.00	0.00	0.04	**0.25**	**0.21**	**0.50**	**0.66**	(((G1,G4),G2),G3,G5)
T_{10}	9.9	0.00	0.00	0.02	**0.14**	**0.18**	**0.43**	**0.59**	(((G1,G4),G3),G2,G5)
T_{11}	12.7	0.00	0.00	0.00	0.00	**0.10**	**0.29**	**0.20**	(((G1,G3),G2),G4,G5)
T_{12}	15.9	0.00	0.00	0.00	0.01	**0.05**	**0.17**	**0.27**	((G1,(G2,G4)),G3,G5)
T_{13}	18.6	0.00	0.00	0.00	0.00	0.03	**0.09**	**0.13**	(G1,((G2,G4),G3),G5)
T_{14}	18.8	0.00	0.00	0.00	0.00	0.02	**0.09**	**0.09**	(((G1,G3),G4),G2,G5)
T_{15}	21.5	0.00	0.00	0.00	0.00	0.01	0.04	**0.10**	((G1,G3),(G2,G4),G5)

Note: Only the 15 candidate tree topologies are considered; the subtree topologies for G1, ..., G5 are specified in Figure 17.1. $\Delta\ell$ denotes the log-likelihood difference from the ML topology. The trees are numbered by increasing order of $\Delta\ell$. PP1 denotes the PP calculated by the MCMCMC using MrBayes with clade constraints, and PP2 denotes the PP calculated by the BIC approximation. p-values ≥ 0.05 are in boldface.

test. For general model selection using AIC, the same idea as the KH test has been proposed independently in Linhart [32] and Vuong [53].

The KH test, like the Cox test, is usually implemented as a normal approximation test. The variance of $\delta(\mathbf{X})$ is estimated by

$$\widehat{\text{var}}(\delta(\mathbf{X})) = \frac{n}{n-1} \sum_{h=1}^{n} \left\{ \ell_h(T) - \ell_h(T') - \frac{1}{n} \sum_{h'=1}^{n} (\ell_{h'}(T) - \ell_{h'}(T')) \right\}^2 .$$

The standardized test statistic $z = \delta(\mathbf{X})/\sqrt{\widehat{\text{var}}(\delta(\mathbf{X}))}$ is compared with the standard normal distribution. If $z > 1.64$, then we conclude T' is better than T at the 5% significance level. The p-value is $1 - \Phi(z)$. For two-sided tests, the p-value should be doubled.

The KH test can best be explained by comparing it with the PB. The p-value of the KH test is interpreted as eq. (17.13), where a replicate $\mathbf{X}^*$ is generated by a modified PB using a weighted sum of the probability functions of the competing trees. The replicates are generated under the additional assumption of $E(\delta(\mathbf{X})) = 0$. This assumption is "least favorable," meaning that it maximizes the frequency of $\delta(\mathbf{X}^*) > \delta(\mathbf{X})$ under the null hypothesis $E(\delta(\mathbf{X})) \leq 0$. The least favorable model for the KH test is sought by changing only $E(\delta(\mathbf{X}))$ but not $\mathrm{var}(\delta(\mathbf{X}))$. Instead of generating replicates directly by the PB under the least favorable assumption, we first generate $\mathbf{X}_1^*, \ldots, \mathbf{X}_B^*$ by the nonparametric bootstrap and calculate a modified test statistic $\delta'(\mathbf{X}^*) = \delta(\mathbf{X}^*) - \bar{\delta}$ for each replicate, where $\bar{\delta} = (\delta(\mathbf{X}_1^*) + \cdots + \delta(\mathbf{X}_B^*))/B$. This centering procedure makes the average of $\delta'(\mathbf{X}^*)$ zero, approximating the replicates from the least favorable model. The p-value is the frequency of observing $\delta'(\mathbf{X}^*) \geq \delta(\mathbf{X})$ in the modified replicates $\delta'(\mathbf{X}_1^*), \ldots, \delta'(\mathbf{X}_B^*)$. This p-value will be indistinguishable from that of the normal approximation for sufficiently long sequences.

Let us consider the simplified working model with $K = 2$. The two regions $\mathcal{R}_T$ and $\mathcal{R}_{T'}$ are separated by the boundary $\partial \mathcal{R}_T = \partial \mathcal{R}_{T'}$. The assumption $E(\delta(\mathbf{X})) = 0$ is represented by the hypothesis that $\boldsymbol{\mu} \in \partial \mathcal{R}_T$, and therefore it does not specify a particular point on the boundary surface. When the KH test is interpreted as the PB, however, the least favorable model must be specified as a point for generating the replicates. It is, in fact, represented approximately by the point $\hat{\boldsymbol{\mu}}(\partial \mathcal{R}_T)$ for the KH test. This explains the similarity between the KH test and the AU test in the case of $K = 2$.

The least favorable model is considered as a consequence of the possibility of misspecification of S. Because only the points on $\mathcal{M}_T$ or those on $\mathcal{M}_{T'}$ represent trees, a general point in the m-dimensional space, including $\partial \mathcal{R}_T$, represents a network. $\partial \mathcal{R}_T$ includes the intersection of $\mathcal{M}_T$ and $\mathcal{M}_{T'}$ as a special case, representing the consensus trees of T and T' (e.g., the star-shaped tree). When $\mathbf{y}$ is in this intersection, the internal branch lengths are estimated to be zero so that $\hat{\boldsymbol{\mu}}(\mathcal{M}_T)$ and $\hat{\boldsymbol{\mu}}(\mathcal{M}_{T'})$ are identical. The regularity condition for justifying the KH test requires, however, that $\boldsymbol{\mu}$ not be in this intersection [53, 46], implying that the least favorable model is strictly a network rather than a tree.

Multiple-comparisons tests

The KH test assumes that the alternative tree topology T' is specified in advance of observing $\mathbf{X}$. In practice, however, we specify $T' = \hat{T}_{\mathrm{ML}}$ after looking at $\mathbf{X}$. In this case, the KH test suffers from a selection bias, meaning that the expected value of $\ell(\hat{T}_{\mathrm{ML}})$ is larger than that of $\ell(T')$ for $T' = \hat{T}_{\mathrm{ML}}$. As K becomes larger, the maximum of $\ell(T')$ over all the alternative tree topologies can easily have a very large value by chance, and $\delta(\mathbf{X})$ of eq. (17.14) tends to be larger than expected from the normal approximation of the KH test. The selection bias often leads to overconfidence in the wrong trees. This may result in conflicting conclusions, each claiming statistical significance.

The selection bias of the KH test has been discussed in the literature, such as Shimodaira and Hasegawa [50], Goldman, Anderson, and Rodrigo [18], and Buckley [4]. We have to modify the KH test in the same way as we did for obtaining the SOWH test from the Cox test. This is implemented as a resampling version of the multiple-comparisons procedure in Shimodaira and Hasegawa and is referred to as the Shimodaira-Hasegawa (SH) test.

486   H. Shimodaira and M. Hasegawa

The statistic to test T_k against $T_1, \ldots, T_K$ is $\delta_k(\mathbf{X}) = \ell(\hat{T}_{\mathrm{ML}}(\mathbf{X}); \mathbf{X}) - \ell(T_k; \mathbf{X})$. This value is compared with the replicates to see if it is significantly large to reject T_k. In the SH test, we first generate $\mathbf{X}_1^*, \ldots, \mathbf{X}_B^*$ by the nonparametric bootstrap and calculate $\ell(T_k; \mathbf{X}_b^*)$ for $k = 1, \ldots, K$, $b = 1, \ldots, B$, using the RELL approximation. The centering procedure is applied to each log-likelihood; $\ell'(T_k; \mathbf{X}_b^*) = \ell(T_k; \mathbf{X}_b^*) - \bar{\ell}_k$, where $\bar{\ell}_k = (\ell(T_k; \mathbf{X}_1^*) + \cdots + \ell(T_k; \mathbf{X}_B^*))/B$. A modified replicate of $\delta_k(\mathbf{X})$ is defined as $\delta_k'(\mathbf{X}_b^*) = \max_{k'=1}^K \ell'(T_{k'}; \mathbf{X}_b^*) - \ell'(T_k; \mathbf{X}_b^*)$. The p-value of T_k is then the frequency of observing $\delta_k'(\mathbf{X}_b^*) \geq \delta_k(\mathbf{X})$, $b = 1, \ldots, B$. This procedure is repeated for $k = 1, \ldots, K$, and the confidence set of tree topologies is obtained by collecting T_k with the p-value not smaller than α.

The test statistic of the SH test can be written as

$$\delta_k(\mathbf{X}) = \max_{k'=1}^K (\ell(T_{k'}; \mathbf{X}) - \ell(T_k; \mathbf{X})).$$

If each term in the maximization is standardized, and the redundant term of $k' = k$ is ignored, the weighted test statistic is obtained as

$$\delta_k(\mathbf{X}) = \max_{k'=1,\ldots,k-1,k+1,\ldots,K} \frac{\ell(T_{k'}; \mathbf{X}) - \ell(T_k; \mathbf{X})}{\sqrt{\widehat{\mathrm{var}}(\ell(T_{k'}; \mathbf{X}) - \ell(T_k; \mathbf{X}))}}. \tag{17.16}$$

The weighted SH (WSH) test uses eq. (17.16) for $\delta_k(\mathbf{X})$. The WSH test often improves the SH test, meaning that the number of nonrejected tree topologies of the WSH test is smaller than that of the SH test, especially for a large K. The WSH test is a resampling version of the Gupta procedure [19, 20] for ranking and selection of normal variables, or equivalently the method of multiple comparisons with the unknown best [26]. The WSH test has been implemented in Shimodaira [45, 47] for a general model selection using the AIC, where $(\ell'(T_1; \mathbf{X}^*), \ldots, \ell'(T_K; \mathbf{X}^*))$ is generated as a normal random vector instead of the RELL approximation. This pseudo-random generation is faster than the RELL method for a relatively small K, but the matrix computation needed for taking account of the covariance structure makes it slower for a large K.

The null hypothesis for testing T_k in the multiple-comparisons tests is that $E(\ell(T_k; \mathbf{X}))$ is the largest among $E(\ell(T_1; \mathbf{X})), \ldots, E(\ell(T_K; \mathbf{X}))$, and it is written as

$$E\left(\ell(T_{k'}; \mathbf{X}) - \ell(T_k; \mathbf{X})\right) \leq 0, \quad k' = 1, \ldots, k-1, k+1, \ldots, K. \tag{17.17}$$

The least favorable model assumes that all the inequalities of (17.17) hold as equalities. In other words, all the tree topologies are assumed to be equally good on average. The centering procedure for the log-likelihoods approximates this assumption. The least favorable model represents the worst case in which the probability of false rejection is maximized due to the selection bias; this probability is not larger than α under the null hypothesis, and it is equal to α in the least favorable model. Typically, the false rejection probability is much smaller than α because only a few of the tree topologies are nearly as good as the best one. As K increases, the number of nonrejected tree topologies can become very large. In order to avoid this conservative behavior of the multiple-comparisons tests, we should make K as small as possible by eliminating extremely unlikely topologies from the candidates. All the possible topologies are not necessarily to be included in the candidates when we are interested in biological hypotheses represented by their typical topologies.

The null hypothesis of (17.17) is represented as the hypothesis that $\boldsymbol{\mu} \in \mathcal{R}_{T_k}$ using the simplified working model. There are K regions $\mathcal{R}_{T_1}, \ldots, \mathcal{R}_{T_K}$ separated by the boundaries $\partial \mathcal{R}_{T_1}, \ldots, \partial \mathcal{R}_{T_K}$. The region $\mathcal{R}_{T_k}$ is in the shape of a polyhedral cone with faces corresponding to the equalities of (17.17). The least favorable model corresponds to the apex of the cone, where the amount of bending of the boundary $\partial \mathcal{R}_{T_k}$ is maximized. The amount of bending indicates the selection bias. This also explains the difference between the BP and the AU test. The MB method will estimate the curvature term $\hat{c}_{T_k}$ as the amount of bending around $\hat{\mu}(\partial \mathcal{R}_{T_k})$; this is the point on the boundary closest to the data set, and it represents the probability function of a typical case. The AU test is adjusting the selection bias in the BP by assuming the typical case, whereas the multiple-comparisons tests are adjusting the selection bias in the KH test by assuming the worst case.

17.5 Concluding Remarks

17.5.1 Two Approaches to Testing

As we have seen, there are several methods for assessing the uncertainty in phylogenetic inference. The uncertainty due to the sampling error is measured by the p-value of statistical testing. There are two very different approaches to testing. They differ in the null hypothesis to be tested.

The standard approach in statistics assumes that the null hypothesis is represented by the parametric model specified by the substitution process S and the tree topology T. This approach includes the hypothesis testing methods (Section 17.4.1) such as the Cox test, the SOWH test, and the LR test. These methods are sensitive to misspecifications of S and may have difficulty comparing nonnested models of the tree topologies. The null hypothesis is denoted by $\mathcal{M}_T$ using the simplified working model; $\boldsymbol{\mu} \in \mathcal{M}_T$ if both S and T are correctly specified, but $\boldsymbol{\mu} \notin \mathcal{M}_T$ for the true T if S is misspecified even slightly.

The second approach is relatively new in statistics. The null hypothesis in the second approach is that the parametric model specified by (S, T) approximates reality better than the other parametric models specified by (S, T'), where T' is one of the other candidate tree topologies. This approach includes the model selection tests (Section 17.4.2) such as the KH test and the multiple-comparisons tests, and also the nonparametric bootstrap methods

(Section 17.3) such as the BP and the AU tests. The second approach is to find the best-approximating model, while the first approach is to find the correct model. The null hypothesis is now denoted by $\mathcal{R}_T$ using the simplified working model; $\boldsymbol{\mu} \in \mathcal{R}_T$ for the true T even if S is misspecified slightly. Thus the methods are robust to misspecifications of S. However, the confidence set of tree topologies is larger than that of the first approach. This is the price we have to pay for the robustness.

The difference between the two approaches is also described as parametric versus nonparametric. The null hypothesis of the first approach is the parametric model itself, whereas the null hypothesis of the second approach is defined not as a parametric model but a family of probability functions in the nonparametric sense. This difference is also apparent in the type of bootstrap procedure. The methods of the first approach are implemented using the parametric bootstrap, while those of the second approach use the nonparametric bootstrap. Parametric models are useful, but we should be careful when parametric methods are used to evaluate a model itself.

The PP of Bayesian inference (Section 17.2.2) is not a frequentist p-value, but it is interpreted similarly as a measure of uncertainty in phylogenetic inference. The calculation of the PP is parametric indeed and belongs to the first approach, but there exists a Bayesian inference belonging to the second approach as well. In fact, the BP can be regarded as the PP for the region $\mathcal{R}_T$, and the bias correction of the AU test can be implemented approximately by giving an elaborate prior distribution (i.e., the matching prior) on the m-dimensional space of the simplified working model as discussed in Efron and Tibshirani [13].

17.5.2 Recommended Methods

Among several methods for p-value calculation, we currently recommend the AU test for general tree selection problems. It satisfies the requirement for unbiasedness at least approximately, and it is not susceptible to an increase in the number of candidate trees. The AU p-values for the branches as well as those for the trees are easily obtained by the MB method. The AU test can be used with any inference methods such as the distance matrix methods, the parsimony method, and the ML method. When used with the ML method, the computational burden is reduced by the MCMC and the RELL techniques; the limited list of candidate tree topologies is prepared by the MCMC simulation, and the ML tree topologies for bootstrap replicates are calculated by the RELL method.

We also recommend the multiple-comparisons tests such as the SH test and the WSH test for obtaining the confidence set of tree topologies when the number of candidate tree topologies is not very large. The p-values of topologies are often larger than those obtained from the AU test, and thus the confidence set becomes larger. The multiple-comparisons tests are conservative, and the conclusions are drawn safely; it is harder for them to miss the true topology in the confidence set than for the AU test. A practical difficulty is that the SH test is very susceptible to an increase in the number of candidate trees. This is alleviated in the WSH test.

The use of Bayesian inference in molecular phylogenetics was advocated by Huelsenbeck et al. [28] and is spreading rapidly (e.g., [35]). However, Bayesian inference is sometimes misleading in phylogenetic tree selection, giving extremely high PP to a wrong tree, as has been pointed out by several authors (e.g., [54]). In spite of this problem, Bayesian inference equipped with MCMC is computationally efficient and may be useful to find candidate trees for large phylogenies, where the standard ML approach is computationally prohibitive.

It should be noted that phylogenetic inference is still a very active research area, and there is still room for improving the methodology. Several inference methods have been developed by introducing new ideas; some of them are, in fact, new to statistics as well. This research subject is also controversial due to the difficulty of the problem. Currently, the field is divided in its opinions regarding the best methodology. Further research is expected to follow.

17.5.3 Biological Discussion

An example of phylogenetic inference is illustrated in Figure 17.1. We analyzed mitochondrial proteins (concatenated sequences of 12 genes encoded on the same strand of mitochondrial DNA) for 28 eutherian mammals with four species from marsupials and monotremes as an outgroup. The mtREV+Γ model was used for amino acid substitutions.

In the tree of Figure 17.1, guinea pig, mouse, and squirrel form a monophyletic clade (rodent monophyly), in contrast with the rodent polyphyly tree suggested by a conventional ML analysis of mitochondrial proteins [9]. The rodent monophyly tree is consistent with the tree obtained from the abundant data set of nuclear DNA [33, 35, 36] and with the ML analyses by some ad hoc method of tree topology search [5, 37]. Although the BP and AU p-values are not high enough to exclude alternative relationships (PP values are high in most of the nodes), the rodents/rabbit clade (Glires) and the primates/Glires clade, both supported by the nuclear DNA data, are suggested. The moonrat/mole clade is also suggested in accord with the nuclear DNA analyses as well as with the traditional morphology, but in contrast with the conventional mitochondrial analyses [37]. Furthermore, it is suggested that the eutherian mammals consist of four major groups; that is, G1: Euarchontoglires (primates + rodents + rabbits), G2: Laurasiatheria (cetartiodactyls + perissodactyls + carnivores + bats + core insectivores), G3: Afrotheria (elephants + dugongs + hyraxes + aardvarks + elephant shrews + golden moles + tenrecs), and G4: Xenarthra (armadillos and their relatives), again consistent with the other evidence [54, 33, 35, 36, 31]. Thus, the combination of the MCMC Bayesian method with the ML method seems to be useful in applying the likelihood approach to phylogenetic problems with many taxa.

It is interesting to mention the analysis of six mammalian species (human, harbor seal, cow, rabbit, mouse, and opossum) of Shimodaira [49]. Both the ML topology and the MAP topology in our analysis of the 32 species correspond to ((human, (rabbit, mouse)), (harbor seal, cow), opossum) (i.e., the tree 7 in Table 3 of Shimodaira [49]), whereas the ML topology for these six species is (((human, (harbor seal, cow)), rabbit), mouse, opossum) (i.e., the tree 1 therein). The PP values are calculated by the BIC approximation: 0.93 for tree 1, 0.07 for tree 2, and almost zero for the others. On the other hand, tree 7 is not rejected by the p-values of the AU, the SH, and the WSH tests.

Acknowledgments

The authors thank Ryota Suzuki, a student programmer at Tokyo Tech, who has performed the calculations for the numerical example and prepared the figures and tables. We also thank the editor, Rasmus Nielsen, for constructive comments. H.S. was supported for this work in part by Grant-in-Aid for Young Scientists (A) KAKENHI-14702061 from MEXT, Japan.

References

[1] J. Adachi and M. Hasegawa. Model of amino acid substitution in proteins encoded by mitochondrial DNA. *J. Mol. Evol.*, 42:459–468, 1996.

[2] H. Akaike. A new look at the statistical model identification. *IEEE Trans. Autom. Control*, 19:716–723, 1974.

[3] S. Aris-Brosou. How Bayes tests of molecular phylogenies compare with frequentist approaches. *Bioinformatics*, 19:618–624, 2003.

[4] T. R. Buckley. Model misspecification and probabilistic tests of topology: Evidence from empirical data sets? *Syst. Biol.*, 51:509–523, 2002.

[5] Y. Cao, M. Fujiwara, M. Nikaido, N. Okada, and M. Hasegawa. Interordinal relationships and timescale of eutherian evolution as inferred from mitochondrial genome data. *Gene*, 259:149–158, 2000.

[6] L. L. Cavalli-Sforza and A. W. F. Edwards. Phylogenetic analysis: Models and estimation procedures. *Evolution*, 32:550–570, 1967.

[7] D. R. Cox. Further results on tests of separate families of hypotheses. *J. R. Stat. Soc. Ser. B*, 24:406–424, 1962.

[8] A. C. Davison and D. V. Hinkley. *Bootstrap Methods and Their Application*. Cambridge University Press, Cambridge, 1997.

[9] A. M. D'Erchia, C. Gissi, G. Pesole, C. Saccone, and U. Arnason. The guinea-pig is not a rodent. *Nature*, 381:597–600, 1996.

[10] J. Dopazo, A. Dress, and A. von Haeseler. Split decomposition: A technique to analyze viral evolution. *Proc. Natl. Acad. Sci. USA*, 90:10320–10324, 1993.

[11] B. Efron. Bootstrap methods: Another look at the jackknife. *Ann. Stat.*, 7:1–26, 1979.

[12] B. Efron, E. Halloran, and S. Holmes. Bootstrap confidence levels for phylogenetic trees. *Proc. Natl. Acad. Sci. USA*, 93:13429–13434, 1996.

[13] B. Efron and R. Tibshirani. The problem of regions. *Ann. Stat.*, 26:1687–1718, 1998.

[14] J. Felsenstein. Evolutionary trees from DNA sequences: A maximum likelihood approach. *J. Mol. Evol.*, 17:368–376, 1981.

[15] J. Felsenstein. Confidence limits on phylogenies: An approach using the bootstrap. *Evolution*, 39:783–791, 1985.

[16] J. Felsenstein and H. Kishino. Is there something wrong with the bootstrap on phylogenies? A reply to Hillis and Bull. *Syst. Biol.*, 42:193–200, 1993.

[17] N. Goldman. Statistical tests of models of DNA substitution. *J. Mol. Evol.*, 36:182–198, 1993.

[18] N. Goldman, J. P. Anderson, and A. G. Rodrigo. Likelihood-based tests of topologies in phylogenetics. *Syst. Biol.*, 49:652–670, 2000.

[19] S. S. Gupta. On some multiple decision (selection and ranking) rules. *Technometrics*, 7:225–245, 1965.

[20] S. S. Gupta and Deng-Yuan Huang. Subset selection procedures for the means and variances of normal populations: Unequal sample sizes case. *Sankhyā Ser. B*, 38:112–128, 1976.

[21] M. Hasegawa and H. Kishino. Confidence limits on the maximum-likelihood estimate of the hominoid tree from mitochondrial-DNA sequences. *Evolution*, 43:672–677, 1989.

[22] M. Hasegawa and H. Kishino. Accuracies of the simple methods for estimating the bootstrap probability of a maximum likelihood tree. *Mol. Biol. Evol.*, 11:142–145, 1994.

[23] M. Hasegawa, H. Kishino, and T. Yano. Dating of the human-ape splitting by a molecular clock of mitochondrial DNA. *J. Mol. Evol.*, 22:160–174, 1985.

[24] M. D. Hendy, D. Penny, and M. A. Steel. A discrete Fourier-analysis for evolutionary trees. *Proc. Natl. Acad. Sci. USA*, 91:3339–3343, 1994.

[25] D. M. Hillis and J. J. Bull. An empirical test of bootstrapping as a method for assessing confidence in phylogenetic analysis. *Syst. Biol.*, 42:182–192, 1993.

[26] J. C. Hsu. *Multiple Comparisons—Theory and methods*. Chapman & Hall, London/New York, 1996.

[27] J. P. Huelsenbeck and F. Ronquist. MRBAYES: Bayesian inference of phylogenetic trees. *Bioinformatics*, 17:754–755, 2001.

[28] J. P. Huelsenbeck, F. Ronquist, R. Nielsen, and J. P. Bollback. Bayesian inference of phylogeny and its impact on evolutionary biology. *Science*, 294:2310–2314, 2001.

[29] H. Kishino and M. Hasegawa. Evaluation of the maximum likelihood estimate of the evolutionary tree topologies from DNA sequence data, and the branching order in Hominoidea. *J. Mol. Evol.*, 29:170–179, 1989.

[30] H. Kishino, T. Miyata, and M. Hasegawa. Maximum likelihood inference of protein phylogeny and the origin of chloroplasts. *J. Mol. Evol.*, 30:151–160, 1990.

[31] Y. H. Lin, P. A. McLenachan, A. R. Gore, M. J. Phillips, R. Ota, M. D. Hendy, and D. Penny. Four new mitochondrial genomes and the increased stability of evolutionary trees of mammals from improved taxon sampling. *Mol. Biol. Evol.*, 19:2060–2070, 2002.

[32] H. Linhart. A test whether two AIC's differ significantly. *S. Afr. Stat. J.*, 22:153–161, 1988.

[33] O. Madsen, M. Scally, C. J. Douady, D. J. Kao, R. W. DeBry, R. Adkins, H. M. Amrine, M. J. Stanhope, W. W. de Jong, and M. S. Springer. Parallel adaptive radiations in two major clades of placental mammals. *Nature*, 409:610–614, 2001.

[34] R. Marcus, E. Peritz, and K. R. Gabriel. On closed testing procedures with special reference to ordered analysis of variance. *Biometrika*, 63:655–660, 1976.

[35] W. J. Murphy, E. Eizirik, W. E. Johnson, Y. P. Zhang, O. A. Ryder, and S. J. O'Brien. Molecular phylogenetics and the origins of placental mammals. *Nature*, 409:614–618, 2001.

[36] W. J. Murphy, E. Eizirik, S. J. O'Brien, O. Madsen, M. Scally, C. J. Douady, E. Teeling, O. A. Ryder, M. J. Stanhope, W. W. de Jong, and M. S. Springer. Resolution of the early placental mammal radiation using Bayesian phylogenetics. *Science*, 294:2348–2351, 2001.

[37] M. Nikaido, Y. Cao, M. Harada, N. Okada, and M. Hasegawa. Mitochondrial phylogeny of hedgehogs and monophyly of Eulipotyphla. *Mol. Phylogenet. Evol.*, 28(2):276–284, 2003.

[38] R. Ota, P. J. Waddell, M. Hasegawa, H. Shimodaira, and H. Kishino. Appropriate likelihood ratio tests and marginal distributions for evolutionary tree models with constraints on parameters. *Mol. Biol. Evol.*, 17:798–803, 2000.

[39] D. Posada and T. Buckley. Advantages of AIC and Bayesian approaches over likelihood ratio tests for model selection in phylogenetics. *Syst. Biol.*, 2004. submitted.

[40] D. Posada and K. A. Crandall. MODELTEST: testing the model of DNA substitution. *Bioinformatics*, 14:817–818, 1998.

[41] A. Rambaut and N. C. Grassly. Seq-Gen: An application for the Monte Carlo simulation of DNA sequence evolution along phylogenetic trees. *Comput. Appl. Biosci.*, 13:235–238, 1997.

[42] Y. Sakamoto, M. Ishiguro, and G. Kitagawa. *Akaike Information Criterion Statistics*. Reidel, Dordrecht, 1986.

[43] G. Schwarz. Estimating the dimension of a model. *Ann. Stat.*, 6:461–464, 1978.

[44] R. Shibata. An optimal selection of regression variables. *Biometrika*, 68:45–54, 1981.

[45] H. Shimodaira. A model search technique based on confidence set and map of models. *Proc. Inst. Stat. Math.*, 41(2):131–147, 1993 (in Japanese).

[46] H. Shimodaira. Assessing the error probability of the model selection test. *Ann. Inst. Stat. Math.*, 49:395–410, 1997.

[47] H. Shimodaira. An application of multiple comparison techniques to model selection. *Ann. Inst. Stat. Math.*, 50:1–13, 1998.

[48] H. Shimodaira. Multiple comparisons of log-likelihoods and combining nonnested models with applications to phylogenetic tree selection. *Commun. in Stat. A–Theory Meth.*, 30:1751–1772, 2001.

[49] H. Shimodaira. An approximately unbiased test of phylogenetic tree selection. *Syst. Biol.*, 51:492–508, 2002.

[50] H. Shimodaira and M. Hasegawa. Multiple comparisons of log-likelihoods with applications to phylogenetic inference. *Mol. Biol. Evol.*, 16:1114–1116, 1999.

[51] H. Shimodaira and M. Hasegawa. CONSEL: For assessing the confidence of phylogenetic tree selection. *Bioinformatics*, 17:1246–1247, 2001.

[52] D. L. Swofford, G. J. Olsen, P. J. Waddell, and D. M. Hillis. Phylogenetic inference. In C. Hillis, D. M. Moritz, and B. K. Mable, editors, *Molecular Systematics*, pages 407–514. Sinauer Associates, Sunderland, MA, 1996.

[53] Q. H. Vuong. Likelihood ratio tests for model selection and non-nested hypotheses. *Econometrica*, 57:307–333, 1989.

[54] P. J. Waddell, H. Kishino, and R. Ota. A phylogenetic foundation for comparative mammalian genomics. *Genome Informatics*, 12:141–154, 2001.

[55] Z. Yang. Among-site rate variation and its impact on phylogenetic analyses. *Trends Ecol. Evol.*, 11:367–372, 1996.

[56] Z. Yang. PAML: A program package for phylogenetic analysis by maximum likelihood. *Comput. Appl. Biosci.*, 13:555–556, 1997.

[57] A. Zharkikh and W. H. Li. Statistical properties of bootstrap estimation of phylogenetic variability from nucleotide sequences. I. Four taxa with a molecular clock. *Mol. Biol. Evol.*, 9:1119–1147, 1992.

Index